RED TRACK

Mathematics for Secondary Schools

NMP

3

RED TRACK

Mathematics for Secondary Schools

This book was written by

Eon Harper
Dietmar Küchemann
Michael Mahoney
Sally Marshall
Edward Martin
Heather McLeay
Peter Reed
Sheila Russell

NMP Director
Eon Harper

NMP Research
Edward Martin

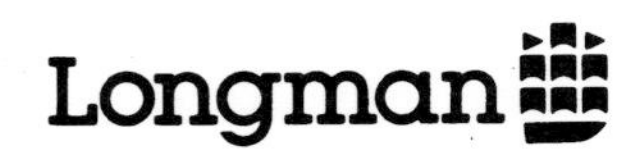

Preface

NMP was founded in 1981 to consider the emerging needs of pupils studying mathematics in secondary schools. Its primary consideration has been the research and development of materials that reflect the recommendations of the Committee of Inquiry into the Teaching of Mathematics in Schools (the Cockcroft Report) and the requirements of the GCSE National Criteria in Mathematics. Careful account has also been taken of the findings of the APU Secondary Surveys in Mathematics, of the outcomes of the CSMS research programme, of the HMI discussion document *Mathematics from 5 to 16*, and of the need to match curriculum and assessment in such areas as practical work, problem solving, investigation, pupil discussion, oral work, written and calculator work, and extended assignments. All of these are important features of the materials, which have undergone extensive trials in schools.

The texts all assume the use of a calculator as a material aid to calculation, and its use should be restricted only when indicated.

This book has been written for Year 3 of secondary schools. Each chapter in the book has three different types of material: link (indicated with a red tab in the margin), ● core (indicated with a black tab) and ▲ enrichment (indicated with a grey tab). All students should complete the core material before going on to the next chapter. Some will find it helpful to complete the support material too, and others will gain much value from the enrichment material. Students should not be alarmed that they seem to be 'missing out' some of the material in the book.

Detailed explanations of all the features of the book are given in the Year 3 Red Track Teachers' Handbook.

NMP is a Charitable Trust for the advancement of mathematics education at school level. The *Mathematics for Secondary Schools* materials were developed from an original research project at the University of Bath.

ADDISON WESLEY LONGMAN LIMITED
Edinburgh Gate, Harlow,
Essex, CM20 2JE, England
and Associated Companies throughout the world.

First published 1988
Fifth impression 1996
ISBN 0 582 20725 8

Set in 12/14pt Univers light

Produced by Longman Singapore Publishers Pte Ltd
Printed in Singapore.

The publisher's policy is to use paper manufactured from sustainable forests.

Contents

Acknowledgements

The materials were researched and evaluated at the University of Bath. NMP thanks the mathematics departments of the following schools for their assistance.

Castell Alun Comprehensive School, Wrexham, Clwyd
The Corsham School, Corsham, Wiltshire
Edgecliff Comprehensive School, Kinver, Staffordshire
Frome College, Frome, Somerset
The George Ward School, Melksham, Wiltshire
Grange Middle School, Keighley, West Yorkshire
Highdown School, Reading, Berkshire
Hope High School, Salford, Manchester
Kingsway County High School, Chester
Leiston High School, Leiston, Suffolk
Manvers Pierrepont High School, Nottingham
St John's School, Marlborough, Wiltshire
St Laurence Comprehensive School, Bradford-on-Avon, Wiltshire
Smestow School, Wolverhampton
Wyke Manor School, Bradford, West Yorkshire

The publishers are grateful to the following for their advice and assistance.

Derek Foxman	Head of Mathematics Department, National Foundation for Educational Research
Arnold Howell	Senior Author, *Mathematics for Schools*
Professor Celia Hoyles	Institute of Education, University of London
John Mason	Faculty of Mathematics, Open University
Professor Ray Ogden	George Sinclair Professor of Mathematics, University of Glasgow
Richard Strong	County Inspector for Mathematics, Somerset

We are also grateful to the following for permission to reproduce photographs and other copyright material;

Bell's Sports Centre, Perth, page 188 (photo Louis Flood Photographers); Camera Press, pages 48 (photo Chip Maury/Orion Press), 51 (photo S. Tihov/Pressphoto-BTA), 165 (photo Peter Mitchell), 172 (photo Ken Lambert), 186 (photo Werner Gorter), 192 right (photo Hartmut Reeh/DPA), 231 (photo Good Year), 251 (photo Barnet Saidman), 255, 271 (photos Jon Blau), 285 athlete (photo Fionnbar Callanan), 291 left and right (photo Homer Sykes); Ford Motor Company, page 143; Sally and Richard Greenhill, page 287 left; Johnson Matthey Jewellery Materials, Birmingham, page 289; Frank Lane Agency, pages 234 (photo M. Nimmo), 285 spider (photo Georg Nystrand), 285 humming bird (photo Leonard Rue), 285 elephant (photo Arthur Christiansen) and 287 right (photo Treat Davidson); Marshall Editions 1984, *Longman Illustrated Animal Encyclopaedia*, page 285 whale; Philips, page 243; Picturepoint, pages 184, 187 and 238; Raleigh, Nottingham, page 97; Rosenthal China, London, page 111; Royal Mint, pages 25 and 121; Safeway Foodstores, page 14 below; Science Museum, page 192 left; Science Photo Library, pages 285 virus (photo M. Wurtz/Biozentrum, University of Basel), 285 molecule (photo Adam Hart-Davis), 285 wasp (photo Sinclair Stammers), 286 above (photo Dr. Jeremy Burgess) and 286 bacteria (photos Dr. Tony Brain); Telegraph Colour Library/Space Frontiers, pages 284 and 285 Mercury.

Photographs on pages 11, 12, 14 above and centre, 15, 50, 142, 169, 189 and 230 are by Longman Photographic Unit.

Illustrations by Oxford Illustrators and cartoons by Paul Dowling.

The front cover shows *Rhythme Colore* by Sonia Delaunay (1885–1979). Sonia Delaunay was a Russian-French painter. She created many designs for fabrics, costumes and book covers and was a powerful influence on the international fashion world of the 1920s.

1 Moving points

A Activities — *Making movies*

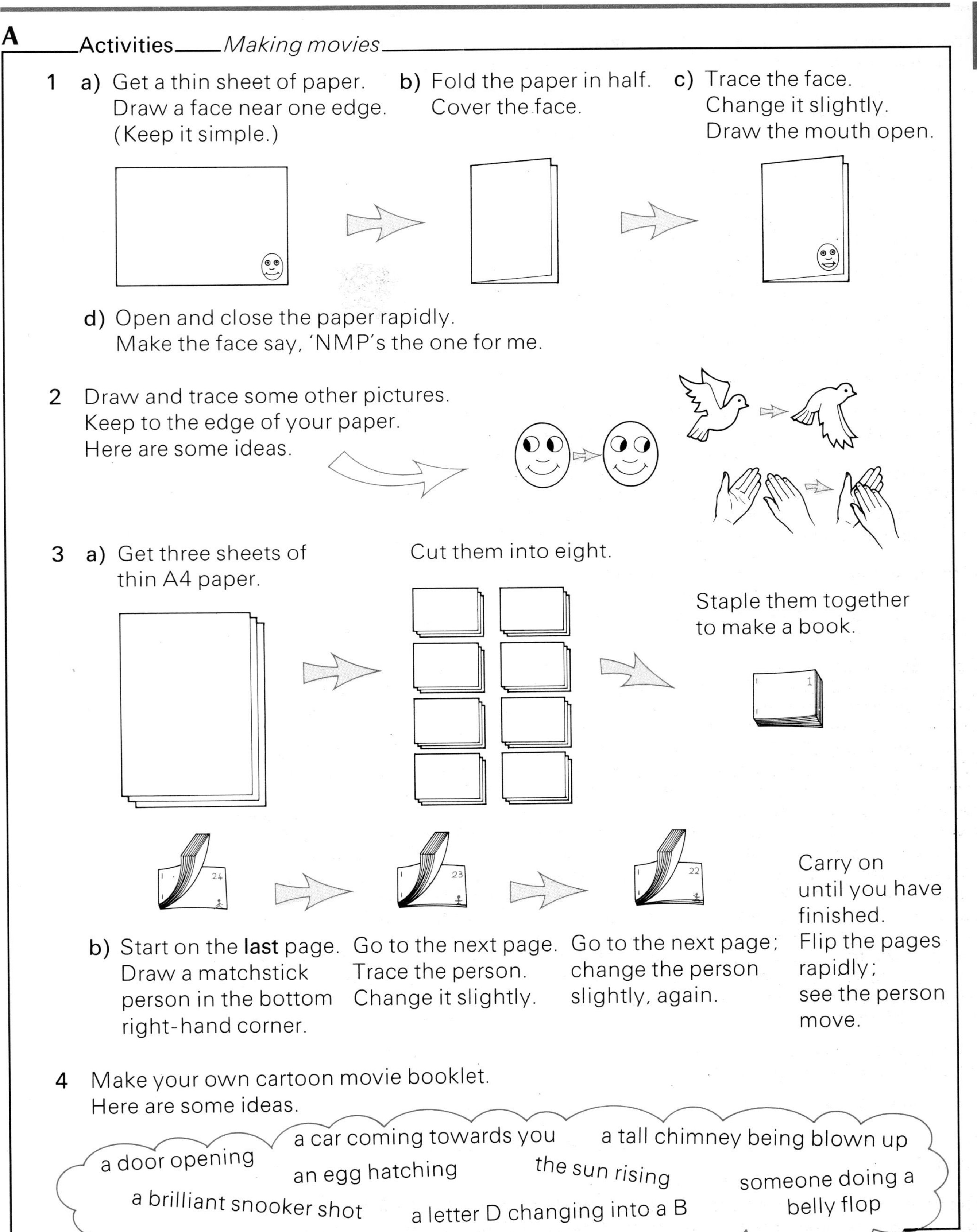

1 a) Get a thin sheet of paper.
Draw a face near one edge.
(Keep it simple.)

b) Fold the paper in half.
Cover the face.

c) Trace the face.
Change it slightly.
Draw the mouth open.

d) Open and close the paper rapidly.
Make the face say, 'NMP's the one for me.

2 Draw and trace some other pictures.
Keep to the edge of your paper.
Here are some ideas.

3 a) Get three sheets of thin A4 paper.

Cut them into eight.

Staple them together to make a book.

b) Start on the **last** page.
Draw a matchstick person in the bottom right-hand corner.

Go to the next page.
Trace the person.
Change it slightly.

Go to the next page; change the person slightly, again.

Carry on until you have finished.
Flip the pages rapidly;
see the person move.

4 Make your own cartoon movie booklet.
Here are some ideas.

a door opening
a car coming towards you
a tall chimney being blown up
an egg hatching
the sun rising
someone doing a belly flop
a brilliant snooker shot
a letter D changing into a B

Using electrical appliances

B You need a pair of compasses for this section.

1 The drawing shows Karen's bedroom floor, covered with red paint spots.
Karen is removing them with an electric sander.

With the sander plugged into the electric socket, Karen can just reach spot A and sand it off.

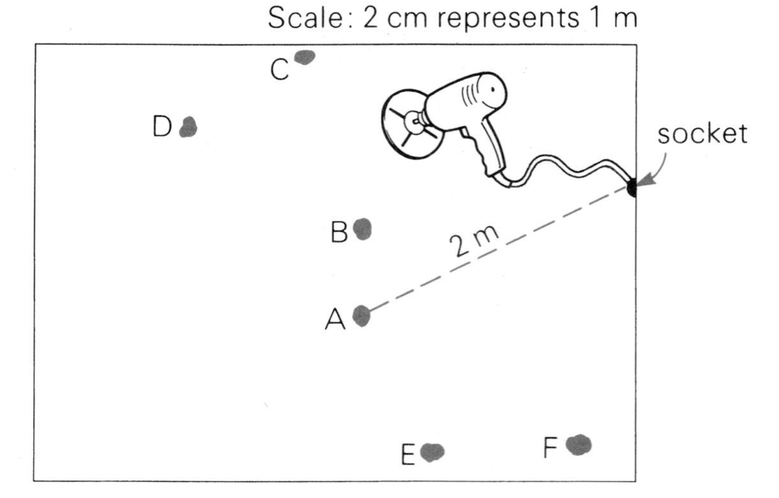

a) Which other spots can the sander reach?

b) Which spots can the sander not reach?

2 Here is another drawing of Karen's bedroom floor.

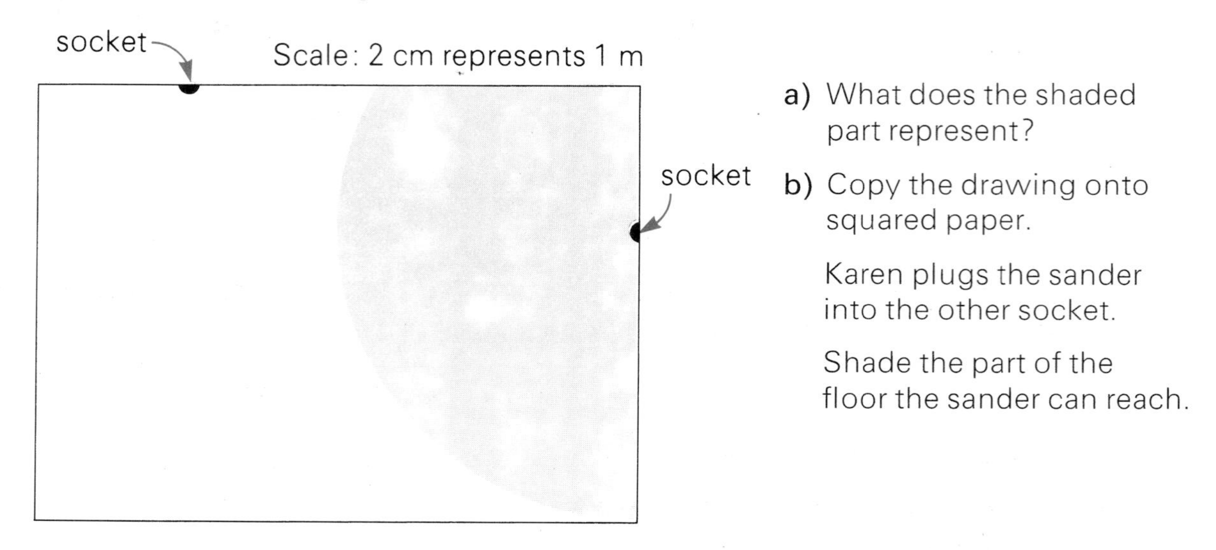

a) What does the shaded part represent?

b) Copy the drawing onto squared paper.

Karen plugs the sander into the other socket.

Shade the part of the floor the sander can reach.

Challenge

3 Karen wishes she had more sockets.

a) She wants to fit a socket so that the sander can reach as much of the floor as possible.
Where should she fit it?

There are two possibilities.

b) Can she now sand the whole floor from these three sockets?
If not, find the smallest number of sockets she would have to add to the original two so that she could.
Where would she have to fit them?

4 Edgar is vacuuming the living room floor. When plugged into this socket, the vacuum cleaner can just reach points A and B.

a) Copy the diagram onto squared paper. Shade the region that Edgar can vacuum.

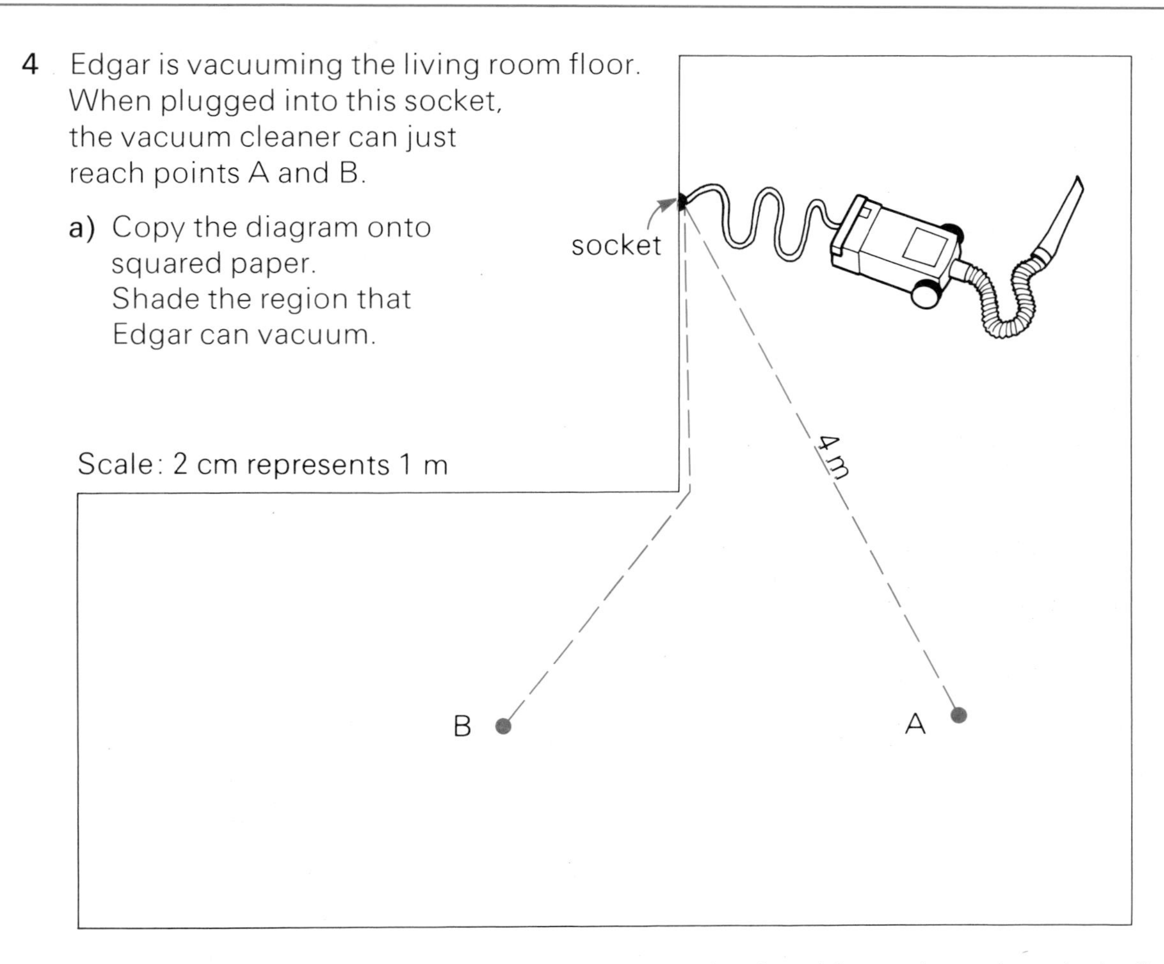

Scale: 2 cm represents 1 m

b) How many more sockets would Edgar need to be able to clean the whole floor? On your diagram, mark where you would put them.

5 Desmond is mowing the back lawn.

From the socket, the mower can just reach point A.

Copy the drawing onto squared paper.

Shade the region that Desmond can mow.

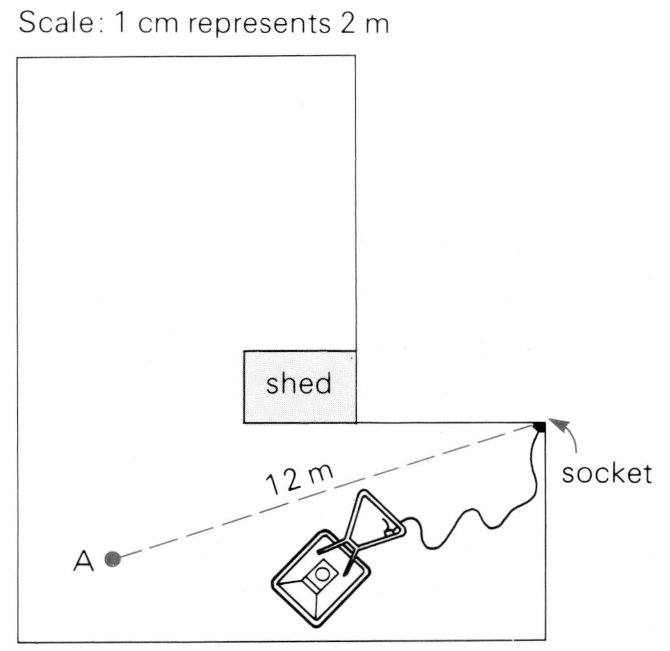

Scale: 1 cm represents 2 m

Swing-ball

C 1 Winston has a swing-ball.
Its string is 4 m long.

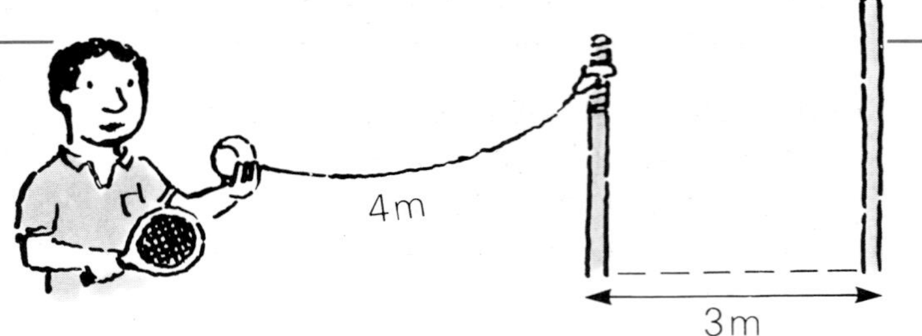

Karen sticks a pole in the ground, 3 m from the swing-ball pole.

This bird's-eye view shows the path of the ball.
Copy the diagram accurately.
Continue the path to show the ball going once around Karen's pole.

Scale: 1 cm represents 1 m

Karen's pole

2 The diagrams show Karen's pole in different positions.
Copy each diagram onto squared paper.

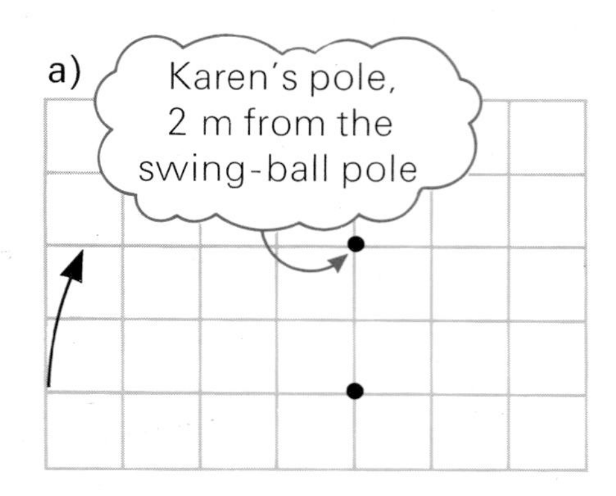

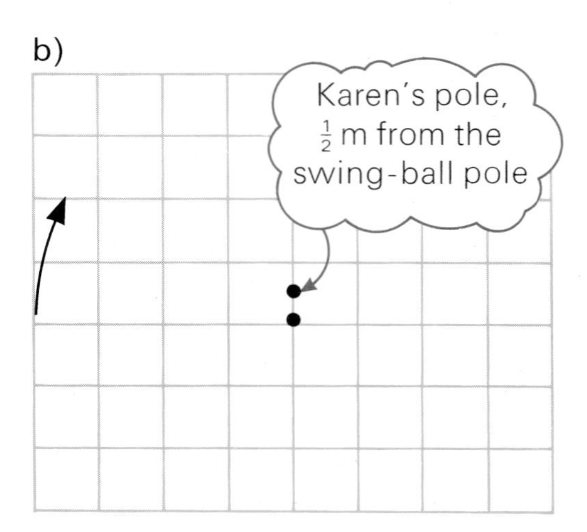

Sketch the path of the ball on each one.
Stop when the ball hits a pole.

3 Make a sketch of this drawing on squared paper.

Draw a dot to show where Karen's pole is.

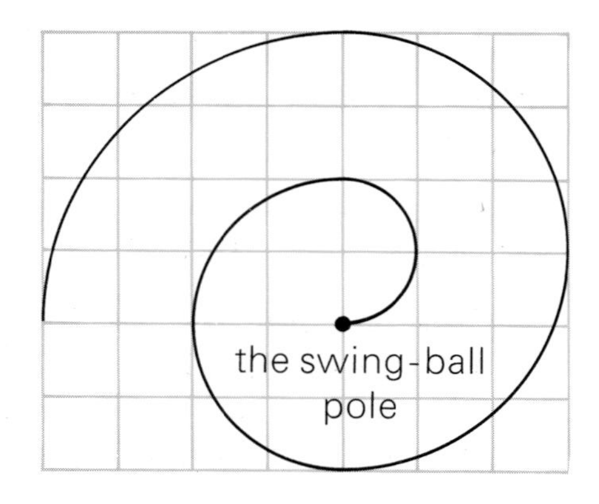

Paths of points on solid objects

D You need a pair of compasses in this section.

1 Rupinder is on a playground swing.
Here are some of the side views.
Draw what you think view **4** should be.

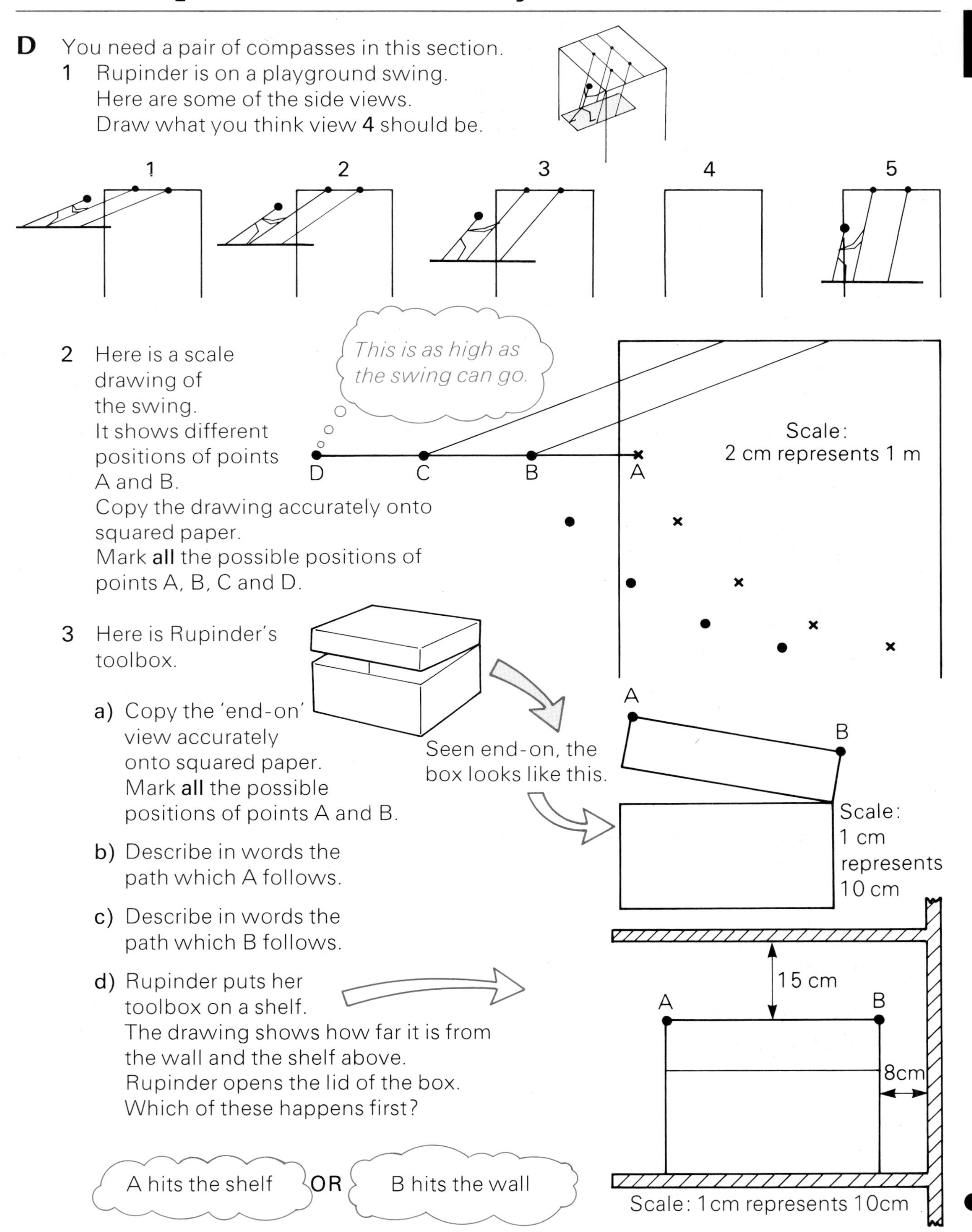

2 Here is a scale drawing of the swing.
It shows different positions of points A and B.
Copy the drawing accurately onto squared paper.
Mark **all** the possible positions of points A, B, C and D.

3 Here is Rupinder's toolbox.

a) Copy the 'end-on' view accurately onto squared paper.
Mark **all** the possible positions of points A and B.

b) Describe in words the path which A follows.

c) Describe in words the path which B follows.

d) Rupinder puts her toolbox on a shelf.
The drawing shows how far it is from the wall and the shelf above.
Rupinder opens the lid of the box.
Which of these happens first?

4 a) The filmstrip shows a car-park barrier opening.
Sketch what you think the barrier should look like in frame **3**.

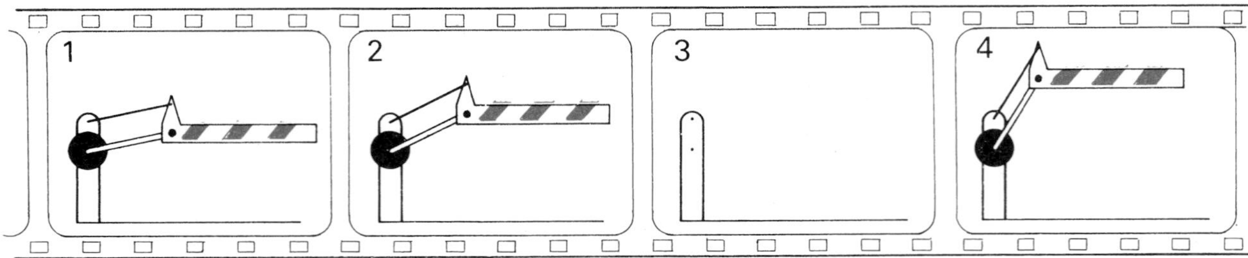

b) This is a larger, simpler version of frame **1**.
It shows the beginnings of the paths of points A and B.
(i) Copy the drawing and continue the paths of A and B.
(ii) Describe the paths of A and B in words.

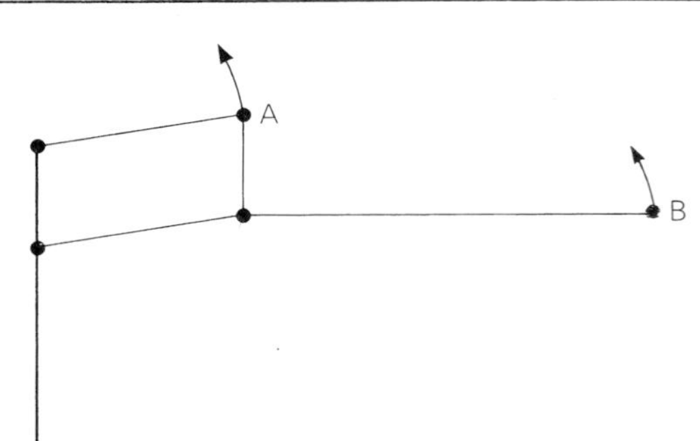

Exploration — *Mad Hatter's Ride*

5 At Disneyworld in Florida there is a Mad Hatter's Tea-party fairground ride.
The base of the ride spins around, and each teacup itself spins around!

The point K shows where Karen is sitting in one of the teacups.

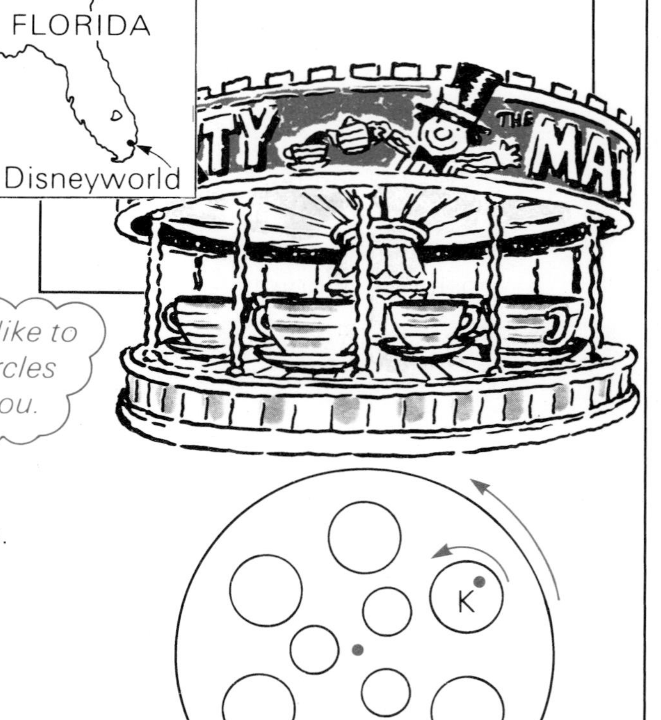

a) Each time the base makes a full turn, each cup makes a full turn.
Sketch Karen's path as the ride spins around.

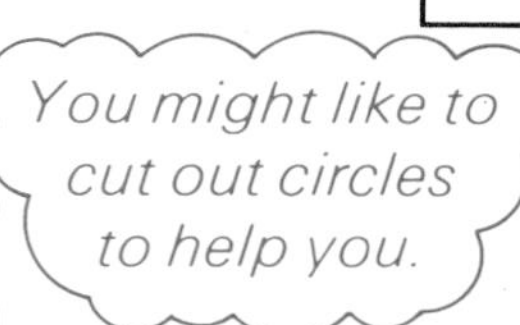

You might like to cut out circles to help you.

b) Investigate what happens for different spinning speeds of the cup.
Draw diagrams for
(i) a fast speed.
(ii) a slow speed.

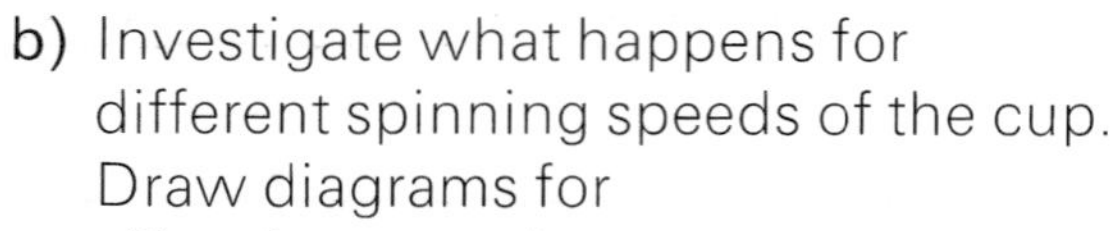

c) Would it make a difference if the cups and the base turned in opposite directions to each other?
Draw diagrams to illustrate your answer.

Activities — *Rolling shapes*

6 Desmond is rolling a wheel along this rocky path.
The coloured line shows the path of the wheel's centre.

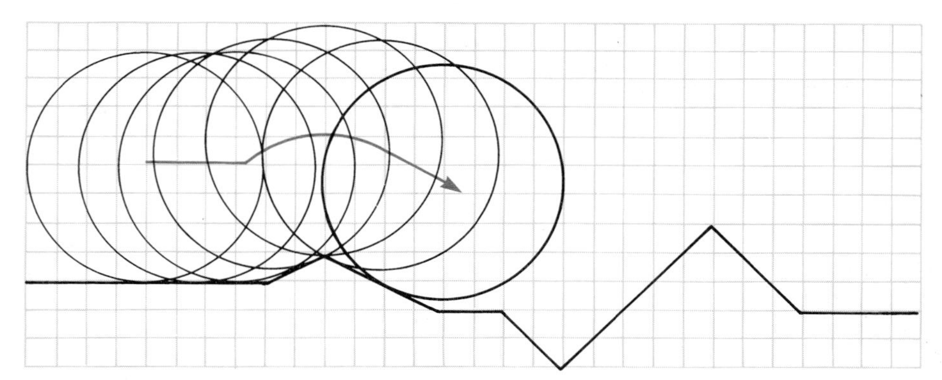

a) Copy the rocky path onto 5 mm squared paper.

b) Copy and continue the path of the wheel's centre.
To help you, make a cardboard disc with radius 2 cm.
Push the point of a pencil through the centre.

c) Make cardboard discs with radius (i) 3 cm. (ii) 4 cm.

d) On your diagram, draw the paths of the centres of these discs.

e) Describe how the paths for different discs are different from each other.

7 a) Draw a circle with radius 2 cm.
Mark three points on the circle.

b) Join the points to make a triangle.
Cut it out.

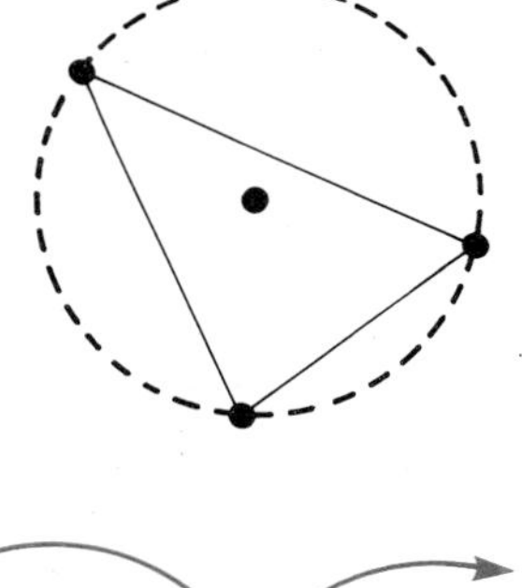

c) Roll the triangle along a flat path.
Plot the path of its 'centre'.

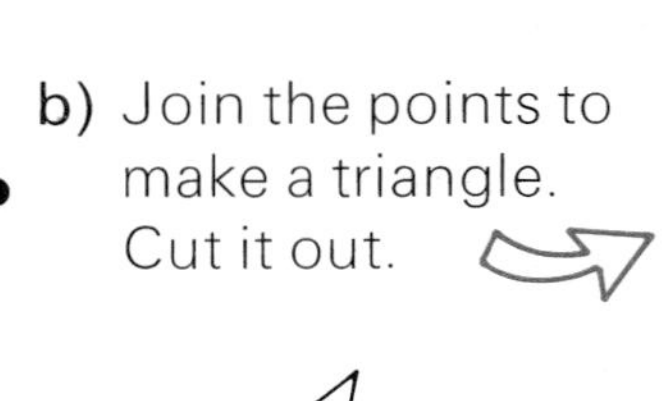

Exploration

8 a) Make a quadrilateral using the idea in questions **7(a)** and **(b)**.
Plot the path of its 'centre'.

b) Repeat for a seven-sided shape.

c) Investigate how the paths for different shapes are different from each other.

Write down what you discover.

Equidistance

1

E 1 Bill and Joyce are planting trees.
They have decided that each tree should be the same distance from A as from B.

a) T is the first tree they have planted.

Check that its distances from A and B are equal.

b) Copy the drawing, checking your measurements carefully.
Find three more places where trees can be planted.

*We say they are **equidistant** from A and B.*

Mark them on your drawing.

c) Joyce decides to mark a line which shows all the places where trees can be planted.
Mark the line yourself on your drawing.

d) The line you should have drawn is called the **perpendicular bisector** of A and B.
Write one or two sentences to explain why you think it is called this.

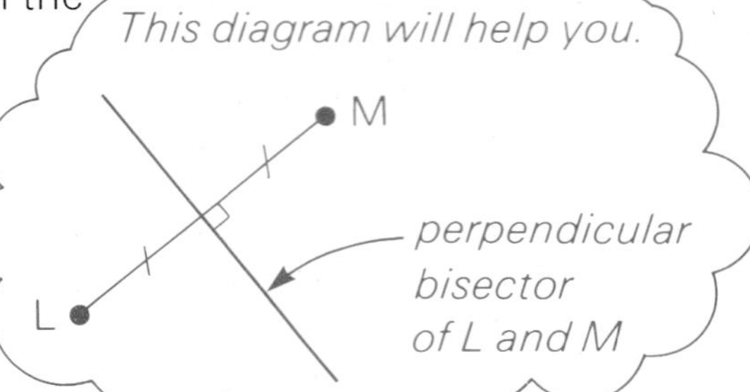

Challenges

2 a) On plain paper, mark two points A and B 6 cm apart.

b) Use compasses to find all the points that lie
(i) 5 cm from A. (ii) 5 cm from B.

c) How many points are 5 cm away from both A and B?
Mark them in colour on your drawing.

on your paper

d) In colour, mark all the points that are equidistant from A and B.

3 a) On plain paper, draw any triangle ABC.
Make it quite large.

use a ruler

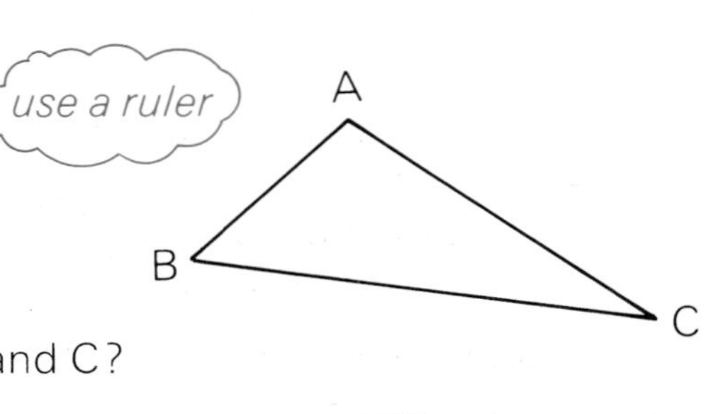

b) Use compasses to help you draw
(i) the perpendicular bisector of A and B.
(ii) the perpendicular bisector of B and C.

c) How many points are equidistant from A, B and C?
Mark them in colour on your drawing.

d) It is possible to draw a circle that passes through A, B and C.
Do it!

Movement relative to fixed points

F Exploration

1 Jenny enjoys flying her Gypsy Moth in circles in the sky. She likes to fly through any little clouds she can find. Investigate how easy it is for her to fly around in circles so that she can pass through all the little clouds

a) if there is only one little cloud.

b) if there are two of them.

c) if there are three or more of them.

Does it matter how the clouds are arranged?

in a straight line; at the corners of a rectangle; at the corners of a tetrahedron; and so on

2 a) Mark two points A and B 9 cm apart in the middle of a blank page.

b) C and D are both twice as far away from A as they are from B.

The curved lines show you how to use compasses to find these points.

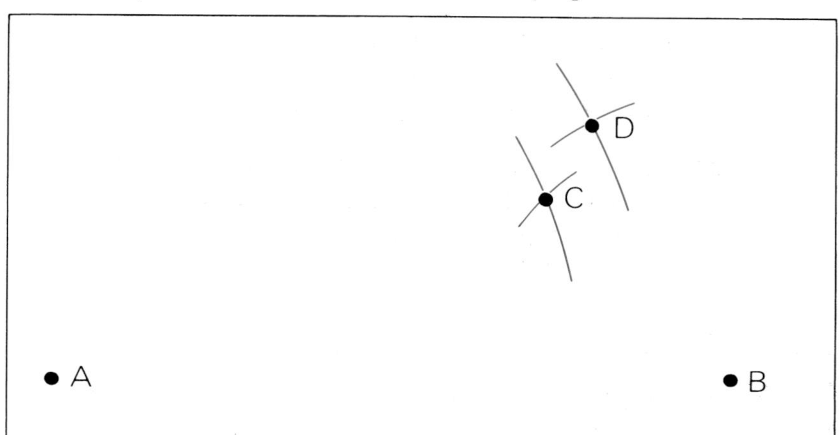

On your diagram mark **six** more points that are twice as far from A as they are from B.

c) Sketch the line which includes all the points that are twice as far from A as from B.

d) Sketch the line which includes all the points that are **three** times as far from A as from B.

3 In this diagram, A and B are 4 cm apart.
The total distance of C from A and B is 6 cm.

The total distance of D from A and B is also 6 cm.
Copy the drawing.
Draw a line to show all the points which are a total distance of 6 cm from A and B.

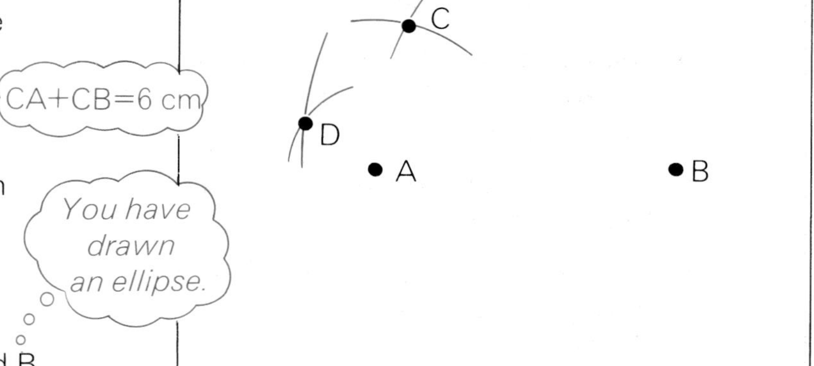

Activity

4 You need a softboard, a sheet of plain paper, two pins, and a piece of thread with a loop at each end.

a) Set up your softboard like this.

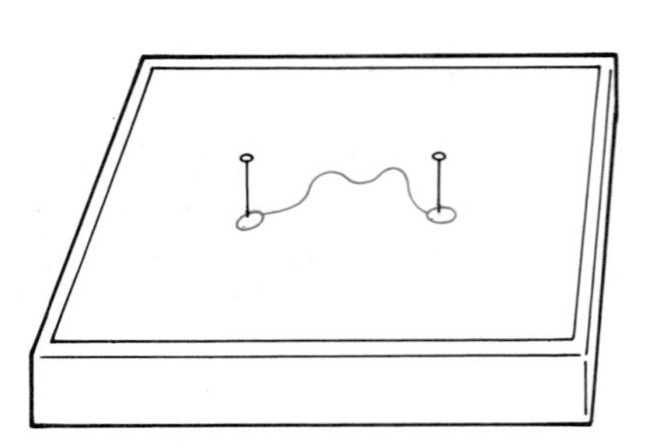

Hold the thread tight with a pencil.
Move the pencil around, keeping the thread tight.

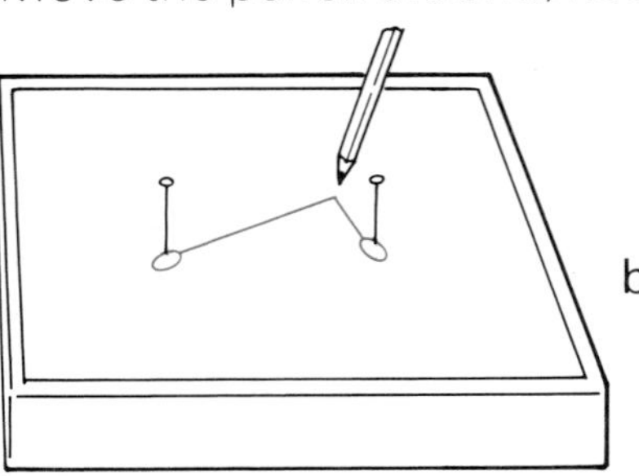

Draw above and below the pins.

b) What does this drawing technique have to do with question **3**?

c) Write one or two sentences and make sketches to show how you could use this technique to mark out an elliptical flower bed or an elliptical table top.

5 Ida the singing fly is in a recording studio.
She has to keep exactly 20 cm from the microphone.
Describe the path she can fly.

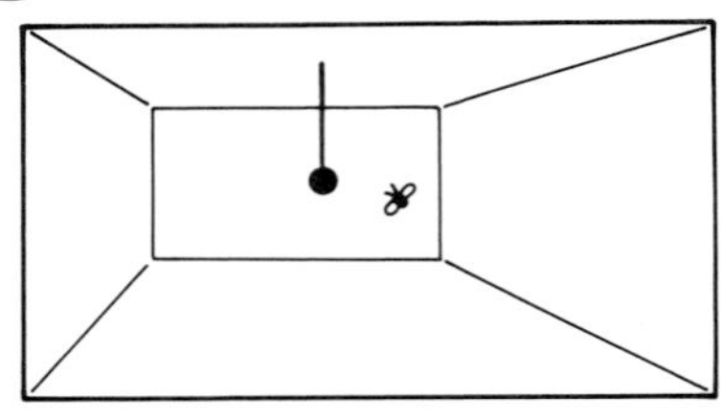

6 Ida is recording in stereo.
Now she has to stay 20 cm from each microphone.
(The microphones are 30 cm apart.)
Describe the path she can fly.

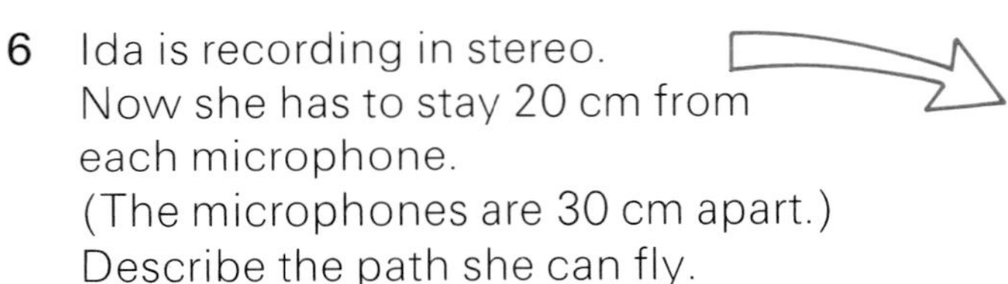

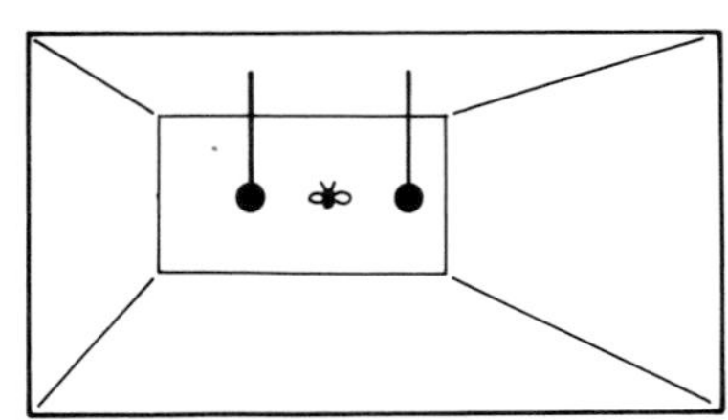

7 Ida is getting dizzy.
The engineer relaxes the 20 cm rule.
But she still has to stay equidistant from both mikes.
Describe the path she can now fly.

8 Ida is reading a fly-sheet.
She can see best if she stays 70 cm from the light.
Describe the path she can fly if she uses

a) this strip-light. b) this strip-light. c) this globe-light.

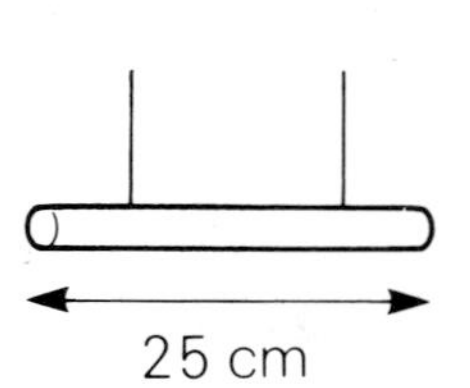

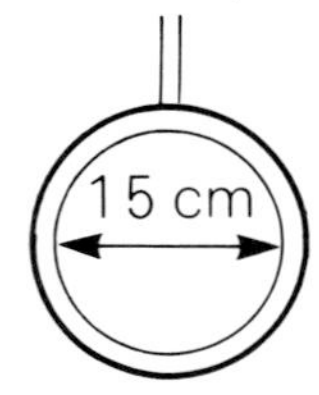

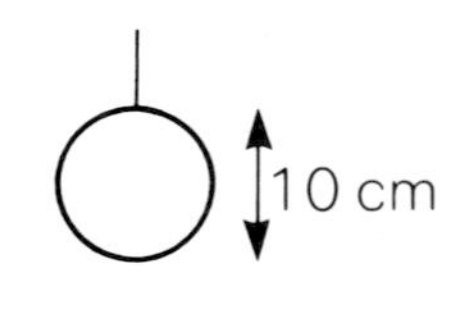

2 Estimating

A

Do you remember . . .?

We often round measurements up or down.

SHE'S ABOUT 2m TALL.

SHE'S ABOUT 1m TALL.

2 m

1 m

1 The pattern says that you need $3\frac{1}{2}$ m of material to make the sarong.

a) Do you think this figure is more likely to have been rounded up or rounded down?

b) Write one or two sentences to explain why you say this.

2 This is an unopened sack of potatoes.

a) What does 'minimum contents 20 kg' mean?

b) Which of these do you think is more likely to be nearer to the true amount of potatoes in the bag?

25 kg

c) Is 20 kg a rounded-up figure or a rounded-down figure?

3 The sign over the bridge means that the maximum height of vehicles should be 4.34 m.

a) Do you think 4.34 m is a rounded-up figure or a rounded-down figure?

b) Write one or two sentences to explain why.

4 Write down whether you think the figures used in these situations are rounded up or rounded down.
For each one, explain why you made this choice.

Find out for yourself

5 By law, labels like these must show the **minimum** amount inside the container . . . so each figure is a rounded-down figure.

Labels like these have an ℮ sign next to the amount.

a) Find out what the ℮ sign means.

b) Which of these two jars do you think is more likely to contain more coffee? Why?

Using rounded figures to estimate

B 1 Make a reasonable estimate for the cost of each of these.
Do each estimate in your head.

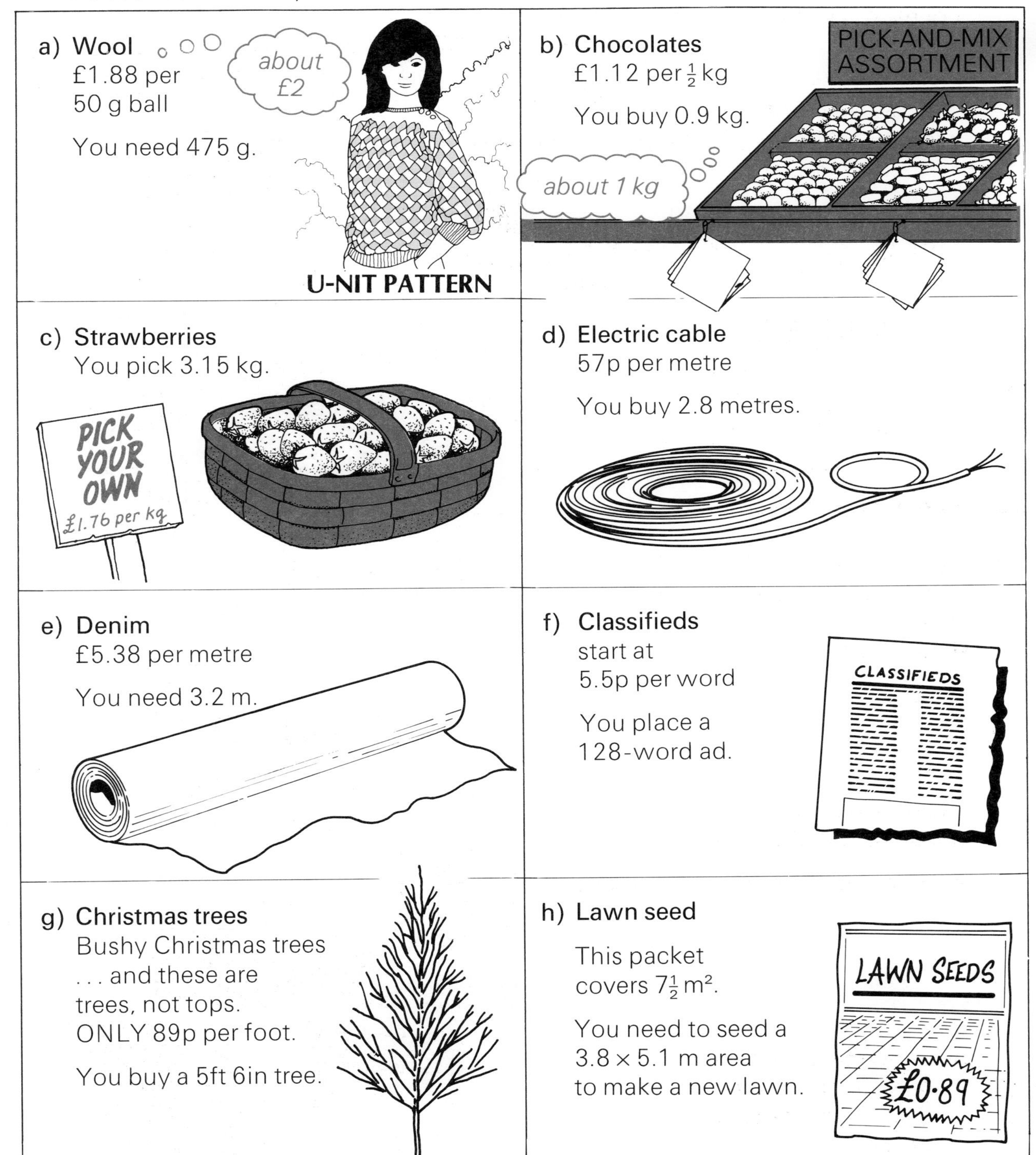

a) **Wool**
£1.88 per
50 g ball

You need 475 g.

b) **Chocolates**
£1.12 per $\frac{1}{2}$ kg

You buy 0.9 kg.

c) **Strawberries**
You pick 3.15 kg.

d) **Electric cable**
57p per metre

You buy 2.8 metres.

e) **Denim**
£5.38 per metre

You need 3.2 m.

f) **Classifieds**
start at
5.5p per word

You place a
128-word ad.

g) **Christmas trees**
Bushy Christmas trees
. . . and these are
trees, not tops.
ONLY 89p per foot.

You buy a 5ft 6in tree.

h) **Lawn seed**

This packet
covers $7\frac{1}{2}$ m².

You need to seed a
3.8 × 5.1 m area
to make a new lawn.

2 Use your calculator to work out the exact costs in question **1**.
Check that your estimates are reasonable.

3 Do these in your head.

Do not use a calculator.
Do not do any written calculation.

Write down only your result each time.

The price charged for each amount is incorrect.
Write either **overcharged** or **undercharged** for each one.

a) 4.2 kg of tomatoes at £0.93 per kilogram.
Price charged: £3.60

b) 9.60 m of wood at £2.30 per metre.
Price charged: £23.00

c) 2.8 kg of grapes at £1.80 per kilogram.
Price charged: £5.40

d) 5 m 60 cm of fabric at £3.25 per metre.
Price charged: £15.50

e) 650 g of liquorice assorted at 43p per 100 g.
Price charged: £2.50

f) 65 cl of ginger beer at 83p per litre.
Price charged: 65p

g) 435 g of cheese at £2.06 per kilogram.
Price charged: £8.70

4 Use your calculator.
Check your results in question **3** by calculating the exact costs.

Estimating using proportion

C

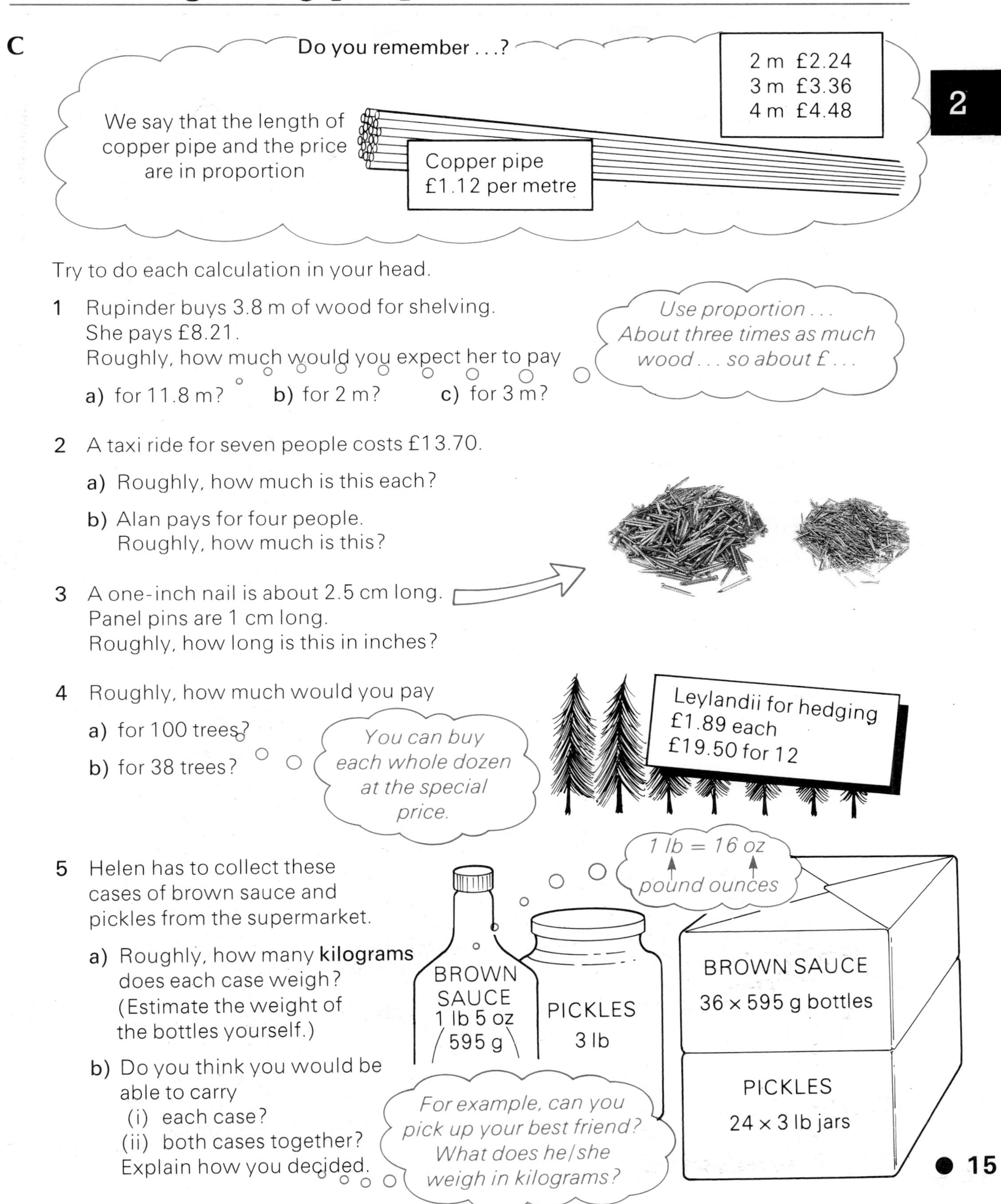

Do you remember . . .?
We say that the length of copper pipe and the price are in proportion
Copper pipe £1.12 per metre
2 m £2.24
3 m £3.36
4 m £4.48
Use proportion . . . About three times as much wood . . . so about £ . . .
You can buy each whole dozen at the special price.
Leylandii for hedging £1.89 each £19.50 for 12
1 lb = 16 oz
pound ounces
BROWN SAUCE 1 lb 5 oz 595 g
PICKLES 3 lb
BROWN SAUCE 36 × 595 g bottles
PICKLES 24 × 3 lb jars
For example, can you pick up your best friend? What does he/she weigh in kilograms?

Try to do each calculation in your head.

1 Rupinder buys 3.8 m of wood for shelving.
She pays £8.21.
Roughly, how much would you expect her to pay

a) for 11.8 m? b) for 2 m? c) for 3 m?

2 A taxi ride for seven people costs £13.70.

a) Roughly, how much is this each?

b) Alan pays for four people.
Roughly, how much is this?

3 A one-inch nail is about 2.5 cm long.
Panel pins are 1 cm long.
Roughly, how long is this in inches?

4 Roughly, how much would you pay

a) for 100 trees?

b) for 38 trees?

5 Helen has to collect these cases of brown sauce and pickles from the supermarket.

a) Roughly, how many **kilograms** does each case weigh?
(Estimate the weight of the bottles yourself.)

b) Do you think you would be able to carry
(i) each case?
(ii) both cases together?
Explain how you decided.

2

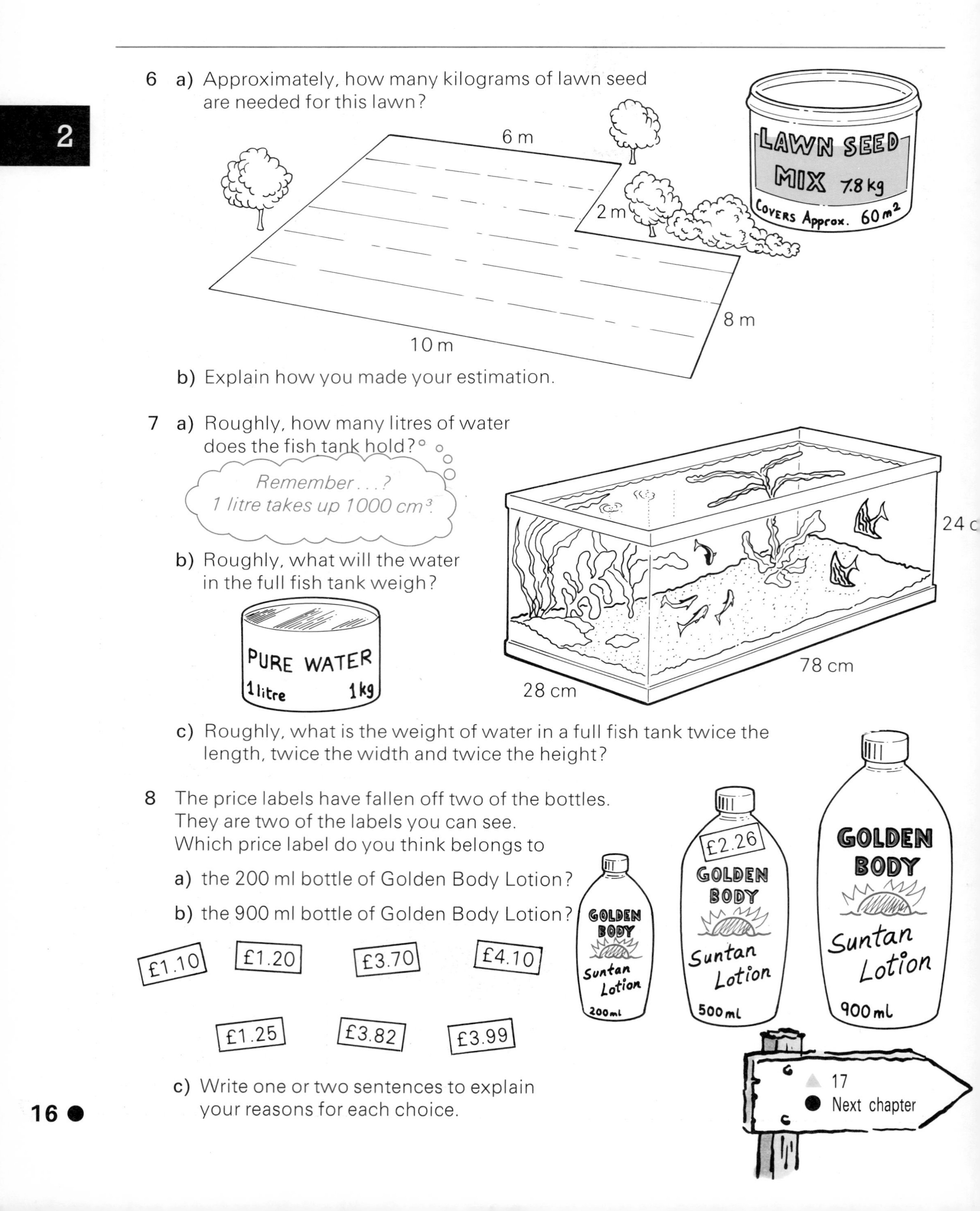

6 a) Approximately, how many kilograms of lawn seed are needed for this lawn?

b) Explain how you made your estimation.

7 a) Roughly, how many litres of water does the fish tank hold?

b) Roughly, what will the water in the full fish tank weigh?

c) Roughly, what is the weight of water in a full fish tank twice the length, twice the width and twice the height?

8 The price labels have fallen off two of the bottles. They are two of the labels you can see. Which price label do you think belongs to

a) the 200 ml bottle of Golden Body Lotion?

b) the 900 ml bottle of Golden Body Lotion?

c) Write one or two sentences to explain your reasons for each choice.

Estimating costs

D **Challenge** *Cardiff to Paris*

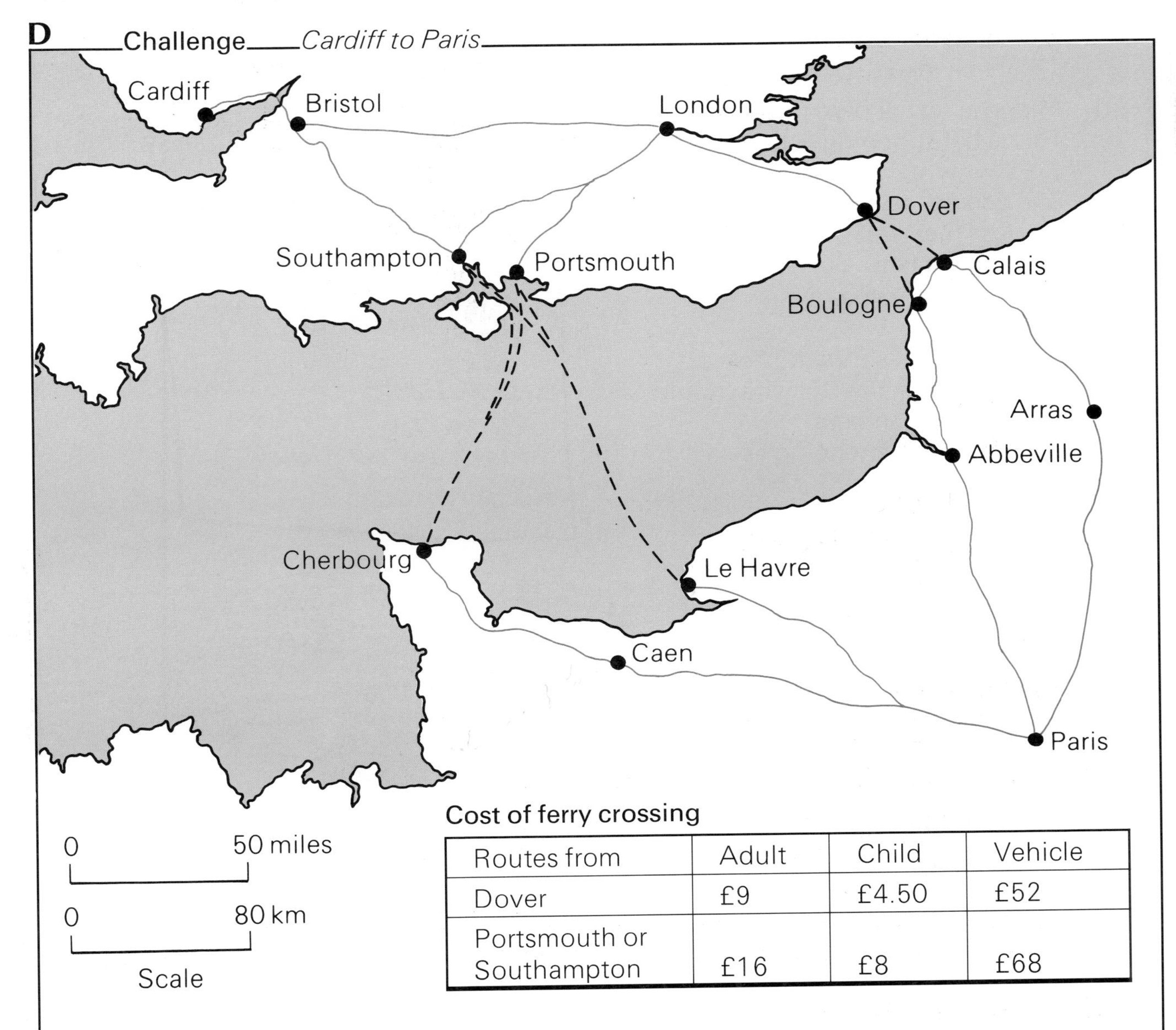

Cost of ferry crossing

Routes from	Adult	Child	Vehicle
Dover	£9	£4.50	£52
Portsmouth or Southampton	£16	£8	£68

1 a) Choose which you think will be the cheapest route from Cardiff to Paris.

b) Estimate the cost for a family of four (two adults and two children), outward journey only.

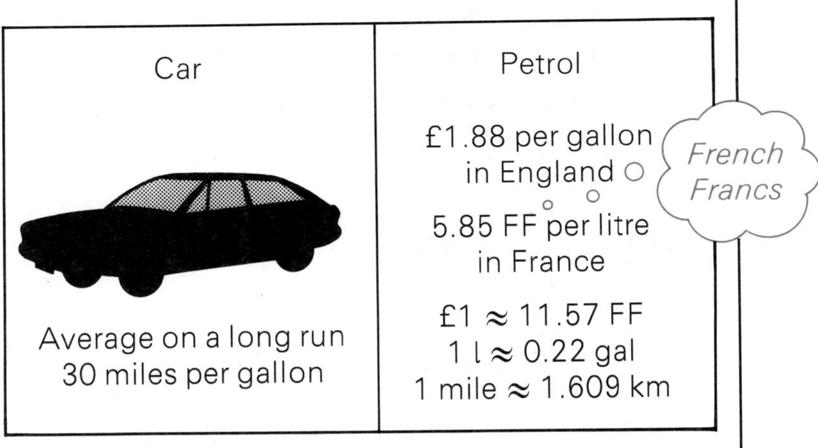

Best buys

E With a friend

1 Midge is on her own this week.
This is her shopping list.

She decides to buy only small sizes
- unless they are very poor value
- unless the large quantities are really good value.

Copy Midge's shopping list.
Decide on the size that she should choose for each item.
Write it on the shopping list.

Tomato Soup
Ketchup
Cornflakes
Shredded Wheat
Sugar Puffs
Rice pudding
Coke
Lemonade
Tuna fish
Washing up liquid
Kitchen rolls

3 Shapes

A

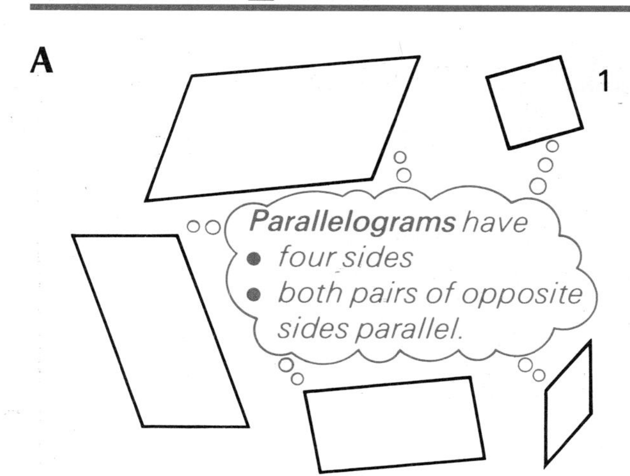

Parallelograms have
four sides
both pairs of opposite sides parallel.

1 a) Each pair of lines shows two sides of a parallelogram.
Copy the lines onto squared paper.
Complete each parallelogram.

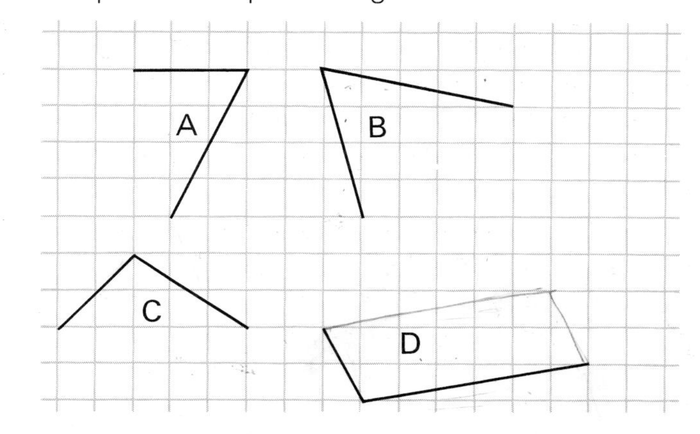

b) What is the area of
  (i) parallelogram **A**?
  (ii) parallelogram **D**?

2 Which of these are special kinds of parallelograms?
Write **yes** or **no** for each one.

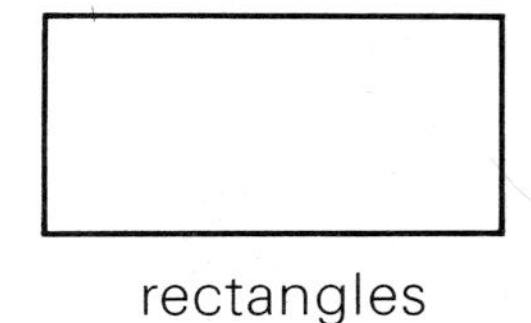
rectangles

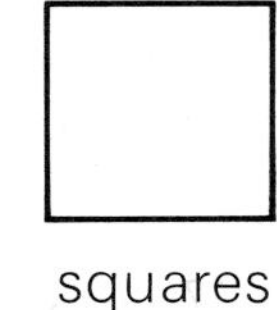
squares

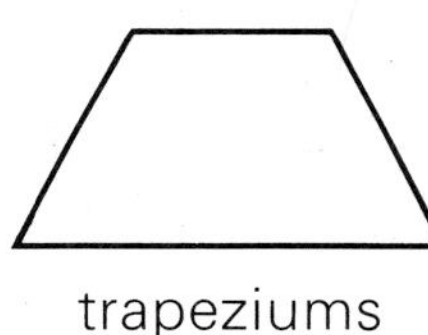
trapeziums

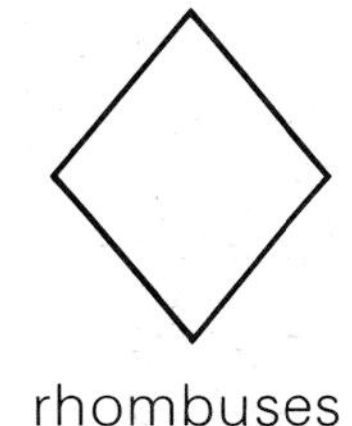
rhombuses

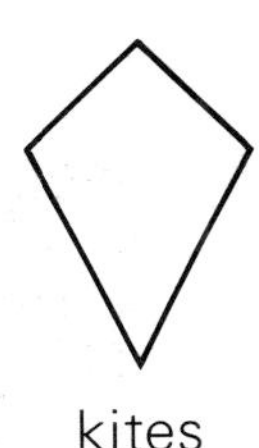
kites

If you say **no**, explain why.

Challenges

3 In the diagram, P, Q, R and S are the midpoints of the sides of the red parallelogram.
The area of PQRS is 15 cm².
What is the area of the red parallelogram?
Explain how you arrived at your result.

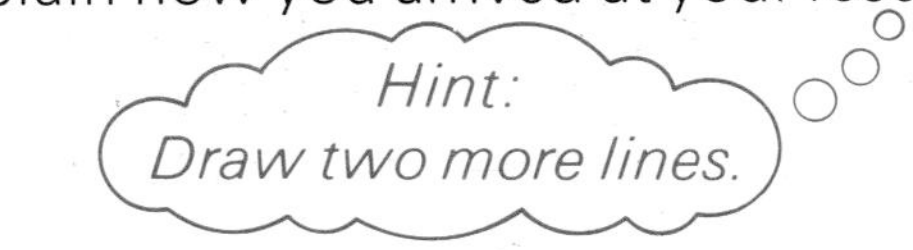

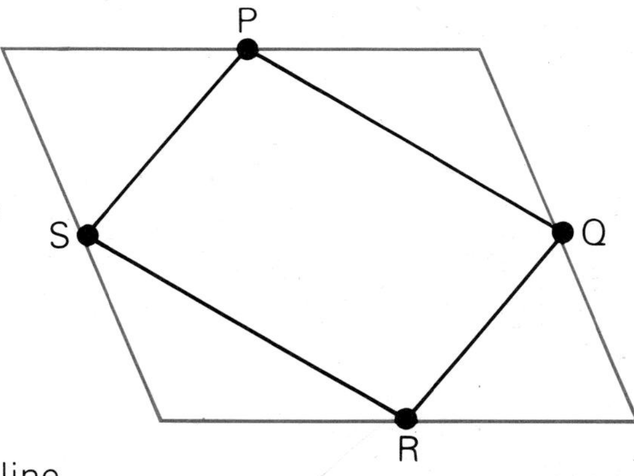

4 a) Mark any three points not in a straight line.
There are three parallelograms which have your three points as three of their corners.
Draw them, using a different colour for each parallelogram.

b) What can you say about the areas of your three parallelograms?
Explain your answer.

B

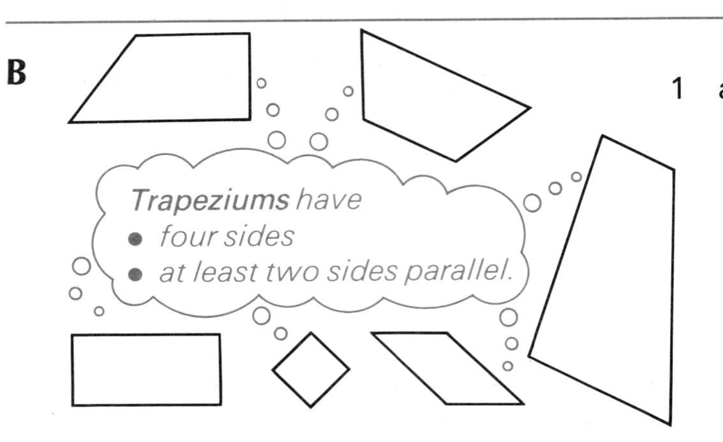
Trapeziums have
• four sides
• at least two sides parallel.

1 a) Draw each shape on squared paper.
Show how to divide each one into two **identical** trapeziums.
Each of your trapeziums must have only one pair of parallel sides.

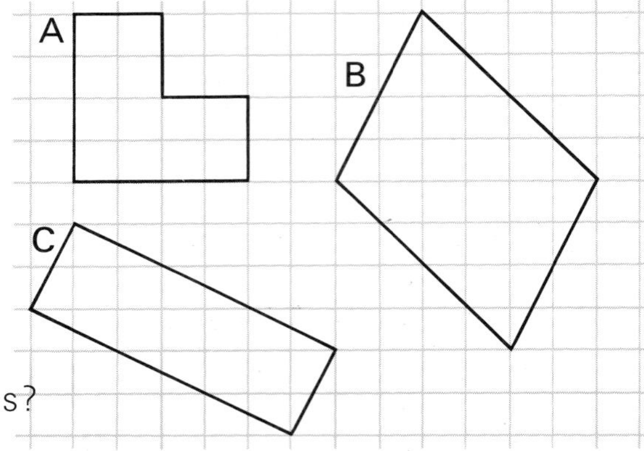
A
B
C

b) What is the area of each of the trapeziums you made in
(i) **A**? (ii) **B**? (iii) **C**?

2 Which of these are special kinds of trapeziums?
Write **yes** or **no** for each one.

a) parallelograms **b)** rectangles **c)** squares **d)** kites **e)** rhombuses

If you say **no**, explain why.

Activities

3 a) Use 1 cm squared paper.
Each of the shapes **A** to **E** can be made by putting together two or more copies of this trapezium.

Make drawings to show how.

flipped over, if necessary

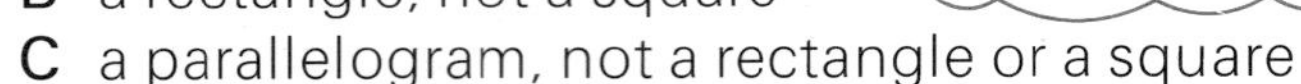

A a square
B a rectangle, not a square
C a parallelogram, not a rectangle or a square
D a bigger trapezium, the same shape as each small one
E an octagon, a bigger version of this:

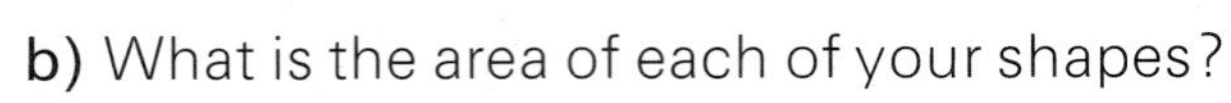

b) What is the area of each of your shapes?

4 Use 1 cm squared paper.

a) It is possible to cut this trapezium in two and rearrange the pieces to make a rectangle.
Make drawings to show how.

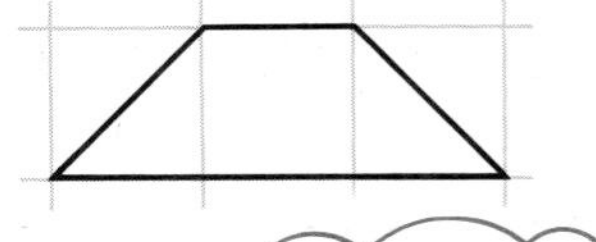

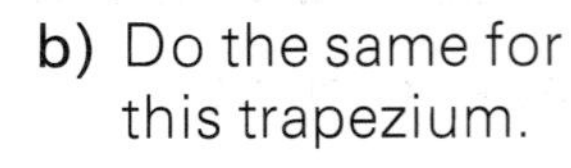

b) Do the same for this trapezium.

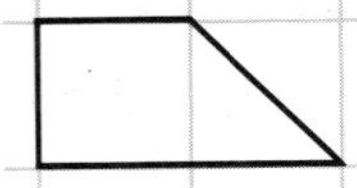

maybe more than two

c) Look for a way to cut **any** trapezium into pieces which can be rearranged to make a rectangle.

Make drawings to explain your method.

for example,

C

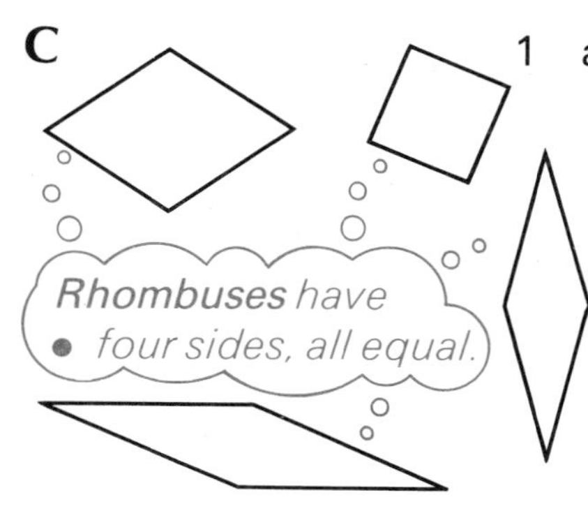
Rhombuses have
• four sides, all equal.

1 a) Copy this arrangement of lines carefully onto squared paper.

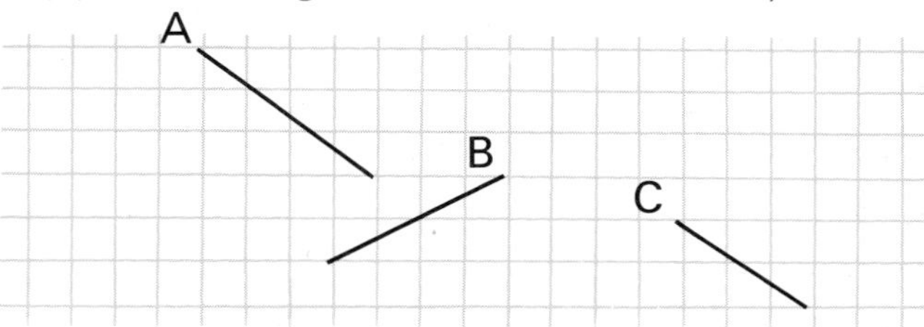
A
B
C

b) Each line is one side of a rhombus.
Complete the rhombuses so that they do not overlap.

c) What is the area of each rhombus?

Activity *Making a rhombus*

2 You need tracing paper.
Draw a line somewhere in the top right-hand part of your tracing paper.
Fold your paper twice, and each time trace what you can see.
Choose where to fold the paper so that, when you finally unfold it, there is a rhombus drawn on it.

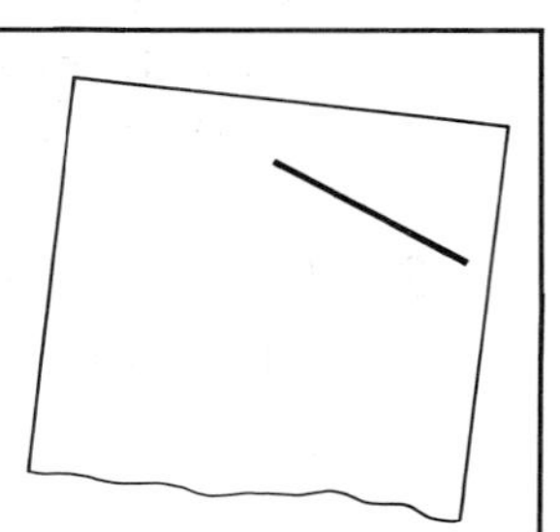

3 Which of these are special kinds of rhombuses?
Write **yes** or **no** for each one.
a) rectangles b) parallelograms c) trapeziums d) squares e) kites
If you say **no**, explain why.

4 Each pair of lines shows two sides of a kite.
Copy the lines onto squared paper and complete each kite.

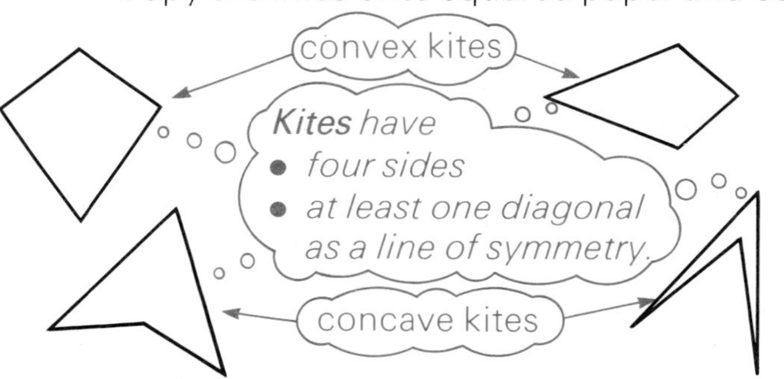
convex kites
Kites have
• four sides
• at least one diagonal as a line of symmetry.
concave kites

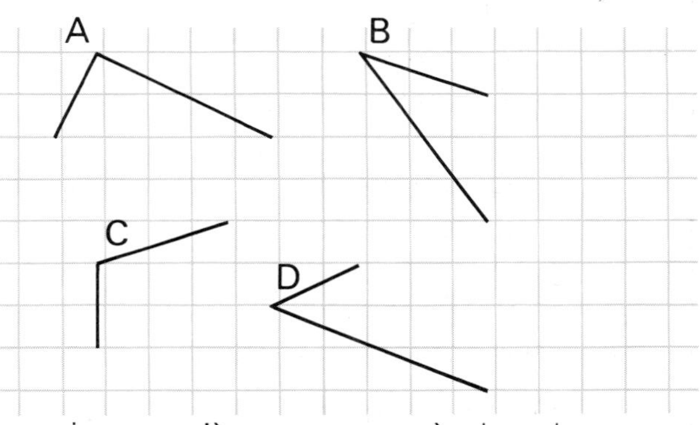
A
B
C
D

5 Which of these are special kinds of kites?
a) rectangles b) parallelograms c) trapeziums d) squares e) rhombuses

Challenge

6 Every kite can be cut into two kites which have the same area as each other.
Investigate how to do this.
Make sure that your method works whether you start with a convex or a concave kite.
Write a short paragraph to explain your method.
Illustrate your explanation with some drawings on squared paper.

D

Exploration — *In good shape*

1 Start with any rectangle that is not a square.

a) With one straight cut, you can separate the rectangle into two parts, one of them a **square**.
Make a drawing to show how you could separate the rectangle into two pieces, one of them a trapezium.

which is not also a parallelogram

b) Investigate the smallest number of cuts you need to produce each of these shapes from a rectangle.

Each cut must separate a shape into two parts.
Each cut must be straight.

(i) a parallelogram which is not also a square or a rectangle

(ii) a convex kite which is not also a square or a rhombus

(iii) a rhombus which is not also a square

(iv) a concave kite

Some shapes may be impossible to make.

Make sketches to show where you would make the cuts each time.

c) This time, start with a trapezium.
Find how many cuts are needed to make each of these.

Some shapes may be impossible to make.

(i) a parallelogram (ii) a convex kite (iii) a concave kite

(iv) a rectangle (v) a square (vi) a rhombus

Your results must be true for **any** starting trapezium.

d) Investigate how to make shapes from other starting shapes.
Is it possible to make all the shapes mentioned on this page, no matter which of the shapes we start with?
If not, write down which ones cannot be made from which starting shapes.

2 Copy and complete these instructions.
Illustrate your instructions with a sketch.

For example:
To change a square into a rectangle, lengthen or shorten two opposite sides by the same amount.

a) To change a rhombus into a kite, . . .

b) To change a kite into a parallelogram, . . .

No cutting allowed; only lengthening, shortening, pushing, and so on.

Quadrilaterals and their diagonals

E 1 This question is all about quadrilaterals.
The diagonals of this rectangle bisect each other.

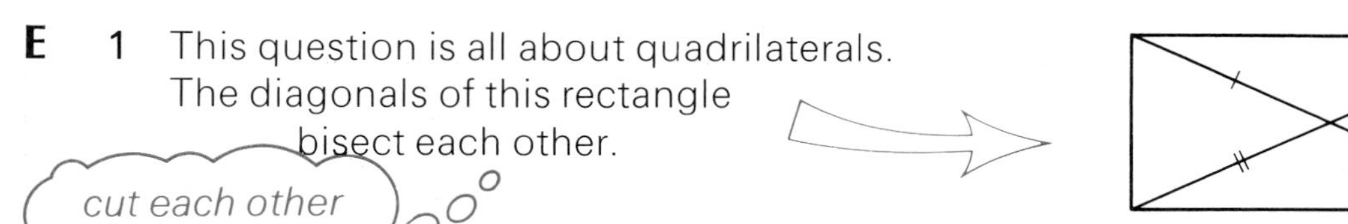

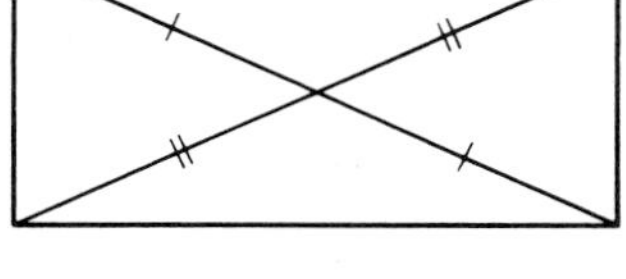

The diagonals of this trapezium do not.

a) Sketch three more shapes whose diagonals bisect each other.

b) The diagonals of a shape bisect each other.
They also meet each other at right angles.
What kind of shape is it?

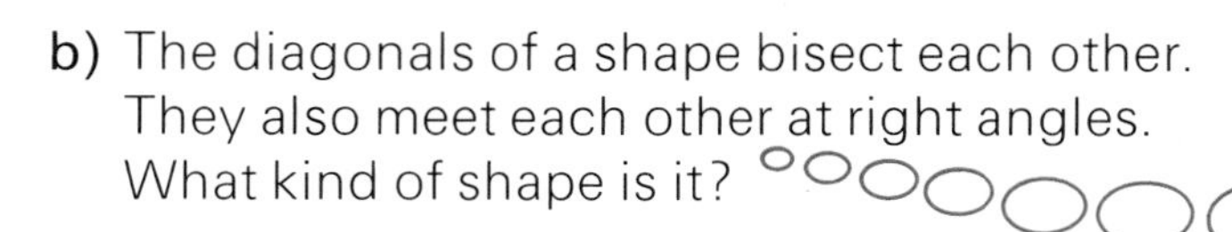

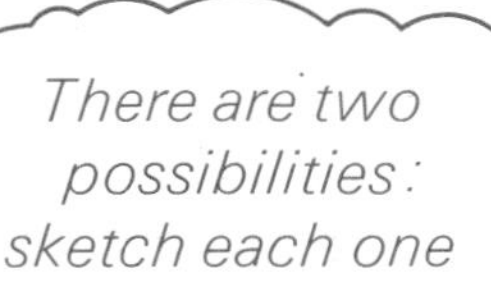

c) The diagonals of a shape meet each other at right angles.
One diagonal bisects the other.
The second diagonal cuts the first in the ratio 2:1.
(i) Sketch the shape. (ii) What kind of shape is it?

d) The diagonals of a shape are the same length.
They cut each other in the ratio 2:1.
(i) Sketch the shape. (ii) What kind of shape is it?

Challenge *Sorting shapes*

2 The flow chart sorts quadrilaterals on the basis of information about their diagonals.

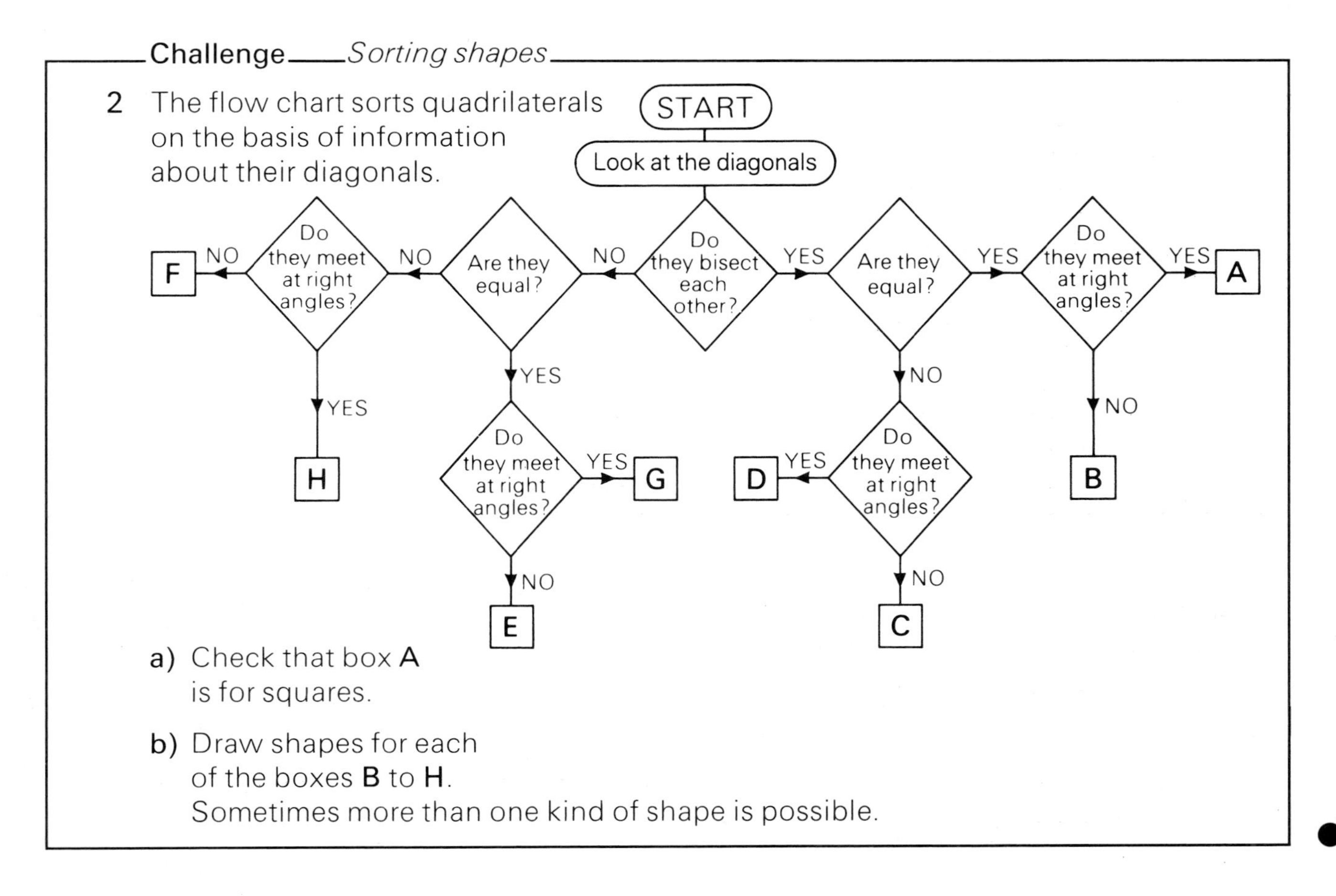

a) Check that box **A** is for squares.

b) Draw shapes for each of the boxes **B** to **H**.
Sometimes more than one kind of shape is possible.

3

Investigating shapes

F

Exploration *Reptiles*

1 Some shapes can be fitted together to make larger copies of themselves.
We will call them **reptiles** . . . because **tiles** of such a shape can be used to **rep**roduce themselves – on a larger scale!
Investigate which of these shapes are reptiles:

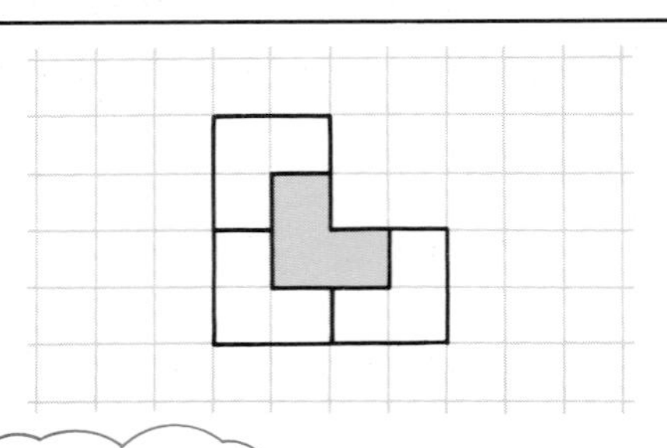

isosceles triangles · squares · rectangles · rhombuses · trapeziums · parallelograms · equilateral triangles · right-angled triangles · scalene triangles · convex kites · concave kites

all sides different lengths

Illustrate what you discover with drawings or sketches.
You may find that only very special shapes of a certain kind are reptiles.

Or you may find that **all** shapes of a certain kind are reptiles.
Mention these discoveries in your report.

Exploration *Unfolding shapes*

2 A shape is folded in half.
Then the folded shape is folded in half again.
Investigate the different possible starting shapes if the final shape is

a) a square.

b) a rectangle.

c) a parallelogram.

d) a rhombus.

e) a convex kite.

f) a concave kite.

Do the dimensions of a particular kind of shape make a difference to the possible starting shapes?
For example, do you get different possibilities for a 3 cm × 1 cm rectangle than you do for a 4 cm × 1 cm rectangle?

Quadrilaterals and circles

G Exploration *Boxing the crown*

1 Midge has an old coin called a **half-crown**.

face value $12\frac{1}{2}$p. but in good condition worth a lot more!

She wants to make a box for it.

She does not want the crown to rattle around in the box.

These boxes would be suitable.

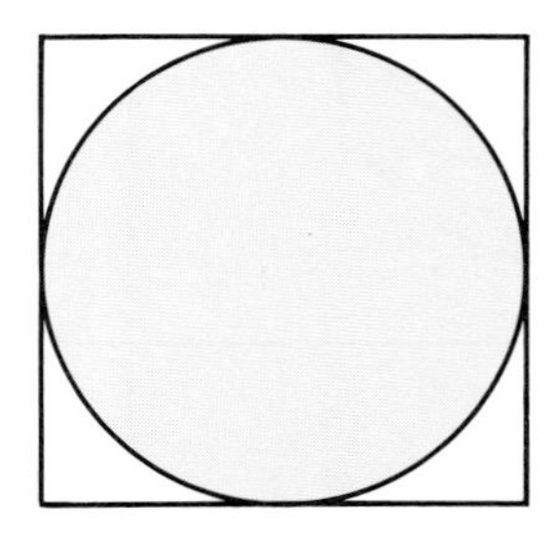

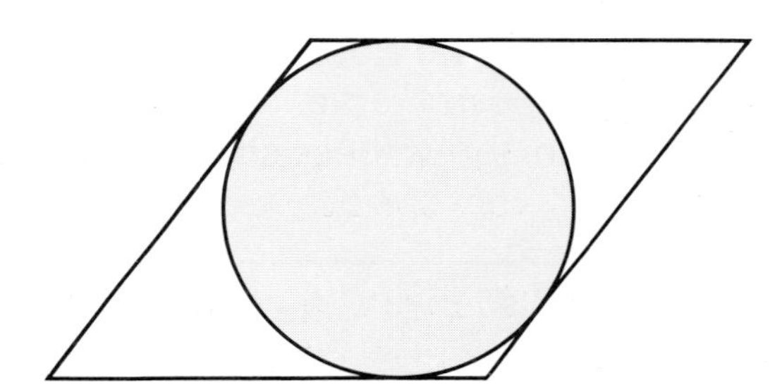

These boxes would not.

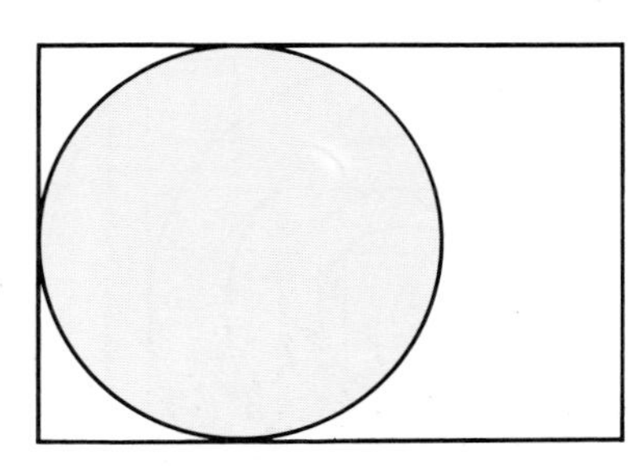

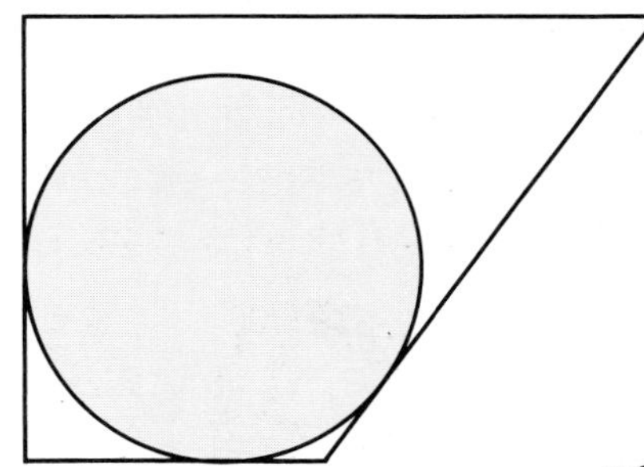

a) Sketch two more boxes which would be suitable.

b) Sketch two more boxes which would not be suitable.

c) Investigate some more quadrilateral shapes for crown boxes.
What is special about the shape of the suitable boxes?
Write down anything you discover.

3

Exploration — *Circling the quad*

2 Winston has a circular plastic tray.
He finds that one of his little sister's toy shapes fits exactly in the tray without rattling.
The toy piece is in the shape of a quadrilateral.

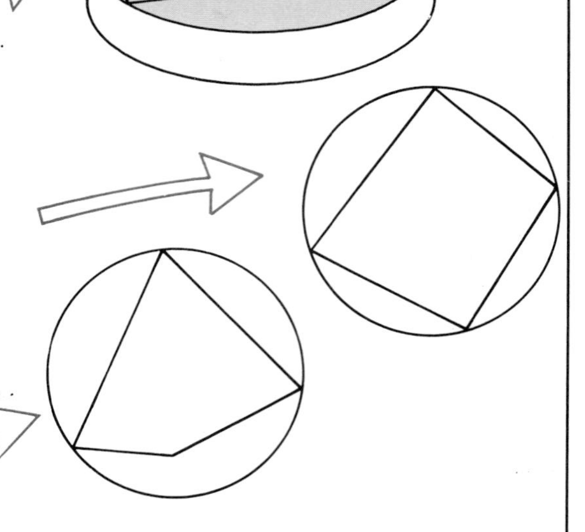

a) Investigate what kind of quadrilateral the toy piece might be.
Write down anything you discover.

b) For some of the possibilities, all four of the shape's corners touch the side of the tray.
Find out all you can about these shapes.

c) For other possibilities, only three of the shape's corners touch the side of the tray . . . but the shape still doesn't rattle.
Find out all you can about these shapes.

Challenge — *Wheels within wheels*

3 In this diagram there are six squares and five circles.

Do not use a ruler.

a) Think out how the areas of the circles relate to each other.

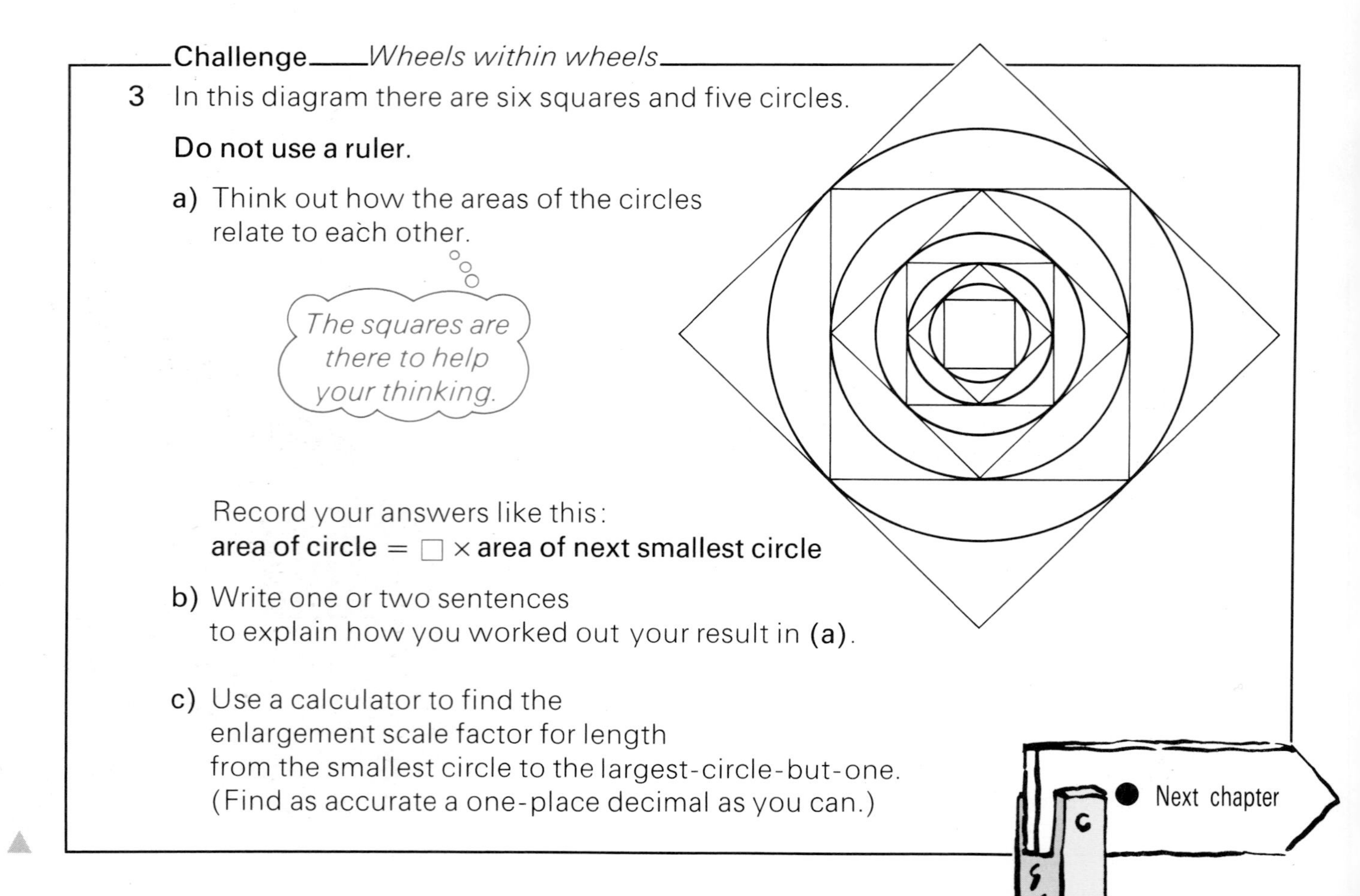

The squares are there to help your thinking.
Next chapter

Record your answers like this:
area of circle = □ × area of next smallest circle

b) Write one or two sentences to explain how you worked out your result in **(a)**.

c) Use a calculator to find the enlargement scale factor for length from the smallest circle to the largest-circle-but-one.
(Find as accurate a one-place decimal as you can.)

4 Planning and calculating

A 1 a) To estimate how much it would cost to carpet the classroom we would need to

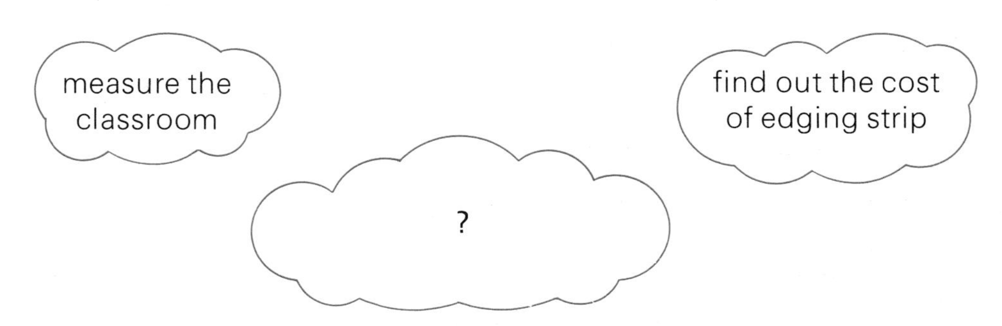

Write down one more thing we would need to do or find out.

b) Measure your classroom.
Choose one of these carpets.
Estimate how much it would cost to carpet the classroom.

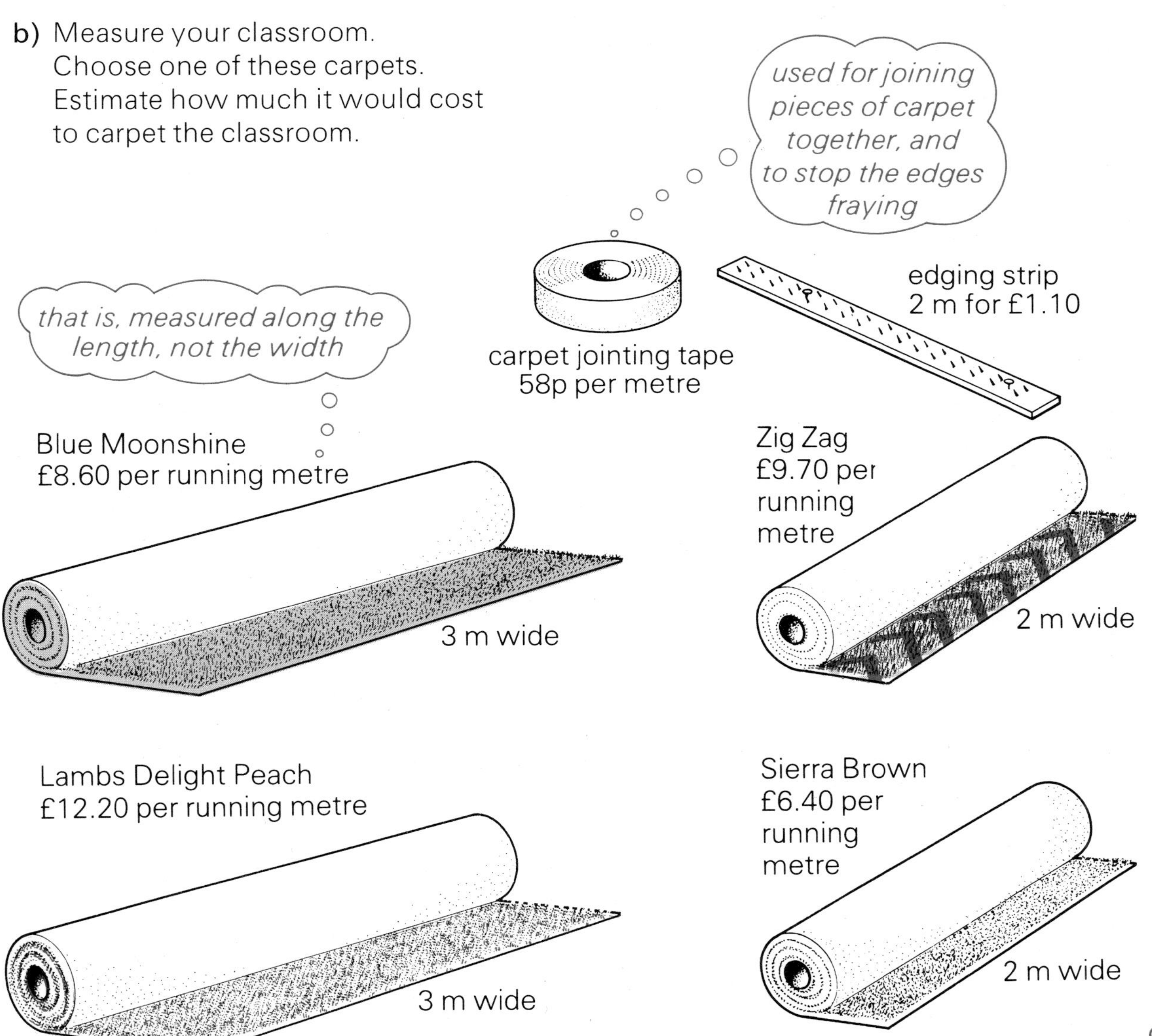

2 Jake is planning to build a brick wall.
He has in mind something like this.

The wall will be two bricks thick.

This is one of the bricks.

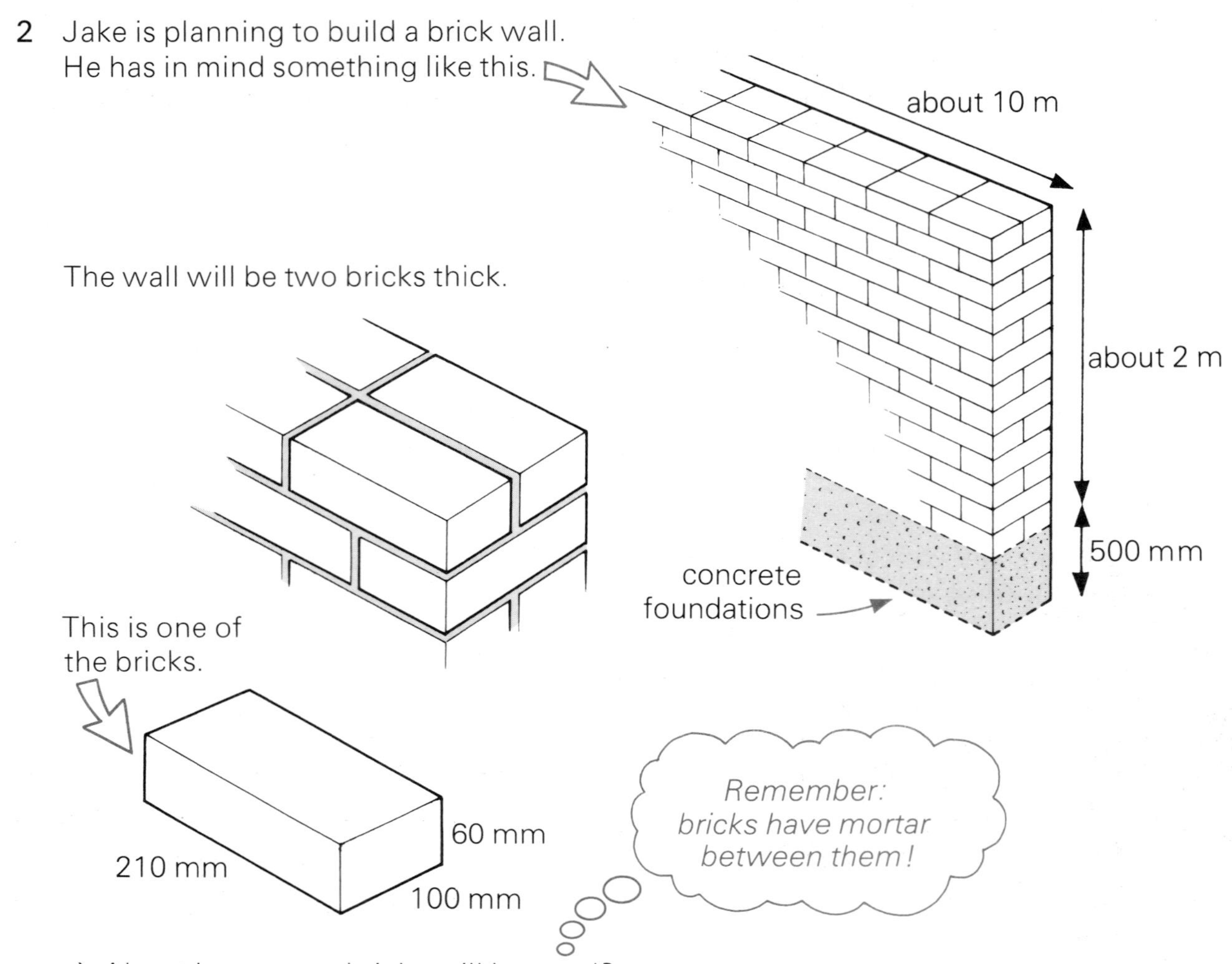

a) About how many bricks will he need?

b) A 10 kg bag of mortar is normally sufficient for 100 bricks.
About how many kilograms of mortar does Jake need?

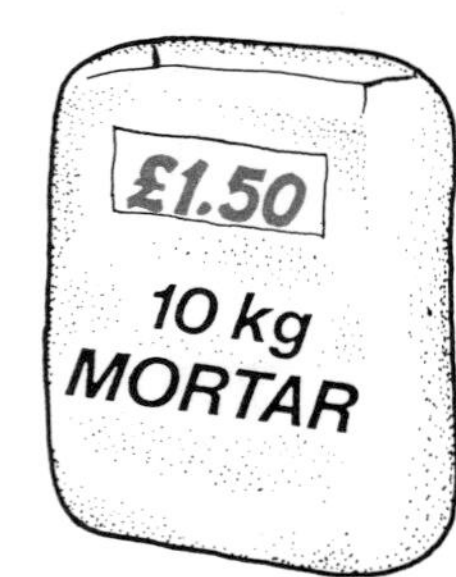

c) Bricks cost £16 per 100.
Roughly, how much will the bricks and mortar cost altogether?

Challenge

d) The wall should have a concrete foundation 500 mm deep.
About how much will the concrete cost?

READYMIX CONCRETE
£66.20 per cubic metre (including delivery)

3 Do **two** of these estimates.
You will need to plan and collect information for each one.
Write down what information you collect, and how you arrive at your estimate.

A Estimate how many hours you have spent sleeping during your life.

B Estimate how many gallons (or litres) of petrol you would need to drive a Metro all the way around mainland Great Britain.

C Estimate how many litres of liquid you drink each year.

D Estimate the area of paper a national daily newspaper uses each week.

E Estimate how many people you talk to each week.

F Estimate how many pupils in your school have a birthday in the same month as yourself.

G Estimate how much it costs to keep an average size dog for a year.

Making sensible estimates

B 1 Medical Services Ltd use cardboard trays like this.

Each tray is filled with 4 cm bandages which are individually shrink-wrapped in cardboard sleeves like this.

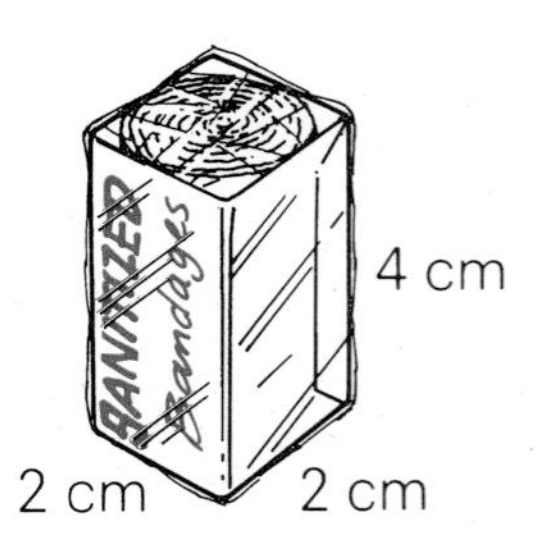

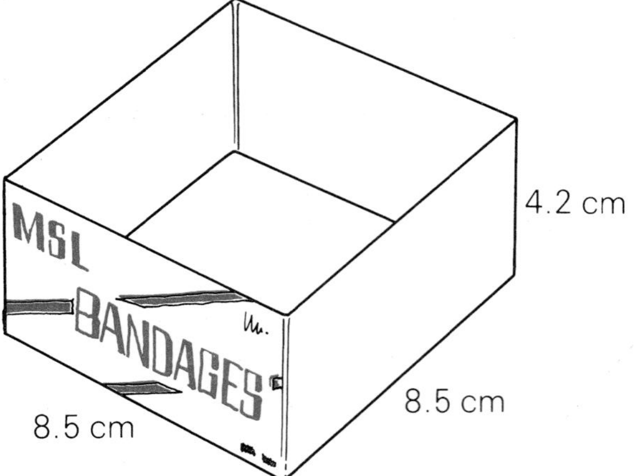

Next year MSL hope to sell 700 000 bandages.

a) About how much card is needed to make all the bandage sleeves?

b) About how much card is needed to make all the trays?

c) Each tray costs £0.92.
About how much will all the trays cost?

When the trays are cut out of sheets of cardboard, some cardboard will be wasted.

With a friend

2 Imagine that you have just been promoted to headteacher!
The Chief Education Officer says that every child must have 924 hours of lessons each year.
Each year there must be 198 school days.
Design your own school day.

Start of registration (Time?)
End of registration (Time?)
Start of lesson one (Time?)

Not the particular lessons – just the times of lessons and breaks, and so on.

Discuss the type of day you would like best.
Remember to include breaks and lunch.
Draw a diagram to show what you decide.

3 Estimate how much it would cost to make the jeans.

Decide yourself what you would have to buy.
List everything you buy.
Write down its cost.

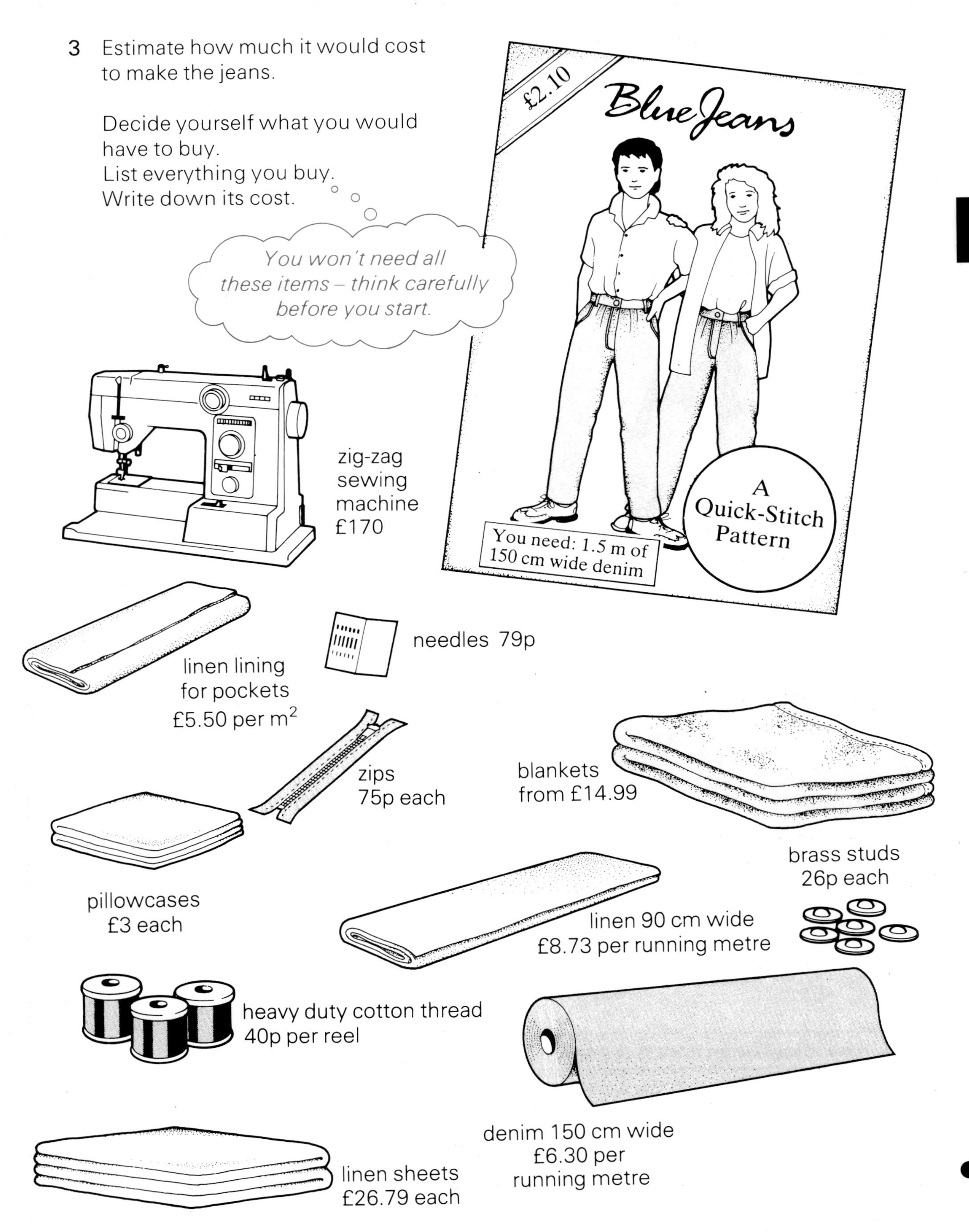

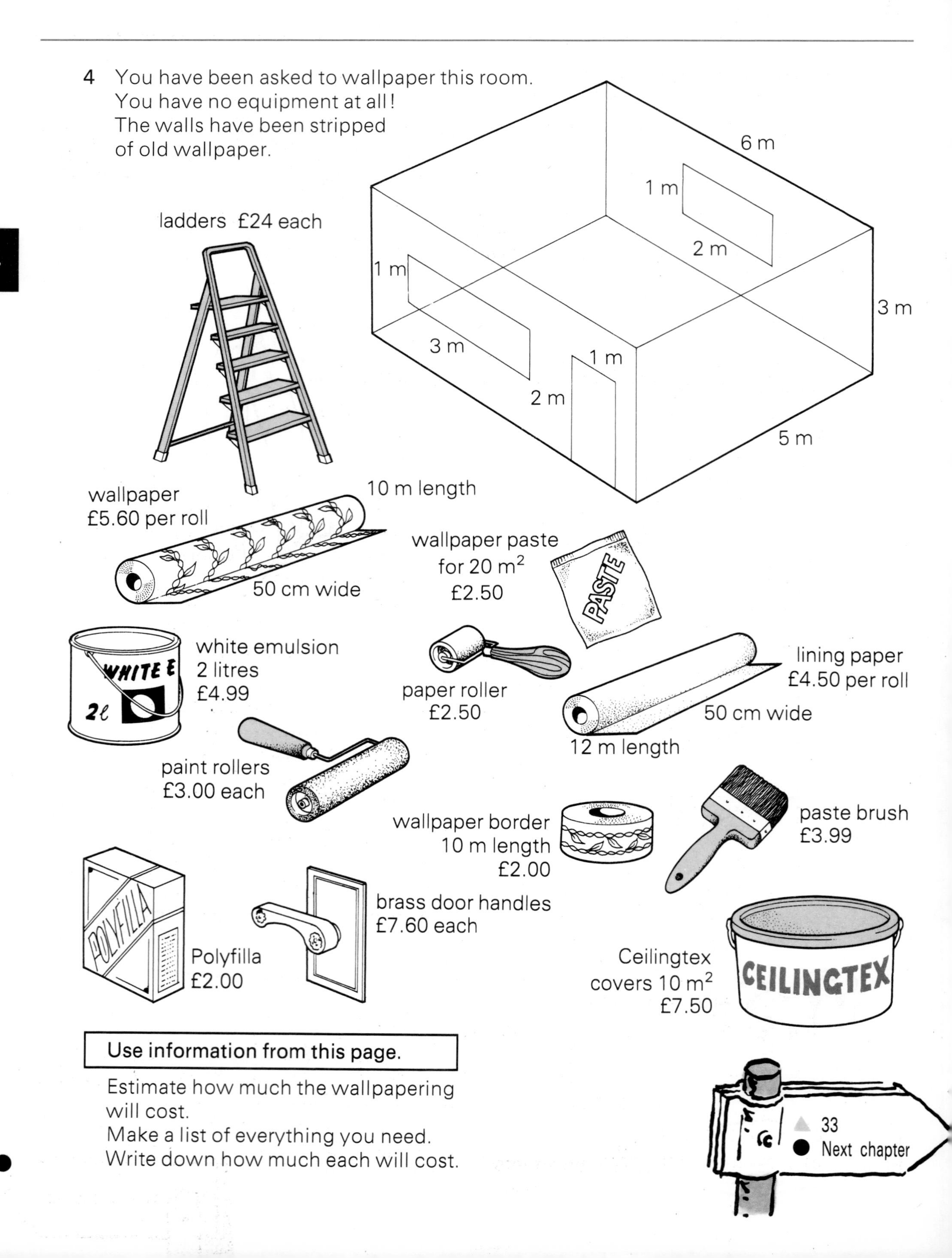
4 You have been asked to wallpaper this room.
You have no equipment at all!
The walls have been stripped
of old wallpaper.
6 m
1 m
2 m
1 m
3 m
3 m
1 m
2 m
5 m
ladders £24 each
wallpaper
£5.60 per roll
10 m length
50 cm wide
wallpaper paste
for 20 m²
£2.50
PASTE
white emulsion
2 litres
£4.99
WHITE E
2ℓ
paper roller
£2.50
lining paper
£4.50 per roll
50 cm wide
12 m length
paint rollers
£3.00 each
wallpaper border
10 m length
£2.00
paste brush
£3.99
POLYFILLA
Polyfilla
£2.00
brass door handles
£7.60 each
Ceilingtex
covers 10 m²
£7.50
CEILINGTEX
Use information from this page.
Estimate how much the wallpapering
will cost.
Make a list of everything you need.
Write down how much each will cost.
33
Next chapter

A day out

C 1 You are planning a day out.
You will go by train from Tadcastle to Bexminster, stopping at Cayton on the way there.

Tadcastle — Cayton — Bexminster

You have £12 to spend.

a) Use the information on this page to organise a possible day for yourself.

TADCASTLE→ROMPTON

Tadcastle	8:20	9:00	9:40	10:20	11:00	11:40
Maypole	8:31	9:12	9:52	10:31	11:10	11:52
Cayton	8:48	9:29	10:09	10:51	11:25	12:10
Seaton	8:53	9:34	10:14	10:55	11:30	12:14
Bexminster	9:15	9:55	10:37	11:18	11:52	12:35
Rompton	9:24	10:05	10:46	11:27	12:04	12:45

Tadcastle – Bexminster £3 single, £5 return

Draw up a timetable and keep a note of what you spend.

At Cayton you visit the castle.

CAYTON CASTLE
Admission £1.20
Guided tours at
9:45 am, 10:45 am, 11:45 am
Tours are FREE
and last about 50 minutes

CAYTON CAFE
Soft Drinks: 25p, 27p, 32p
Ice Creams: 30p, 40p, 50p
and a wide selection of confectionary.

At Bexminster there are three interesting places to visit and two cafes.

BEXMINSTER CAVES
Entrance: £1.75
Program: 70p
Tour takes 90 minutes.

BEXMINSTER ROLL-A-DROME
£1 Entrance
75p Skate hire
15p Locker fee

BEXMINSTER ART AND HISTORY MUSEUM
Entrance free
(Donation welcome)

ROSY'S CAFE

EGG – CHIPS	£1·25
BURGER – CHIPS	£1·55
PIZZA – CHIPS	£1·75
PIE – CHIPS	£1·35
COLA · MILK · JUICE	£0·45

THE CHIPPY

CHIPS	40p	SAUSAGES	35p
PIES	55p	PIZZA	£1·85p
FISH	85p	CHICKEN	£1·05
		PEAS	25p

Don't forget to keep an eye on time, and remember £12 doesn't go a long way.

ROMPTON→TADCASTLE

Rompton	14:00	14:40	15:20	16:00	16:40	17:20	18:
Bexminster	14:10	14:49	15:31	16:10	16:52	17:29	18:
Seaton	14:31	15:01	15:54	16:33	17:13	17:49	18:
Cayton	14:36	15:07	16:00	16:38	17:19	17:54	18:
Maypole	14:54	15:25	16:17	16:55	17:37	18:13	18:5
Tadcastle	15:05	15:37	16:28	17:05	17:49	18:24	19:0

b) How long is your day? (from Tadcastle station and back again)

c) How much money do you have left at the end of the day?

5 Thinking in three dimensions

A 1 Horace and his friends meet at the local cafe.
On their table is a ketchup bottle, a sugar bowl and a salt pot.

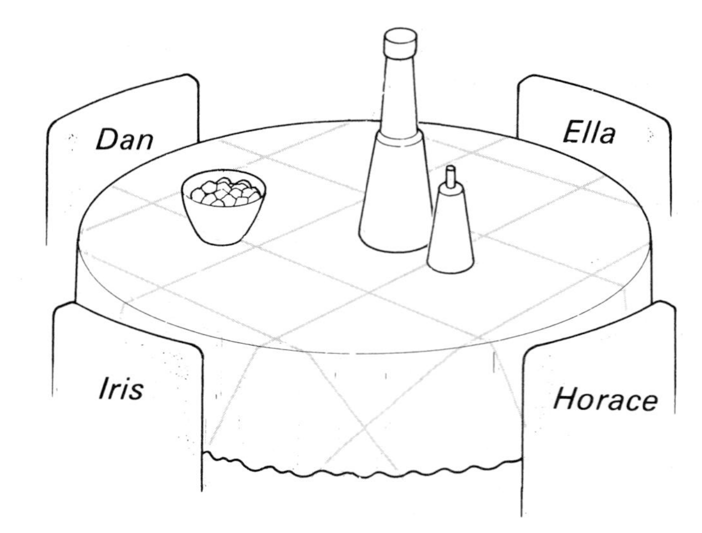

a) Look at these **elevations** of the ketchup, sugar and salt.

View **A** is Horace's elevation.

Whose elevation is
(i) **B**?
(ii) **C**?

A

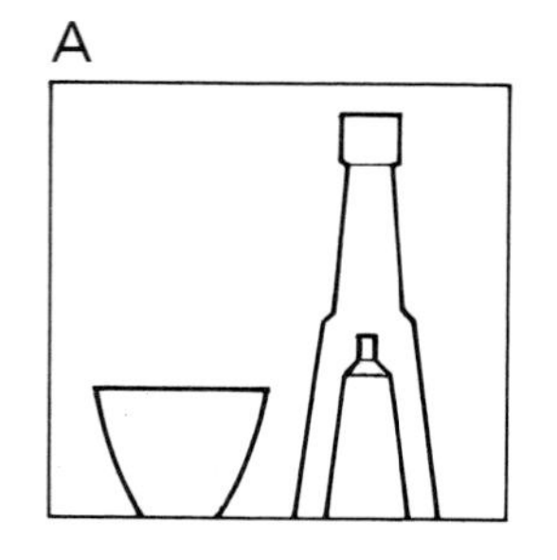

B

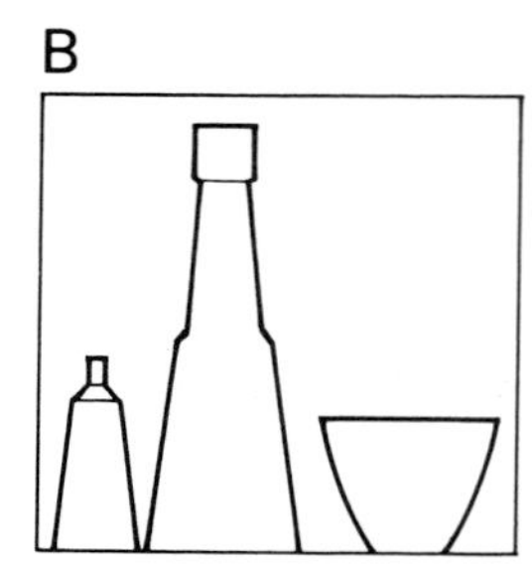

C

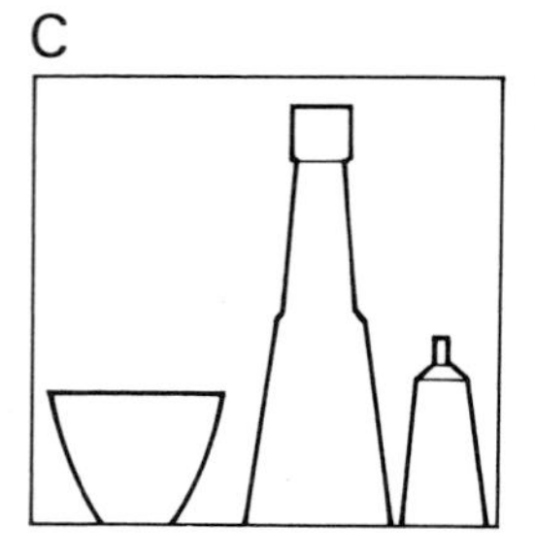

b) Sketch the fourth person's elevation.

c) Look at these **plans** of the ketchup, sugar and salt.

View **A** is Horace's plan.

Whose plan is
(i) **B**?
(ii) **C**?

A

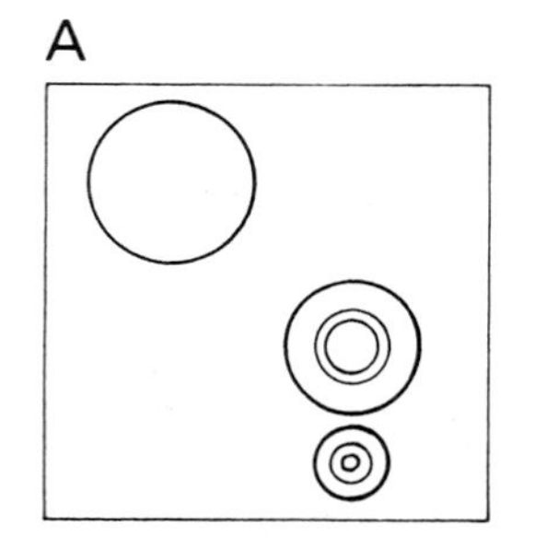

B

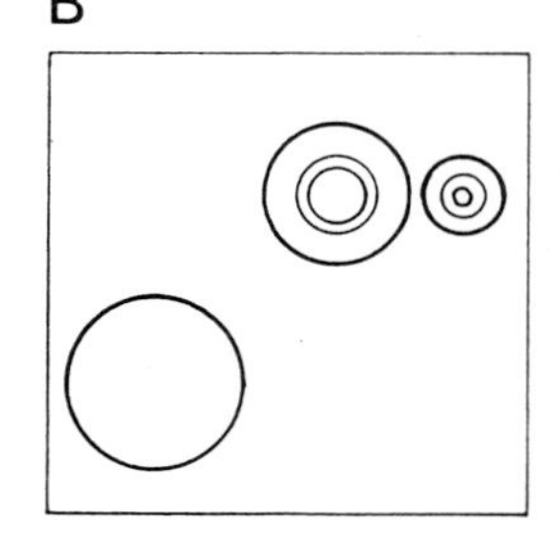

C

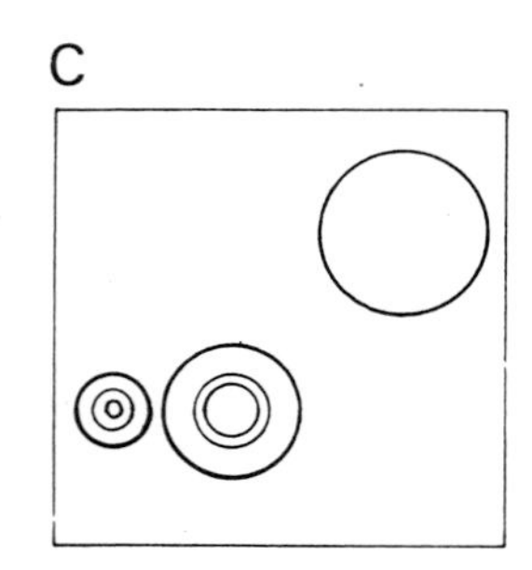

d) Sketch the fourth person's plan.

2 Horace moves the ketchup bottle towards Iris.
His elevation now looks like this.

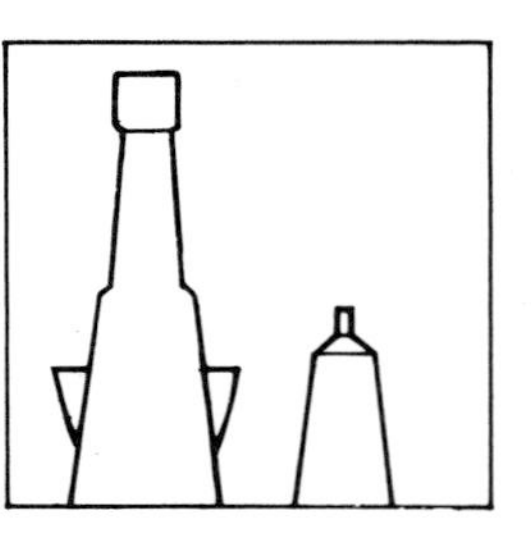

Sketch Dan's

a) plan.

b) elevation.

3 Horace now moves the sugar towards Ella.
His plan now looks like this.

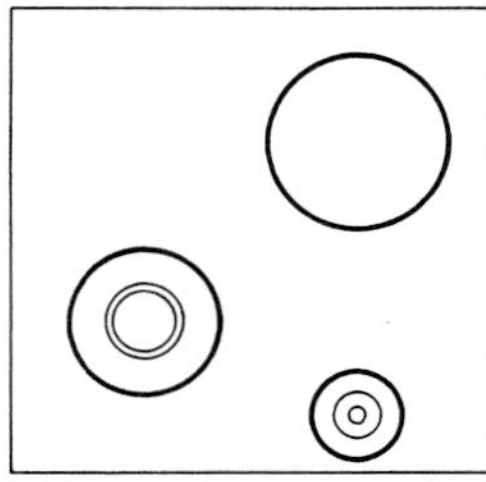

a) Sketch Horace's elevation.

b) Sketch Dan's
  (i) plan.
  (ii) elevation.

Think it through

4 Penny arrives and pulls up a chair.
Her elevation looks like this.

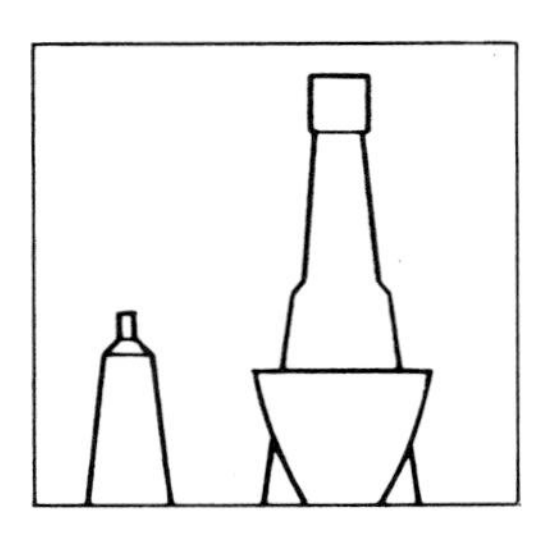

Between which two people is she sitting?

5 Horace moves the ketchup bottle again.
His elevation now looks like this.

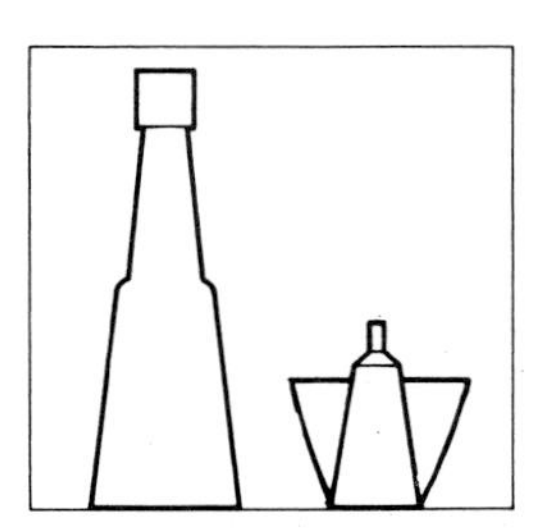

Which of these could be
Ella's elevation?

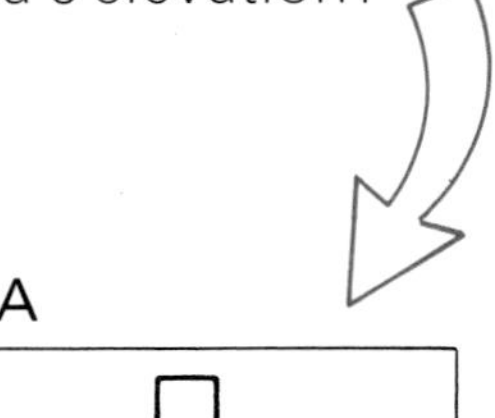

A

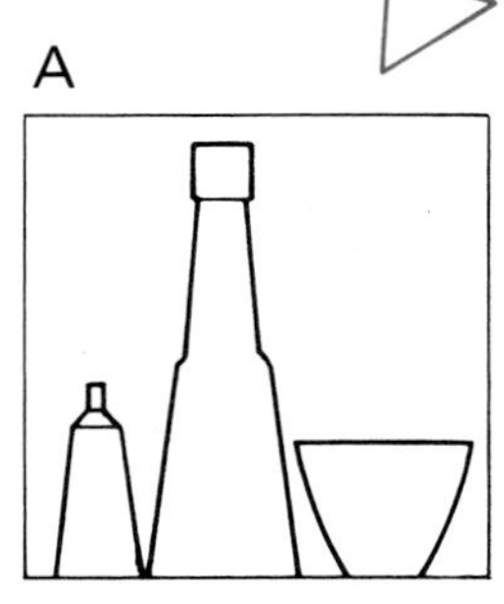

B

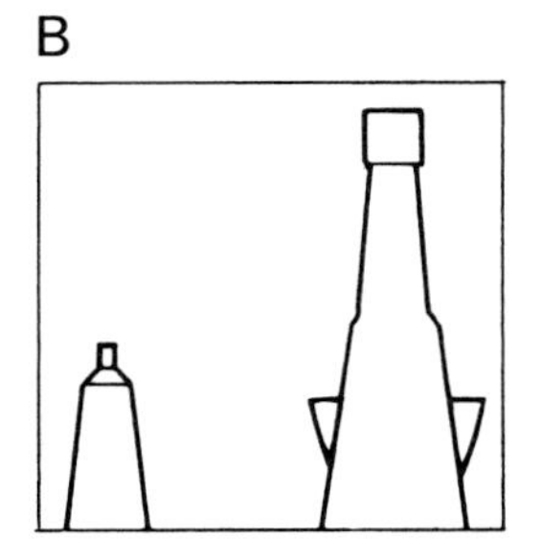

C

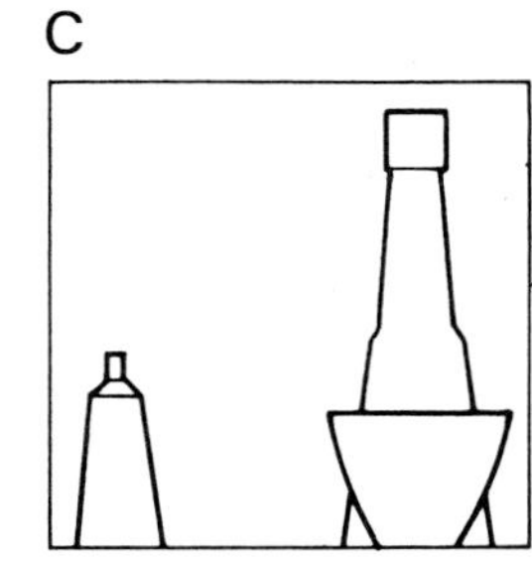

D

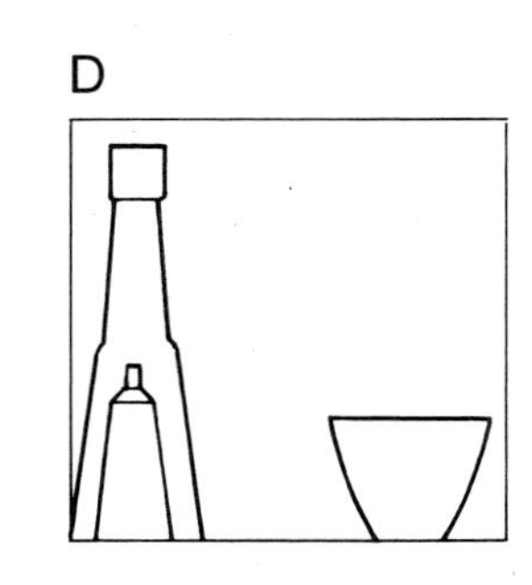

6 Horace moves the salt.
Here are his and Ella's new elevations.
Draw Dan's plan.

Horace's elevation

Ella's elevation

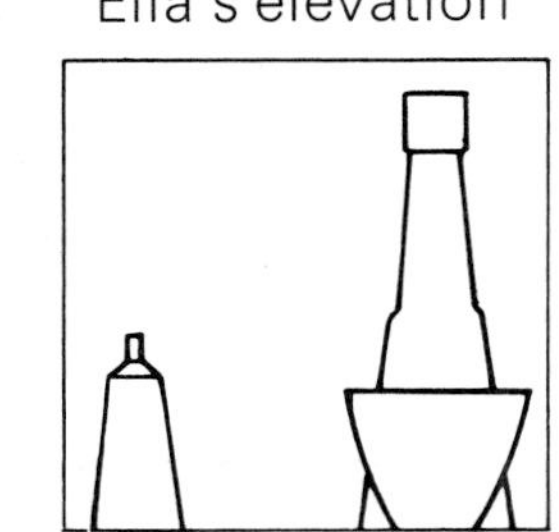

Plans and elevations

B You need 1 cm dotted isometric paper,
1 cm dotted squared paper,
five 1 cm cubes,
and a 5 cm square marked out like this.

C
B D
A

1 a) Build this object on your 5 cm square.
(Use all your cubes.)

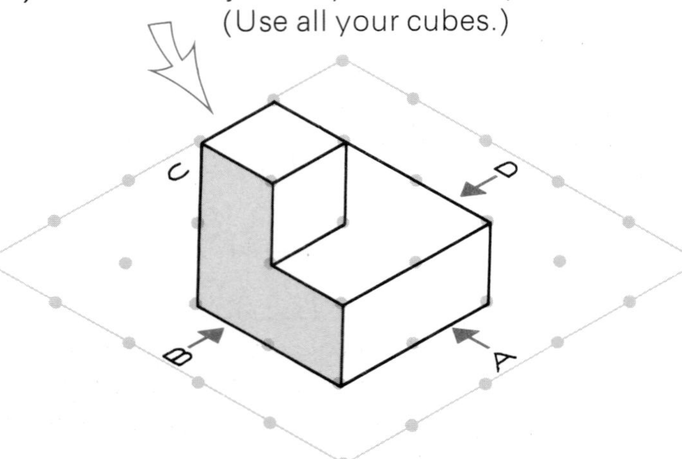

b) Here are the object's plan and elevation from B.
Draw the plan and elevation from C.

plan from B

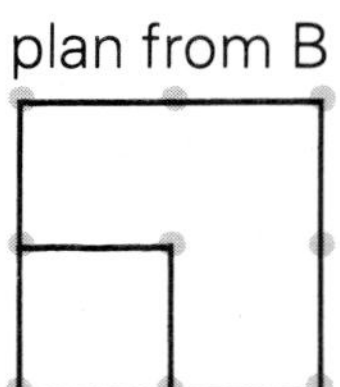

elevation from B

c) Compare these elevations from A and from B.

elevation from A

elevation from B

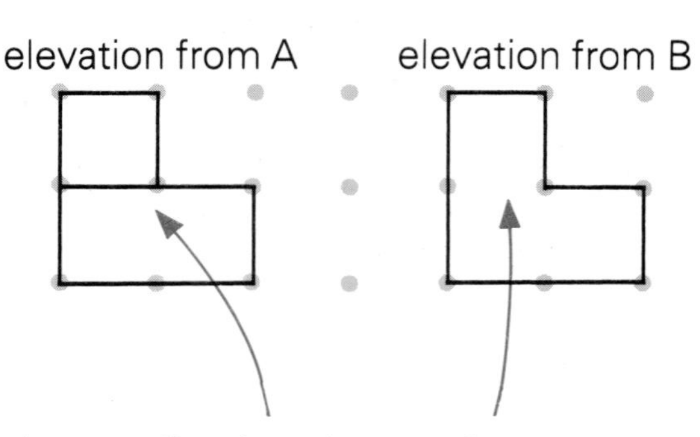

Explain why we have drawn a line here but not here.

d) Look at these two elevations.
Which is from C and which is from D?

elevation 1

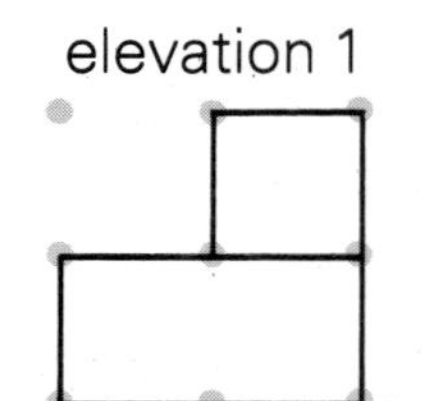

elevation 2

2 a) Build this object on your 5 cm square.
(You will need all your cubes.)

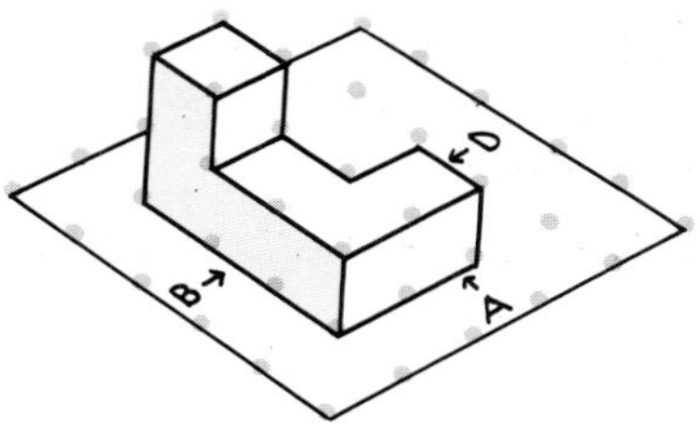

b) Here are the plan and elevation from A.

plan from A — elevation from A

Draw the
(i) plan from B.
(ii) elevation from B.

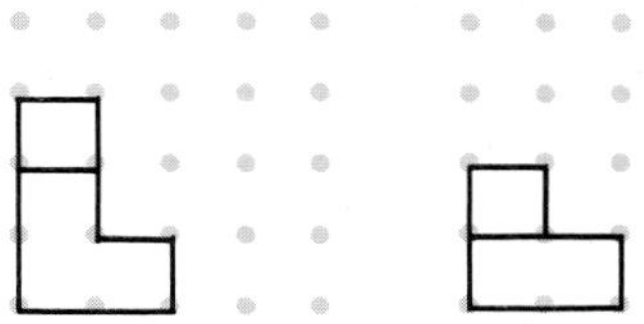

c) Look at these drawings.

1 2 3 4

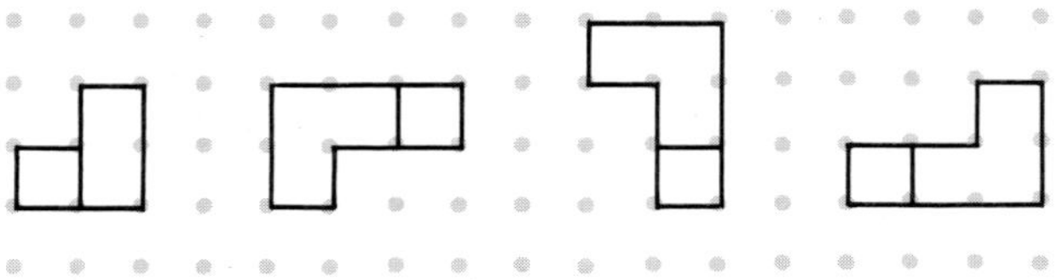

(i) Which drawings could be plans of the object in part **(a)**?
(ii) Which ones could be elevations?

3 a) Make the object that has these views.
(You will need all your cubes.)

b) Think of another five-cube object with the same plan view from A.

Draw its elevation from A.

plan from A — elevation from A

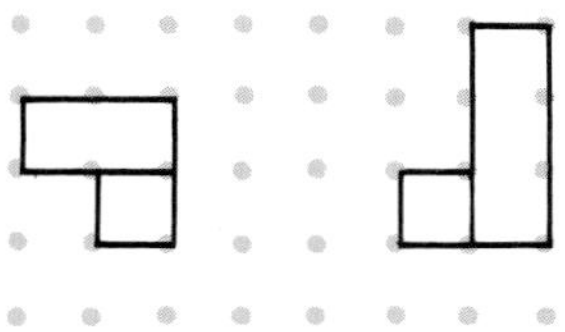

4 These views are of a five-cube object.

Draw its elevation from B.

plan from A — elevation from A

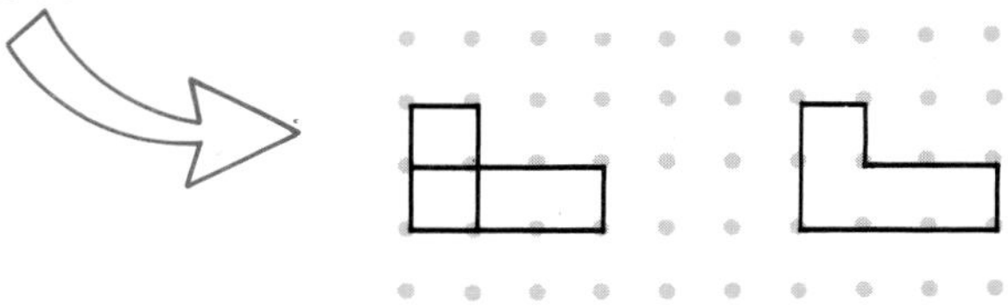

5 Think of a five-cube object whose plan from A and elevation from A are the same.

Draw your object on isometric paper.

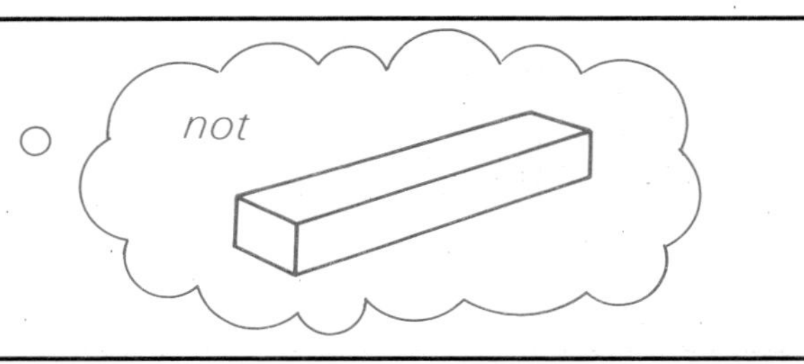

6 a) Three of these shapes can be posted through this slot.

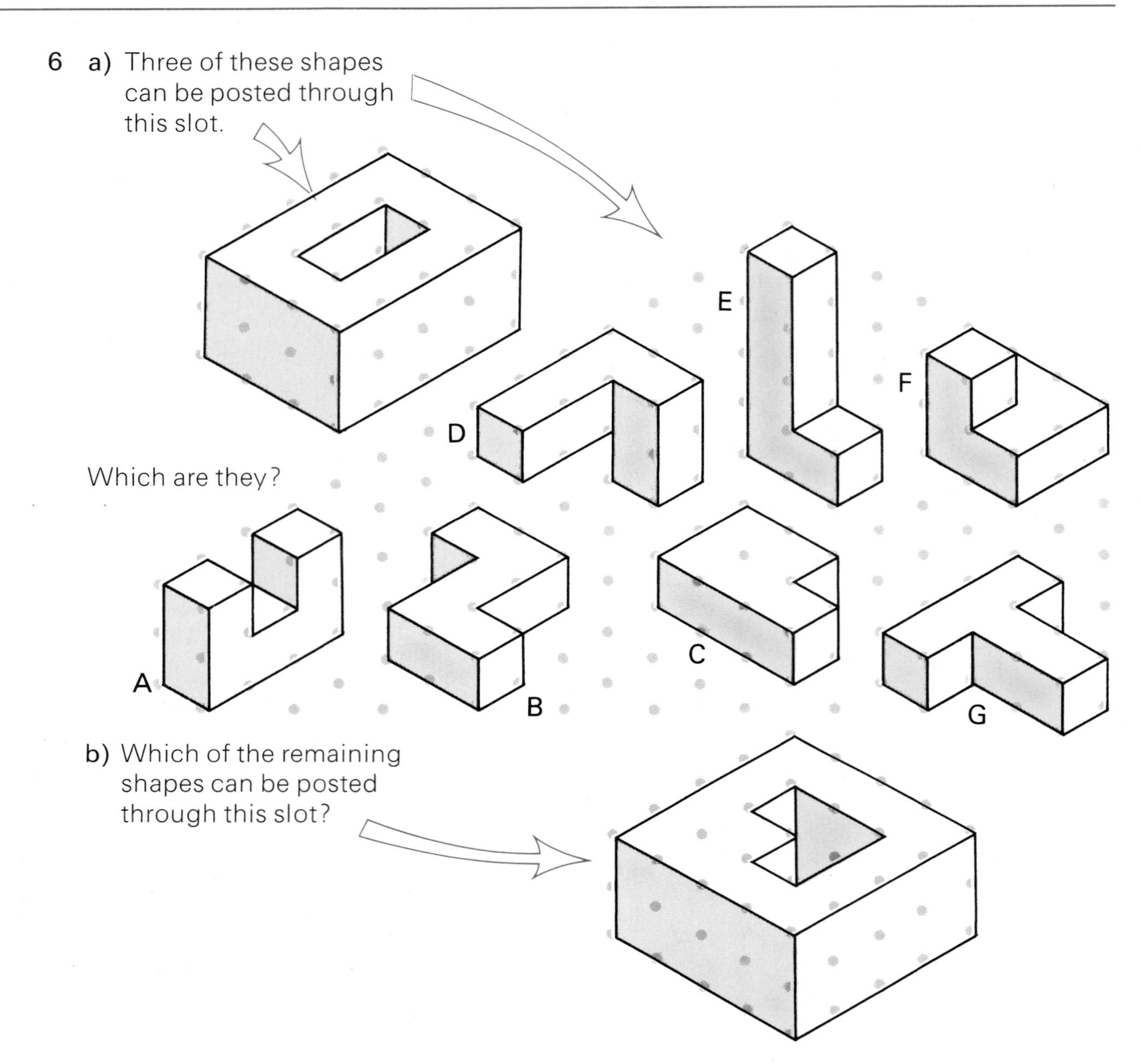

Which are they?

b) Which of the remaining shapes can be posted through this slot?

Think it through

c) Two of the shapes fit together to make this shape. Which two?

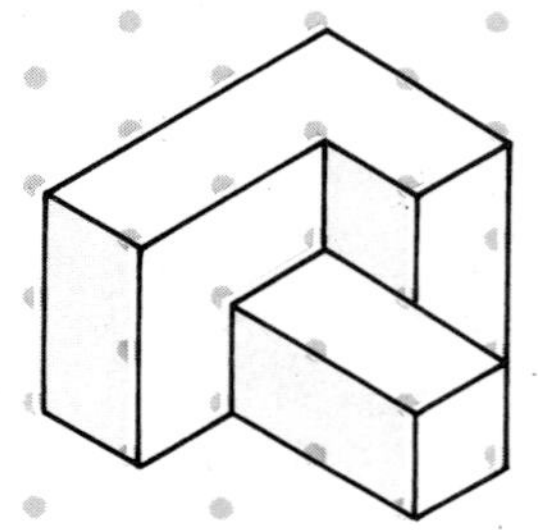

d) Two of the shapes fit together to make this shape. Which two?

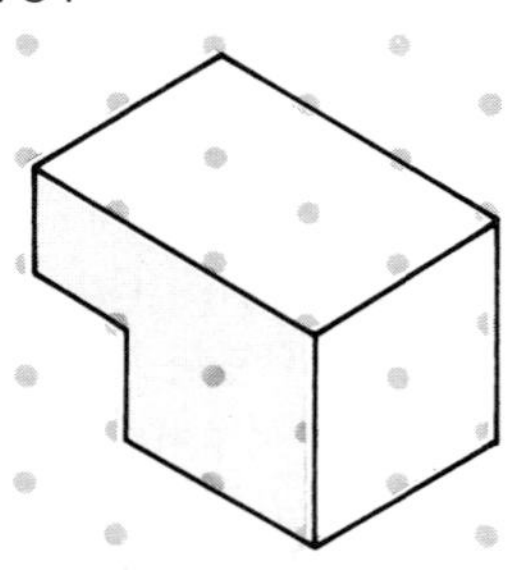

Model houses

C 1 Horace has two rectangular blocks and three triangular blocks.
He puts them together to make this house.

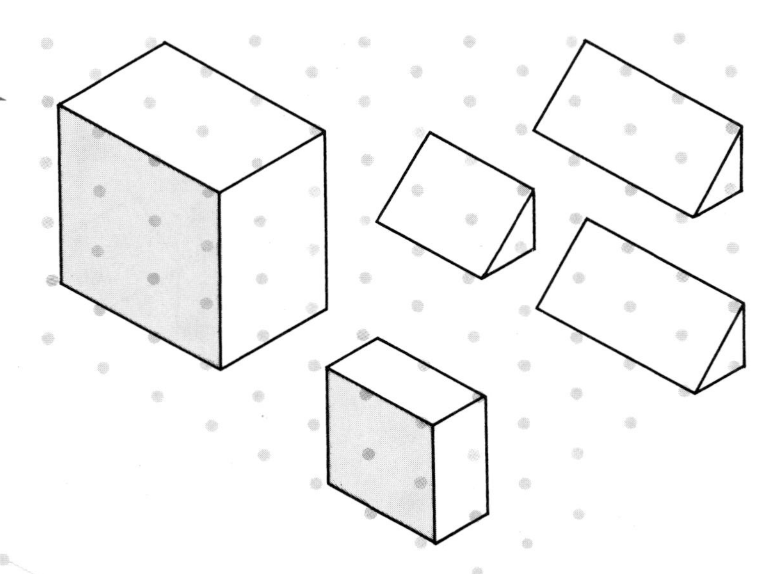

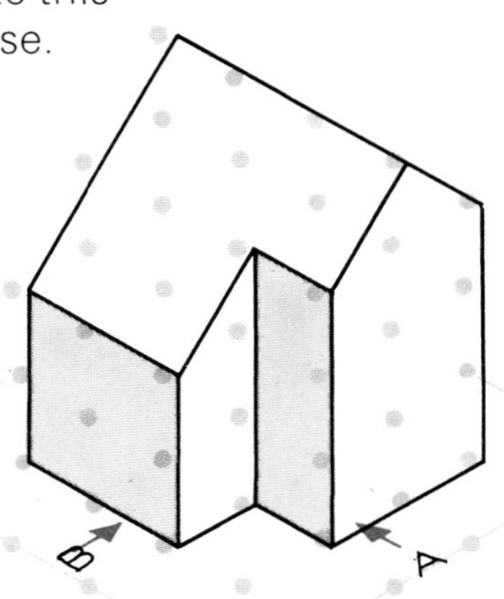

Here are the plan and elevation from A.

a) Draw
   (i) the plan from B.
   (ii) the elevation from B.

b) Draw the elevation
   (i) from C.
   (ii) from D.

plan from A

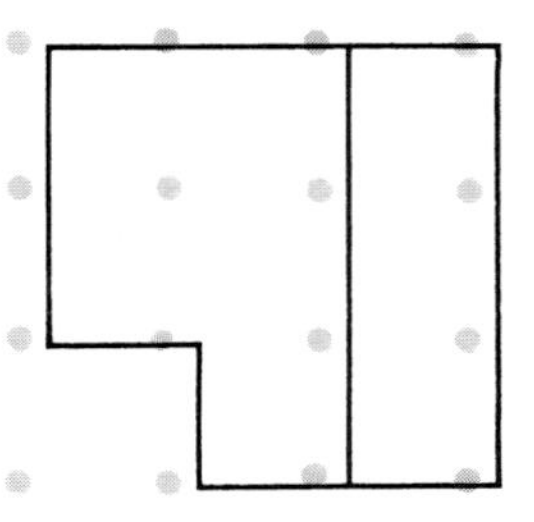

elevation from A

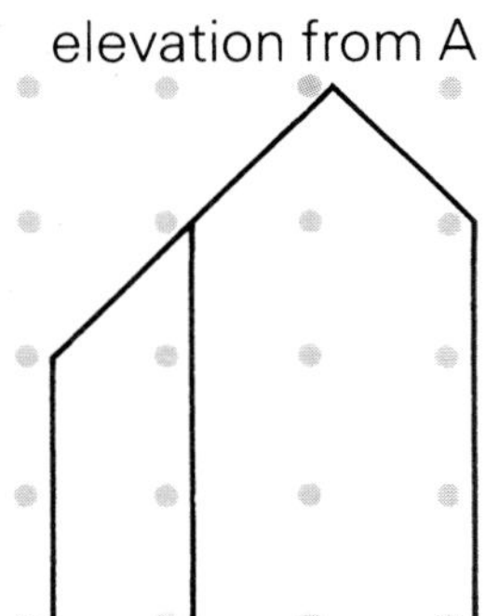

2 Horace rearranges the pieces like this

. . . and then like this.

a)

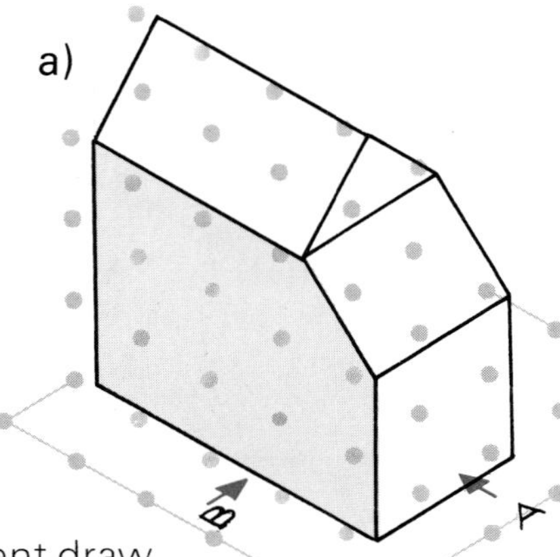

b)

For each arrangement draw
 (i) the plan from A.
 (ii) the elevation from A.

3 Here are some more arrangements of the pieces.

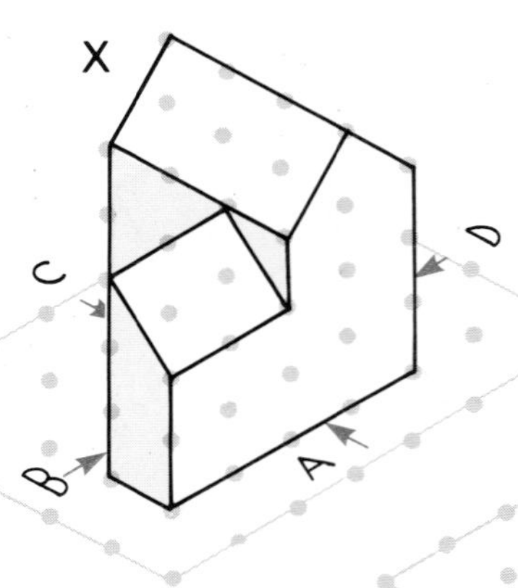

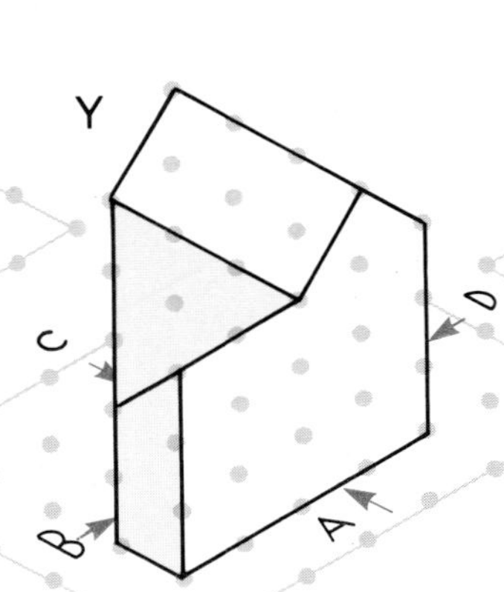

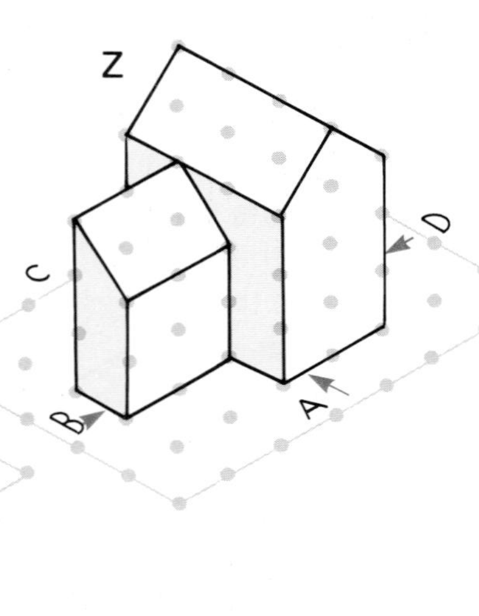

a) These are the elevations from A.
Match **1**, **2**, **3** with **X**, **Y**, **Z**.

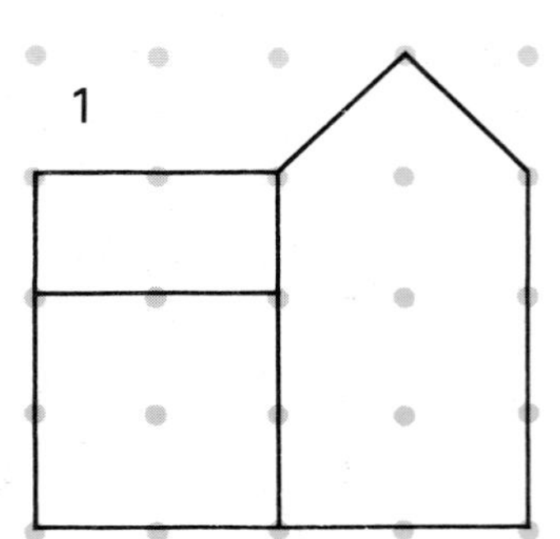

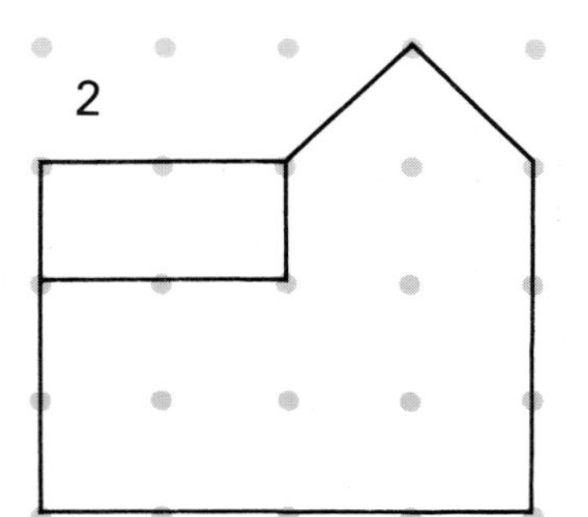

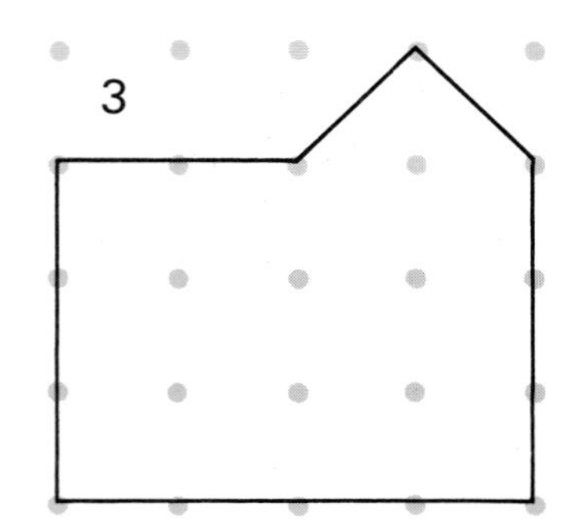

b) (i) These are elevations from B. Match them with **X**, **Y** or **Z**.

(ii) Draw the missing elevation from B.

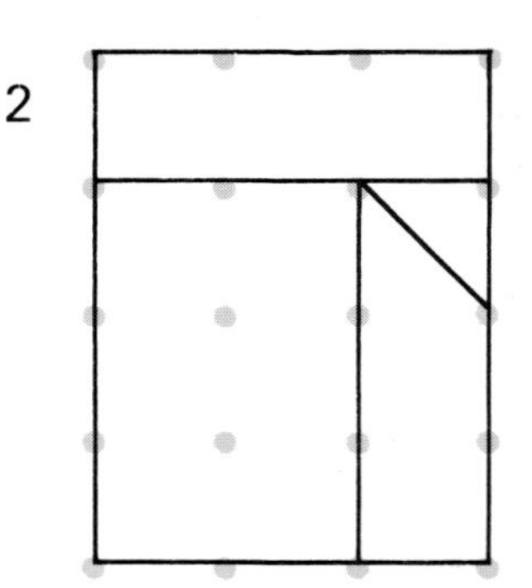

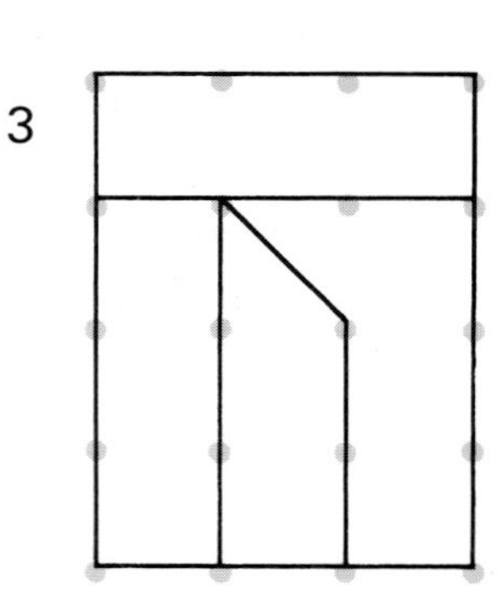

c) Two of the arrangements have the same plan from B. Which two?

Think it through

d) Which arrangements have
(i) the same elevation from C?
(ii) the same elevation from D?

Slanting planes

D 1 Here are some blocks.

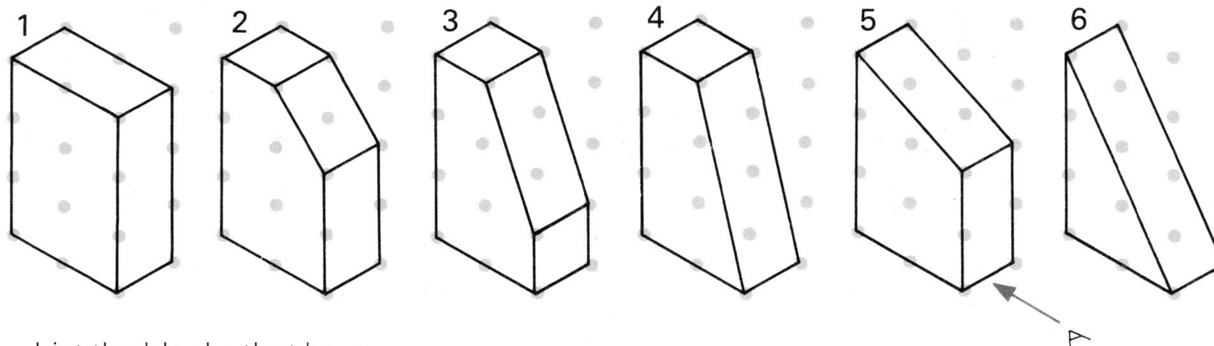

List the blocks that have

a) this plan from A.
b) the same elevation from A as block **1**.
c) the same elevation from A as block **2**.

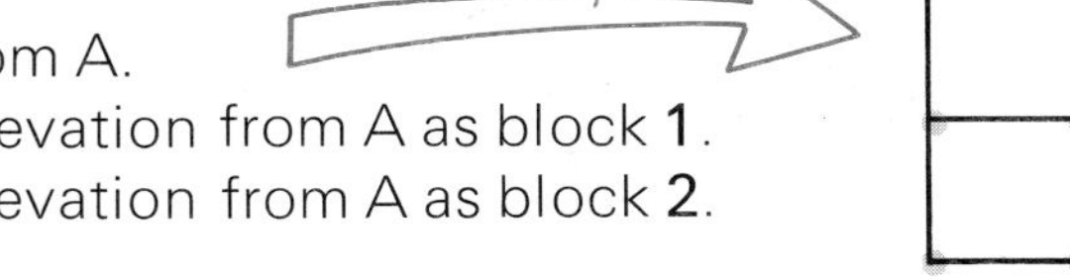

2 Here are some more blocks.

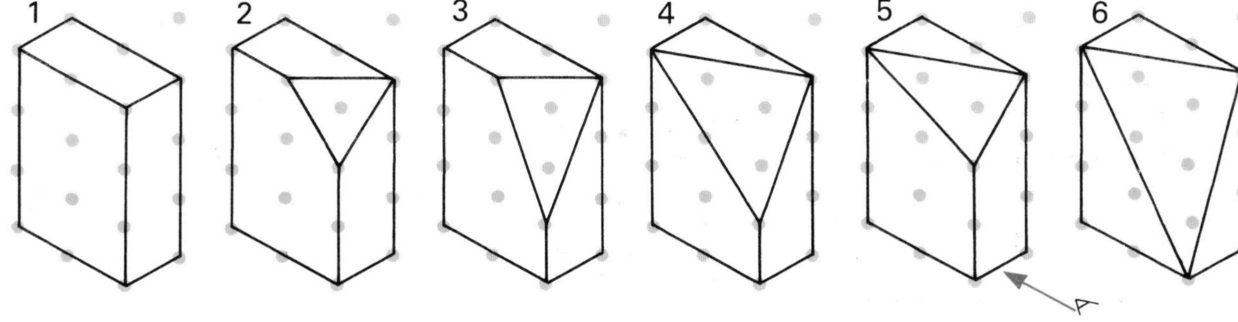

a) List the blocks that have (i) this plan from A. (ii) this elevation from A.

b) Draw the plan and elevation from A of
 (i) block **5**.
 (ii) block **6**.

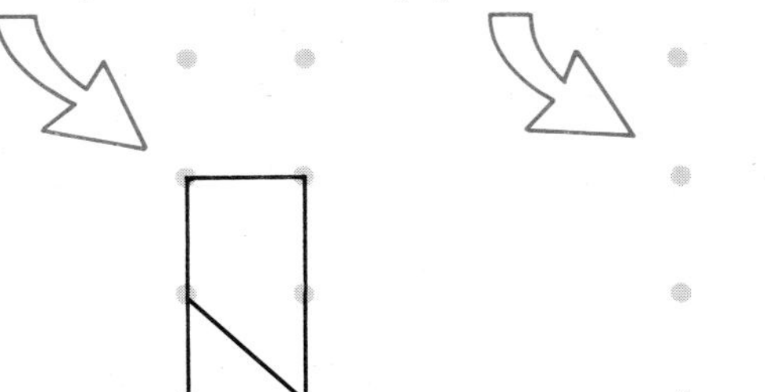

Think it through

3 Horace has a block of wood.

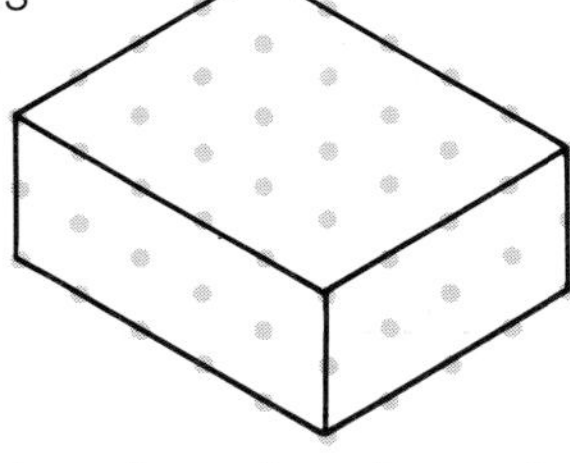

He saws a bit off.

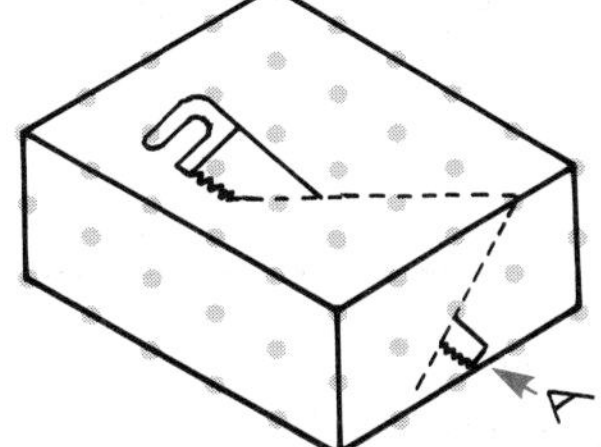

The plan from A now looks like this.

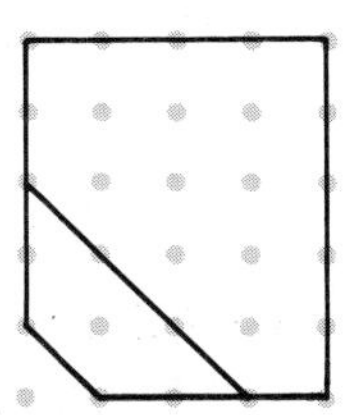

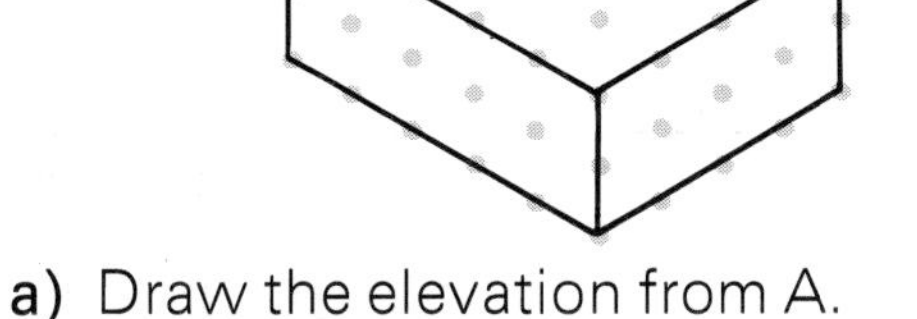

a) Draw the elevation from A.
b) Draw, on isometric paper, the piece of wood that is left.

Pyramids

E 1 In this rectangular-based pyramid, this vertex is directly above the point X of the base.

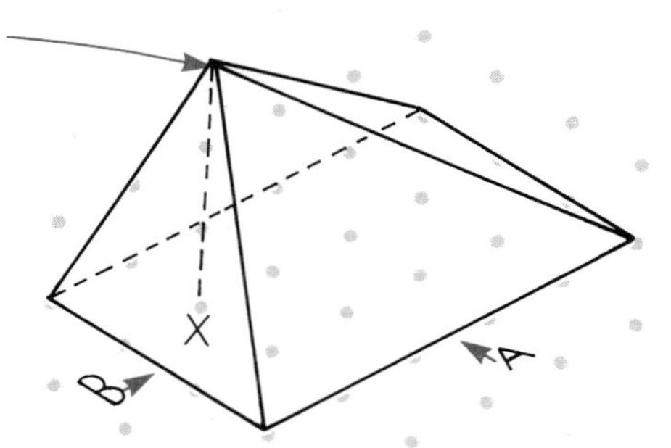

Here are the plan and elevation from A.

plan from A

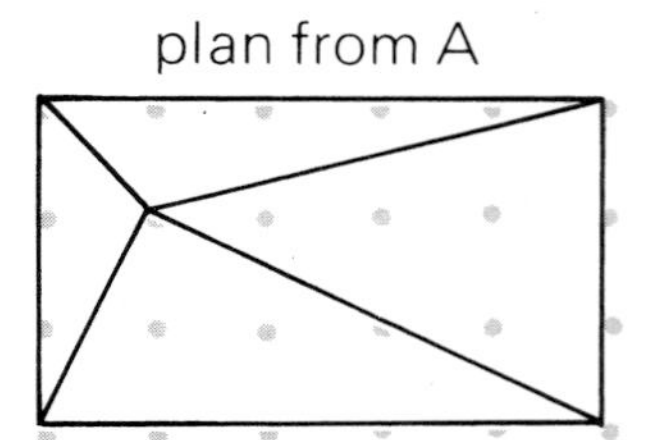

elevation from A

Draw the plan and elevation from B.

2 Here are some more rectangular-based pyramids.

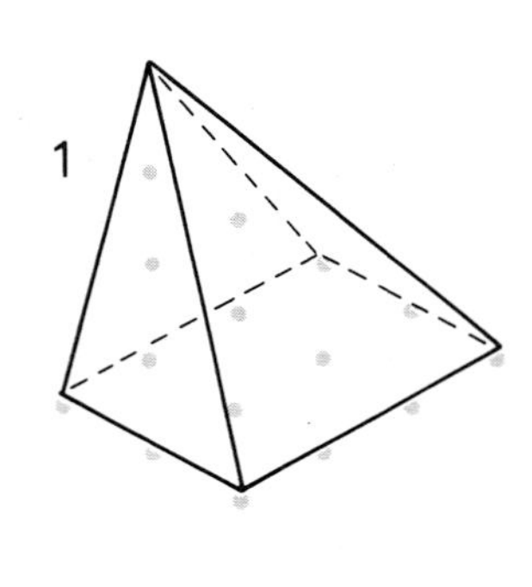

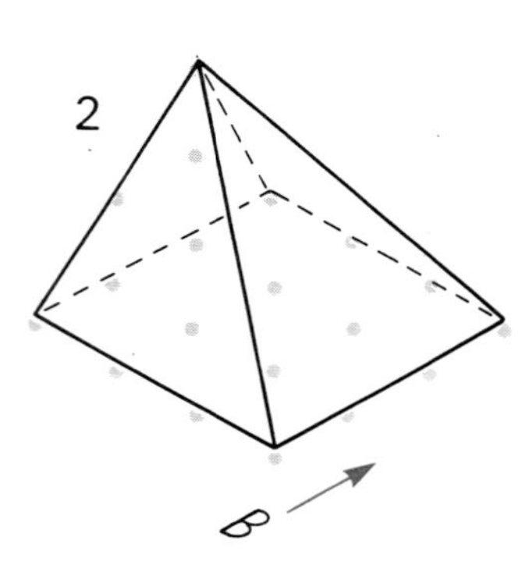

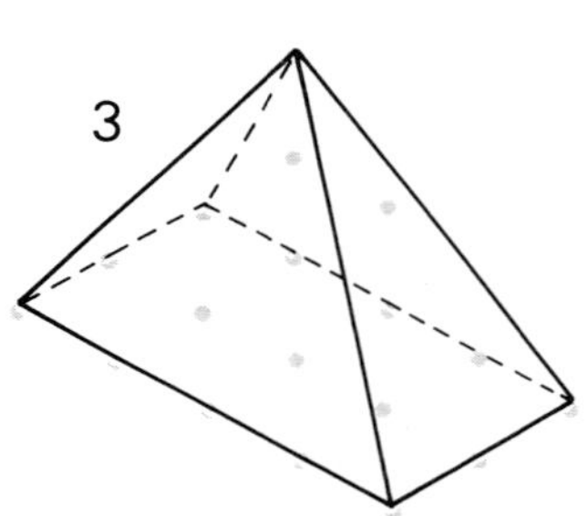

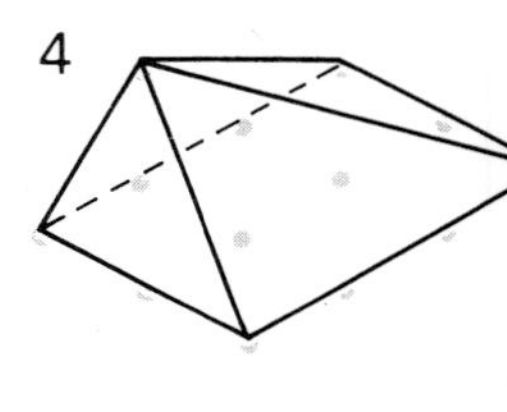

These are some of their plans and elevations from B.
Draw the missing views.

	1	2	3	4
plan from B		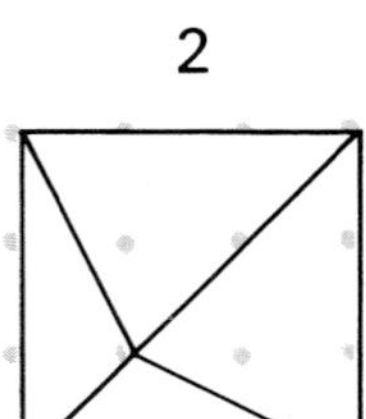		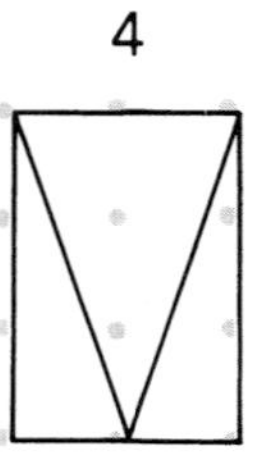
elevation from B	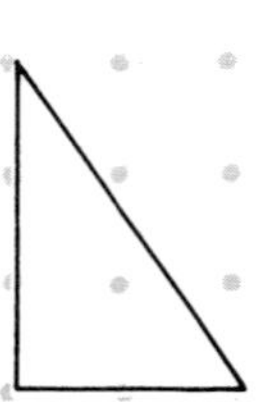		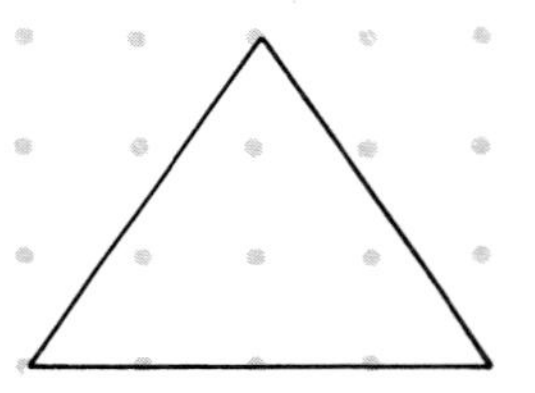	

Sloping roofs

F 1 Here is a drawing of a house.
Point X is in the horizontal plane EFGH.
V is directly above X.

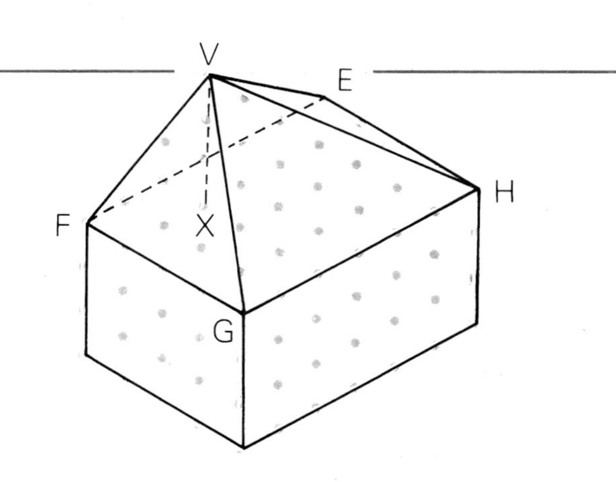

The roof has four faces:
VEF, VFG, VGH and VHE.

a) Which face is
   (i) the steepest?
   (ii) the least steep?

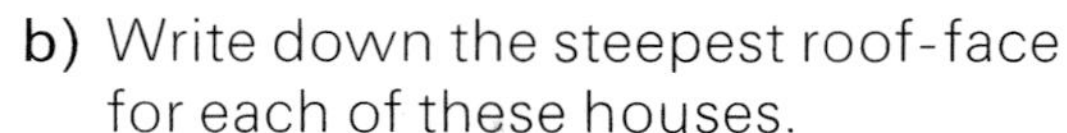

b) Write down the steepest roof-face for each of these houses.

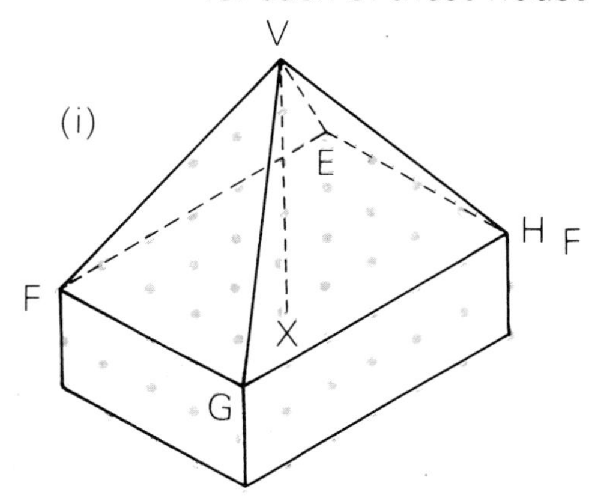

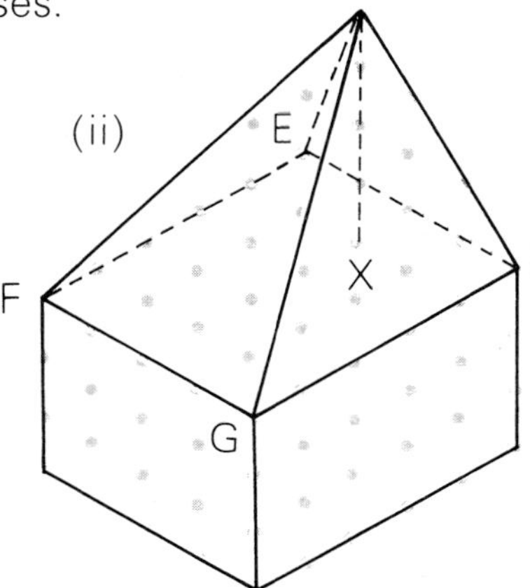

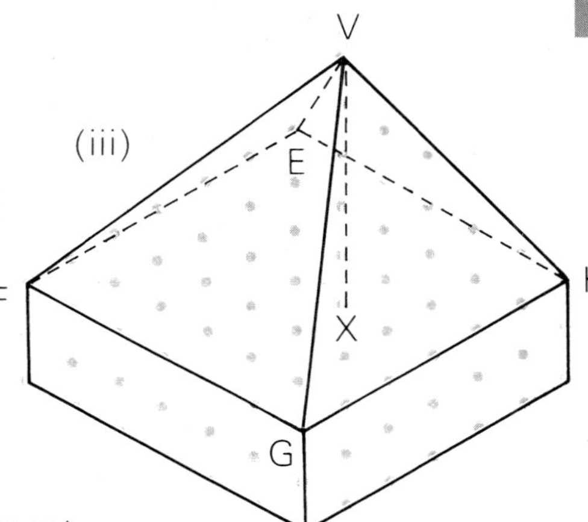

Explain why you think these roof-faces are the steepest.

5

2 Here is another house and its elevation from A.

a) Which is steeper, VFG or VHE?

b) Which of the four roof-faces is the steepest?

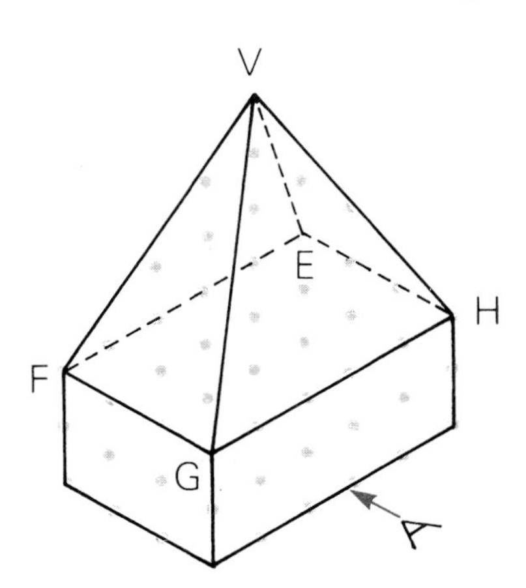

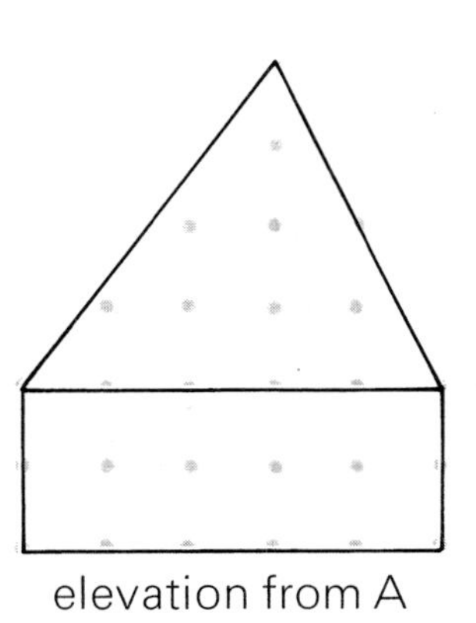

elevation from A

3 Here is another house and its elevation from A.

a) Which is steeper, VFG or VHE?

b) Which of the four roof-faces is the steepest?

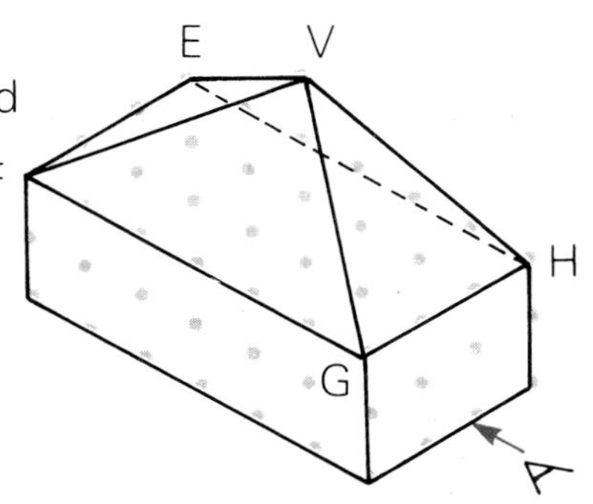

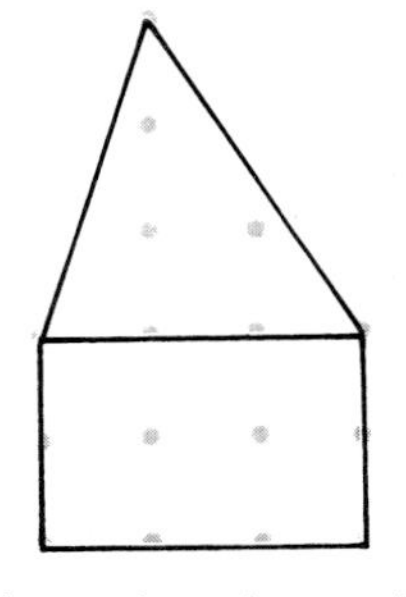

elevation from A

4 The plan from A of another house looks like this.

List the roof-faces in order, steepest first.

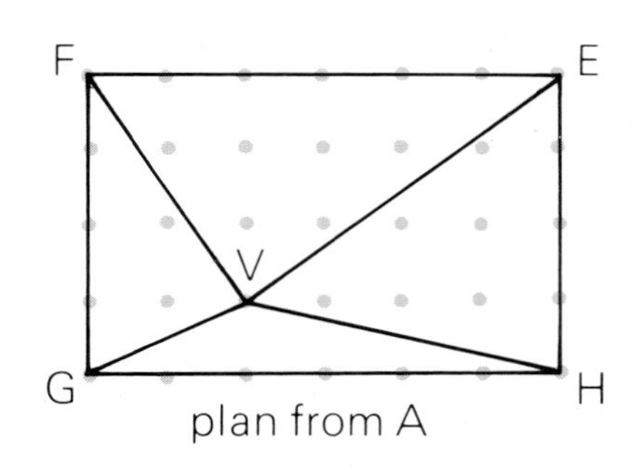

5

Think it through

5 Look at this tent.
The pole supporting V is vertical.
But you cannot tell from the drawing where the pole touches the ground.

a) Draw four possible **plans** from A, each showing V in a different position.

b) For each of your plans, write down which sloping face of the tent is steepest.

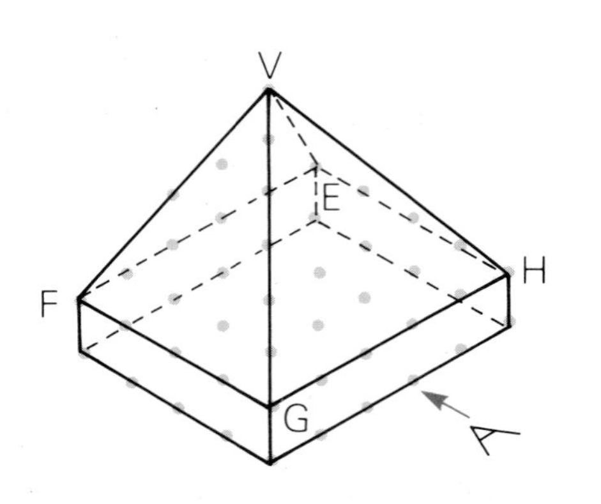

Challenge

6 Look at this triangular-based pyramid.

This is the plan from A.

F V G E

V E F X G 7 cm 5 cm 10 cm A

Draw the elevation from this direction.

Draw it full-size on 1 cm dotted squared paper.

Taking some measurements from a full-size plan may help.

Next chapter

6 Using ratio

A 1 a) Each card of Pearl buttons has large and small buttons.
There is something the same about the mixture of buttons on each card.
What is it?

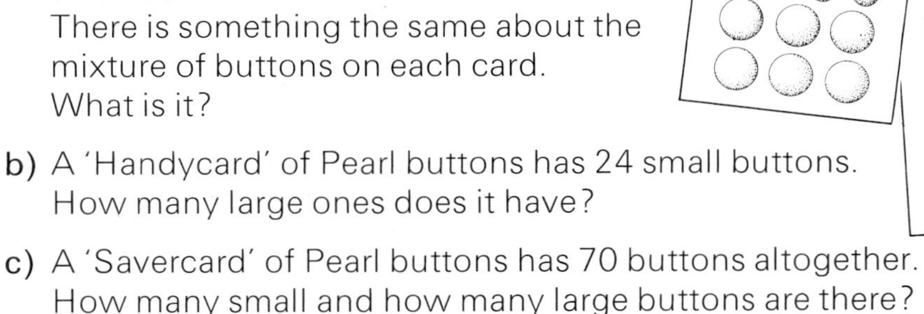

b) A 'Handycard' of Pearl buttons has 24 small buttons.
How many large ones does it have?

c) A 'Savercard' of Pearl buttons has 70 buttons altogether.
How many small and how many large buttons are there?

2 Glory button cards have 5 small buttons for every 4 large buttons.

a) How many large buttons are there on a card of Glory buttons with 15 small buttons?

b) How many small buttons are there on a card of Glory buttons with 16 large buttons?

c) How many buttons of each size are there on a card of Glory buttons with 45 buttons altogether?

Take note

Comparisons such as '3 large buttons **for every** 5 small buttons' are called **ratio** comparisons.
We say that the **ratio** of large buttons to small buttons is '3 to 5'.
We write this as
number of large buttons : number of small buttons = 3:5.

d) What is the ratio of large to small buttons on a Glory button card?

3 The scale on a map is given as 1:25 000.
What do you think this means?

4 a) These pots of paint are all mixed together to make orange.

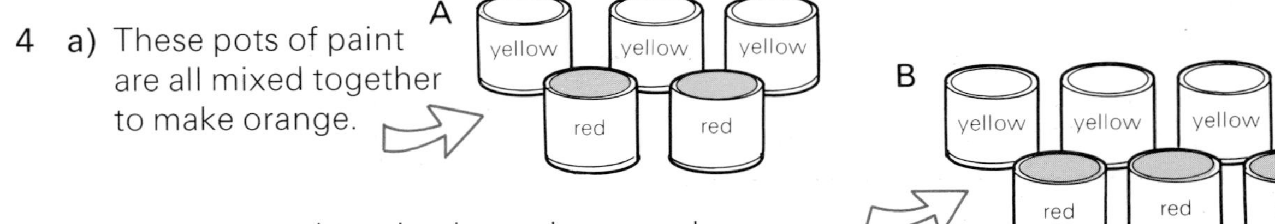

These pots are also mixed together to make orange.

Are the two orange shades identical?

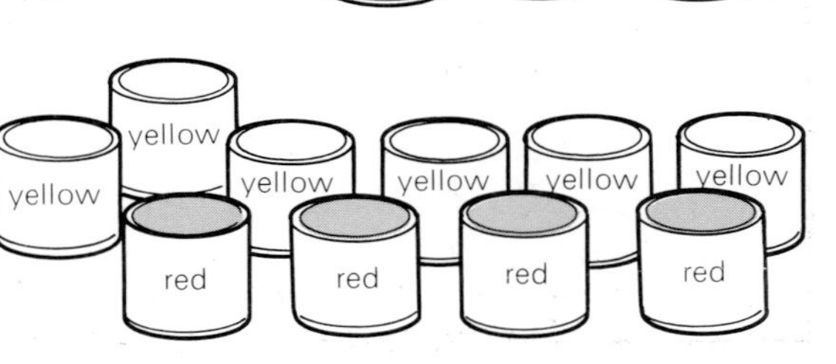

b) These pots produce a shade of orange identical to one of the oranges in A and B.
Which one?

6

5 Jake is planting beds of tulips.
He is using mixtures of red and yellow flowers.

a) In one bed he plants 5 red tulips for every 3 yellow tulips.
In another he plants 3 red tulips for every 2 yellow tulips.
Which bed of tulips will look more yellow from a distance?

b) Which of these mixtures will look more yellow?

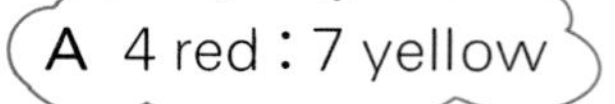

A 4 red : 7 yellow

B 7 red : 10 yellow

c) Which of these mixtures will look more red?

A 3 red : 10 yellow

B 5 red : 16 yellow

6 To make mortar, sand is mixed with cement and water.

a) Helen uses a ratio of 5 sand : 2 cement.
She has thrown 15 shovelfuls of sand into the mixer.
How much cement should she add?

b) Next time, Helen uses 42 shovelfuls altogether of sand and cement.
How many of these are sand, and how many are cement?

7 a) What is the **ratio** of red material to white material in each of these patchwork designs?

A

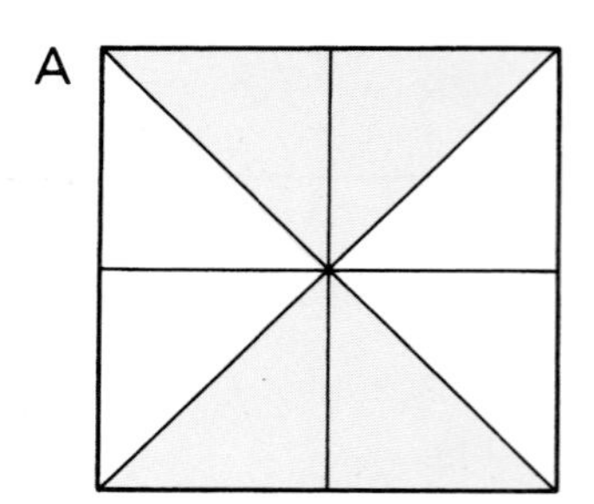

B

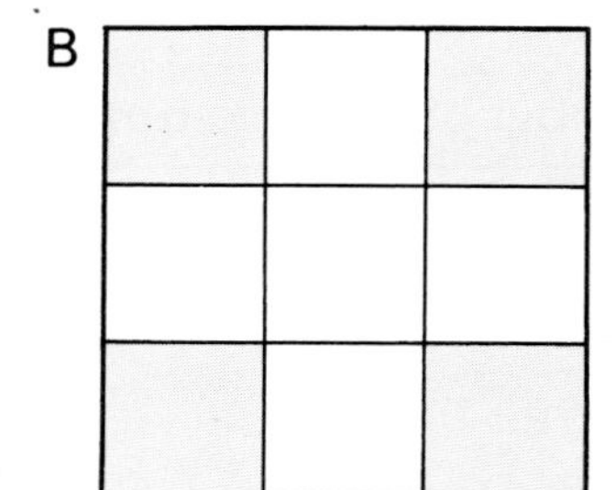

C

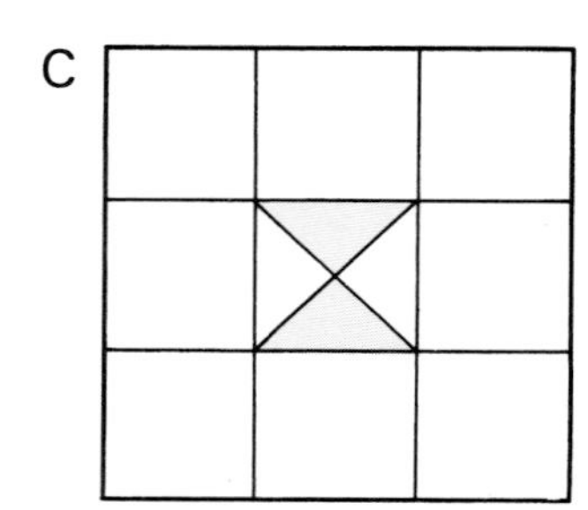

b) What **fraction** of each design is red?

c) Design your own patchwork for which the ratio red area : white area is 1 : 2.

Challenge

8 Here are some ratios of sand to cement.
Arrange them in order, from 'sandiest mixture' to 'least sandy mixture'.

sand cement

A 5:2

B 4:3

C 3:2

D 7:3

E 2:1

9 Estimate each of these ratios.
Write your estimates like this: □ : □.

a) The approximate ratio
the time I spend sleeping : the time I am awake

b) The approximate ratio
my height : the length of my arm

c) The approximate ratio
the distance I walk on my way to school : the distance I ride on my way to school

d) The approximate ratio
the weight of food and drink I take in each day : my weight

e) The approximate ratio
the number of days I have lived : the number of days my teacher has lived

f) The approximate ratio
the number of days per year I spend at home : the number of days per year I spend at school

g) The approximate ratio
the number of times per year I get up before 8.00 am : the number of times per year I get up after 8:00 am

6

Find out for yourself

10 Find out these exact ratios.

a) the number of female MPs in Parliament : the number of male MPs in Parliament

b) the number of kings of England since 1065 : the number of queens of England since 1065

Rugs, crosses, prisoners and teachers

B 1 The ratio length : width of all Wainwright rugs is 3 : 5.

a) Sketch three different rugs which satisfy the ratio.
On each sketch write down the dimensions of the rug.

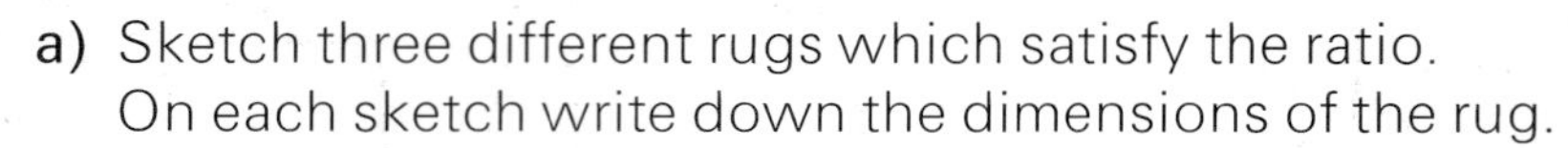

b) The width of a particular Wainwright rug is 3.5 m.
What is its length?

c) The length of a particular Wainwright rug is 12.5 m.
What is its width?

Challenge

d) The area of a particular Wainwright rug is 2.4 m^2.
What is its length and width?

2 The numbers of crosses and circles in each collection are in the same ratio.

a) How many crosses are missing from collection **C**?

b) Write the ratio of crosses to circles in each collection in the form □ : 1.

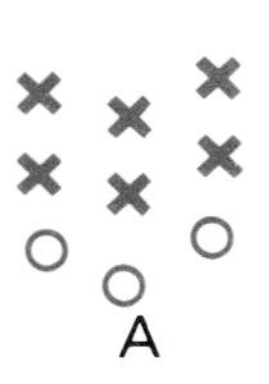

A

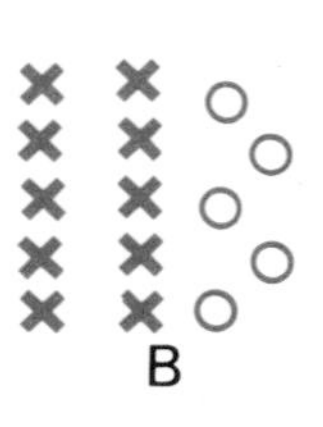

B

C

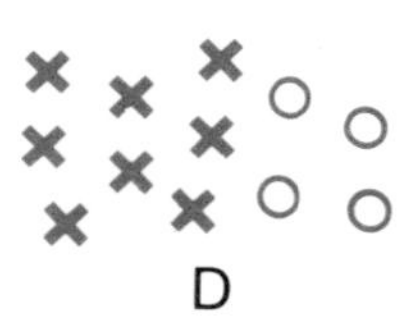

D

3 Write each of these ratios in the form □ : 1.

a) 10 : 5 b) 25 : 5 c) 12 : 3 d) 8 : 5

4 The ratio

males sentenced to prison in 1985 : **females sentenced to prison in 1985**

was about 29 : 1.
About 34 300 men were sent to prison.
Approximately, how many women were sent to prison?

5 a) Approximately, what is the pupil to teacher ratio in your school?

number of pupils to each teacher

b) The ratio of girls to teachers in a school is 10.1 : 1.
The ratio of boys to teachers in the school is 10.2 : 1.
Are there more boys or more girls in the school?

c) There are 505 girls in the school.
How many teachers are there?

d) How many boys are there in the school?

Ratio in practice

Activity *Paper sizes*

6 Standard paper sizes are coded like this:

A0, A1, A2, A3, A4, A5, . . .

The system is very simple.
A0 is the largest.
Each of the other sizes is obtained by cutting the previous size in half, like this:

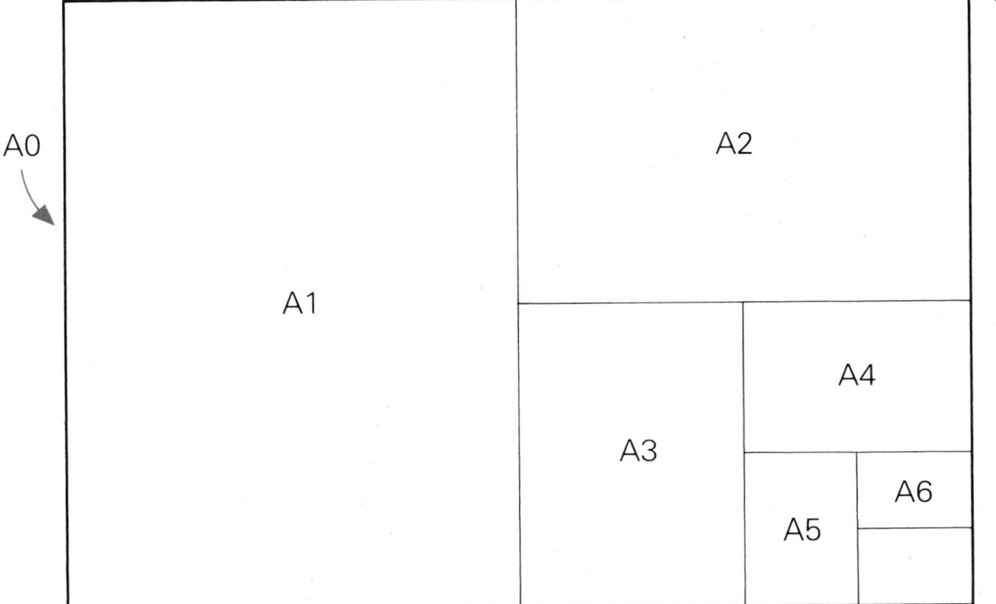

a) Get a sheet of A4 paper.
Use it to help you copy and complete this table.

the longer of the two dimensions

Paper size	A0	A1	A2	A3	A4	A5	A6
Length (cm)							
Width (cm)							

b) There is something special about the length : width ratio of the sheets of paper.
What is it?

7 The **gear ratio** of a bicycle gear is

the number of teeth on the cog at the pedals : the number of teeth on the cog at the back wheel

A ten-speed bicycle has two cogs at the pedals, one with 40 teeth and one with 52 teeth.
It also has five cogs at the back wheel with 14, 17, 20, 24 and 28 teeth.

a) The bike is in top gear when the chain is round a front cog and a back cog so that one turn of the pedals makes the back wheel turn more than with any other combination.
What is the gear ratio of top gear?

b) The bike is in bottom gear when one turn of the pedals makes the back wheel turn less than any other combination of front and back cogs.
What is the gear ratio of bottom gear?

c) List the gear ratios of the ten gears in order from top gear to bottom gear.

8 The amount of gold in jewellery and coins is measured in **karats**.
Pure gold is 24 karat, written '24K'.
If a ring is 18K gold, then

the mass of gold in the ring : the total mass of the ring = 18 : 24

a) Meg has a 14K gold bracelet.
Write the gold content of her bracelet as a ratio.

b) The mass of Meg's bracelet is 60 g.
How many grams of gold are there in it?

c) In June 1987 1 g of pure gold was worth £9.50.
What was the gold in Meg's bracelet worth in June 1987?

Challenge

9 The highest that a $1\frac{1}{2}$ mm high common flea has been known to jump is 197 mm.
If humans were as athletic, what would the world high-jump record be?

Make your own estimate of the height of a world record high-jumper.

Dividing things up

C 1 In 1980, 40% of UK adults smoked cigarettes.

a) Two of these are correct for the ratio
'UK adult smokers : UK adult non-smokers' in 1980.
Which two?
A 40:100 B 40:60 C 2:5 D 2:3

b) By 1987 the ratio of adult smokers to adult non-smokers was 17:33.
Had the proportion of smokers to non-smokers increased or decreased?
Explain how you arrived at your result.

2 This bar of metal is to be sawn into two parts, in the ratio 3:5.
How long is each part?

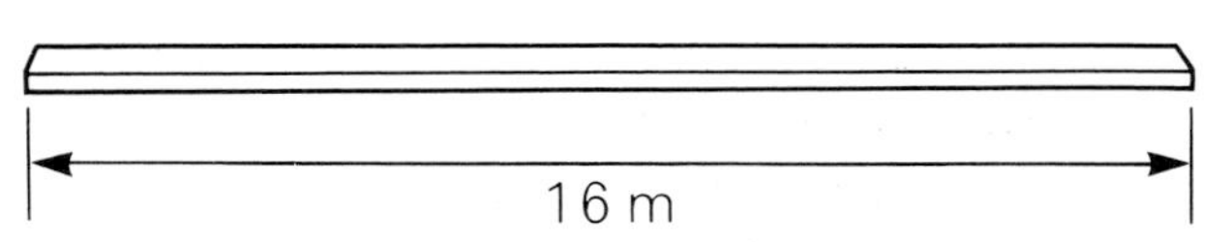

3 Two friends fill in a football pools coupon.
The total cost of the entry is 80p; one pays 20p and the other 60p.
They win £800 000!
They decide to share the money in the same ratio as their original stake.
How much should each get?

4 The number of women per 100 men employed in the textile industry in 1987 was 140.

a) Write this as a ratio of women : men.

b) Approximately 6 000 people were employed in the textile industry altogether.
How many were men, and how many were women?

5 A sports club wants to keep its membership ratio 'under 30 years old : over 30 years old' to about 3:2.
The maximum number of members is 280.

a) How many members under 30 years old should there be?

b) At one point in time there are 46 members over 30 in the club.
How many under-30s should there be?

6 a) What is the ratio

houses without structural faults : houses with structural faults ?

b) In a town there are 12 000 houses.
How many would you expect to have structural faults?

c) 32 houses in a village have structural faults.
How many houses do you think there are altogether in the village?

The golden ratio

D **Exploration** *Find a golden rectangle*

1 Start with this rectangle.

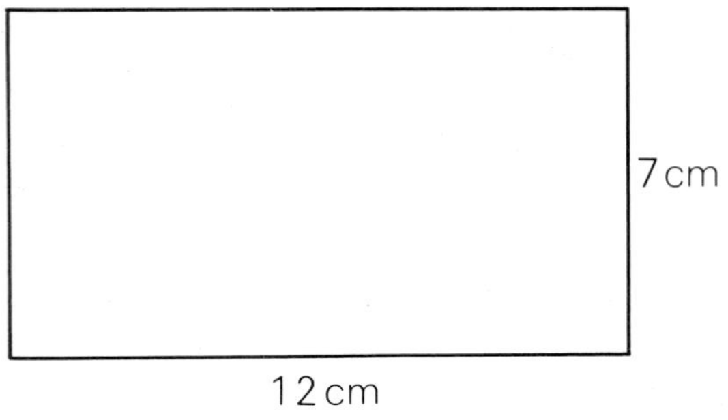

7 cm

12 cm

a) Cut the biggest possible square from one end, like this

Write down the approximate length : width ratio for the new rectangle.

In this question, always take the length to be the longer of the two dimensions.

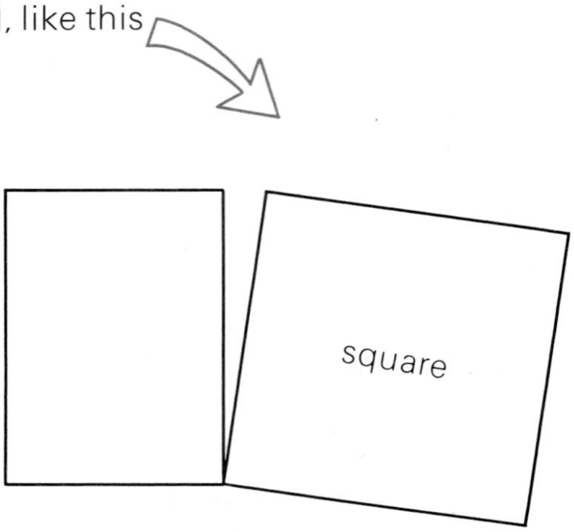

square

b) Repeat **(a)** for the new rectangle:

square

c) ... and then again for this new rectangle.

d) Try to find a starting rectangle for which this is true:

Each rectangle you make has the same length : width ratio.

Your rectangle is called a **GOLDEN RECTANGLE**.

Write down the dimensions of your starting rectangle.

Photoreduction

E Use your calculator on this page.

1 Some kinds of photocopying machines can produce reduced-size copies.
One machine has four buttons labelled

1

2

3

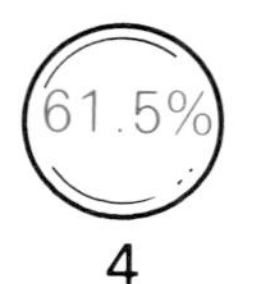

4

Button **1** produces copies so that

a reduced distance : the original distance
= 85:100

The other buttons work in a similar way.

An original rectangular picture measures 20 cm × 15 cm.

20 cm

15 cm

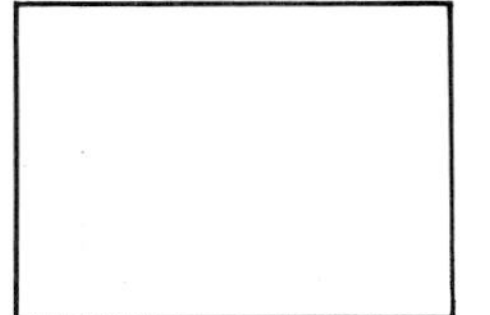

a) What size drawing does a button **1** reduction produce?

b) A reduction using button **1** is reduced a second time using button **1**.
What size drawing is produced?

c) On another machine there is a button that has the same effect as using button **1** twice in a row.
What percentage is shown on the button?

Give lengths to the nearest millimetre.

Give percentages to the nearest 1%.

d) With what percentage would a button be labelled that had the same effect as
   (i) using button **2** twice in a row?
   (ii) using button **1** and then button **4**?
   (iii) using button **3** three times in a row?

e) Is there a succession of buttons on the original machine which produces half-size copies?
If so, what is it?

*that is, reduces **distances** to half-size*

f) The percentages on the buttons on the original machine tell us what happens to **distances**.
Suppose we now want the buttons to show **area** reductions.
What should be printed on button **1**?

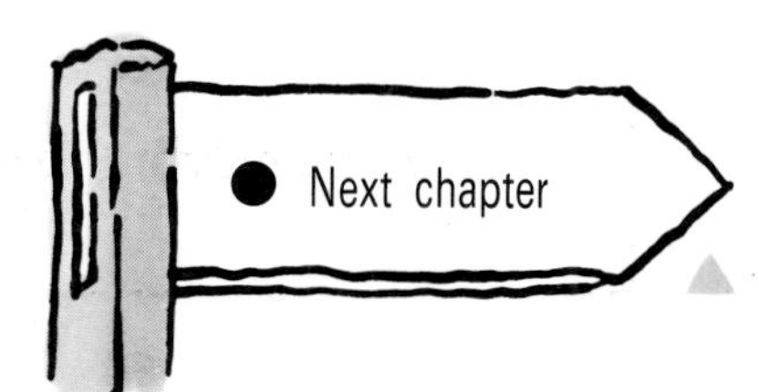

7 Dealing with shapes

A 1 What is the area of

a) parallelogram **A**?

b) parallelogram **B**?

c) parallelogram **C**?

The dotted-line divisions will help you.

2 Only two sides of parallelogram **D** have been drawn.

a) Copy and complete it on 1 cm squared paper.

b) What is its area?

3 Use squared paper.

a) Draw three different parallelograms.
All three must have
- base length 4 cm

and
- area 12 cm².

b) As well as the base length and area, what other measurements are the same in your three parallelograms?

c) Now draw three different parallelograms which have
- height 4 cm

and
- area 20 cm².

As well as height and area, what other measurements are the same in your three parallelograms?

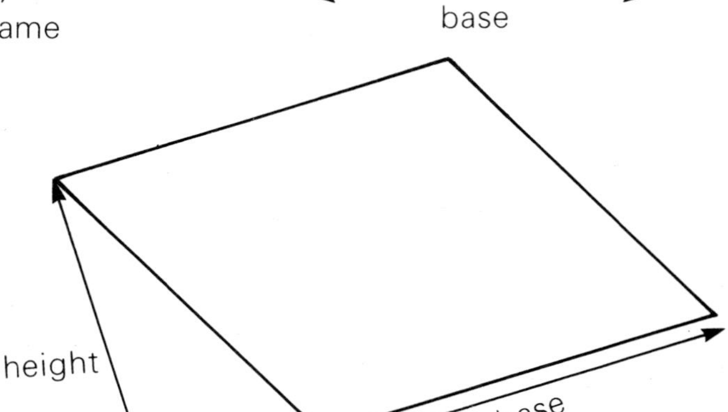

Challenge

4 How many different parallelograms are there whose area is 36 cm², and whose base length and height are whole numbers of centimetres?

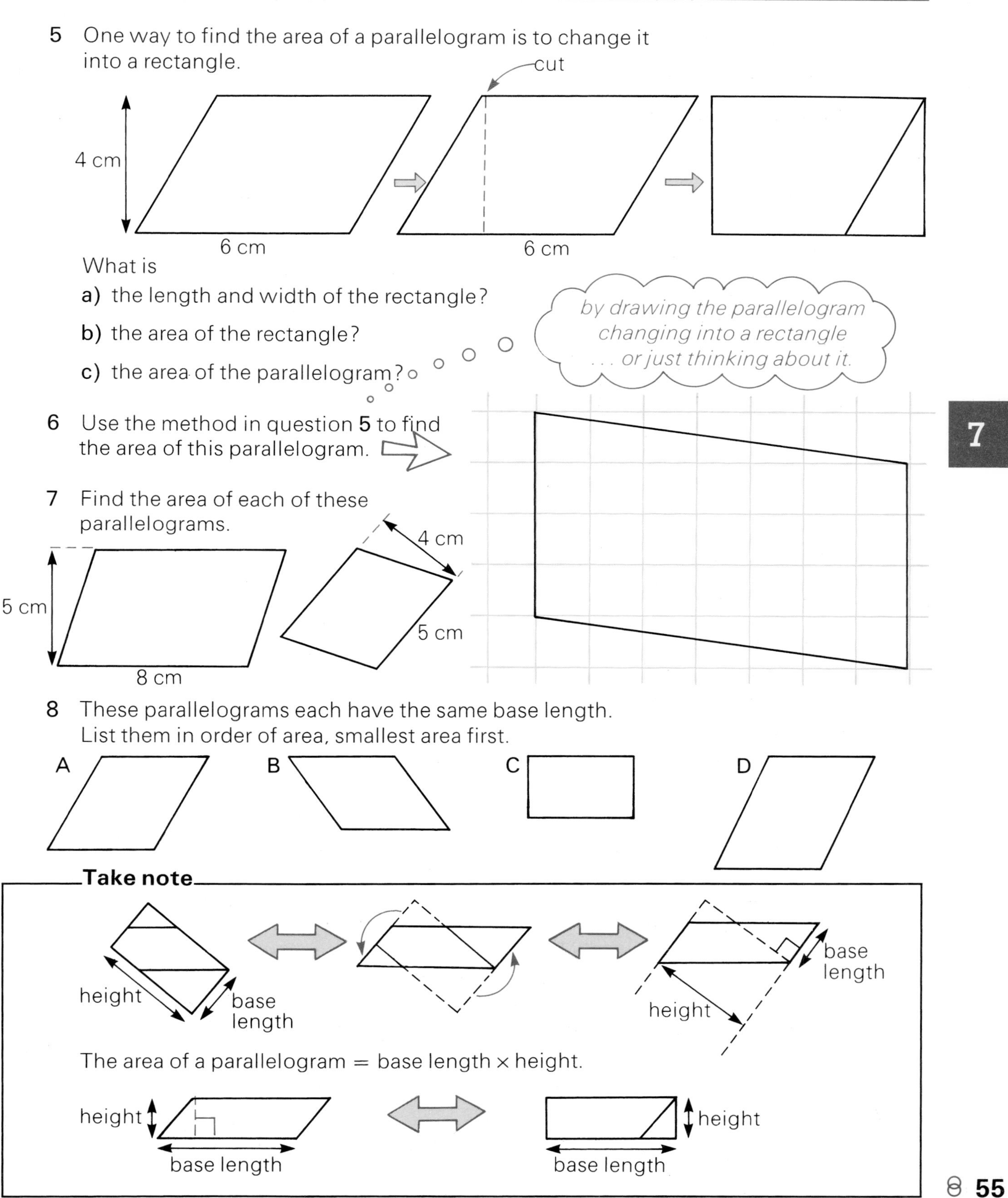

5 One way to find the area of a parallelogram is to change it into a rectangle.

What is

a) the length and width of the rectangle?

b) the area of the rectangle?

c) the area of the parallelogram?

6 Use the method in question **5** to find the area of this parallelogram.

7 Find the area of each of these parallelograms.

8 These parallelograms each have the same base length. List them in order of area, smallest area first.

Take note

The area of a parallelogram = base length × height.

Triangles

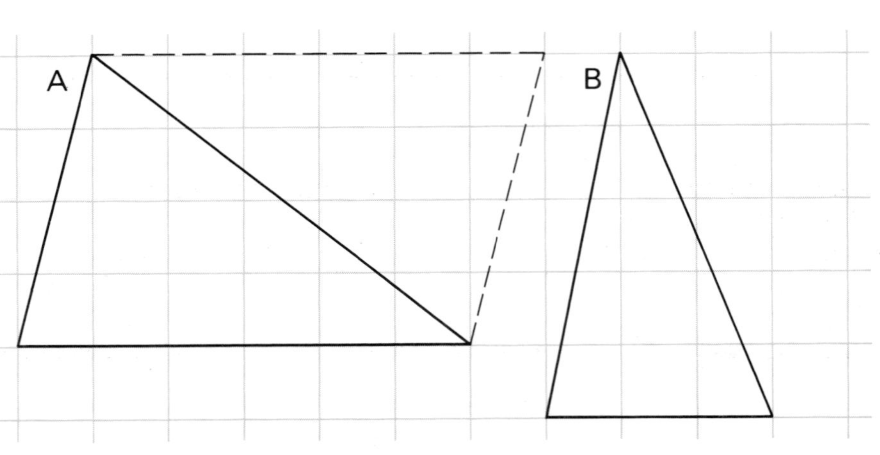

B 1 a) Find the area of triangle **A**.

b) Find the area of triangle **B**.

c) Draw a triangle with the same area as triangle **B**, but a different shape.

2 The drawings show the same parallelogram cut into two identical triangles in two different ways.
Copy and complete the **Take note**.

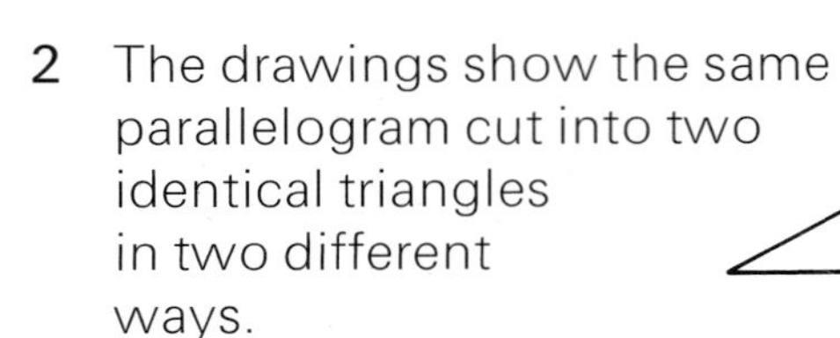

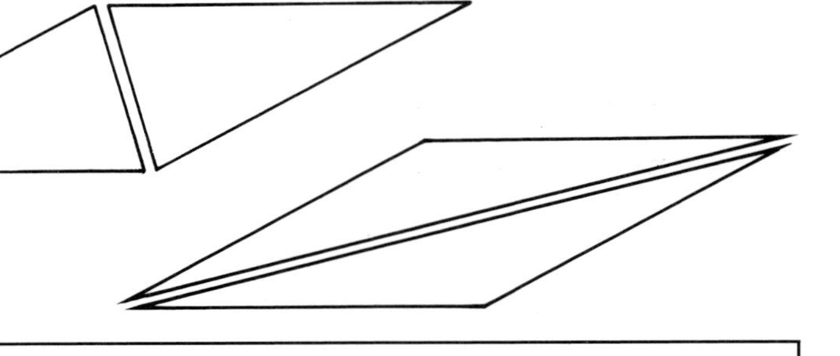

7

Take note

Every triangle is half a parallelogram.

height

base length

height

base length

The area of a triangle
=
□ × (base length × height).

3 This is a pin board. ○ ○ ○ ○

The pins are at the corners of 1 cm squares.

An elastic band can be stretched over the pins to make triangles.

a) What is the area of the triangle in the drawing?

b) What is the greatest area a triangle on this board can have?
Draw a triangle with the greatest possible area.

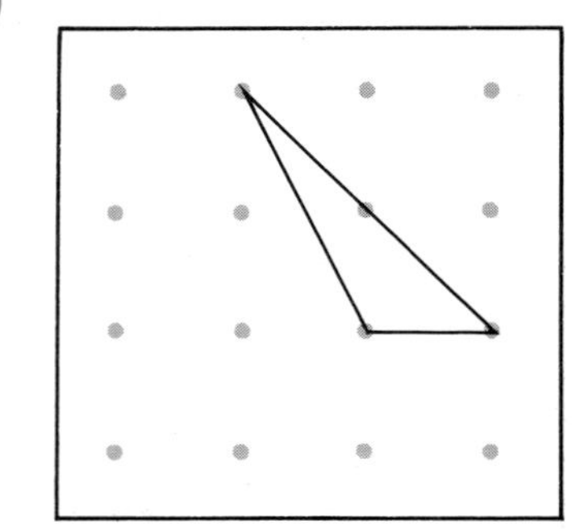

c) What is the smallest area a triangle on the board can have?
Draw a triangle with the smallest possible area.

Challenge

4 This pin board also has 1 cm squares.
What is the area of the shape?

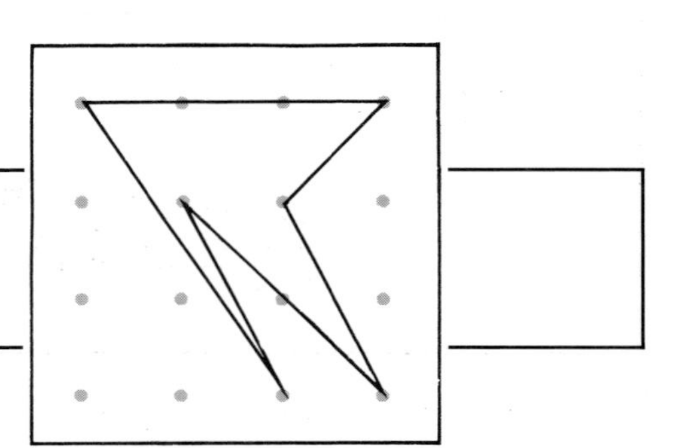

5

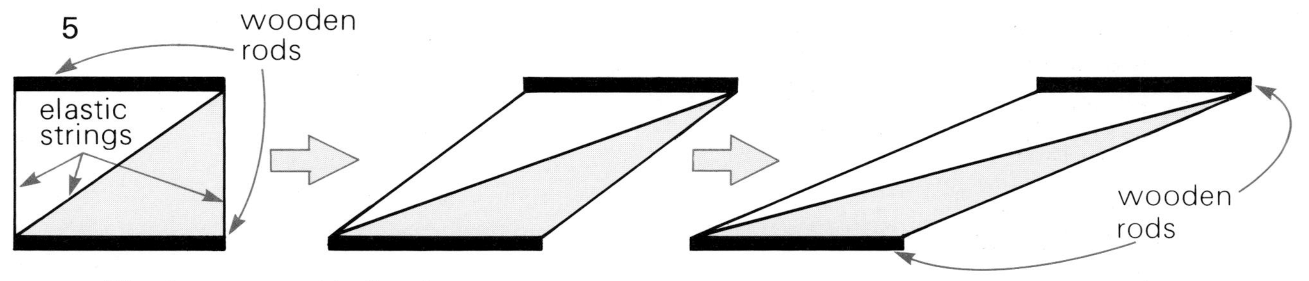

The bottom rod is fixed.
Think about the top rod moving right and left along a horizontal line.
The elastic strings stretch.
Describe what happens to

a) the height of the shaded triangle.

b) the length of its base.

c) its area.

6 a) List these triangles in order of area, smallest first.

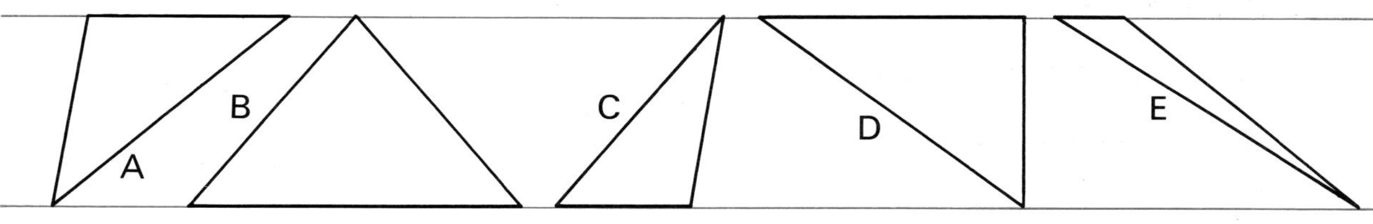

b) List these triangles in order of area, smallest first.

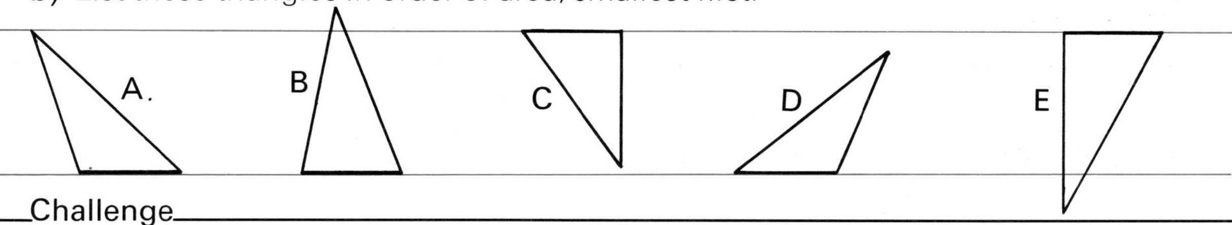

Challenge

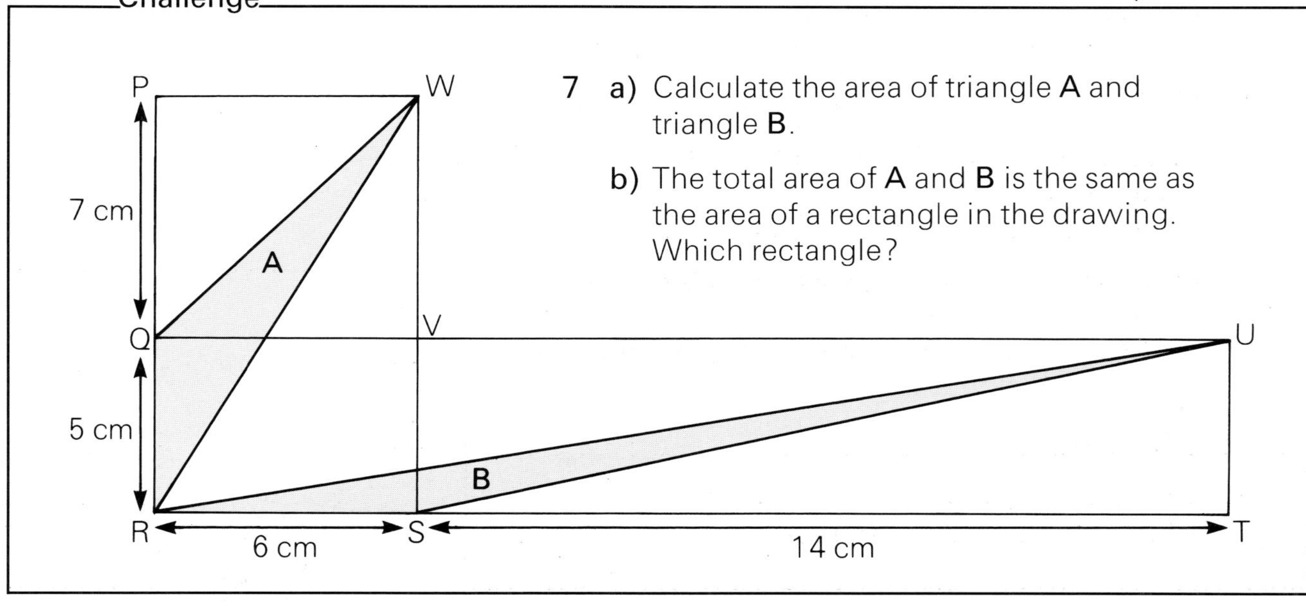

7 a) Calculate the area of triangle **A** and triangle **B**.

b) The total area of **A** and **B** is the same as the area of a rectangle in the drawing. Which rectangle?

Trapeziums

C 1 a) These are two identical trapeziums.
Use 1 cm squared paper.
Draw the trapeziums fitted together to make a parallelogram.

b) Now draw them fitted together to make a different parallelogram.

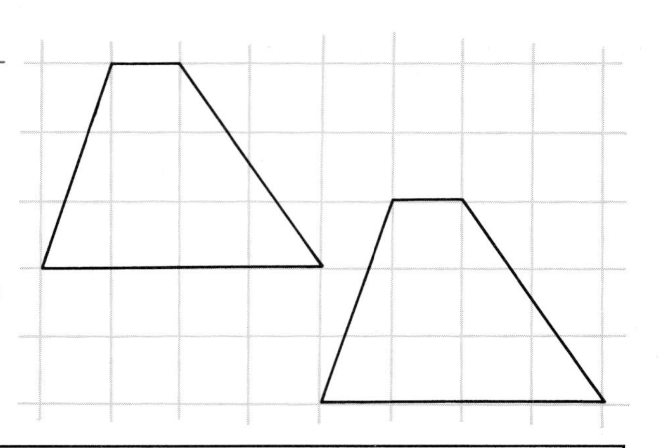

Activity

2 Draw a parallelogram on squared paper.
Cut it into two identical trapeziums.
Stick them in your book.

four sides; at least two of them parallel to each other

What is the area of

a) your parallelogram?

b) each of your trapeziums?

7

3 Every parallelogram can be cut into a pair of identical trapeziums

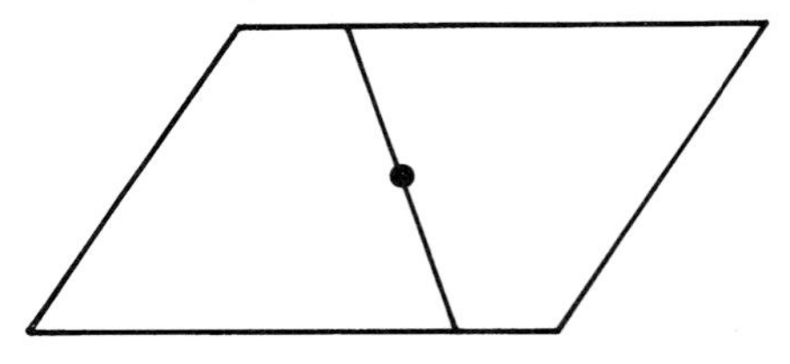

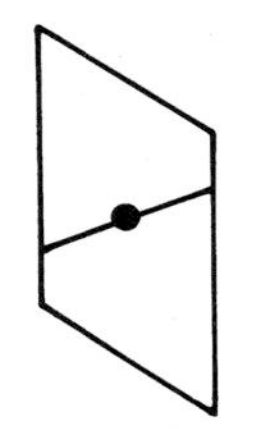

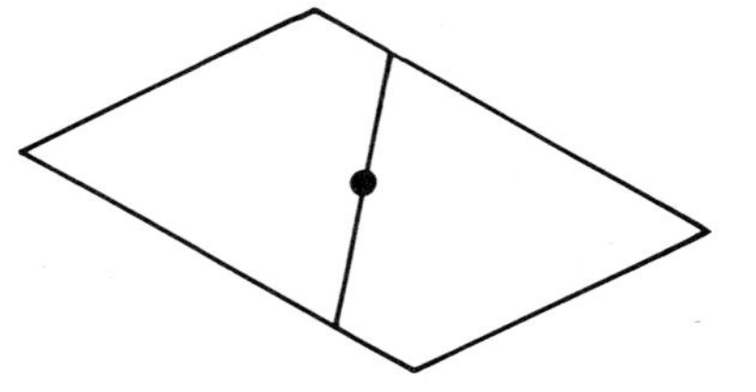

. . . and every trapezium is half of a parallelogram.

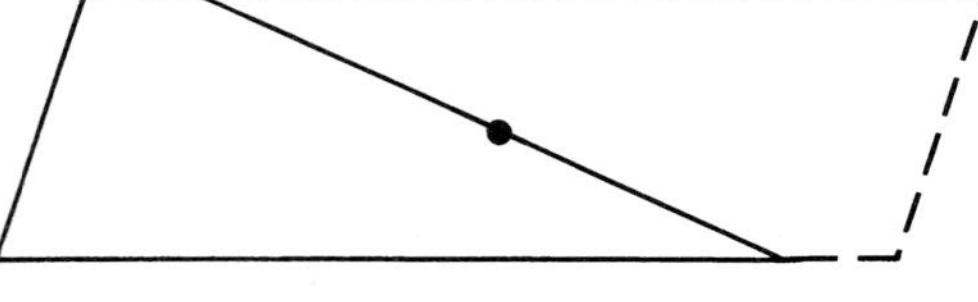

a) Copy and complete the **Take note**.

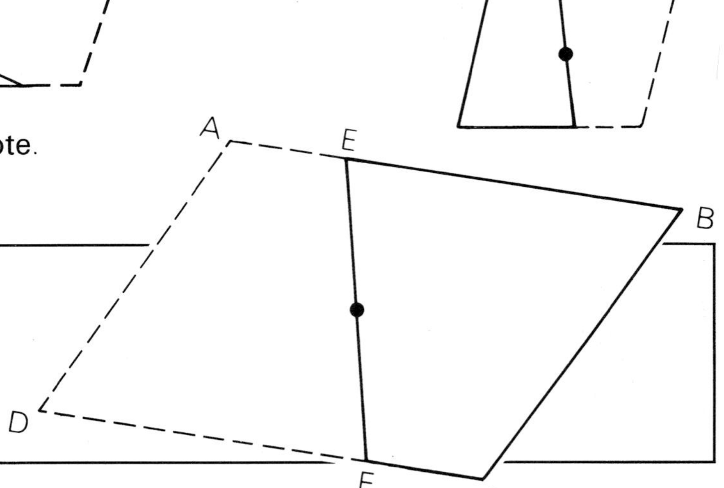

Take note

Area of trapezium EBCF
=
□ × area of parallelogram ABCD.

b) In all the drawings in this question, the marked dot is in the same position relative to the corners of the parallelogram.
Write one or two sentences to describe the position of the marked dots.

Think it through

4 What is the area of trapezium ABCD?

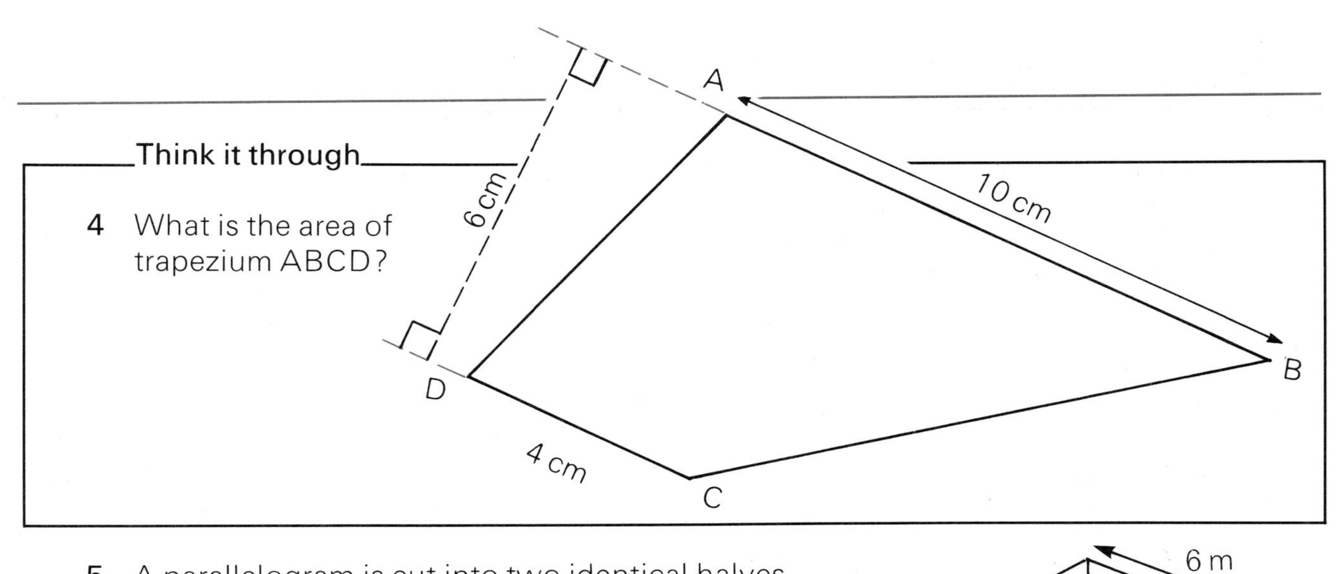

5 A parallelogram is cut into two identical halves.
Does each half have to be a trapezium?
Draw parallelograms to help you explain your answer.

6 Use 1 cm squared paper.
Draw a trapezium with an area of 24 cm².

7 Flat roof material for this Police Station
costs £24 per square metre (including labour).
How much will it cost to put a roof on the building?

6 m
90°
10 m
10 m
90°

8 Here are two trapeziums.

a) What is the value of d?

b) What is the value of b?

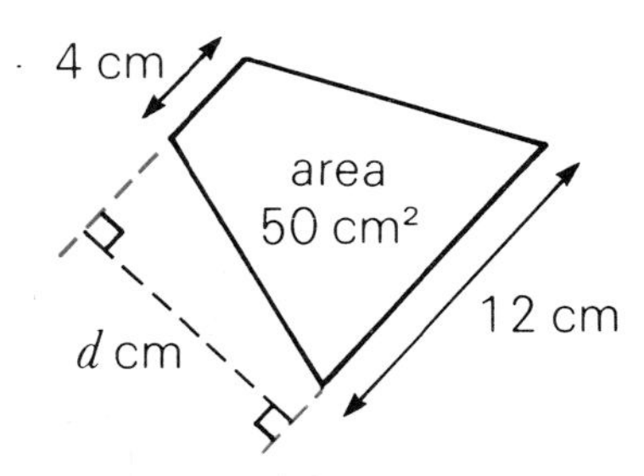

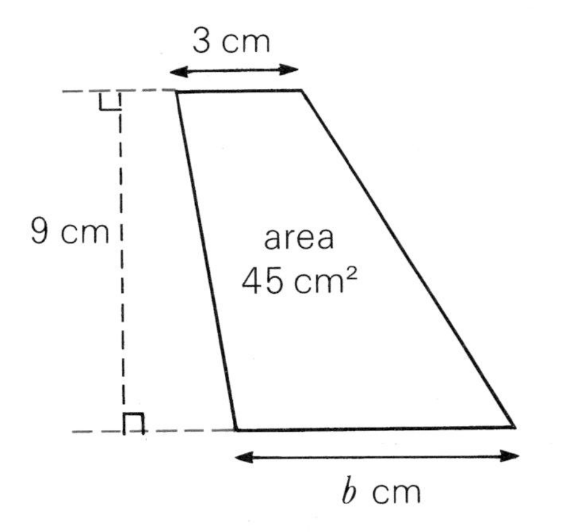

Challenge

9 This is a folded parallelogram.

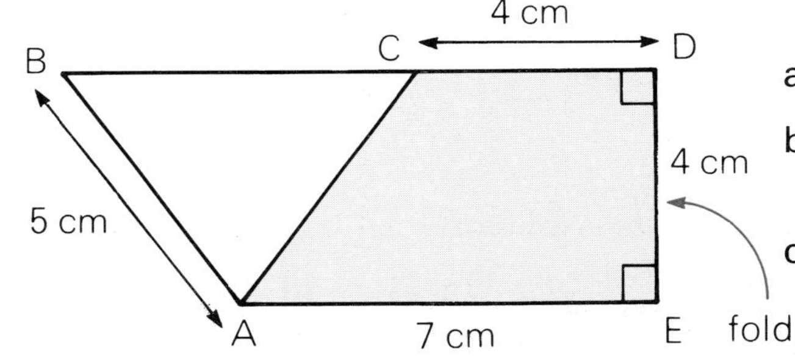

a) What is the area of triangle ABC?

b) What was the area of the parallelogram before it was folded?

c) What is the area of trapezium AEDC?

7

Kites and rhombuses

D Activity

1 You need 1 cm squared paper.
Cut out two copies of each of these triangles.

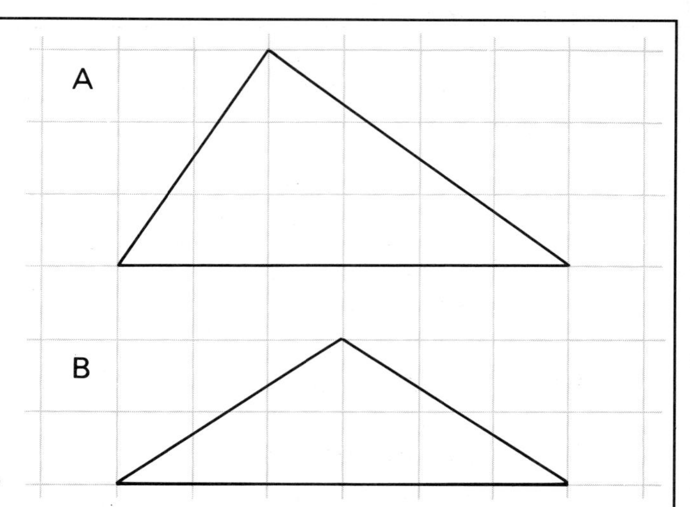

a) Fit your two copies of **A** together to make a kite:
What is its area?

b) Fit your two copies of **B** together to make a rhombus:
What is its area?

c) Copy and complete the **Take note.**

Take note

A kite can be made from two identical triangles.
Its area is twice the area of each triangle.

A rhombus can be made from two identical __________ triangles.
Its area is ________ the area of each triangle.

d) Draw a kite which has an area of 24 cm^2.

e) Draw a rhombus which has an area of 24 cm^2.

Challenge

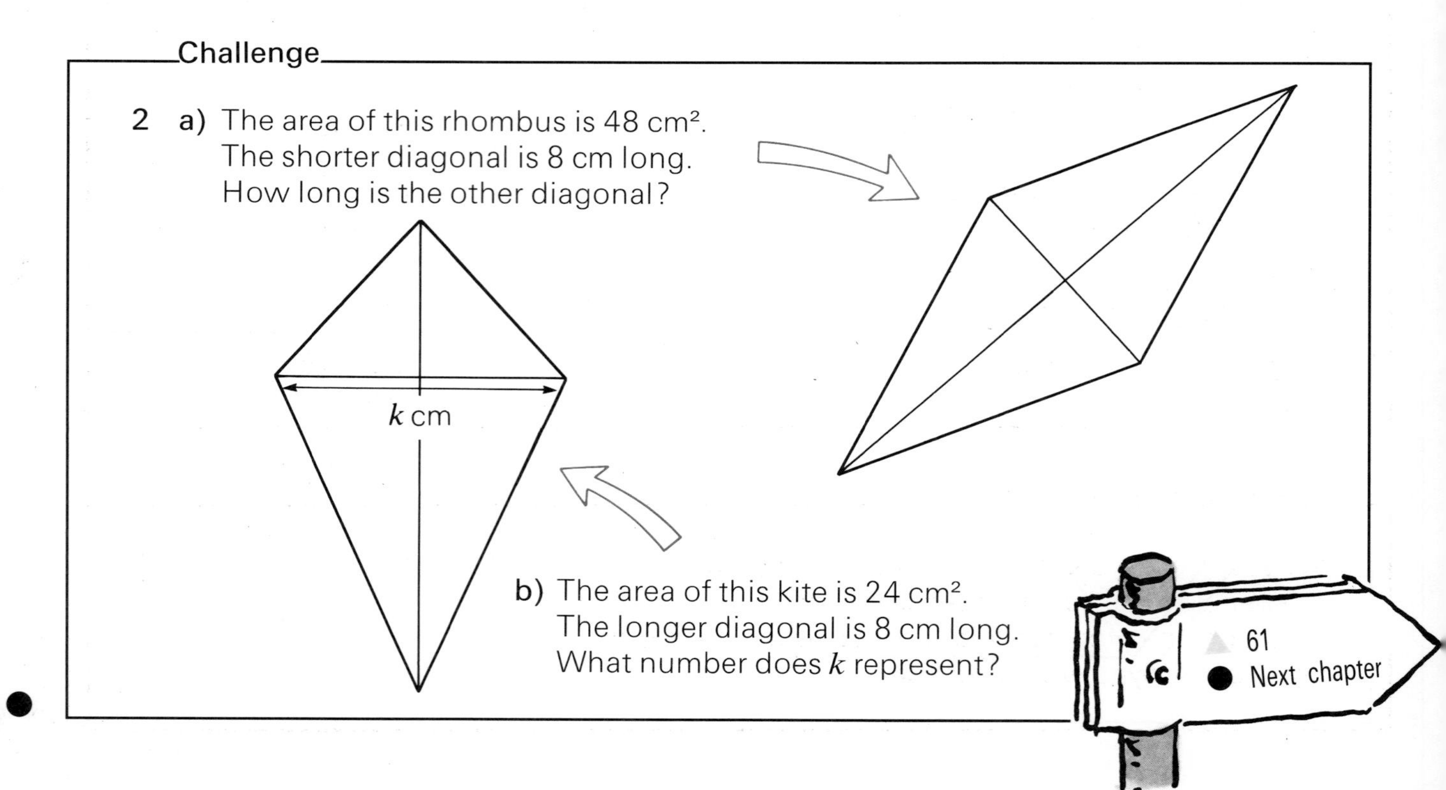

2 a) The area of this rhombus is 48 cm^2.
The shorter diagonal is 8 cm long.
How long is the other diagonal?

b) The area of this kite is 24 cm^2.
The longer diagonal is 8 cm long.
What number does k represent?

Right-angled triangles and squares

E

Exploration

1 a) Squares have been drawn on the sides of this triangle.
What is the area of
(i) square **A**?
(ii) square **B**?
(iii) square **C**?

The dotted lines will help.

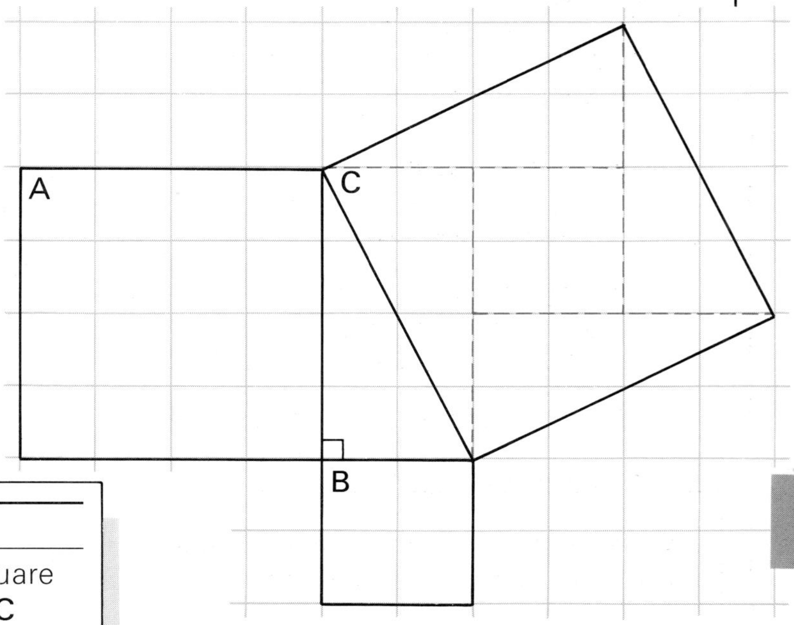

A
C
B

b) Investigate some more triangles.
Make a table of your results like this

Area (cm^2) of		
Square A	Square B	Square C
16	4	

c) Look for a connection between the areas of the squares.
Write down what you discover.

d) A square has been drawn on the longest side of this right-angled triangle.

called the ***hypotenuse***

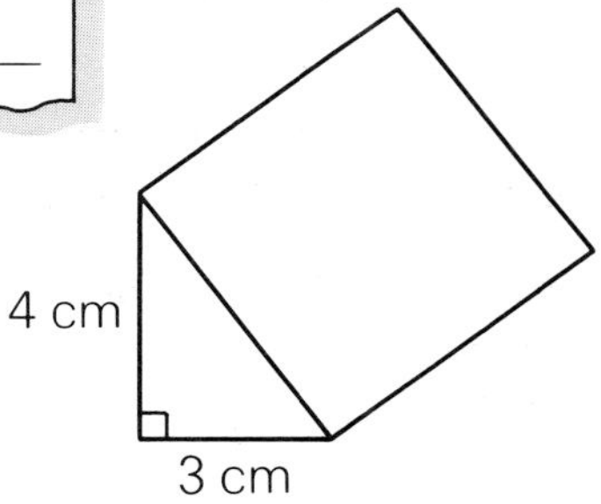

4 cm
3 cm

Use what you discovered in **(c)** to find its area.

e) What is the area of the square on the side AC of this triangle?

f) What is the length of AC?

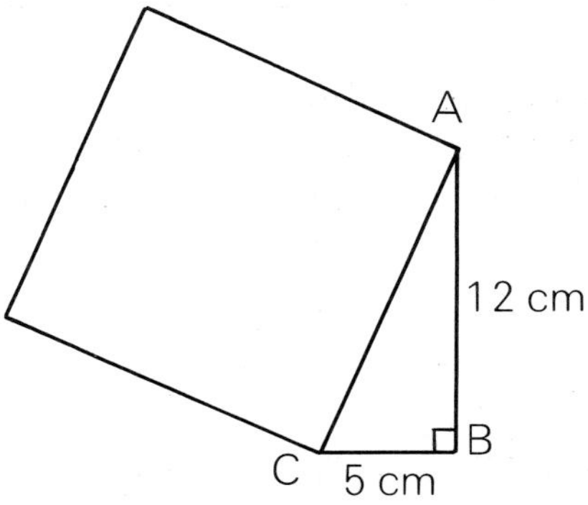

A
12 cm
B
C
5 cm

Find out for yourself

What you have discovered on this page is linked with the famous Greek mathematician Pythagoras.
Find out what you can about him, and his 'theorem'.
Write about what you discover.

2 a) Find the areas of the squares drawn on the sides of these triangles. Write each area to the nearest 1 cm².

b) Find the length of the third side of each triangle, to the nearest millimetre.

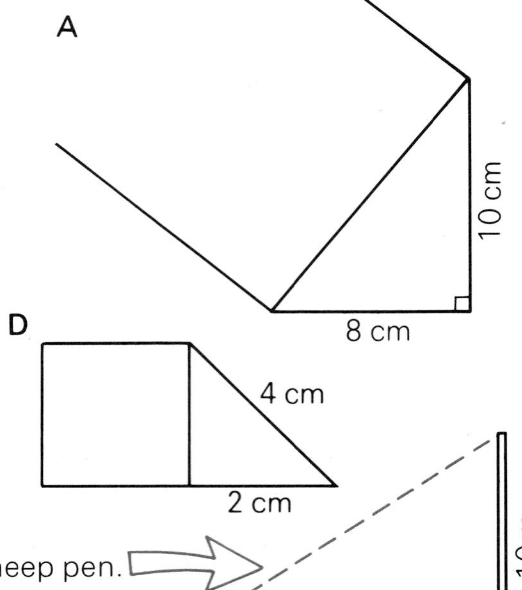

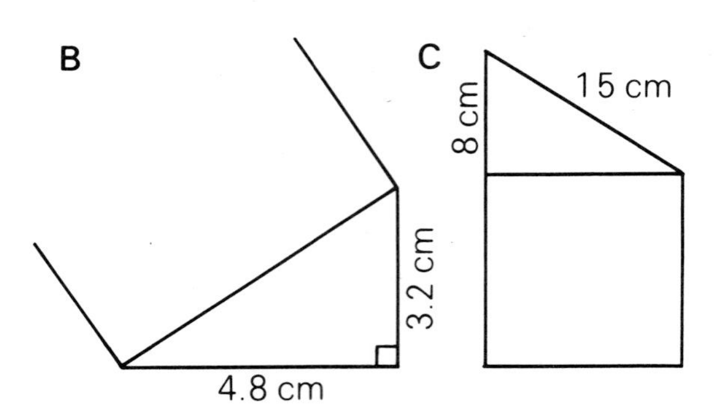

3 The drawing shows two sides of a triangular sheep pen. What length of fencing is needed for the third side, to the nearest metre?

15 m

4 a) Find the length of each of the sides marked **?** in this triangle spiral.

b) What is the length of the hypotenuse of the 50th triangle?

c) Sketch the 100th triangle, and mark in the lengths of each of its sides.

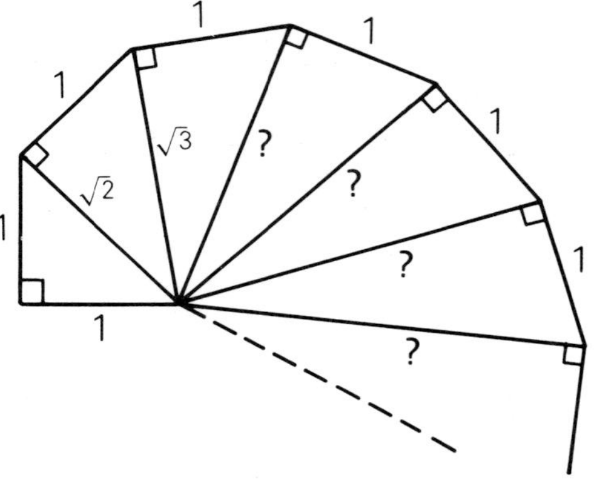

5 In this tiling pattern you will see three different sized squares. The smallest squares have sides 6 cm long. How long are the sides of each of the other two sizes of squares.

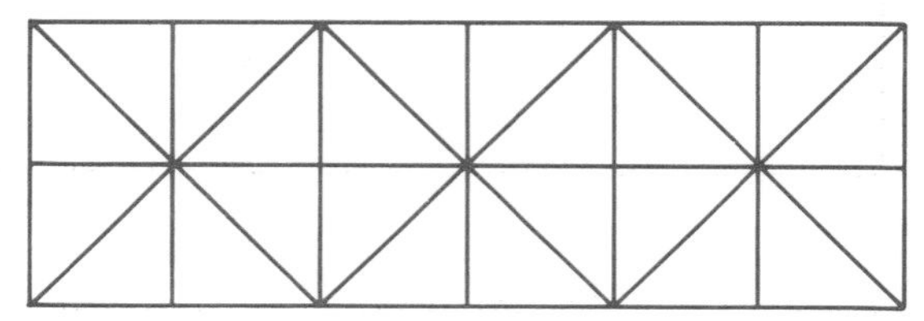

Challenge

6 a) Are there any right-angled triangles with hypotenuse 45 cm long which have each of their other two sides a whole number of centimetres long? If so, write down the lengths of the sides.

b) Sketch the right-angled triangle with hypotenuse $\sqrt{50}$ cm long which has the largest possible area.

7

Area and diagonals

F **Exploration** *Stretching kites*

1 Draw the kite ABCD on 1 cm squared paper.

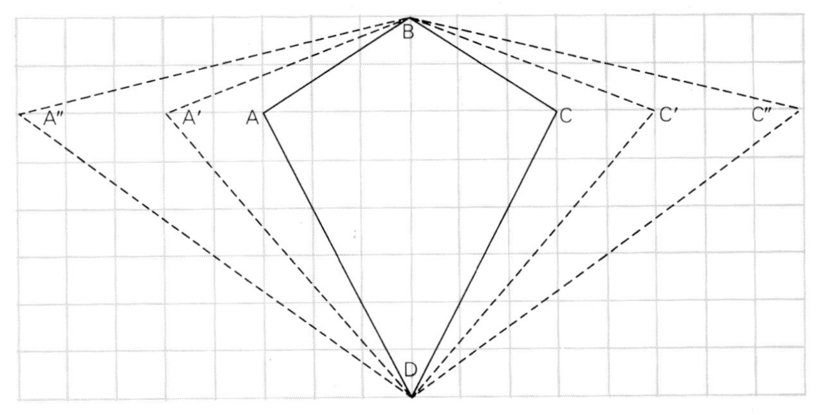

a) Check that its area is 24 cm².

b) Stretch its 'shoulders' to A′ and C′.
What is its new area?

c) Stretch its shoulders to A″ and C″.
What is its new area?

d) Try some more stretches of its shoulders.
Write one or two sentences to explain how its area changes.

e) Now investigate what happens when only one shoulder is stretched.
Write one or two sentences to explain how the shape's area changes.

f) Think about stretching its tail

. . . and its head

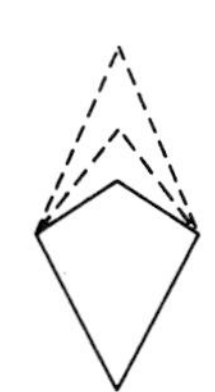

. . . and its tail, head and shoulders!
Investigate what happens to its area.
Write a report about what you discover.

Challenge

2 a) There is something special about the diagonals of all the shapes you got in question **1** by stretching a kite.
What is it?

b) There is an easy way to find the area of quadrilaterals whose diagonals are like this, provided you know the lengths of the diagonals.
Write down how to do it.

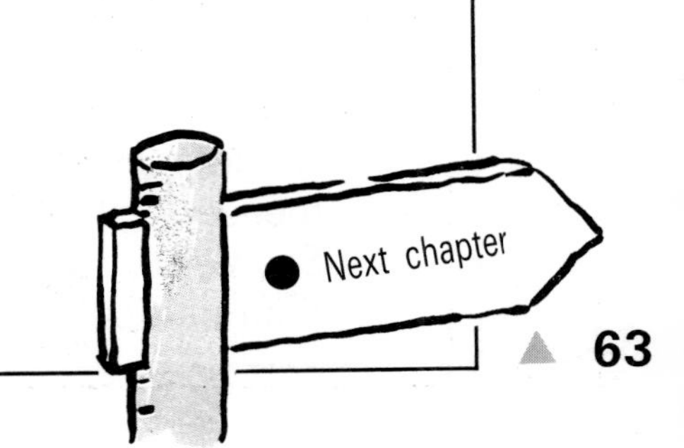

8 Accuracy and approximation

A 1 Fiona has written an airmail letter to her uncle in Australia.
She is weighing it on the kitchen scales to find out how much postage to put on the envelope.
Airmail letter postage to Australia is 34p for a letter weighing up to 10 g and 15p for every additional 10 g up to 250 g.

a) How much postage do you think Fiona should put on the envelope?

b) Write one or two sentences to explain any difficulty you might have had answering (a).

c) Fiona decided to have the letter weighed at the Post Office.
What do you think is the most important difference between the Post Office scales and Fiona's kitchen scales?

2 a) At international athletics meetings the 100 m race is timed by an electronic timer.
An ordinary stopwatch or a watch with a second hand is not good enough.
Write down at least two reasons why not.

b) For the 10 000 m race, an ordinary stopwatch is just about as good as an electronic timer.
Why do you think this is?

Think it through

3 Winston's mum is an industrial chemist.
At work, she mixes liquids, making her measurements with the help of a burette and measuring cylinders.

At home, in the kitchen, she uses a measuring jug to measure liquids
. . . or, when the children need medicine, she uses a medicine glass.

Write a short paragraph explaining the differences between these methods of measuring liquids.
Say why you think they are suited to their particular purposes.

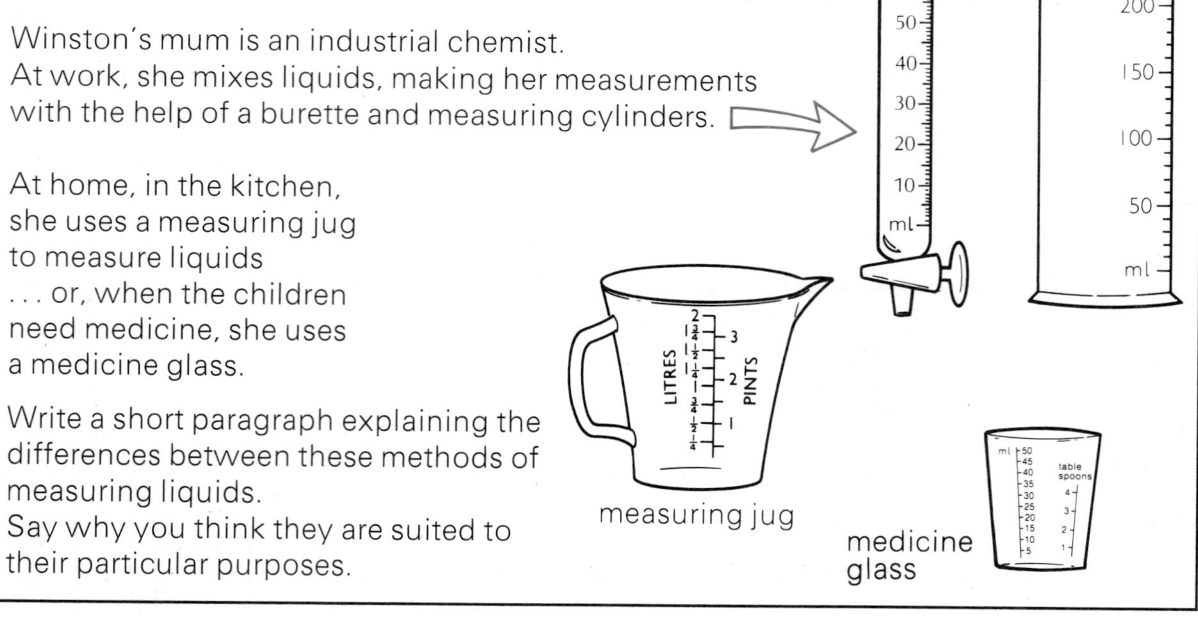

4 Winston's little sister Molly made this centimetre ruler at school out of card.

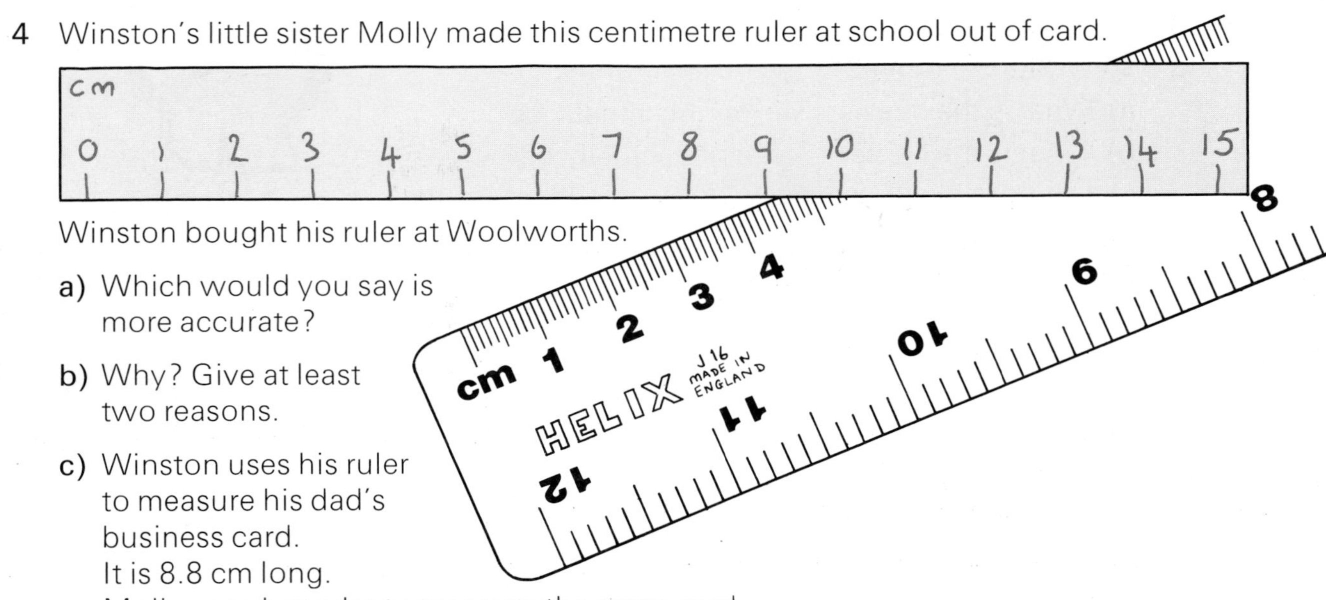

Winston bought his ruler at Woolworths.

a) Which would you say is more accurate?

b) Why? Give at least two reasons.

c) Winston uses his ruler to measure his dad's business card.
It is 8.8 cm long.
Molly uses her ruler to measure the same card.
She writes, 'This card is 9 cm long, to the nearest centimetre.'
   (i) What does 'to the nearest centimetre' mean?
   (ii) What is it about Molly's ruler that means she has to add this phrase to all her measurements?
   (iii) Winston ought really to add a similar phrase to all his measurements. What phrase?

5 Draw a centimetre ruler 10 cm long.
Make just enough marks on it for it to be useful if you had to measure lengths to the **nearest $\frac{1}{2}$ cm**.

up to 10 cm

Take note

We can never measure with total accuracy.
Accuracy depends partly upon how good our measuring instruments are.
That is why we give measurements 'to the nearest ...'

rulers, stopwatches, weighing scales, etc

Think it through

6 We usually choose our measuring instruments according to how accurate our measurements need to be.
For each measurement, say how accurate you think it needs to be. Choose from these.

a) the distance between two towns
b) the amount of water in a swimming pool
c) the amount of flour in a bread mixture

Write one or two sentences to explain your choices.

To the nearest ...
1 mm, 1 cm, 1 m,
500 m, 1 km, 5 km,
1 ml, 1 cl, 10 cl, 1 l,
100 l, 1 m³, 10 m³,
1 g, 10 g, 50 g, 100 g,
500 g, 1 kg, 10 kg

8

7 A zoo-keeper weighs a hippopotamus, and finds that she weighs 4 tonnes to the nearest tonne.

a) What is the heaviest she might actually be?

b) What is the lightest she might actually be?

c) Is it possible to give her weight

(i) to the nearest 5 tonnes?

(ii) to the nearest 500 kg?

d) Explain your answers in **(c)**.

You might like to draw scales to help you.

8 Try to answer these questions about yourself. You might have to take some measurements . . . or ask a friend to help you.
What is

a) your height, to the nearest 10 cm?

b) your weight, to the nearest 5 kg?

c) your age, to the nearest 6 months?

d) the distance you travel to school, to the nearest 10 km?

e) the amount of milk you have drunk so far today, to the nearest litre?

f) the amount of sleep you had last night, to the nearest hour?

g) your age, to the nearest century?

9 These two road signs give advance warning of restrictions ahead.

Sign **A** gives the maximum load a bridge can carry.
Sign **B** gives the height of a low bridge.

A

B

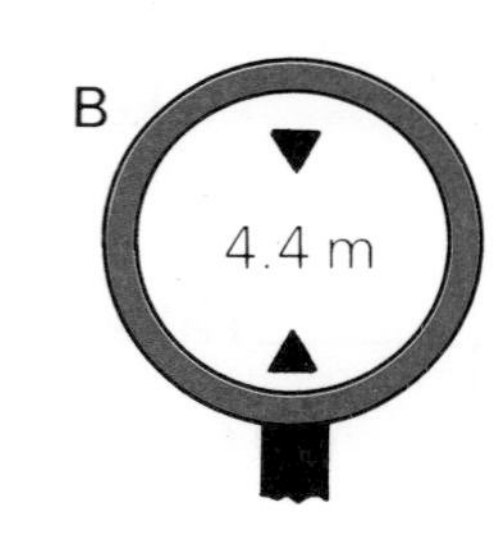

a) How do you think the weight restriction has been approximated?

- to the nearest tonne

or • rounded up to the next tonne

or • rounded down to the next tonne

b) How do you think the height restriction has been approximated?

- to the nearest 0.1 m

or • rounded up to the next 0.1 m

or • rounded down to the next 0.1 m

c) Explain your choices in **(a)** and **(b)**.

10 Think of **two** examples of measurements you see around you which you think are

a) measured to the **nearest** unit.

b) rounded **up** to the next unit.

c) rounded **down** to the next unit.

Write down all six of your examples.

cm, kg, 10 l, £1 million, 1000 people, . . .

in the newspapers, on road signs, on TV, on food packets, . . .

Working with approximations

B 1 We can use 'rounding to the nearest . . .' to help us make estimations.
Here is an example:

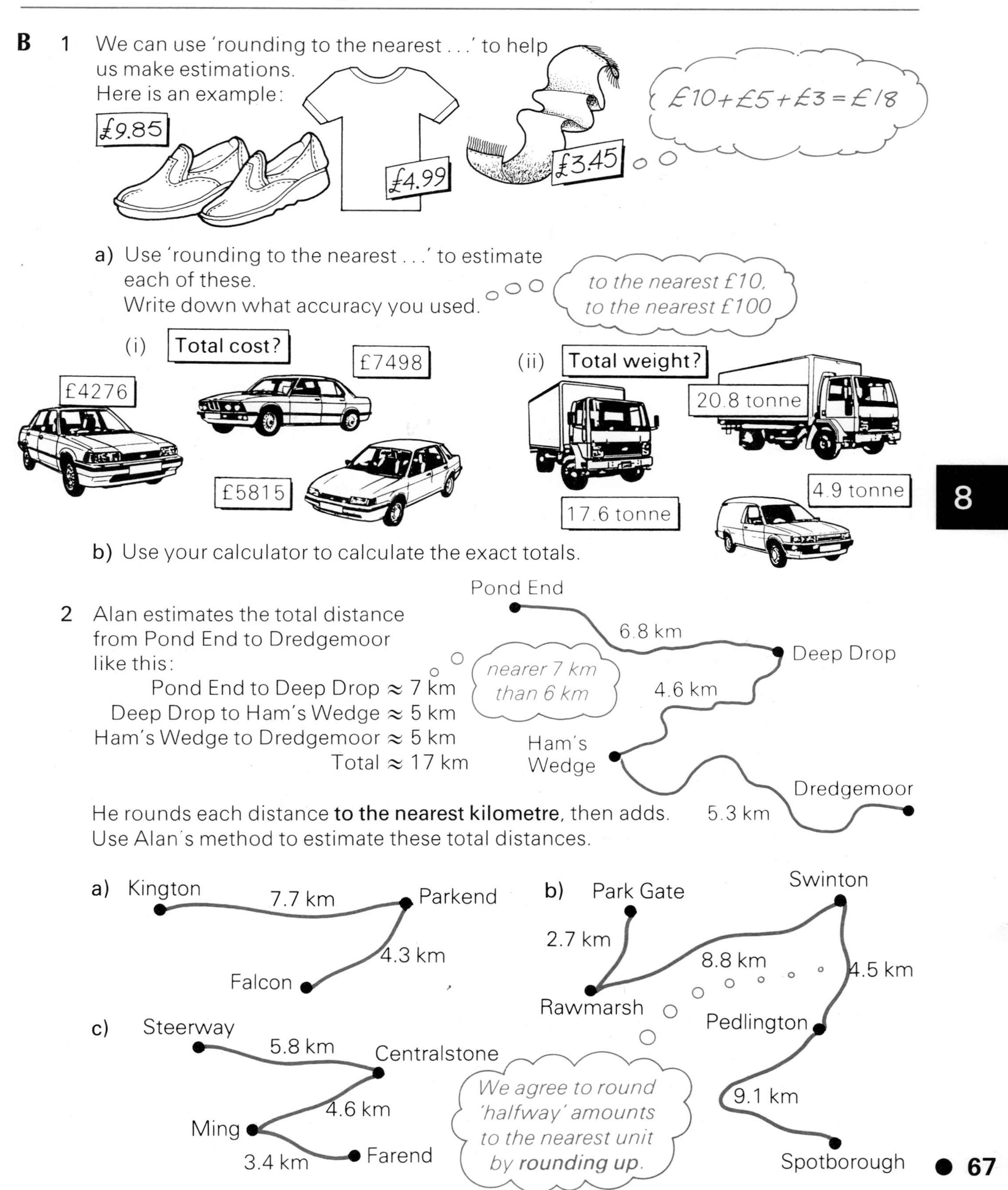

a) Use 'rounding to the nearest . . .' to estimate each of these.
Write down what accuracy you used.

(i) Total cost?

(ii) Total weight?

b) Use your calculator to calculate the exact totals.

2 Alan estimates the total distance from Pond End to Dredgemoor like this:

Pond End to Deep Drop $\approx$ 7 km
Deep Drop to Ham's Wedge $\approx$ 5 km
Ham's Wedge to Dredgemoor $\approx$ 5 km
Total $\approx$ 17 km

He rounds each distance **to the nearest kilometre**, then adds.
Use Alan's method to estimate these total distances.

a) Kington – Parkend – Falcon

b) Park Gate – Rawmarsh – Swinton – Pedlington – Spotborough

c) Steerway – Centralstone – Ming – Farend

3 Use 'rounding to the nearest . . .' to estimate the cost of each of these.
For each one, write down what approximations you made.

a) 2.85 kg of tomatoes at £0.69 per kilogram

b) 2.175 kg of grapes at £1.37 per kilogram

c) 380 g of cheese at £1.92 per kilogram

d) 3.585 kg of cashew nuts at £2.21 per kilogram

For example, to the nearest 1 kg and to the nearest 10p.

With a friend

4 Here are three estimates for the total cost of the three plants.

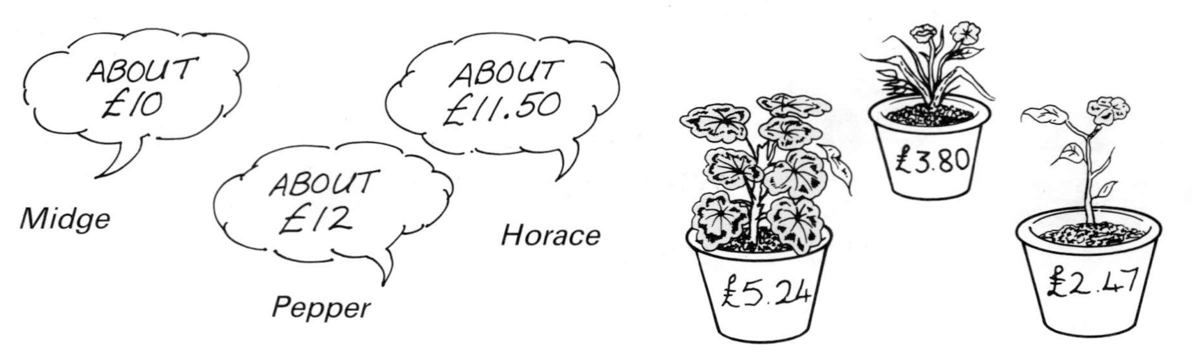

ABOUT £10
Midge
ABOUT £12
Pepper
ABOUT £11.50
Horace
£5.24
£3.80
£2.47

Discuss with your friend how you think each estimate was made.
Write down what you decide.

Exploration

5 a) Here are the prices of two tins of cat food.
Which estimate gives the nearest value to their true total cost:
A rounding to the nearest 10p and adding?
B rounding to the nearest 5p and adding?

b) We might expect rounding to the nearest 5p always to give a more accurate estimate than rounding to the nearest 10p
. . . but it does not.
Find some more examples in which rounding to the nearest 10p gives a more accurate estimate.
Try to find a general rule which tells us which will give the more accurate estimate when we are adding two amounts.

c) Does your rule work when we are adding three amounts?
If it does not, try to find a rule for three amounts.

d) What about four amounts?
. . . five amounts?

Decimal places

C 1 The produce department at the Sav-it Supermarket has a price labelling machine.

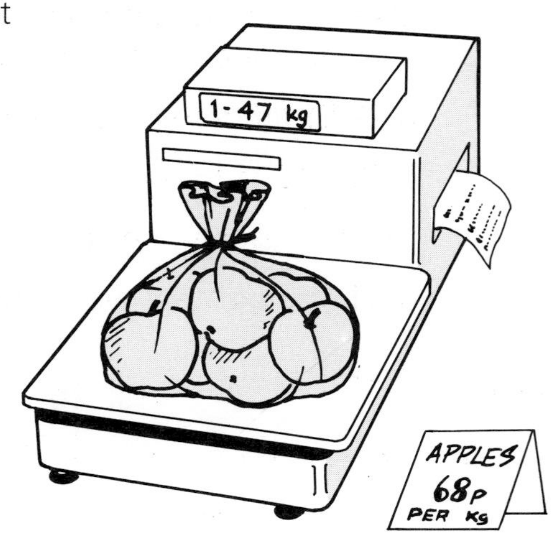

a) The assistant puts this bag of apples on the machine.
It calculates:

$$1.47 \times 0.68 = 0.9996$$

But it does **not** print a price label for £0.9996. Why not?

b) Instead, the machine always rounds prices to the nearest 1p before printing a price label. What is the amount on the label it prints for the bag of apples?

c) Use your calculator.
Write down what price label the machine prints for each of these:
(i) 6.75 kg of potatoes at 47p per kilogram.
(ii) 1.15 kg of asparagus tips at 39p per 100 g.
(iii) 2.64 kg of tomatoes at 63p per kilogram.

Take note

We say that the labelling machine rounds prices in pounds **to 2 decimal places.**

We write: 2 DP.

Rounding to 1 decimal place means to the nearest tenth.
to 2 decimal places means to the nearest hundredth.
to 3 decimal places means to the nearest thousandth.
⋮ ⋮

2 A bee humming-bird weighs about 0.0016 kg.
Scientists studying the growth of bee humming-birds would not record their results in kilograms rounded to 2 DP. Why not?

3 Barnard's star is about 56 573 960 000 000 km away from earth.
Astronomers do not record distances in kilometres rounded to 2 DP.
Why not?

not involving money

4 Give two examples where it **would** be sensible to round results to 2 DP.

Significant figures

D 1 The newspaper report of Saturday's football match gives the gate as 14 800.

a) The number of spectators has been rounded
. . . but definitely **not** to some number of decimal places.
How can we be so sure?

b) To describe this kind of rounding, we can use **significant figures**.
The actual gate was 14785.

The newspaper had rounded to 3 SF.
What is the gate, rounded to
(i) 1 SF? (ii) 2 SF? (iii) 4 SF?

8

2 Significant figures can also be used with decimals.

a) Weighed accurately, a certain 5 kg bag of potatoes contained 5.063 kg of potatoes.
What is this amount, rounded to
(i) 1 SF? (ii) 2 SF? (iii) 3 SF?

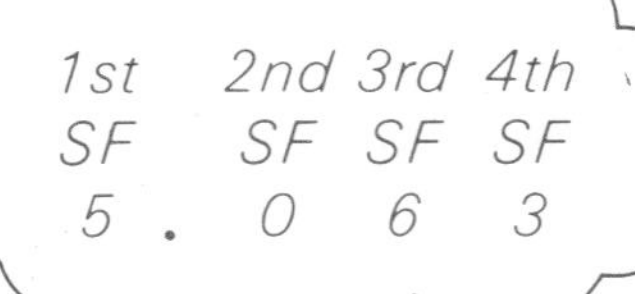

b) We don't start counting significant figures until we reach one that is not zero.
At the end of 1986, 1 Italian lira was worth £0.000 501 002.

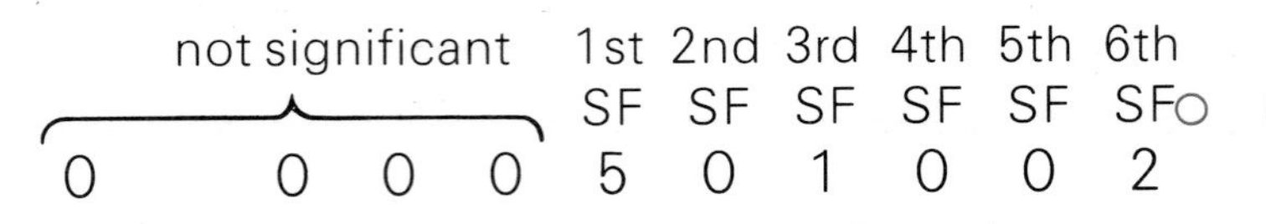

Notice: Some zeros are significant; others are not.

Rounded to 1 SF, this amount is £0.0005
What is it, rounded to
(i) 2 SF? (ii) 3 SF?

3 A three-toed sloth can average about 0.158 km/h on the ground.

a) Round this speed to
(i) 2 SF. (ii) 1 DP. (iii) 1 SF.

b) What has to be true about a number if it makes no difference whether you round it to 1 SF or 1 DP?

Exploration

4 The table gives the results of rounding two numbers to 1 DP and to 1 SF.

Number	Rounded to 1 DP	Rounded to 1 SF
193.07	193.1	200
0.0123	0.0	0.01

a) For each number, which type of rounding gives the more accurate approximation?

that is, closer to the original number

b) Investigate other numbers.
Try to discover answers to these questions.
For which numbers
 (i) is rounding to 1 DP more accurate than rounding to 1 SF?
 (ii) is rounding to 1 SF more accurate than rounding to 1 DP?
 (iii) are rounding to 1 DP and rounding to 1 SF equally accurate?
Write a report of what you find out.

5 a) Approximate the floor areas of the bedroom and the dining room by first rounding each distance to
 (i) 1 DP. (ii) 1 SF.

bedroom: 5.06 m × 3.48 m

dining room: 4.17 m × 3.84 m

b) Suppose that the distances given on the drawings are accurate.
For each room, write down which kind of rounding gives the more accurate approximation to the floor area.

Challenge

c) Try to find another size of rectangular room so that rounding the distances to 1 SF leads to a more accurate approximation to the floor area than rounding them to 1 DP.

Challenge

6 Mr Hornblower has a rectangular allotment which is between four and six times as long as it is wide. He has calculated its area.

Mr H: Round the length and breadth to 1 SF; you get an area of 80 m².
Round the length and breadth to 2 SF; you get 54 m².
Round the true area to 2 SF; you get 55 m².

What could the length and breadth of the allotment be?
Find one possibility.

Accurate measurement

E 1 This is a special kind of ruler.
It is made from perspex so that you can see through it.
It can be used to measure in inches correct to 2 DP.

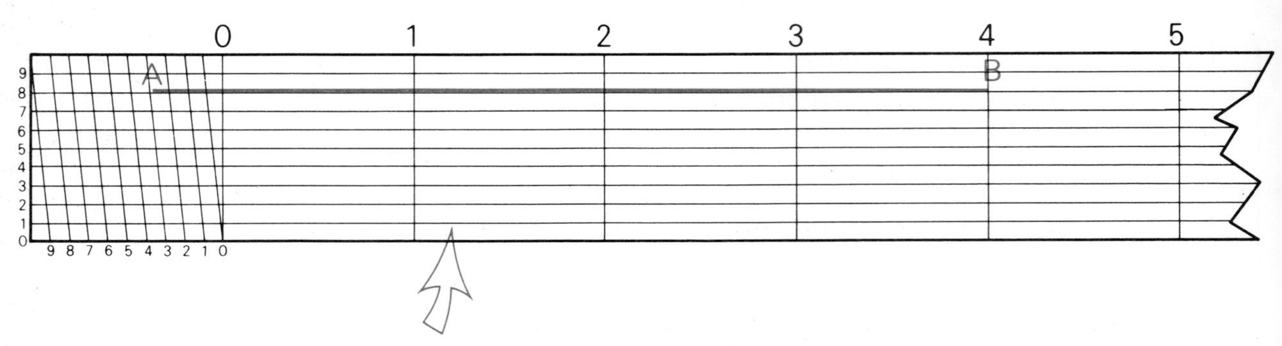

a) The line being measured here (line AB) is 4.37 inches long.
What is the length of the line XY being measured here?

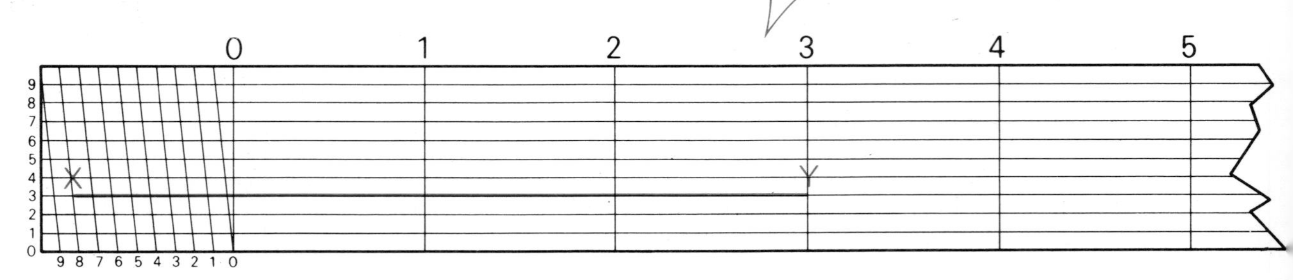

b) Copy this enlarged drawing of the end of the ruler. Use it to help you explain how the ruler works.

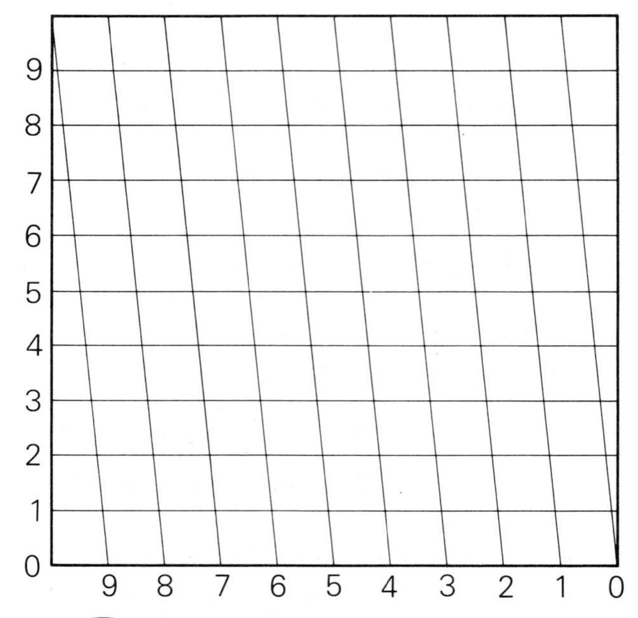

Think it through

c) You need some 2 mm squared paper.
Draw a line which is 9.46 cm long.

You will need a very sharp pencil!

Keeping track of error

F 1 'Slam-dunk' McIntyre is a professional basketball player. His height, rounded to 1 SF, is 2 m. Based on this information alone, there is a whole range of possible values for Slam-dunk's height.

a) Copy and complete this sentence to show **all** the possible heights:

$\square$ m $\leqslant$ Slam-dunk's height $<$ $\square$ m

b) Why is one inequality sign '$\leqslant$' and the other '$<$'?

2 For each of these, copy and complete the sentences to give the whole range of possible values.

a) Time for 1500 m race = 213.3 s (to 1 DP).

$\square$ s $\leqslant$ race time $<$ $\square$ s

b) Population of the world = 5 000 000 000 (to 1 SF).

□ ≤ world population ≤ □
Careful!
Both signs are '≤'.

c) Length of arrow = 0.77 m (to 2 DP).

$\square$ m $\leqslant$ arrow's length $<$ $\square$ m

d) Mass of wedge of cheese = 350 g (to 2 SF).

$\square$ g $\leqslant$ mass of cheese $<$ $\square$ g

e) Number of telephones in the Soviet Union in 1982 = 23 707 000 (to 5 SF).

$\square$ $\leqslant$ number of phones in USSR in 1982 $\leqslant$ $\square$

3 Why do you think some parts of question **2** have two '$\leqslant$' signs, while the other parts have one '$\leqslant$' sign and one '$<$' sign?

4 Look at question **1** again. Slam-dunk McIntyre's height is **at least** 1.5 m. By saying he is 2 m tall, we overestimate Slam-dunk's height by at most 0.5 m.

a) Check that we underestimate his height by less than 0.5 m.

or 50 cm

b) We say that the approximate height of 2 m has a **maximum error** of 0.5 m.

Find the maximum error for each of the approximate amounts in question **2**.

5 The amounts of liquid in these two containers are rounded to 1 DP.

a) Write sentences giving the range of possible amounts in each container.

b) Use your results in **(a)**.
Write sentences giving the range of possibilities for the **total** amount of liquid in the two containers.

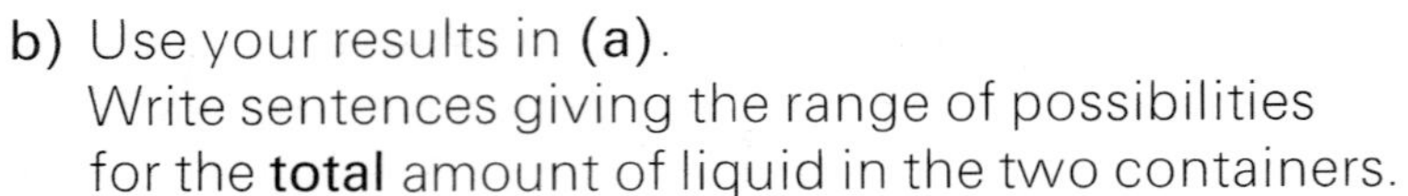

c) 1.4 l + 2.7 l = 4.1 l
What is the maximum error made if the total amount of liquid is given as 4.1 l?

d) Can we be sure that rounding the total amount to 1 DP will give 4.1 l?
Explain your answer.

It may help to give examples of possibilities for the ***actual*** *amounts of liquid in the containers.*

6 In 1982 there were 65 000 000 TV sets in Japan (rounded to 2 SF) and 160 000 000 TV sets in the USA (rounded to 3 SF).

8

a) Write sentences giving the range of possible numbers of TV sets in the two countries.

b) There were at least 94 000 001 more TV sets in the USA than in Japan.
Explain how this number was calculated.

c) Write sentences giving the range of possibilities for the difference between the numbers of TV sets in the two countries.

d) 160 000 000 − 65 000 000 = 95 000 000
What is the maximum error made if we say that in 1982 there were 95 000 000 more TV sets in the USA than in Japan?

Think it through

7 A garden shed has a rectangular floor measuring 3.4 m by 4.6 m.
Both these measurements have been rounded to 1 DP.
Find the maximum error made if we say that the floor area of the shed is
$3.4 \times 4.6\ \text{m}^2 = 15.64\ \text{m}^2$.

Challenge

8 Subtracting approximate amounts can lead to unexpected problems.
Find two amounts for which this is true:

If the amounts are rounded to 1 SF, and the approximations are subtracted, then the maximum error in the result is greater than the result itself.

9 Enlarging and reducing

A 1 Glenda visits Hull in the school minibus. She buys a small sticker for herself, and a larger one for Frankie. Frankie's sticker is an enlargement of Glenda's.

a) Copy and complete Frankie's sticker on 5 mm squared paper.

b) What is the scale factor of enlargement?

2 Ben buys this sticker. It is larger than Glenda's, but is **not** an enlargement of it. Explain why.

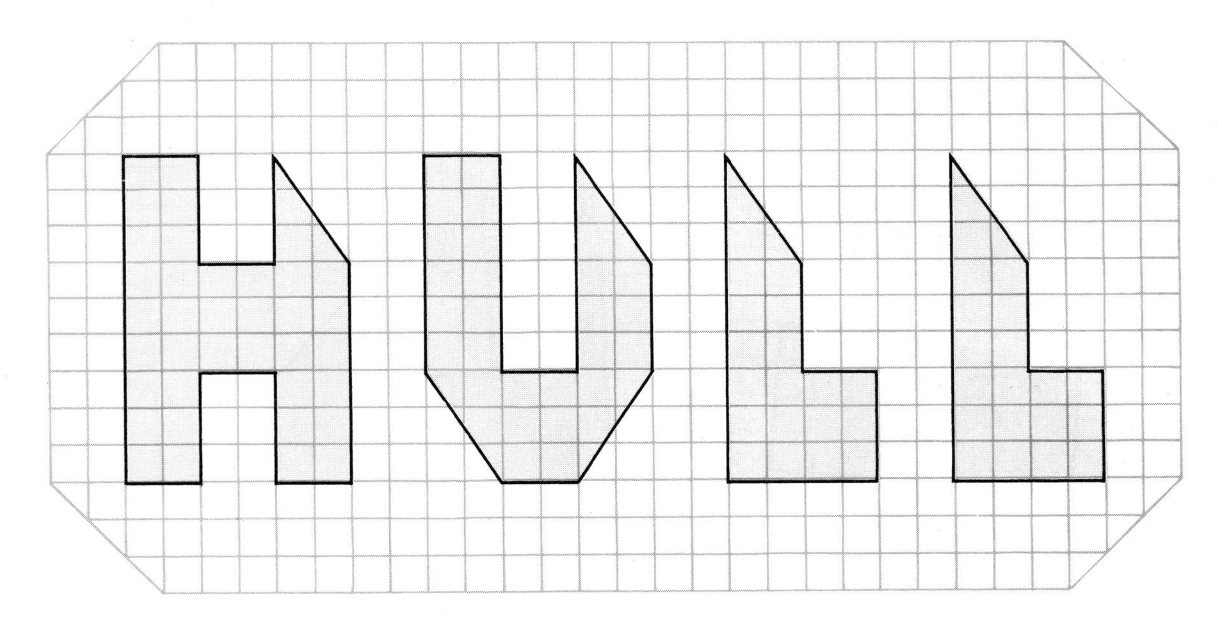

Think it through

3 Pepper buys a sticker which **is** an enlargement of Glenda's. It is the same height as Ben's.

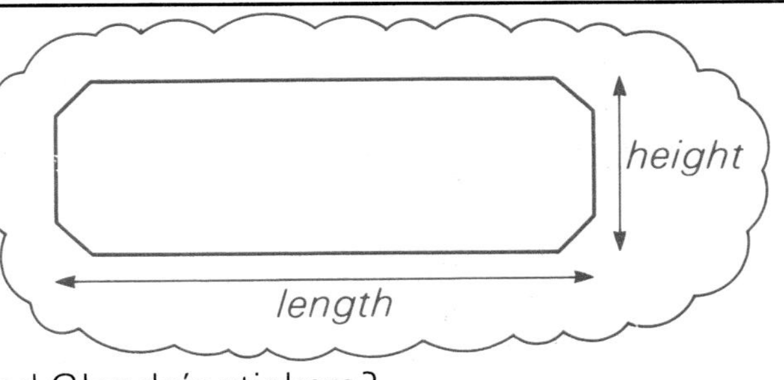

a) Is it longer or shorter than Ben's?

b) How long is it?

c) What scale factor connects Pepper's and Glenda's stickers?

9

4 Rupinder buys all these Hull stickers.

a) Six of them are **not** enlargements of Glenda's. Which are they?

b) For the three stickers which **are** enlargements, what is the enlargement scale factor?

Centre of enlargement

B 1 a) You need some 1 cm squared paper. Copy these two **L**s. Draw them this size, and in exactly the same position.

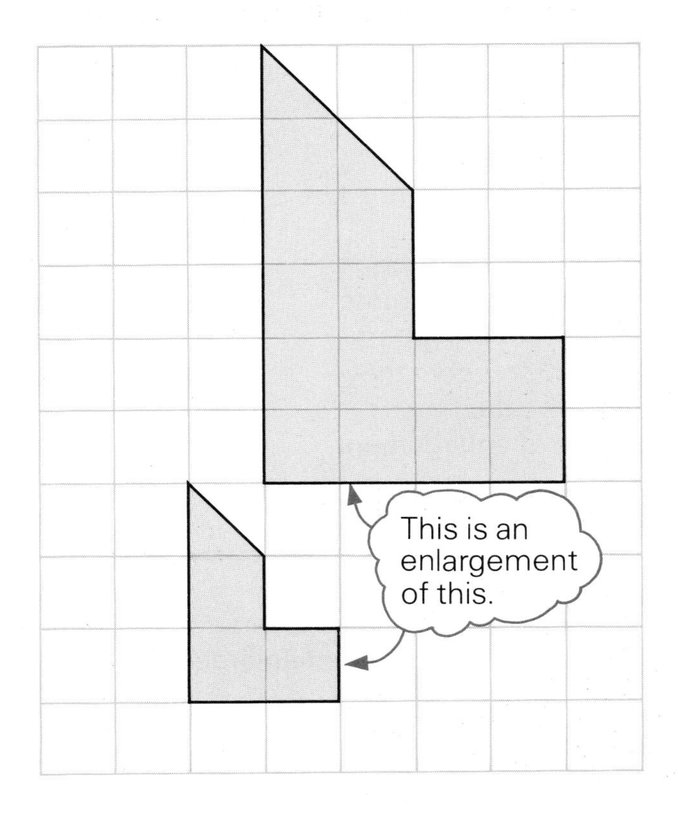

This is an enlargement of this.

9

b) Mark this pair of **corresponding points** on your letters.

Draw a straight line through the pair of corresponding points.

Repeat for another pair of corresponding points. Call the point where the lines meet point O.

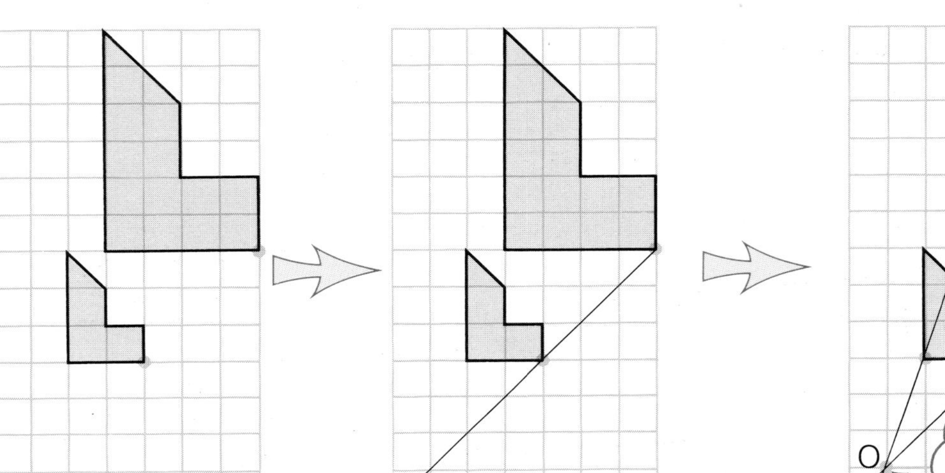

O
The lines meet here.

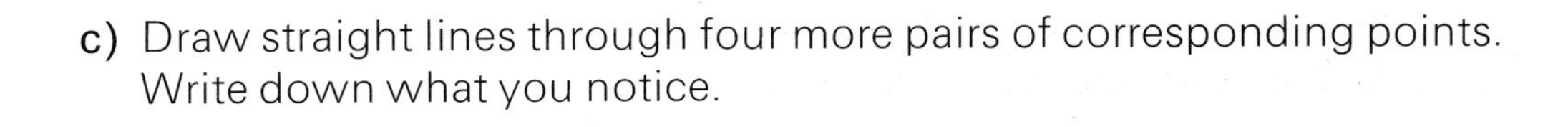

c) Draw straight lines through four more pairs of corresponding points. Write down what you notice.

Take note

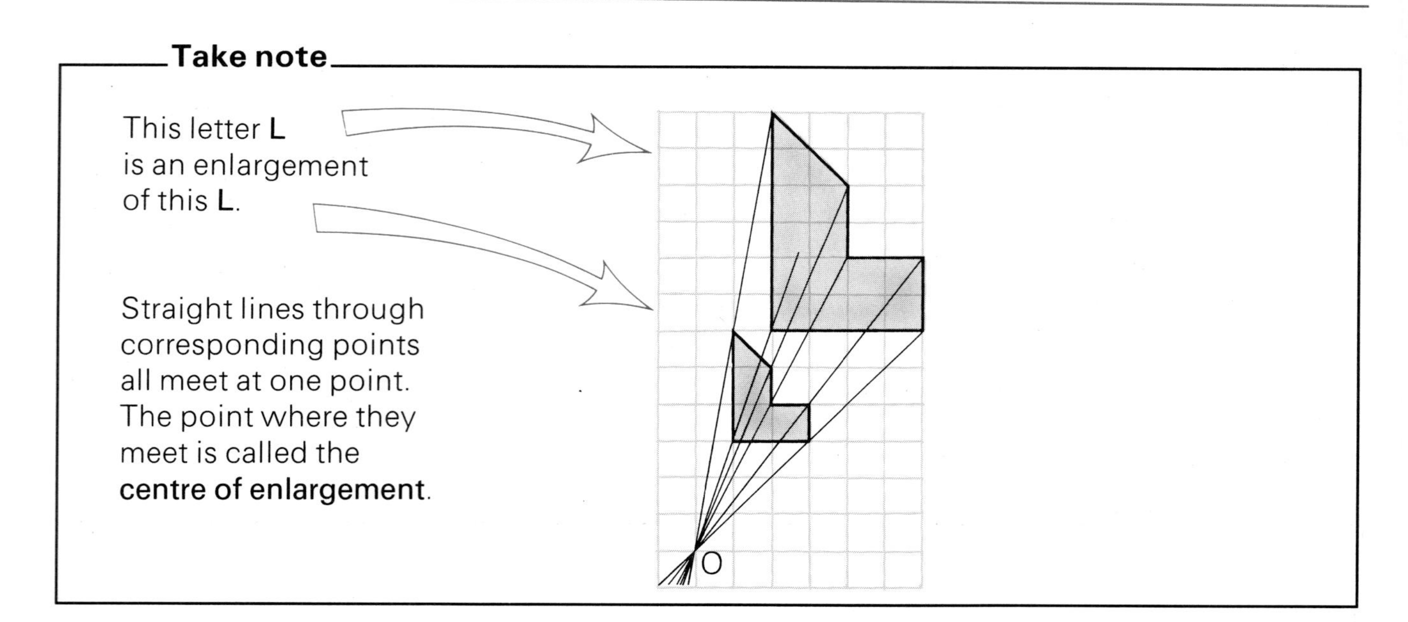

This letter **L** is an enlargement of this **L**.

Straight lines through corresponding points all meet at one point. The point where they meet is called the **centre of enlargement**.

2 a) Copy these pairs of letters onto 1 cm squared paper.
Find the centre of enlargement for each pair.

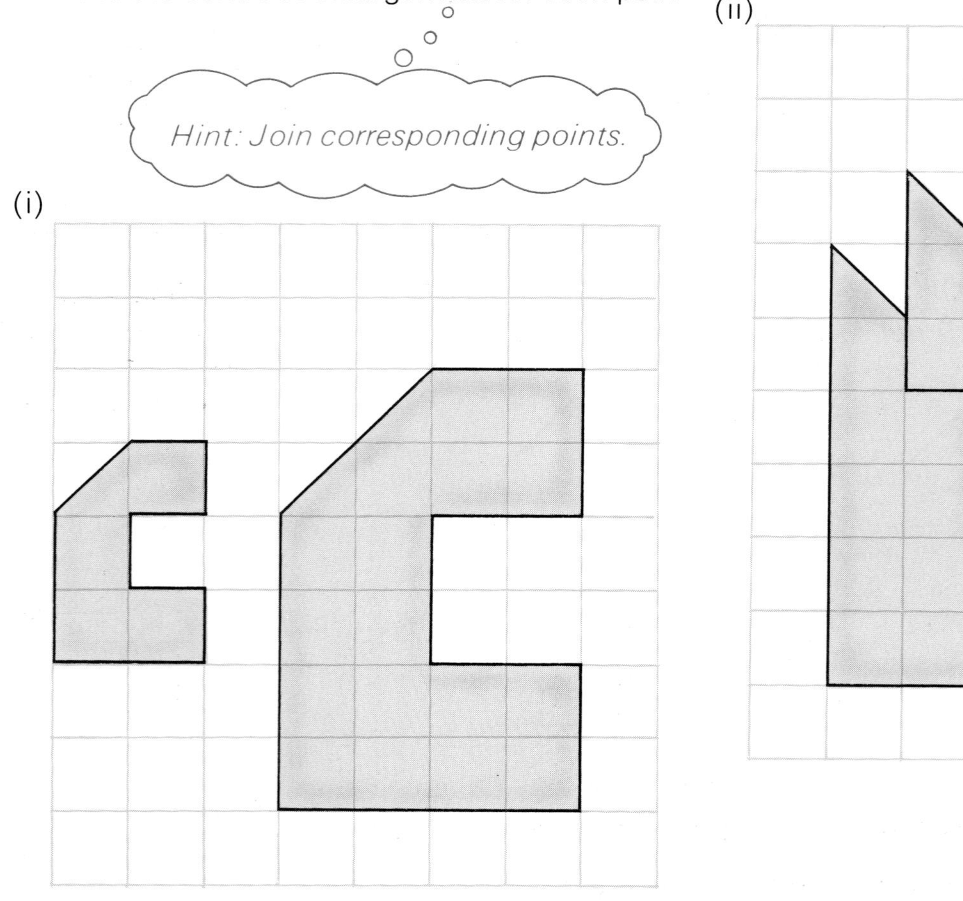

b) What is the scale factor of enlargement for
(i) the letter **C**s? (ii) the letter **L**s?

9

3 a) On your diagram for question **2(a)** (i) label the centre of enlargement O.

b) Call this pair of corresponding points A and A′.

c) How many times as long as OA is OA′?

d) Choose another pair of corresponding points. Label them B and B′.
How many times as long as OB is OB′?

e) What is the scale factor of the enlargement?

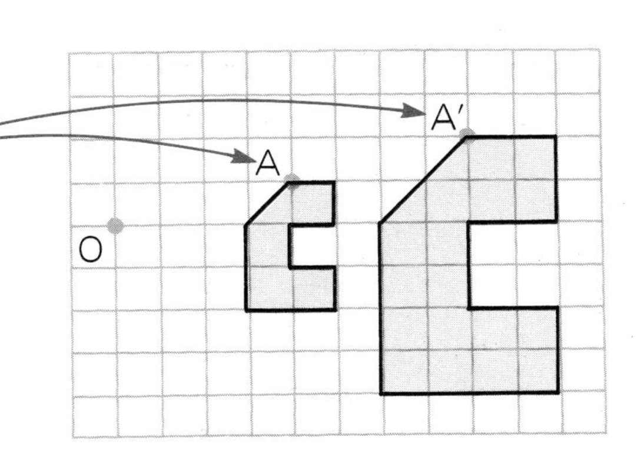

4 Copy and complete the **Take note**.

Take note

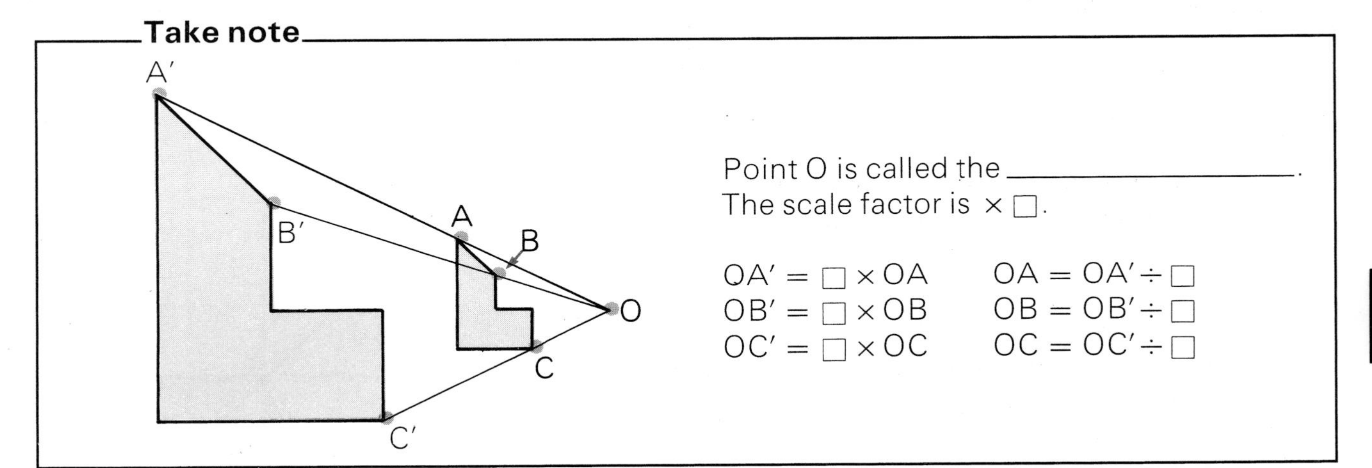

Point O is called the ______________.
The scale factor is × □.

OA′ = □ × OA OA = OA′ ÷ □
OB′ = □ × OB OB = OB′ ÷ □
OC′ = □ × OC OC = OC′ ÷ □

5 a) On plain paper draw a 5 cm × 2 cm rectangle. Label the corners A, B, C and D.

b) Mark a point O near the rectangle.

c) Draw a line from O to A. Continue the line until it is **three** times as long as OA. Call the endpoint A′.

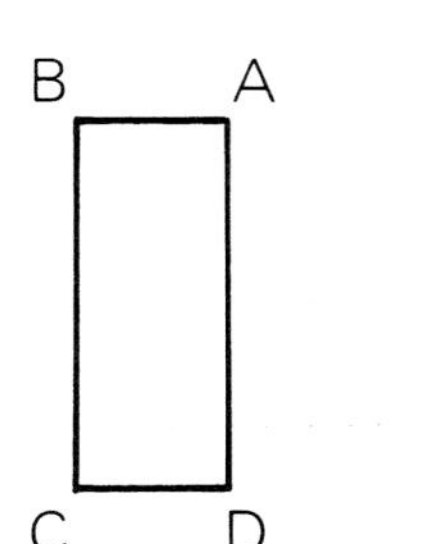

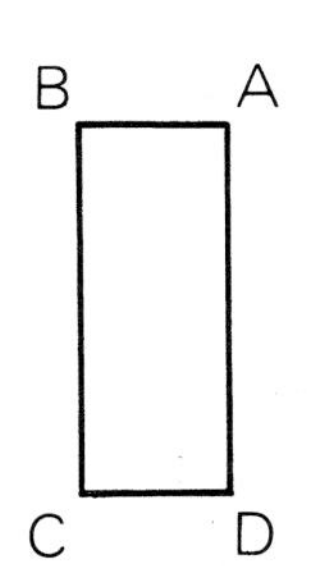

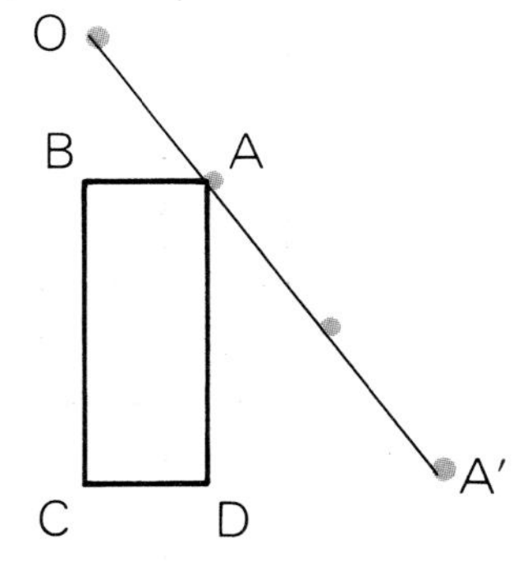

d) In the same way, find points B′, C′, D′.

e) Join the four new points. You should get a rectangle.

f) Measure your new rectangle's sides.
Are they three times as long as the original sides?

g) ABCD is a reduction of A′B′C′D′. Write down its scale factor, like this: ÷ □.

9

6 The quadrilateral ABCD is being enlarged by a scale factor of $\times 4$.

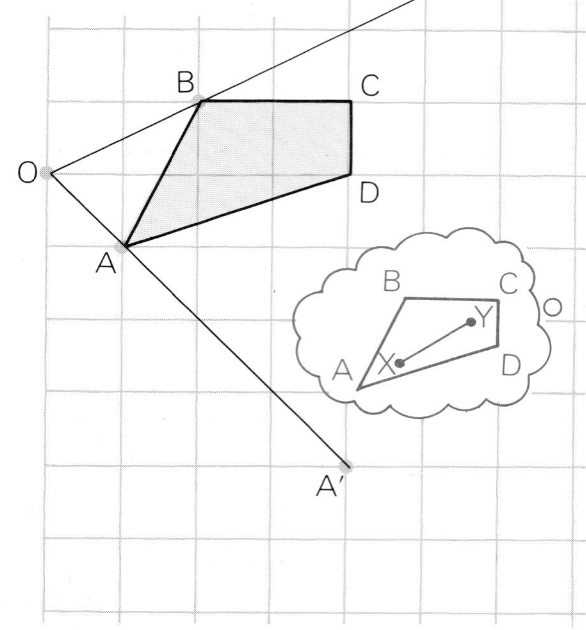

Point O is the centre of the enlargement.

a) Copy the diagram onto 1 cm squared paper. Complete the enlargement.

b) BC is 2 cm long. Check that B′C′ is four times as long.

c) How many times as long as CA is C′A′?

d) Choose any two points on or inside ABCD. Call them X and Y. Write down what you can about the length of X′Y′.

7 Quadrilateral ABCD is again being enlarged by a scale factor of $\times 4$.

a) Copy and complete the enlargement.

b) Compare this enlarged figure with the one in question **6**. Write down what you notice.

B′
B
C
O
D
A
A′

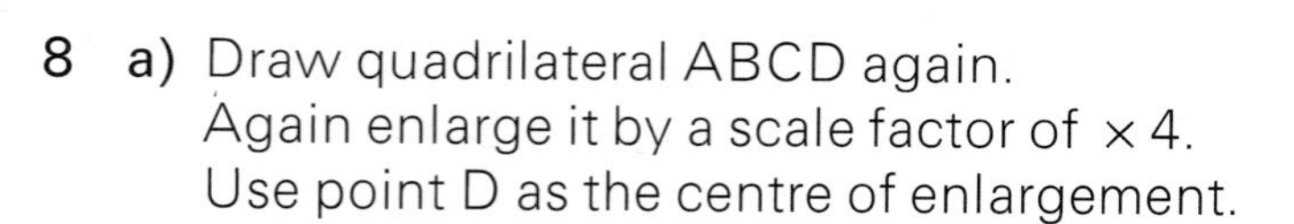

Think it through

8 a) Draw quadrilateral ABCD again. Again enlarge it by a scale factor of $\times 4$. Use point D as the centre of enlargement.

b) Choose a point M anywhere inside ABCD.

Write down what you can about the length of DM′.

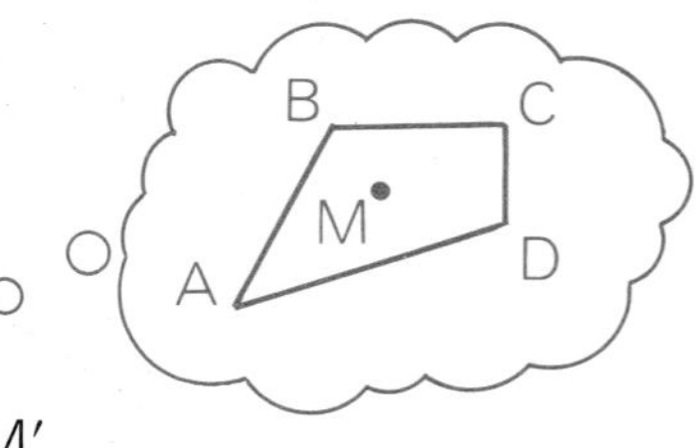

Take note

In an enlargement, all lengths are enlarged by the same scale factor.

9

Making larger but not enlarging

C 1 Horace starts with this rectangle.

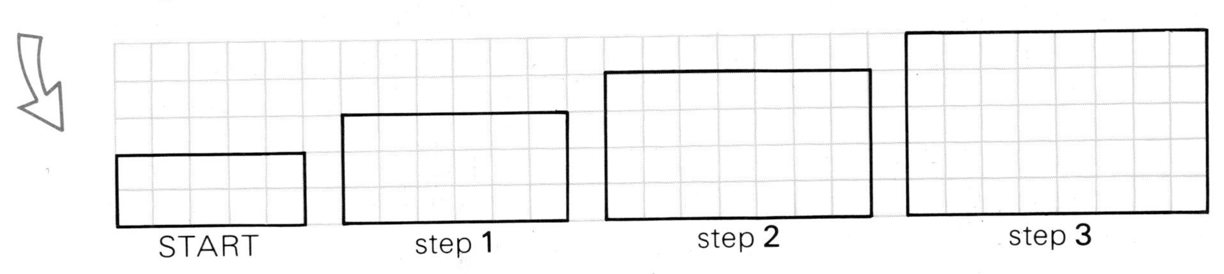

He keeps adding 1 unit to each side.

a) Draw his rectangle after
   (i) step **4**.
   (ii) step **10**.

b) Imagine Horace's rectangle after step **100**.
The rectangle will look like a blown-up version of one of these.
Which one?

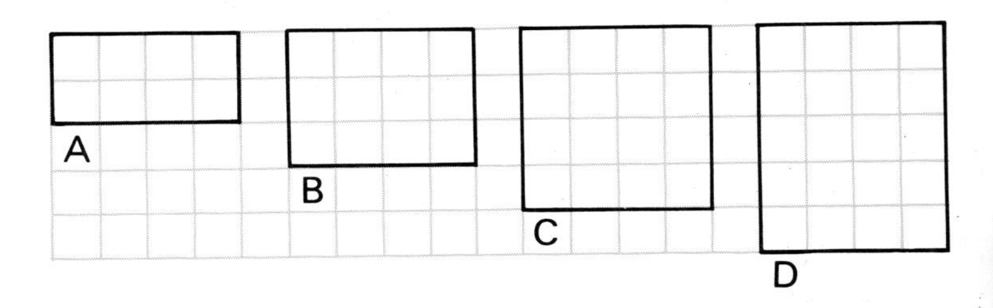

9

2 Horace starts with this rectangle.

O
START
step **1**
step **2**
step **3**

He keeps adding 3 cm to the distance between each corner and point O.

a) Copy the diagram.
Draw the rectangle after
   (i) step **3**.
   (ii) step **4**.

b) Describe the shape of the rectangle after step **100**.

Reducing

D 1 a) The new **T** is being drawn using a scale factor of $\times \frac{1}{2}$.

(i) Write this as $\div \square$.

(ii) Copy and complete the new **T**.

b) Points M and N are inside the original **T**.
The distance MN is 6 cm.
What is the distance M′N′?

2 a) What scale factor is being used to draw the new letter **L**?

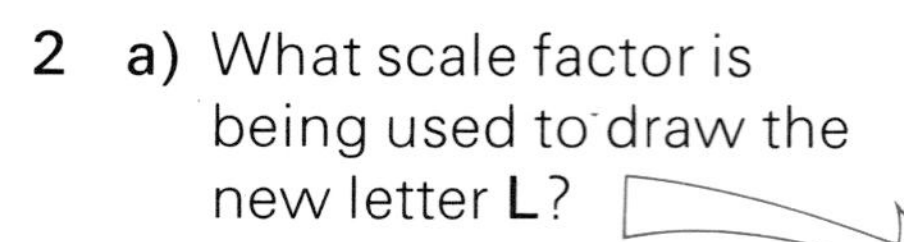

b) Copy and complete the new **L**.

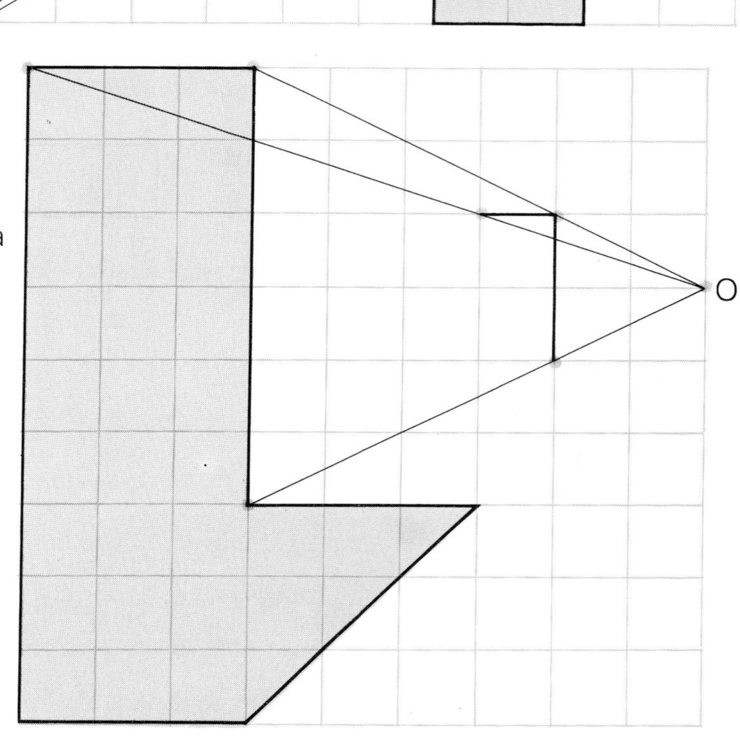

3 Use a scale factor of $\times \frac{1}{4}$ to draw a smaller version of this triangle.
Use O as the centre of enlargement.

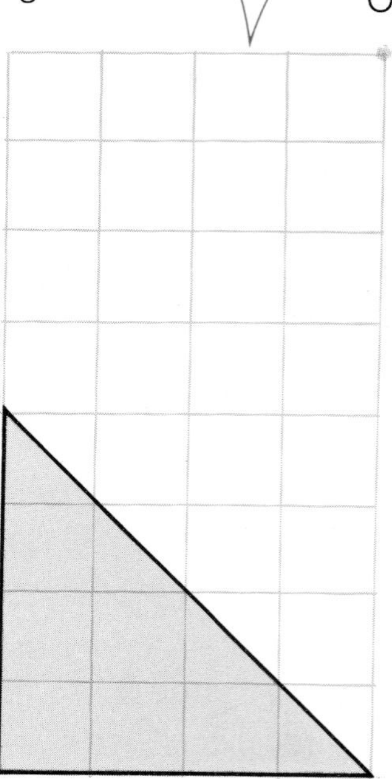

Activity *Self-portrait*

4 Use squared paper.
Measure with a ruler or tape measure.
Draw a smaller version of yourself.
Use a scale factor of $\times \frac{1}{10}$.

5 a) Reduce the triangle.
Use A_1 as the centre and use a scale factor of $\times \frac{1}{2}$.

b) Enlarge your new triangle.
Use A_2 as the centre and use a scale factor of $\times 2$.

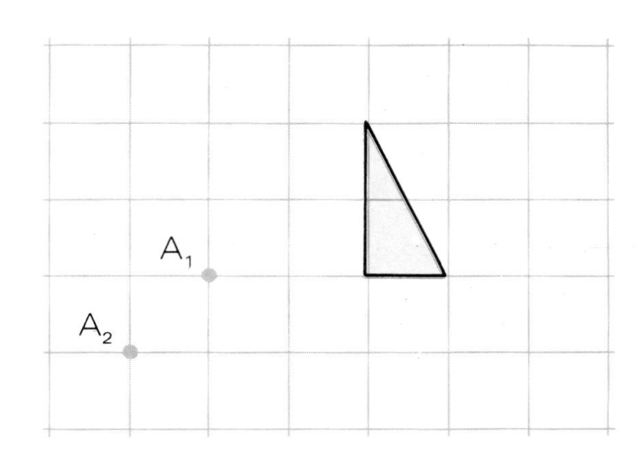

6 Repeat question **5**, but this time use B_1 and B_2 as the centres.

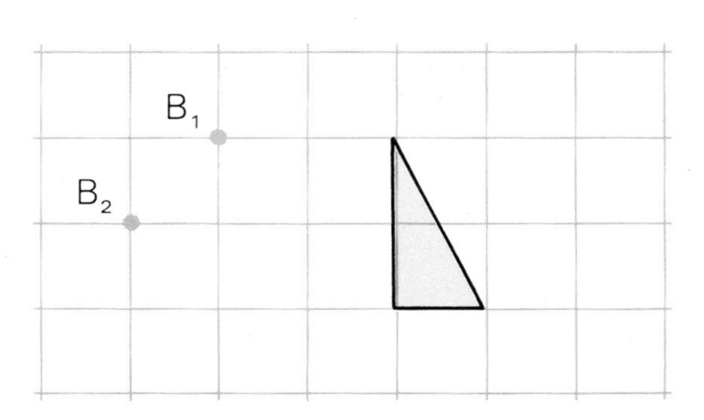

7 Glenda repeats question **5**, but uses C_1 and C_2 as the centres.
She says the final triangle will end up here.

a) Do you think she is right?

b) Check by reducing and enlarging.

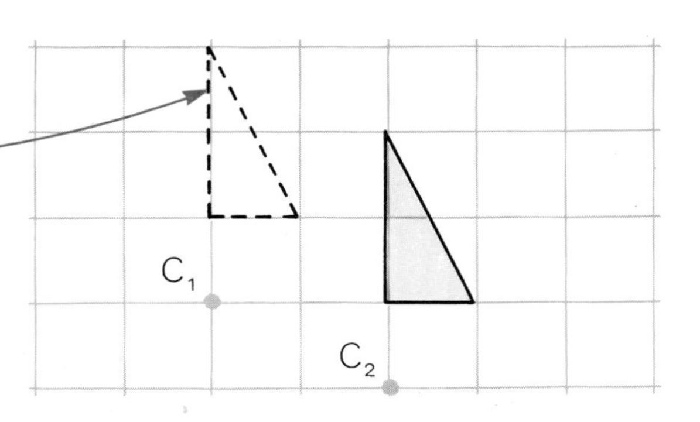

Think it through

8 Horace repeats question **5**, but uses D_1 and D_2 as the centres.

a) Copy the triangle.

b) **Without** reducing and enlarging draw the position of the final triangle.

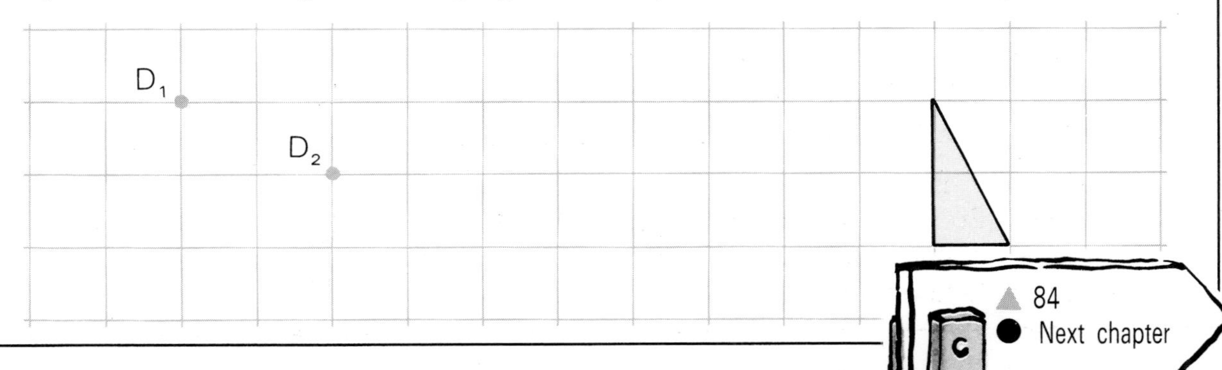

▲ 84
● Next chapter

9

Enlargements and lengths

E 1 Triangle A′B′C′ is an enlargement of ABC.
O is the centre of enlargement.
The scale factor is × 3.

Work out the lengths of

a) OC′.

b) B′C′.

c) AA′.

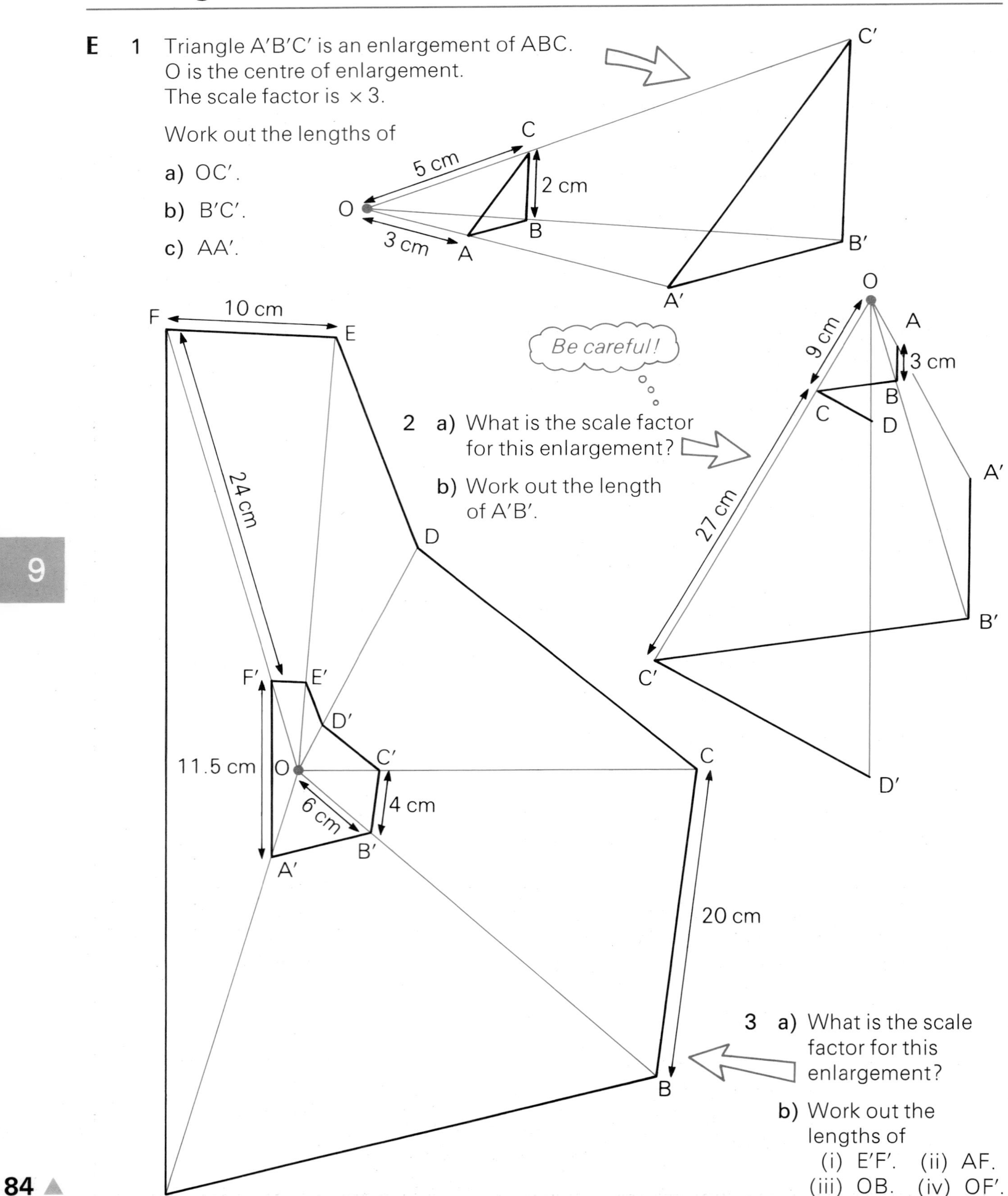

2 a) What is the scale factor for this enlargement?

b) Work out the length of A′B′.

3 a) What is the scale factor for this enlargement?

b) Work out the lengths of
(i) E′F′. (ii) AF.
(iii) OB. (iv) OF′.

9

Enlarging and reducing solids

F

Explorations

1 You need some dotted isometric paper.

a) Copy the diagram, near the right-hand side of your sheet of isometric paper.

b) Enlarge the 1 cm cube with centre of enlargement O and scale factor $\times 6$.

c) Reduce the enlarged cube with centre of enlargement A and scale factor $\times \frac{1}{3}$.

d) Your final cube should be an enlargement of the original 1 cm cube.
Write down the scale factor, and describe where the centre of enlargement is.

A
B
O

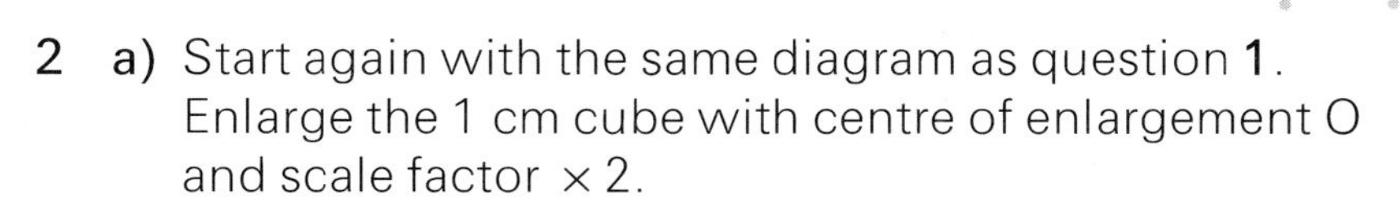

2 a) Start again with the same diagram as question **1**.
Enlarge the 1 cm cube with centre of enlargement O and scale factor $\times 2$.

b) Enlarge the enlarged cube with centre of enlargement B and scale factor $\times 3$.

c) Your final cube should be an enlargement of the original 1 cm cube.
Write down the scale factor, and describe where the centre of enlargement is.

3 Repeat question **2**, but this time use $\times 3$ first and $\times 2$ second.

4 Investigate this situation.
Look for a way to predict in advance what the final scale factor will be, and where the centre of enlargement is.
Try doing

- an enlargement followed by an enlargement.
- an enlargement followed by a reduction.
- a reduction followed by an enlargement.
- a reduction followed by a reduction.

Avoid choosing a reduction and an enlargement whose scale factors multiply together to give 1.

For example, avoid $\times 2$ followed by $\times \frac{1}{2}$.

Always use O as the first centre of enlargement.

5 Why did we suggest you avoided choosing a reduction and an enlargement whose scale factors multiply together to give $\times 1$?

Hint: Look back at page 83.

Shadows

G 1 Glenda has a screen, a piece of card, and a table lamp.

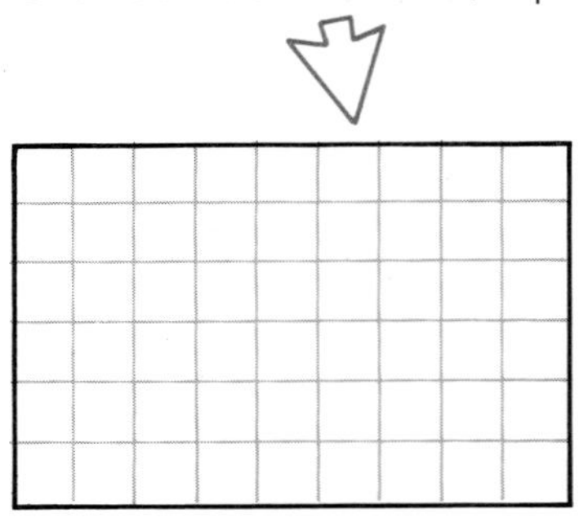
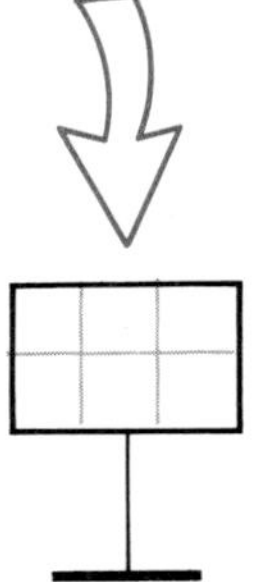

She stands the card midway between the screen and the lamp, like this.

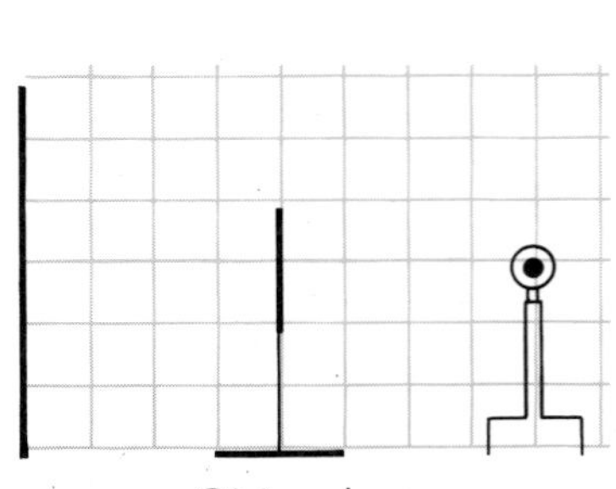
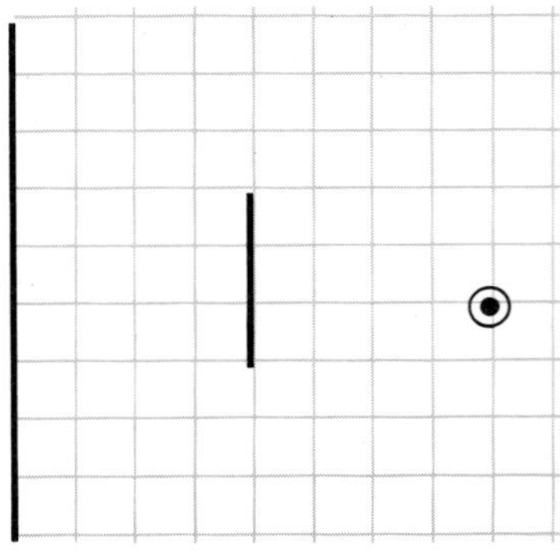

Side view Top view

The shadow of the card on the screen looks like this.

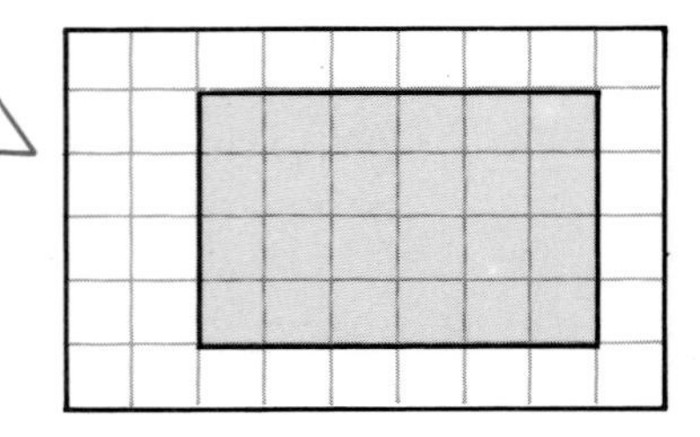

a) What happens to the size of the shadow when
   (i) the lamp is moved away from the card?
   (ii) the card is moved nearer the lamp?
   (iii) the card and lamp stay the same distance apart but are moved away from the screen?

b) Draw the shadow of the card on the screen when the top view looks like this.

(i)

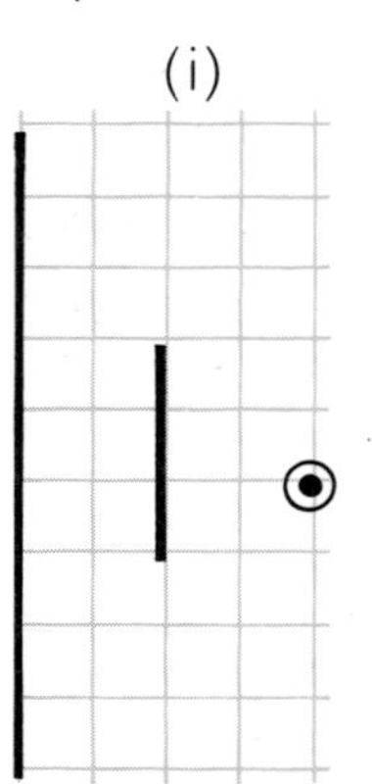

(ii)

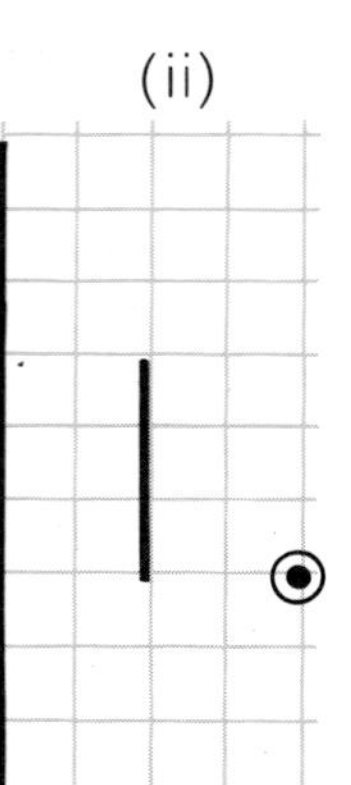

(iii)

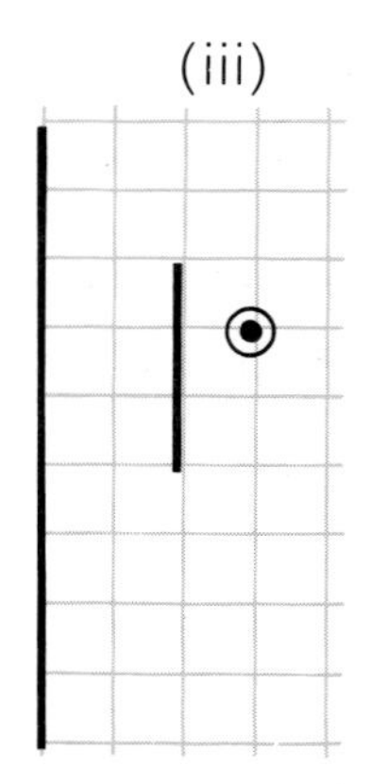

(iv)

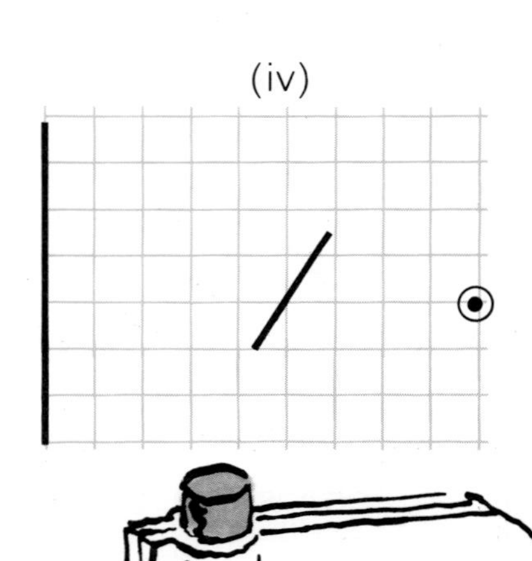

10 Using letters

A

Challenge

1 Jake is building a wooden W to strengthen the roof of a shed. This is his drawing.

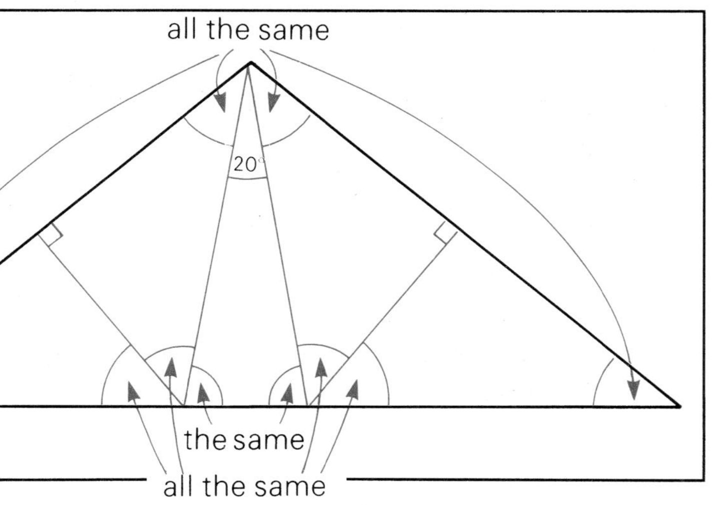

Now he needs to work out the angles. Try to work out the angles for him. Make a sketch and mark the size of each angle on it.

2 This is another W inside a roof space.

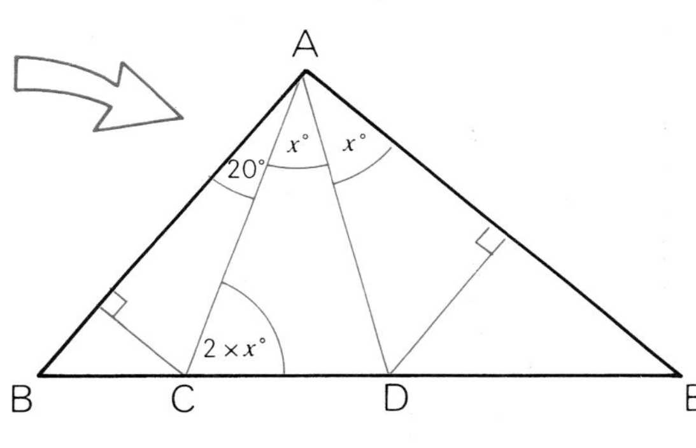

This time the builder has marked some angles using letters.

For example, angle CAD and angle DAE are equal; angle ACE is twice angle CAD . . .

a) Which of these is correct for the size of angle AEC?

A 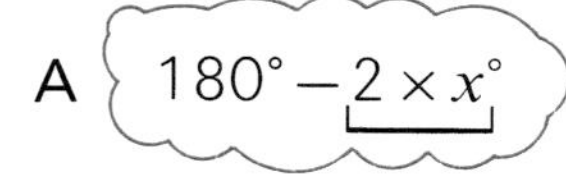 $180° - 2 \times x°$ B $4 \times x°$ C 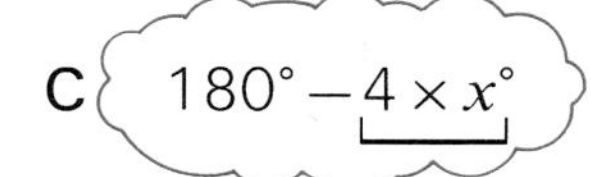 $180° - 4 \times x°$

b) Angle BAE is a right angle.
Which of these is correct for x?

A  20 B 30 C 35 D  40

c) How many degrees is angle AEC? *Use your results in **(a)** and **(b)**.*

d) Sketch the roof space and mark in the size of each angle.

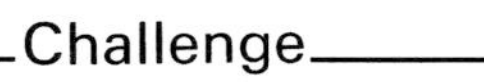

Challenge

3 Sketch this roof space. On your sketch mark in the sizes of all the angles.

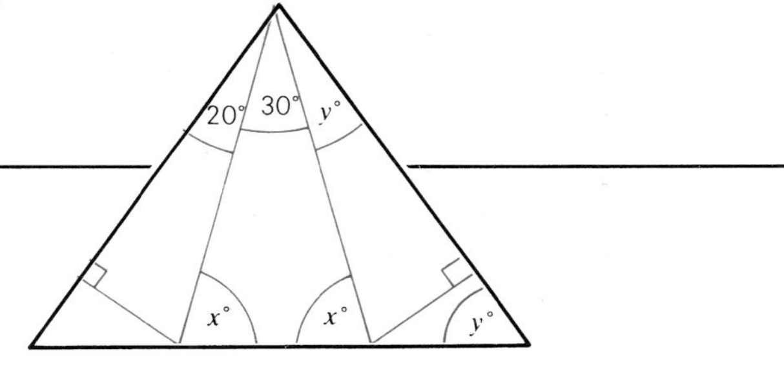

4 **Do you remember . . .?**

To help us to solve problems we can use letters; and write number sentences.
For example, this number sentence is for the outer triangle in Jake's problem:

$$x + x + 20 + x + x = 180$$

So $4 \times x + 20 = 180$

Jake's problem
This is Jake's problem from question **1**, with letters used for the angles.

Jake is building a wooden W to strengthen the roof of a shed.
He wants the top angle to be 20°.
He wants the central triangle, and the two side triangles to be isosceles.

each divided in half by part of the W

a) Test numbers in the number sentence until you find what x stands for.
b) Write a number sentence for the central triangle. Find out which number y stands for.
c) Write a number sentence and find out which number z stands for.
d) Check that the angles you have found agree with your solution in question **1**.

5 Pepper is making a set of picture frames.
The frames are 12 cm longer than they are wide.
For the first frame Pepper uses 84 cm of wood.
This number sentence will help her find the width and length:

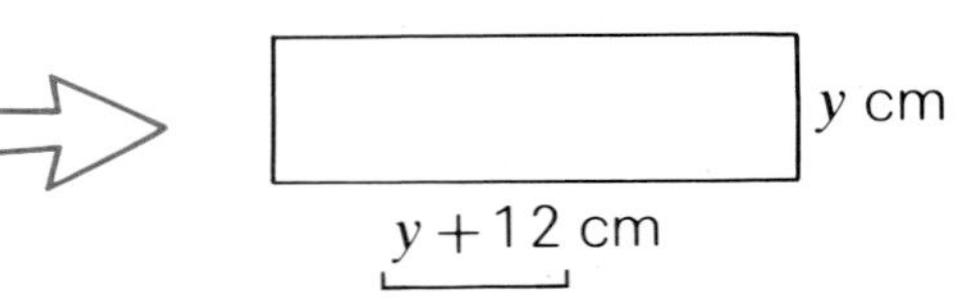

$$y + y + 12 + y + y + 12 = \square$$

a) Copy and complete it.
b) Write it more simply using one $\times$ and one $+$.
c) Think and test until you find the number which y represents.
Write down the length and width of the frame.

Challenge

6 This is one of Pepper's more complicated designs.
The frame has two lines of symmetry.

a) Pepper uses 96 cm of wood to make one of the frames.
Write a number sentence and use it to find the lengths of all the parts of the frame.
b) Pepper finds that whatever frame she makes using this design, she always needs at least 40 cm of wood. Explain why.

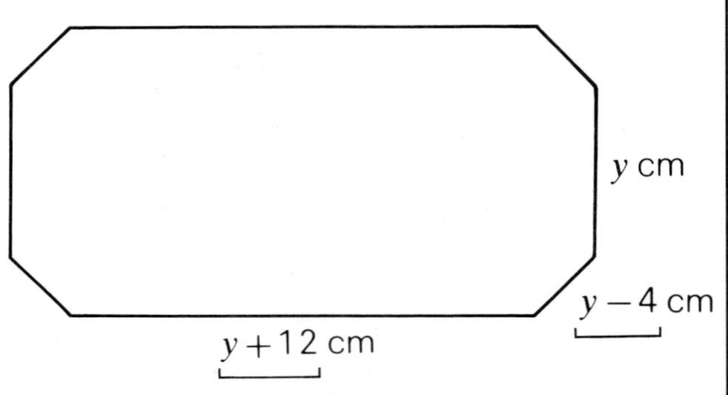

Equations

B 1 Here are lots of number sentences.
Think and test numbers for each one until you find out which number the letter represents.
Use your calculator if it helps.
Write your solutions like this.

Example:
$\frac{x}{4} = 5.$
So x is 20.
Check: $\frac{20}{4} = 5.$

Think: What number divided by 4 gives 5?

Always write a check like this.

Always remember to check that your solution fits the number sentence.

a) $y+8 = 15$

b) $y-8 = 15$

c) $6 \times w = 24$

d) $\frac{w}{6} = 24$

e) $9 \times y = 81$

f) $\frac{y}{30} = 30$

g) $x-14 = 28$

h) $28-x = 16$

i) $x+2.7 = 7.2$

j) $1.2 \times x = 6$

k) $\frac{2.5}{w} = 0.5$

l) $2.4+y = 3.15$

m) $56 \div w = 7$

n) $w \div 5 = 7$

o) $w \div 14 = 3$

p) $7 \div w = 2$

q) $1.6 \times y = 8$

r) $y \times 2.4 = 48$

s) $w \times 4 = 0.4$

t) $\frac{10}{y} = 2.5$

u) $\frac{x}{20} = 1.4$

v) $12 \div k = 1.5$

w) $15 = w-11$

x) $8 = \frac{4}{x}$

y) $\underbrace{2 \times y}+16 = 40$

z) $10-\underbrace{4 \times k} = 1$

Think it through

2 These number sentences are tougher than those in question **1**.
Try to find out which number each letter represents.
Think and test until you find it.
Write each of your solutions like this.

Example:
$9-p = 2 \times p.$
So p is 3.
Check: $9-3 = 6$
$2 \times 3 = 6.$

a) 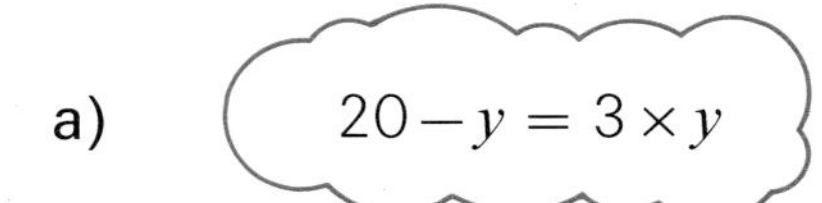

b) $\frac{t}{3} = t-3$

c) $25-\frac{2 \times n}{3} = n$

d) 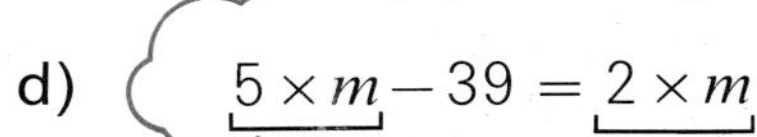

10

Take note

Number sentences like these are also called **equations**.

a) $x+87=104$ b) $\frac{y}{12}=7$ c) $35-\underline{2\times w}=\underline{5\times w}$ d) $x+y=10$

Each number which the letter represents is called a **solution** of the equation.
When we have found all the possible numbers, we say we have **solved** the equation.

3 Solve each of the equations in the **Take note**. Think and test until you find all the possible solutions.

In part (d) find all the whole-number solutions, and one solution for which x and y are not whole numbers.

4 a) Write down an addition equation like this whose solution is 28. $y+47=82$ (*Solution: $y=35$*)

b) Write down a subtraction equation like this whose solution is 53. $y-19=37$

c) Write down a subtraction equation like this whose solution is 35. $59-y=16$

d) Write down a multiplication equation like this whose solution is 41. $3\times y=111$

e) Write down a division equation like this whose solution is 135. $w\div 12=144$

f) Write down a division equation like this whose solution is 84. $\frac{w}{6}=16$

g) Write down a division equation like this whose solution is $1\frac{1}{2}$. $\frac{4}{w}=8$

5 Solve each of the example equations in question **4**. Think and test until you find all the possible solutions.

Challenge

6 a) Write down an addition and subtraction equation with a letter on both sides of the = sign like this, and whose solution is 12. $y+6=14-y$

b) Write down a multiplication and addition equation with a letter on both sides of the = sign like this, and whose solution is 18. $3\times p=p+9$

10

7 For each of these situations,

a) form an equation.

b) solve the equation.

c) write down the information asked for in the question.

Example:

total length = 30 cm
How long is each section?

Answers: a) $x + 6 + 3 \times x = 30$

b) $4 \times x + 6 = 30$
So $x = 6$.

c) *The sections are 12 cm and 18 cm long.*

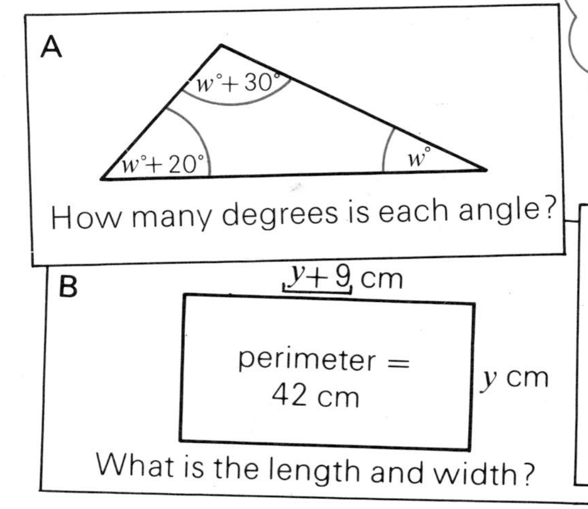

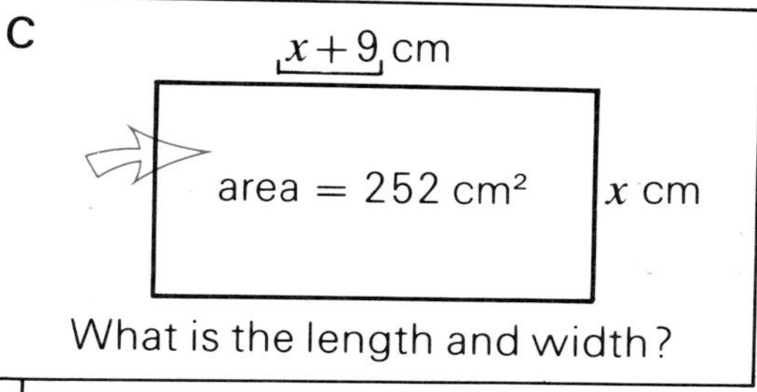

8 A haulage company has an order to deliver 540 tonnes of coal. It has one large lorry which can carry 60 tonnes, and lots of smaller lorries which can carry 15 tonnes each.

a) Use y for the number of small lorries needed.
   (i) Write down an expression using y for the total amount of coal the small lorries can carry.
   (ii) Copy and complete this equation for the order: $\square \times y + \square = 540$.

b) Find the solution to the equation.

c) How many small lorries does the haulage company need for the delivery?

Think it through

9 Form your own equation for this problem.

Use n to represent how much each of the younger nephews and nieces receives.

Use your equation to help you to solve the problem.

Rich Uncle Harry has lots of nephews and nieces.
Some of them are older than 12 years, and some are younger.
At the family get-together Uncle Harry gives £8 to each nephew and niece who is 12 years old or more. There are seven of them.
He also gives each nephew or niece who is less than 12 years old the same amount as each other. There are nine of them.
He gives away £101 altogether.
How much does he give each of his younger nephews and nieces?

Using two letters in the same equation

C 1 Annie is planning a vegetable patch.
She decides she needs 20 m², and she would like the patch to be a rectangle or a square.
This equation tells us what she wants: $w \times y = 20$

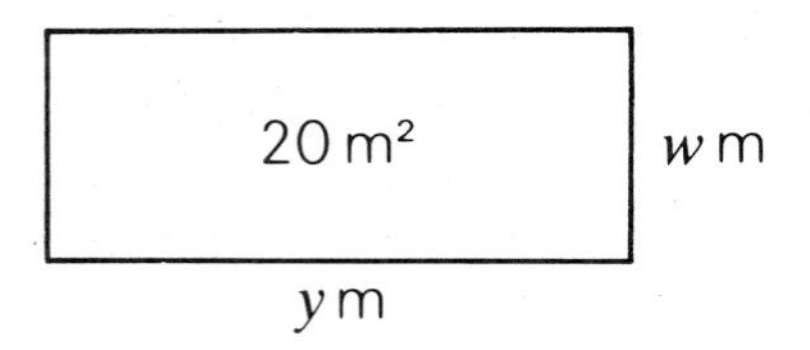

One possible solution is $w = 4, y = 5$.

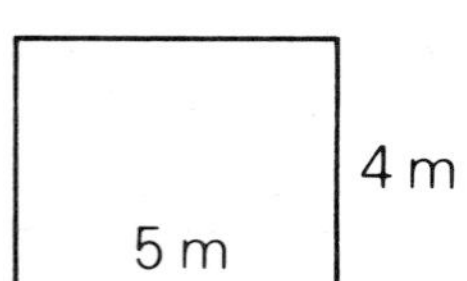

Another is $w = 2, y = 10$.

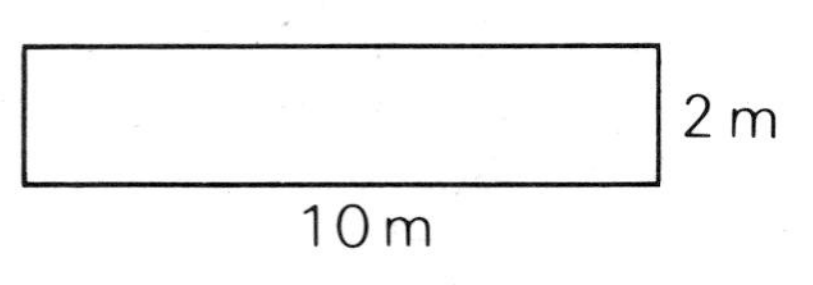

a) Find all the possible solutions for which w and y are whole numbers.

including the 'silly' ones which correspond to long, skinny vegetable patches

b) Annie chooses w to be 2.5.
How long and how wide is her vegetable patch?

2 Horace is designing T-shirts.
He is going to print large **T**s like this all over them.
He decides that each **T** should cover an area of 150 cm².

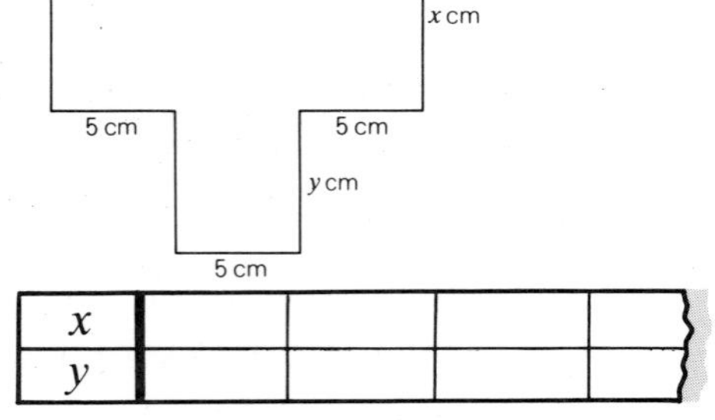

a) Find a number for x and a number for y which Horace can use.

b) Find all the possible whole-number pairs for x and y which Horace can use.
Record them in a table, like this:

x				
y				

c) Horace is not going to want to use some of these possibilities.
Which ones, and why not?

d) Use this 'exploded' diagram to help you to write an equation, using x and y, for the area of Horace's **T**s.

15 cm

x cm

e) Copy and complete these solutions of the equation:

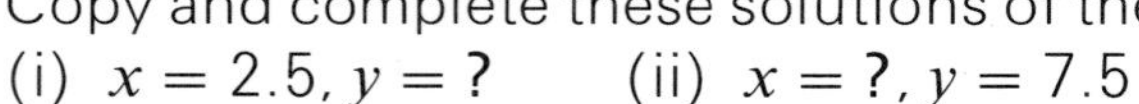
(i) $x = 2.5, y = ?$ (ii) $x = ?, y = 7.5$

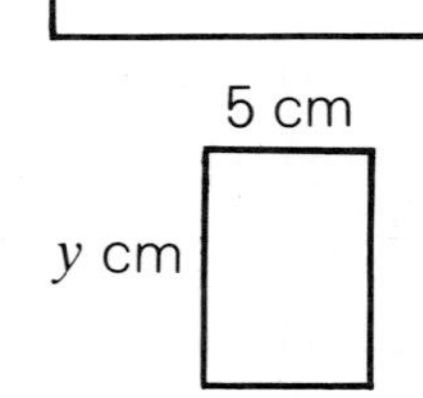

3 a) For this equation find all the whole-number solutions for p and q, where p is less than 25.

b) What is p when q is chosen to be 1.5? $\frac{p}{q} = 4$

Challenge

4 a) Find an equation which has these pairs for p and q as some of its solutions.

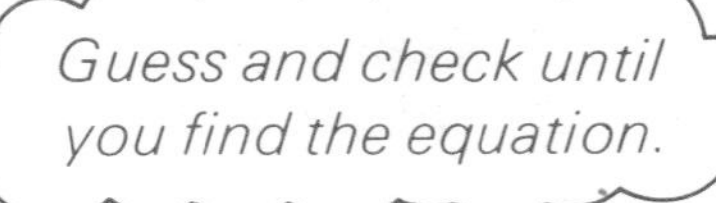

Guess and check until you find the equation.

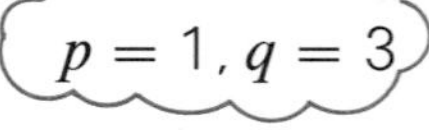

$p = 3, q = 7$ $p = 7, q = 15$ $p = \frac{1}{2}, q = 2$

b) Find three more pairs for p and q which are solutions of the equation.

Solving equations

D 1 a) Try to find the solution to this 'flour-scoop' problem. If you don't succeed after five minutes go on to **(b)**.

Flour-scoop problem

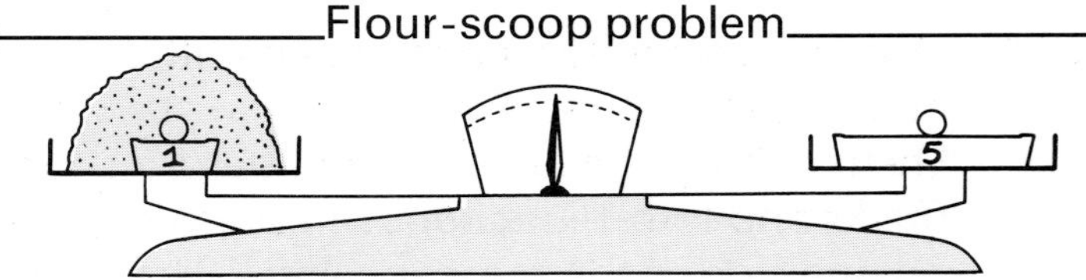

Twelve scoops of flour and a 1 kg weight on one pan of an old kitchen scales exactly balance a 5 kg weight on the other. How many kilograms of flour does a scoop hold?

b) Fiona attacks the problem by writing this equation.

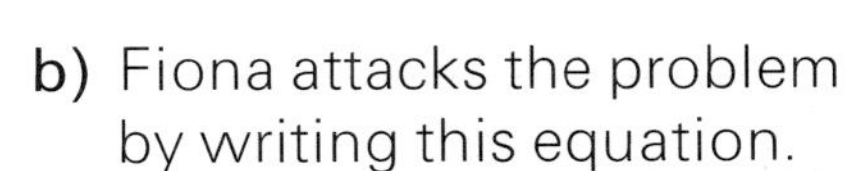

$\square \times x + \square = 5$

Copy and complete it.

c) Next she thinks, 'I can take 1 kg from each side of the scale and it will still balance.'
Copy and complete her equation for the new situation:

$\square \times x = 4$

d) Then Fiona thinks, 'My new equation tells me that 12 scoops of flour weigh 4 kg.
(i) What does 1 scoopful weigh?
(ii) Copy and complete this equation:

$x = \frac{\square}{12} = \frac{1}{\square}$

e) This is how Fiona writes down the whole solution to the problem.

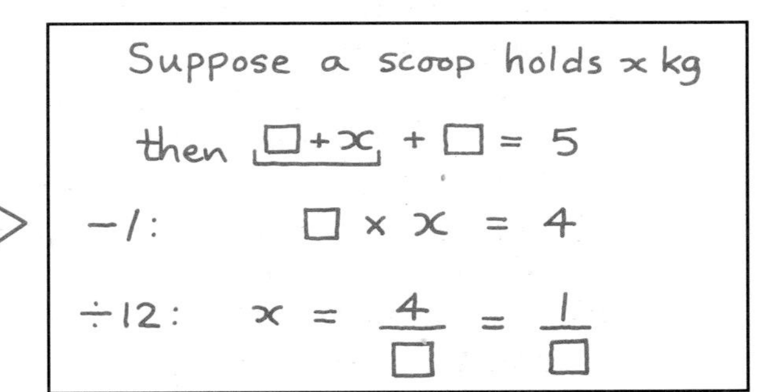

Copy and complete it.

f) Why do you think she writes
(i) '−1'? (ii) '÷12'?

2 a) Try to find the solution to this 'lemonade' problem. Use any method you wish.

b) Imagine how Fiona would solve the problem.
Write down what you think she would write.

Lemonade problem

Jim pours 36 glassfuls of lemonade into the washing-up bowl.
He scoops out exactly 2 litres of lemonade with a measuring jug.
He discovers that there is just enough left in the bowl to fill a 7-litre bucket.
How many litres of lemonade does a glass hold?

3 Solve these equations, using Fiona's method.

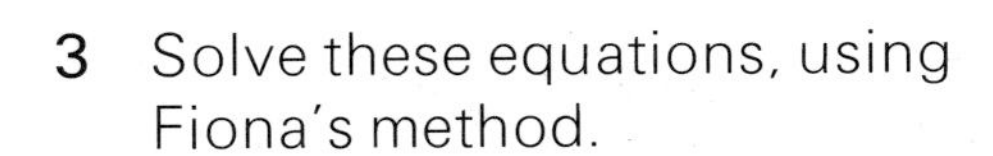

a) $15 \times x + 2 = 12$

b) $8 \times x - 10 = 8$

c) $6 \times x + 2 = 15$

4 For each equation in question **3**, write out a problem which goes with it.

10

5 Horace has been experimenting with building bricks and a see-saw.
He finds that, if he puts 15 standard bricks and two 10 kg bags of potatoes on one side of the see-saw, and 23 bricks and two 1 kg bags of sugar on the other, then the see-saw balances.
Fiona helps Horace work out the weight of a brick.

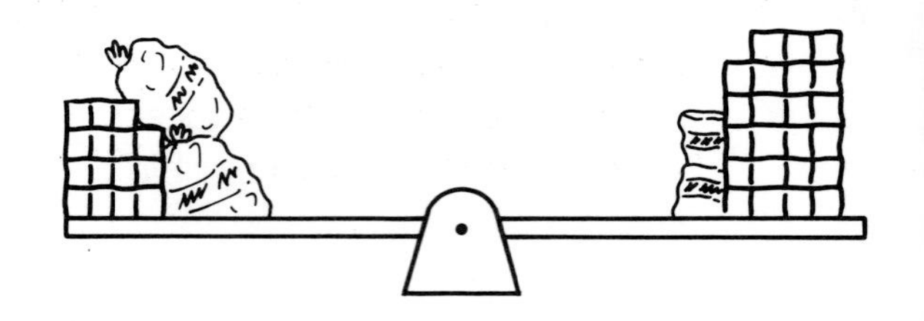

a) Here is part of what she writes:

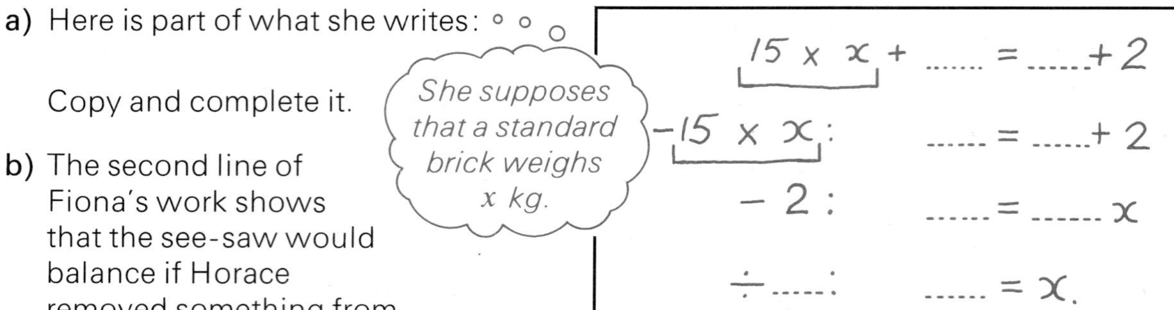

She supposes that a standard brick weighs x kg.
15 × x + ... = ...+ 2
−15 × x: ...... = ...+ 2
− 2 : ...... = ... x
÷......: ...... = x.

Copy and complete it.

b) The second line of Fiona's work shows that the see-saw would balance if Horace removed something from each side. What?

c) Explain what Fiona does in each of the last two lines of her work.

d) What is the weight of a standard brick?

6 Imagine how Fiona would solve the 'corridor' problem.
Write down what you think she would write.
Then write down the solution to the problem.

Corridor problem

> Lenny has a 1 metre long metal rod. By using the rod and lying down, he discovers that the length of the corridor outside the maths room is 2 metres longer than 16 times his height.
> It is also 10 metres longer than 11 times his height.
> How tall is Lenny?

7 Solve these equations, using Fiona's method.

a) $18 \times x + 5 = 6 \times x + 20$

b) $17 \times x - 2 = 7 \times x + 2$

c) $3 - 2 \times x = 4 \times x - 6$

8 For each equation in question **7**, write out a problem which goes with it.

A gallery of equations

E 1 Here is a collection of equations for you to solve.
Some are easy to solve; some are difficult.
Some have no solution; some have one solution; and others have more than one.
Solve as many as you can. Use any method you wish.

a) $15-\underline{2\times x}=3$ b) $30-\underline{5\times x}=\underline{4\times x}+12$

c) $13-\frac{48}{x}=5$ d) $x+\frac{3}{x}=3$

e) $5-\underline{2\times x}-\frac{3}{x}=0$ f) $\underline{23\times x}+17=\underline{8\times x}+26$

g) $\underline{19\times x}-14=100$ h) $3-x=\frac{3}{x}-1$

i) $\frac{108}{x+3}=9$ j) $2\times 5+\underline{3\times x}=\underline{6\times x}+10$

k) $28-\underline{3\times x}=55-\underline{6\times x}$ l) $\frac{5}{x}+3=\frac{3}{x}+11$

m) $\underline{3\times x}+\frac{13}{x}=\frac{4}{x}+\underline{2\times x}$ n) $x\times\underline{11-x\times\underline{6-x}}=6$

2 Choose two of the equations you could solve in question **1**.
For each one, write out a problem which goes with it.

3 Choose one of the two equations you could not solve in question **1**.
Change one of the numbers in the equation so that it can be solved.
Then solve your modified equation.

Challenge

4 Messy Maddox wrote down this information about a rectangle.

He has managed to smudge the last digit of each measurement.
The length (x cm) and the width (y cm) are a whole number of centimetres long.

Enough of this digit is readable for us to see that it is either 6 or 8.

a) Write down two equations, using x and y in each one, for the rectangle.
Use your equations to show that there are four possible rectangles which agree with the information.

b) Messy Maddox remembers that the actual rectangle is very close to being a golden rectangle.

What are its dimensions?

Remember . . . ?

Square

For a golden rectangle, the original rectangle and this part of it have the same length : width ratio.

11 Using matrices

A 1 Information is often given in a table.
Here are some examples.

Fat and energy contents of ices and yoghurt

	Serving (ml)	Total fat (g)	Saturated fatty acids (g)	Cholesterol (mg)	Food energy (calories)
Ice milk	250	7	4	26	200
Ice cream	250	14	8	53	255
Plain yogurt	250	4	2	17	125

TABLE OF OLYMPIC MEDAL WINNERS, WINTER GAMES, 1924–84

	Gold	Silver	Bronze	Total
USSR	68	48	50	166
Norway	54	57	52	163
USA	40	46	31	117
Finland	29	42	32	103
Austria	25	33	30	88
Sweden	32	25	29	86

How much would it cost to rebuild your home?
Terraced house, Pre-1920 (September 1983 costings)

Region	Large (about 1650 ft²)	Medium (about 1350 ft²)	Small (about 1050 ft²)
1	£48/ft²	£47/ft²	£47/ft²
2	£42/ft²	£41.50/ft²	£41.50/ft²
3	£40/ft²	£39.50/ft²	£39.50/ft²
4	£38/ft²	£37.50/ft²	£37/ft²

Stockmarket performance (%)

	Sept. 1982 to end 1985	1985
Europe	+197.0	+48.8
U.K.	+ 95.7	+19.7
Japan	+ 89.8	+13.6
U.S.A	+ 72.6	+28.4

Clothing sizes

Dresses and Suits (Women)						
British	36	38	40	42	44	46
American	34	36	38	40	42	44
Continental	42	44	46	48	50	52
Men's Suits and Overcoats						
British and American	36	38	40	42	44	46
Continental	46	48	50	52	54	56

a) How much cholesterol is there in a 250 ml serving of ice cream?

b) (i) About how much would it have cost in 1983 to rebuild a pre-1920, 1400 ft² terraced house in region 3?
(ii) The four regions are
A Scotland, Wales and Northern England
B East Anglia, Midlands, South West, Yorkshire, Humberside and Northern Ireland
C London (Greater London area)
D South East and North West England.
Which do you think is which? Why?

c) Meg bought a size 44 dress while on holiday in Frankfurt. What is this in British sizes?

We don't count the column and row headings.

d) The Olympic Medals table has 6 rows and 4 columns. How many rows and how many columns are there in
(i) the stock performance table?
(ii) the men's suits and overcoats table?

2 This table has 2 rows and 3 columns.
We call it a 2 × 3 table.

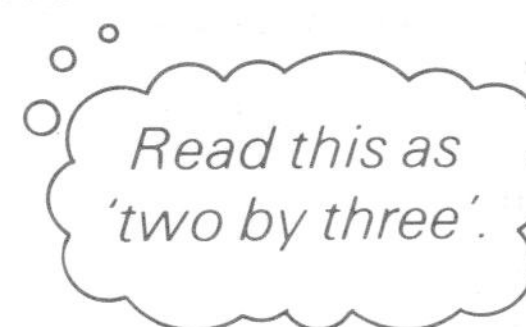
Read this as
'two by three'.

Corner Peninsular Base Unit Code	Rembrandt	Barley	Fresh Olive
CPB5A	£137.75	£75.28	£73.47
CPB6A	£155.46	£76.77	£74.86

a) Each row in the table gives information about a different size of base unit.
What does each column give information about?

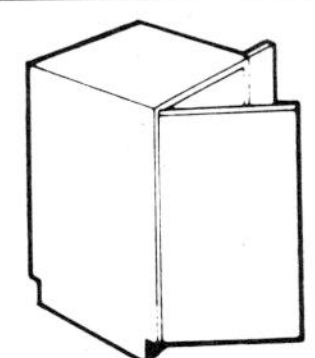

b) Here is some information about bicycles.
Arrange it into a 5 × 2 price table.
Remember to give each row and column a heading.

SALE

18″ frame
Raleigh Avanti
Normally £142.60
SALE PRICE £120.00

16″ frame
Raleigh Streetwolf
Normally £88.00
SALE PRICE
£74.00

18″ frame
Raleigh Shopper
Normally £107.80
SALE PRICE
£90.00

20″ frame
Raleigh
Rhapsody
Normally £120.00
SALE PRICE £87.60

20″ frame
Raleigh Flyer
Normally £154.80
SALE PRICE
£138.00

c) Write one or two sentences to say what advantages you think there are in using a table to give the information.

3 A knitting pattern gives instructions for knitting a sweater to fit chest sizes 81 cm, 86 cm, 91 cm and 97 cm.
It says how many balls of double-knitting wool of various kinds are needed.

A sweater in 'Balmoral' wool needs 8, 9, 9 or 10 50 g balls, depending on the size of the sweater.
A 'Beehive' wool sweater needs 6, 6, 7 or 7 50 g balls.
A 'Trident' wool sweater needs 17, 18, 19 or 20 25 g balls.

Arrange this information into a 3 × 4 'You will need' table.
Give each row and column a heading.

'Parcelling' numerical information

B 1 The Park Hotel chef buys some food items locally each day. These are his shopping lists for Sunday and Monday.

Sunday	
Bread	10 loaves
Milk	20 pints
Eggs	5 dozen
Sugar	4 kg
Margarine	10 kg

Monday	
Bread	9 loaves
Milk	18 pints
Eggs	6 dozen
Sugar	3 kg
Margarine	7 kg

Write the information as a 5×2 table.

2 a) To save time, the chef writes his shopping list for Tuesday like this. What do you think the 2 means?

$$\begin{pmatrix} 7 \\ 19 \\ 2 \\ 5 \\ 8 \end{pmatrix}$$

b) To save even more time, he writes the shopping lists for Wednesday, Thursday, Friday and Saturday in one table, like this:

first row →

$$\begin{pmatrix} 8 & 5 & 10 & 8 \\ 16 & 15 & 22 & 19 \\ 4 & 7 & 4 & 6 \\ 6 & 2 & 4 & 3 \\ 9 & 11 & 10 & 7 \end{pmatrix}$$

first column ↑

(i) What do you think the second column tells us?
(ii) What do you think the fourth row tells us?
(iii) What do you think the number 22 in the table tells us?

11

Take note

We call a 'parcel' of numbers like the chef's table a **matrix**.

Matrices are a simple and compact way of representing information.

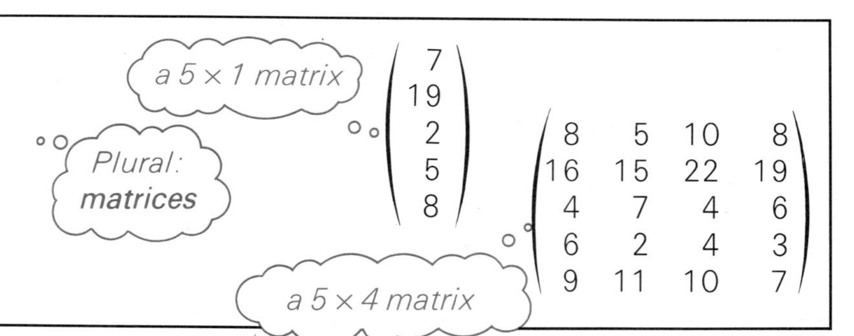

$$\begin{pmatrix} 7 \\ 19 \\ 2 \\ 5 \\ 8 \end{pmatrix} \qquad \begin{pmatrix} 8 & 5 & 10 & 8 \\ 16 & 15 & 22 & 19 \\ 4 & 7 & 4 & 6 \\ 6 & 2 & 4 & 3 \\ 9 & 11 & 10 & 7 \end{pmatrix}$$

3 Write the chef's shopping list for the whole week as a 5×7 matrix.

4 Write a short paragraph about the advantages and disadvantages of the chef's use of matrices.

Think about what might happen if one day the chef called in sick and one of the under-chefs had to use the chef's shopping list.

5 Write out a matrix for the number of minutes you spend in each type of school lesson each day of the week.

Think it through

6 Four teams take part in a hockey tournament.
They all play each other once.
Here are the results:

Crusaders 2	Wasps 2
Wasps 1	Terrapins 3
Giants 3	Crusaders 2
Terrapins 0	Giants 0
Terrapins 1	Crusaders 1
Wasps 0	Giants 2

a) This matrix tells us how many goals each team scored.
Copy and complete it.

$$\begin{pmatrix} 5 \\ 4 \\ 5 \\ \_ \end{pmatrix}$$

b) This matrix tells us how many points each team gained.
Copy and complete it.

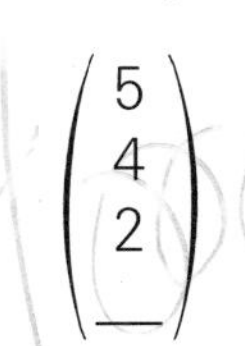

$$\begin{pmatrix} 5 \\ 4 \\ 2 \\ \_ \end{pmatrix}$$

You will need to decide for yourself how the points system works.

c) This matrix gives us a lot of information about the hockey tournament.
(i) What do you think the number 5 in the third row and fifth column means?
(ii) Copy the matrix.
Write in headings for the rows and columns so that anyone could understand it.

$$\begin{pmatrix} 3 & 2 & 1 & 0 & 5 & 2 & 5 \\ 3 & 1 & 2 & 0 & 4 & 2 & 4 \\ 3 & 0 & 2 & 1 & 5 & 6 & 2 \\ 3 & 0 & 1 & 2 & 3 & 7 & 1 \end{pmatrix}$$

Assignment

7 a) Find a league table in the sports pages of a newspaper.
Find out what the abbreviations at the top of the columns stand for.
Cut the table out and stick it in your book. Write a short newspaper article which compares the progress of two of the teams.

b) Compare your table with tables for the same league in other newspapers.
Do all the tables have the same matrix of results?
Explain the differences, if there are any.

11

Doing arithmetic with matrices

C 1 The nurse is completing the births record for two maternity wards.

Breckenhall Hospital births	Canning Ward		Twerton Ward		Totals	
	Girls	Boys	Girls	Boys	Girls	Boys
Monday	2	1	0	2	2	3
Tuesday	3	2	4	1	7	
Wednesday	2	4	0	3		
Thursday	3	1	4	2		
Friday	0	2	1	3		
Saturday	6	2	0	4		
Sunday	2	2	2	2		

a) This addition will show the same information in matrix form. Copy and complete it.

$$\begin{pmatrix} 2 & 1 \\ 3 & 2 \\ 2 & 4 \\ 3 & 1 \\ 0 & 2 \\ 6 & 2 \\ 2 & 2 \end{pmatrix} + \begin{pmatrix} 0 & \_\_ \\ 4 & \_\_ \\ \_\_ & \_\_ \\ \_\_ & \_\_ \\ \_\_ & \_\_ \\ \_\_ & \_\_ \\ \_\_ & \_\_ \end{pmatrix} = \begin{pmatrix} 2 & 3 \\ 7 & \_\_ \\ \_\_ & \_\_ \\ \_\_ & \_\_ \\ \_\_ & \_\_ \\ \_\_ & \_\_ \\ \_\_ & \_\_ \end{pmatrix}$$

b) (i) This matrix subtraction also tells us something about the births in the two wards. Copy and complete it.

$$\begin{pmatrix} 2 & 3 \\ 7 & \_\_ \\ 2 & \_\_ \\ \_\_ & 3 \\ \_\_ & \_\_ \\ \_\_ & \_\_ \\ 4 & \_\_ \end{pmatrix} - \begin{pmatrix} 2 & 1 \\ 3 & 2 \\ 2 & 4 \\ 3 & 1 \\ 0 & 2 \\ 6 & 2 \\ 2 & 2 \end{pmatrix} = \begin{pmatrix} 0 & \_\_ \\ 4 & \_\_ \\ \_\_ & \_\_ \\ \_\_ & \_\_ \\ \_\_ & \_\_ \\ \_\_ & \_\_ \\ \_\_ & \_\_ \end{pmatrix}$$

(ii) Underneath your subtraction, explain what each matrix tells us.

Challenge

2 a) Think of a real-life situation which this subtraction might represent, then write the information in a table.

$$\begin{pmatrix} 2 & 1 \\ 3 & 2 \\ 4 & 6 \end{pmatrix} - \begin{pmatrix} 1 & 1 \\ 1 & 2 \\ 3 & 2 \end{pmatrix} = \begin{pmatrix} \_\_ & \_\_ \\ \_\_ & \_\_ \\ \_\_ & \_\_ \end{pmatrix}$$

b) Complete the matrix subtraction, and your table.

Think it through

3 On Saturdays, Meg works in the stockroom of the DIY store. She is in charge of hardware. Here are two of her stock lists:

Round wire nails			
Length (mm)	Small packs	Large packs	Loose
20	360	210	25 kg
30	470	150	12 kg
40	250	220	18 kg
50	1080	600	8 kg
60	640	560	32 kg

Screws			
Type	Boxes of 100	Boxes of 200	Boxes of 500
25 mm 6 gauge	75	40	35
25 mm 8 gauge	60	55	35
38 mm 6 gauge	20	80	70
38 mm 8 gauge	110	40	40
38 mm 10 gauge	85	65	75
50 mm 8 gauge	80	105	45

a) At 10:00 am a delivery van arrives, bringing (among other things) these amounts of nails and screws:

Round wire nails

Small packs: 200 × 20 mm, 100 × 30 mm, 300 × 40 mm.
Large packs: 200 × 20 mm, 300 × 30 mm, 200 × 40 mm, 100 × 60 mm.
Loose: 5 kg of 20 mm, 15 kg of 30 mm, 10 kg of 40 mm, 25 kg of 50 mm.

Screws

100s: 50 × 25 mm/6 g, 100 × 25 mm/8 g, 150 × 38 mm/6 g, 50 × 38 mm/10 g, 50 × 50 mm/8 g.
200s: 50 × 25 mm/6 g, 50 × 25 mm/8 g, 50 × 38 mm/8 g, 50 × 38 mm/10 g.
500s: 50 × 25 mm/6 g, 50 × 25 mm/8 g, 50 × 38 mm/8 g, 50 × 50 mm/8 g.

Write matrix calculations to show the effect this delivery has on Meg's nail stocks.

b) Sales during the morning are good, and at 1:30 pm Meg has to re-stock some items in the shop.
Here is what she adds to the goods displayed for sale in the shop:

Round wire nails

Small packs: 40 × 20 mm, 10 × 30 mm, 65 × 40 mm, 50 × 50 mm, 45 × 60 mm.
Large packs: 55 × 20 mm, 40 × 30 mm, 15 × 40 mm, 35 × 50 mm, 50 × 60 mm.
Loose: 15 kg of 20 mm, 5 kg of 30 mm, 5 kg of 40 mm, 10 kg of 50 mm, 15 kg of 60 mm.

Screws

100s: 30 × 25 mm/6 g, 30 × 25 mm/8 g, 15 × 38 mm/6 g, 40 × 38 mm/8 g, 10 × 38 mm/10 g, 25 × 50 mm/8 g.
200s: 10 × 25 mm/6 g, 25 × 25 mm/8 g, 30 × 38 mm/6 g, 15 × 38 mm/8 g, 20 × 38 mm/10 g, 30 × 50 mm/8 g.
500s: 10 × 25 mm/6 g, 10 × 25 mm/8 g, 20 × 38 mm/6 g, 25 × 38 mm/8 g, 15 × 38 mm/10 g, 20 × 50 mm/8 g.

Write matrix calculations to show the effect this re-stocking of the shop has on Meg's screw stocks.

Matrices and coordinates

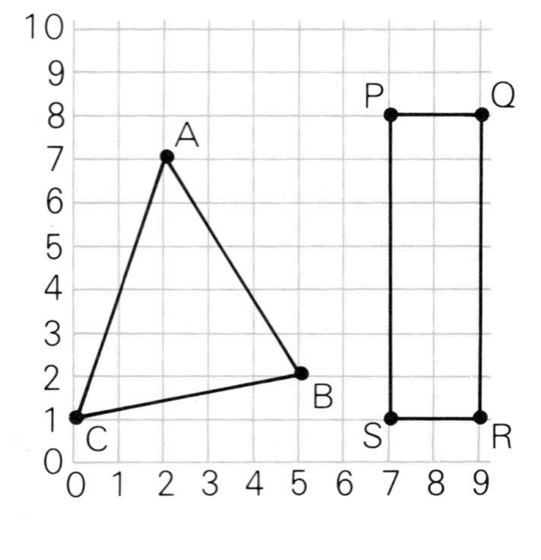

D 1 For triangle ABC, we can record the position of C like this: (0, 1).

a) Record the position of A in the same way.

b) We can record the position of the whole triangle like this:

$\begin{pmatrix} 2 & \text{—} & 0 \\ 7 & 2 & 1 \end{pmatrix}$ or like this: $\begin{pmatrix} 2 & \text{—} \\ \text{—} & 2 \\ 0 & 1 \end{pmatrix}$

Copy and complete each matrix

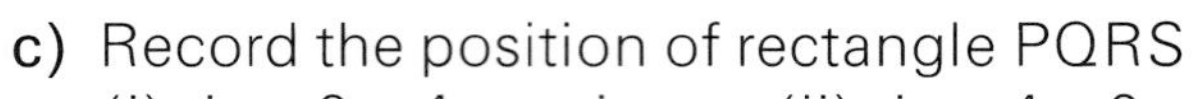

It doesn't matter which we use ... as long as we are consistent ... and we remember what each row and column means.

c) Record the position of rectangle PQRS
(i) in a 2×4 matrix. (ii) in a 4×2 matrix.

2 This matrix records the position of a pentagon. Draw the pentagon on squared paper.

$$\begin{pmatrix} 0 & 4 & 5 & 2 & 2 \\ 0 & 1 & 2 & 3 & 1 \end{pmatrix}$$

3 We can write the journey F→F′ as a 2×1 matrix like this: $\begin{pmatrix} 3 \\ 1 \end{pmatrix}$

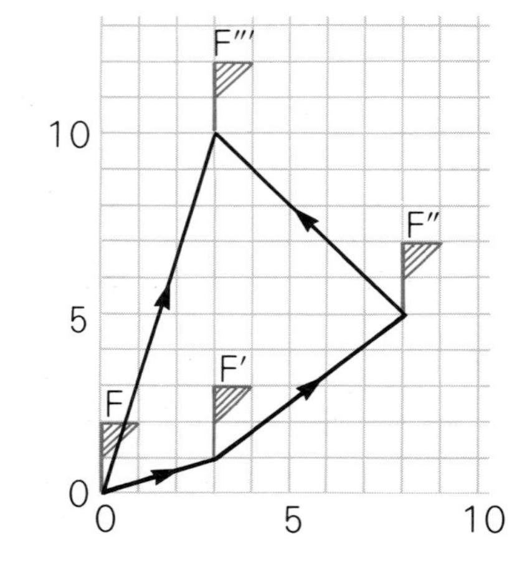

a) Write these journeys in the same way:
(i) F′→F″ (ii) F″→F‴. (iii) F→F‴.

b) This represents the two-stage journey F→F′, F′→F″:

$$\begin{pmatrix} 3 \\ 1 \end{pmatrix} + \begin{pmatrix} 5 \\ 4 \end{pmatrix} = \begin{pmatrix} 8 \\ 5 \end{pmatrix}$$

Which single journey does $\begin{pmatrix} 8 \\ 5 \end{pmatrix}$ represent?
Write it like this: □→□.

Take note

$\begin{pmatrix} 8 \\ 5 \end{pmatrix}$ is called the **resultant**.
It is what results from the addition of $\begin{pmatrix} 3 \\ 1 \end{pmatrix}$ and $\begin{pmatrix} 5 \\ 4 \end{pmatrix}$

4 a) Write these two-stage journeys using +s and $\begin{pmatrix} \; \\ \; \end{pmatrix}$s:
(i) F′→F″, F″→F‴ (ii) F″→F‴, F‴→F.

b) Which two-stage journey does this represent? $\begin{pmatrix} ^-3 \\ ^-1 \end{pmatrix} + \begin{pmatrix} 3 \\ 10 \end{pmatrix} = \begin{pmatrix} 0 \\ 9 \end{pmatrix}$

11

5 a) Write this four-stage journey using +s and $\begin{pmatrix} \\ \end{pmatrix}$s:

F′→F, F→F‴, F‴→F″, F″→F′.

b) Write down a five-stage journey whose resultant is $\begin{pmatrix} 0 \\ 0 \end{pmatrix}$.

Start with F″→. . .

c) Write your five-stage journey using +s and $\begin{pmatrix} \\ \end{pmatrix}$s.

6 a) This drawing is made on a computer screen.
The computer was given these instructions using 2 × 1 matrices.

START: $\begin{pmatrix} 5 \\ 1 \end{pmatrix} + \begin{pmatrix} 0 \\ 4 \end{pmatrix} + \begin{pmatrix} ^-3 \\ 0 \end{pmatrix} + \begin{pmatrix} 0 \\ 1 \end{pmatrix} + \begin{pmatrix} ^-1 \\ 0 \end{pmatrix} + \begin{pmatrix} 0 \\ ^-5 \end{pmatrix} + \begin{pmatrix} ^-1 \\ ^-1 \end{pmatrix}$ END

What is the resultant of the addition?

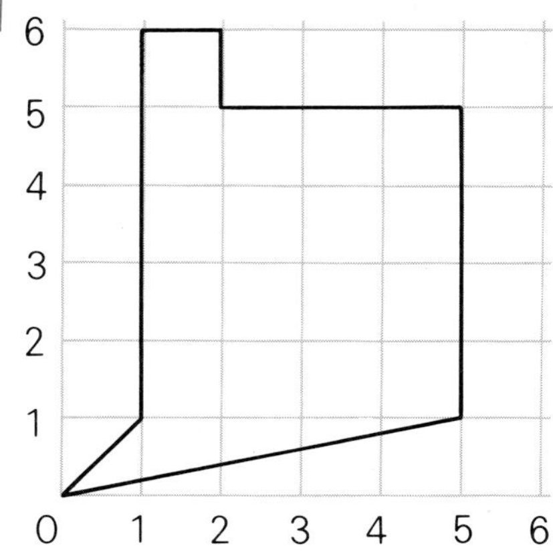

b) Write out instructions for the computer to draw this shape.

Challenge

7 The same kinds of instructions as those in question **6** can be given to an embroidery machine to draw patterns like these.

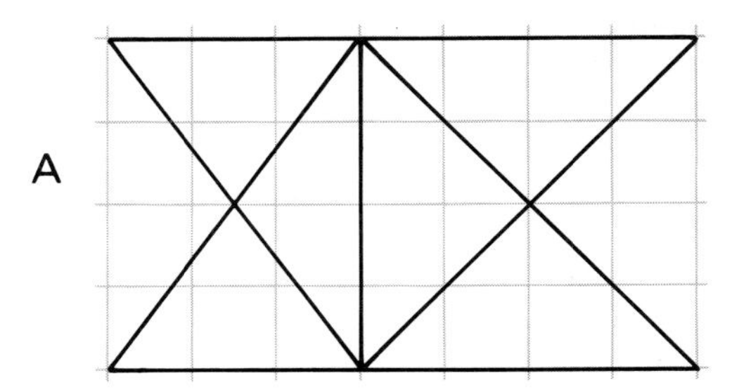

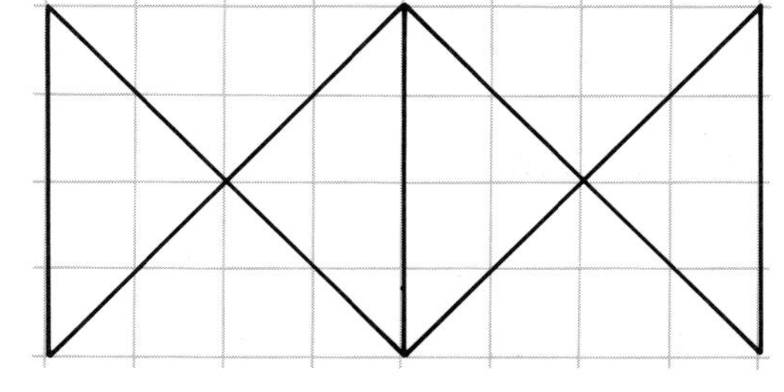

The instructions must be written so that the machine doesn't embroider the same line twice.

a) Write down a set of instructions for each pattern.

b) Design your own embroidery pattern which can be made by the machine.
Write out machine instructions for your pattern.

Scaling matrices

E 1 This table gives the ingredients for making 100 kg of each of three alloys.

mixtures of metals

Chemical symbols:
Al – aluminium
Ni – nickel
Fe – iron
Cu – copper

	Alloy **1**	Alloy **2**	Alloy **3**
Al	70 kg	50 kg	65 kg
Ni	20 kg	10 kg	30 kg
Fe	3 kg	30 kg	0 kg
Cu	7 kg	10 kg	5 kg

a) What are the ingredients for making 100 kg of alloy **3**?

b) Write the ingredients information as a 4×3 matrix.

c) What are the ingredients for making **500** kg of alloy **1**?

d) Write a 4×3 matrix giving the ingredients information for **500** kg of each alloy.

e) This calculation shows the ingredients matrix being multiplied by 3:

$$3 \times \begin{pmatrix} 70 & 50 & 65 \\ 20 & 10 & 30 \\ 3 & 30 & 0 \\ 7 & 10 & 5 \end{pmatrix} = \begin{pmatrix} 210 & \_\_ & \_\_ \\ 60 & 30 & \_\_ \\ \_\_ & 90 & \_\_ \\ 21 & \_\_ & \_\_ \end{pmatrix}$$

(i) Copy and complete the calculation.
(ii) Write one or two sentences to explain what the '3 ×' ingredients matrix tells you.

f) Use **M** to represent the original ingredients matrix. Explain what these represent:
(i) $\mathbf{M}+\mathbf{M}$. (ii) $5 \times \mathbf{M}$. (iii) $10 \times \mathbf{M}$.

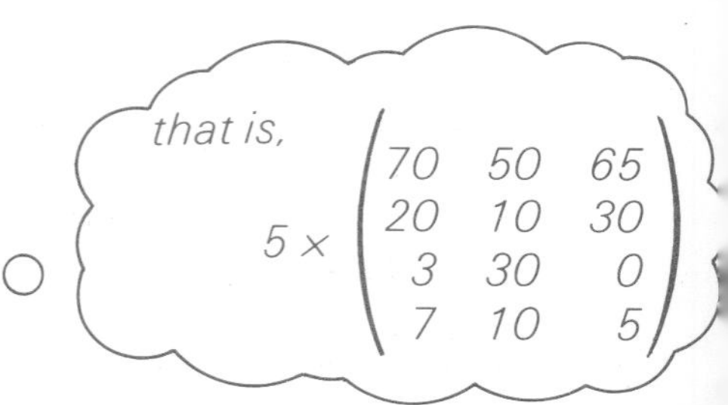

Think it through

2 The grid shows a shape and its enlargement.

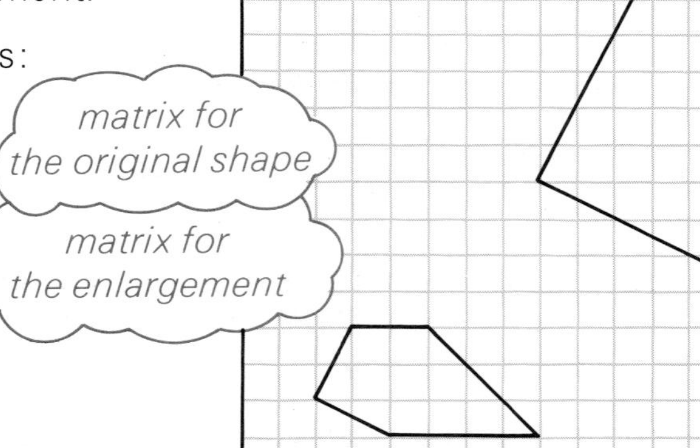

a) Copy and complete the two matrices:

$$\mathbf{K} = \begin{pmatrix} 2 & 4 & 8 & \_\_ & \_\_ \\ 2 & 1 & 1 & \_\_ & \_\_ \end{pmatrix}$$

$$\mathbf{N} = \begin{pmatrix} 8 & 16 & \_\_ & \_\_ & \_\_ \\ 8 & 4 & \_\_ & \_\_ & \_\_ \end{pmatrix}$$

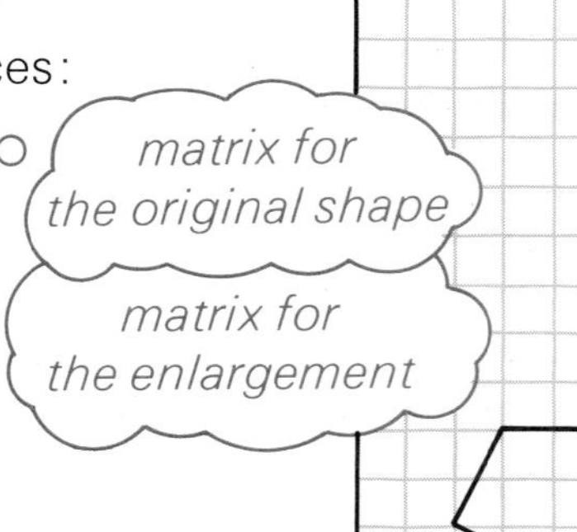

b) What is the scale factor for the enlargement?

c) Copy and complete: $\mathbf{N} = \square \times \mathbf{K}$.

d) Draw your own shape and a reduction of it with scale factor ÷ 3. Write down a matrix for the shape and a matrix for its reduction. Write an equation connecting the two matrices, like that in **(c)**.

Matrices, translations and enlargements

F 1 a) Record the position of triangle PQR in a 2×3 matrix.

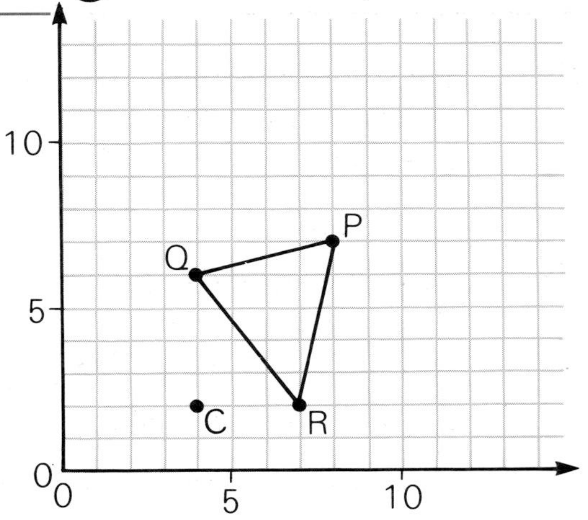

b) Add $\begin{pmatrix} ^-4 & ^-4 & ^-4 \\ ^-2 & ^-2 & ^-2 \end{pmatrix}$ to your matrix and draw the triangle P′Q′R′ whose position is given by the resulting matrix.

c) How are the two triangles related: by an enlargement, a translation or a reflection?

d) The triangle P″Q″R″ is an enlargement of triangle P′Q′R′ with scale factor $\times 2$ and centre of enlargement at the origin. This matrix calculation gives the position of triangle P″Q″R″. Copy and complete it.

$$2 \times \begin{pmatrix} 4 & 0 & 3 \\ 5 & 4 & 0 \end{pmatrix} = \begin{pmatrix} 8 & \_ & \_ \\ 10 & \_ & \_ \end{pmatrix}$$

The opposite journey to the one which took PQR to P′Q′R′.

e) The triangle P″Q″R″ now slides so that each vertex makes the journey $\begin{pmatrix} 4 \\ 2 \end{pmatrix}$.

The resulting triangle is P‴Q‴R‴. Draw triangle P‴Q‴R‴.

f) Write a matrix calculation which gives the position of triangle P‴Q‴R‴.

g) Look at triangles PQR and P‴Q‴R‴. Describe how they and the point C are related.

Challenge

2 a) Copy and complete this matrix calculation to show how A′B′C′ can be obtained from ABC.

$$\begin{pmatrix} 1 & 2 & 5 \\ 5 & 1 & 5 \end{pmatrix} + \begin{pmatrix} \_ & \_ & \_ \\ \_ & \_ & \_ \end{pmatrix}$$

$$= \begin{pmatrix} 3 & \_ & \_ \\ 7 & \_ & \_ \end{pmatrix}$$

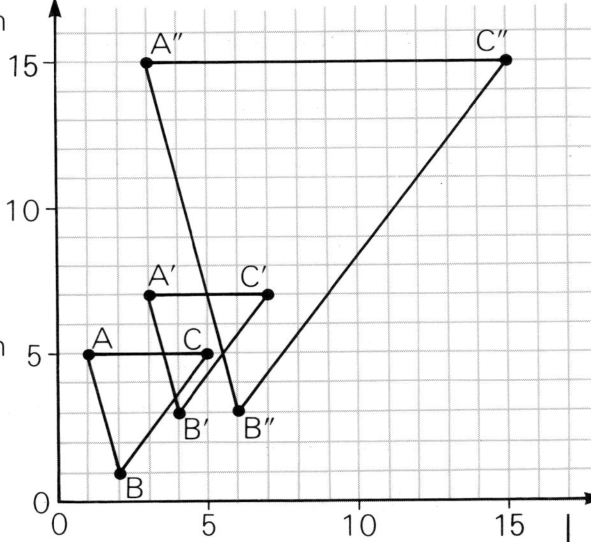

b) Copy and complete this matrix calculation to show how A″B″C″ can be obtained from ABC.

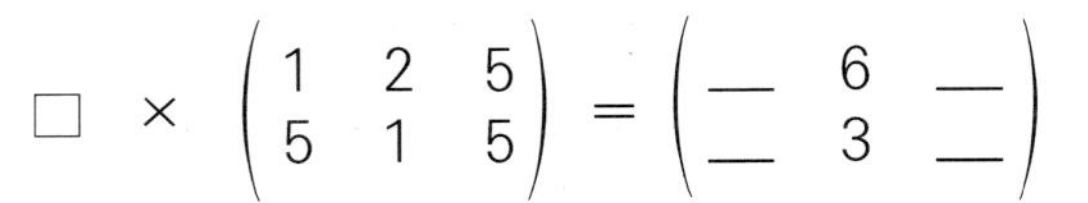

$$\square \times \begin{pmatrix} 1 & 2 & 5 \\ 5 & 1 & 5 \end{pmatrix} = \begin{pmatrix} \_ & 6 & \_ \\ \_ & 3 & \_ \end{pmatrix}$$

c) Find a matrix calculation which shows how
(i) A″B″C″ can be obtained from ABC.
(ii) A′B′C′ can be obtained from ABC.
(iii) A″B″C″ can be obtained from A′B′C′.

3 a) This matrix gives the position of the cuboid:

$$\begin{pmatrix} 0 & 1 & 1 & 0 & 0 & 1 & 1 & \_ \\ 0 & 0 & 0 & 0 & 2 & 2 & 2 & \_ \\ 0 & 0 & 2 & 2 & 0 & 0 & 1 & \_ \end{pmatrix}$$

Copy and complete it.

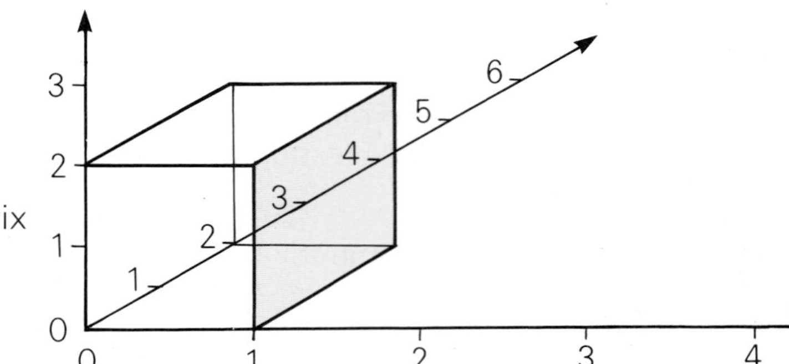

b) Add the matrix in **(a)** and the matrix

$$\begin{pmatrix} 2 & 2 & 2 & 2 & 2 & 2 & 2 & 2 \\ 1 & 1 & 1 & 1 & 1 & 1 & 1 & 1 \\ 3 & 3 & 3 & 3 & 3 & 3 & 3 & 3 \end{pmatrix}$$

c) The resulting matrix in **(b)** gives a new position of the cuboid. Is the position the result of a slide, an enlargement, a reflection or a rotation?

d) The original cuboid is enlarged, centre the origin, with scale factor $\times 3$.

(i) Write down the matrix which describes the new position of the cuboid.

(ii) Copy and complete this matrix calculation to show how the new position may be obtained:

$$\square \times \begin{pmatrix} 0 & 1 & 1 & 0 & 0 & 1 & 1 & \_ \\ 0 & 0 & 0 & 0 & 2 & 2 & 2 & \_ \\ 0 & 0 & 2 & 2 & 0 & 0 & 1 & \_ \end{pmatrix} = \begin{pmatrix} \_ & \_ & \_ & \_ & \_ & \_ & \_ & \_ \\ \_ & \_ & \_ & \_ & \_ & \_ & \_ & \_ \\ \_ & \_ & \_ & \_ & \_ & \_ & \_ & \_ \end{pmatrix}$$

Challenge

4 This matrix gives the position of the pyramid:

$$\begin{pmatrix} 3 & 7 & 7 & 3 & 5 \\ 1 & 1 & 5 & 5 & 3 \\ 2 & 2 & 2 & 2 & 8 \end{pmatrix}$$

S is the point (1, 0, 2).

Pyramid A′B′C′D′E′ is an enlargement of pyramid ABCDE with scale factor $\times 3$ and centre of enlargement S.
Find the coordinates of the points A′B′C′D′E′ on the enlarged pyramid.

Next chapter

12 Thinking with brackets

A 1 a) This is the tarrif for the Captain's Table.

Work out the cost of these orders in your head.

b) Tom works in the fish and chip shop. Here he is working out an order. Write down what you think was ordered.

c) Tom usually works out the total cost on paper. Here are two of his calculations.

(i)

2×50 1.00
3×30 90
£1.90

(ii)

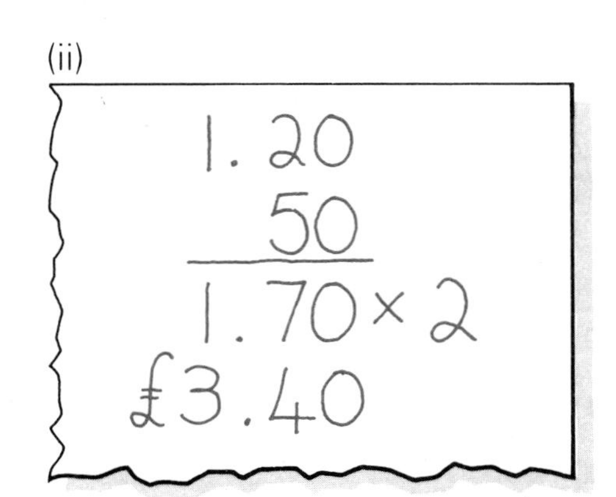

What do you think was ordered each time?

d) Write down as many different orders as you can which have a total cost of £3.00.

12

2 a) Moira works with Tom.
For one order she wrote:

$$£1.20 + 50 \times 2 = £2.20$$

What do you think was ordered?

b) Later, she wrote:

$$£1.20 + 50 \times 2 = £3.40$$

What do you think was ordered this time?

c) For another order she wrote:
This could be for two possible orders.
What are they?
Write down the cost for each.

d) The working for these orders uses brackets.

$$(£1.20 + 30) \times 2 = £3.00$$

$$£1.20 + (30 \times 2) = £1.80$$

Write one or two sentences to explain what the brackets tell us.

Think it through

3 Copy these sentences.
Insert your own brackets so that the sentences are correct.

a) $£1.20 + 60p \times 2 = £3.60$
b) $£1.20 + 60p \times 2 = £2.40$
c) $£4.00 - £1.50 \times 2 = £5.00$
d) $£4.00 - £1.50 \times 2 = £1.00$
e) $18 + 2 \times 4 = 80$
f) $18 + 2 \times 4 = 26$
g) $28 - 8 \times 3 = 60$
h) $28 - 8 \times 3 = 4$
i) $24 \div 12 \div 2 = 1$
j) $24 \div 12 \div 2 = 4$

Why use brackets?

B 1 a) **RILEY SPORTS**

DISCOUNT
£1.00 off per sweater for orders over 6 sweaters.

Here is an order for sweaters.

Write out one calculation, using brackets, to find the cost of the order.
Write down the total cost.

b) This is the total cost of an order. What was ordered?

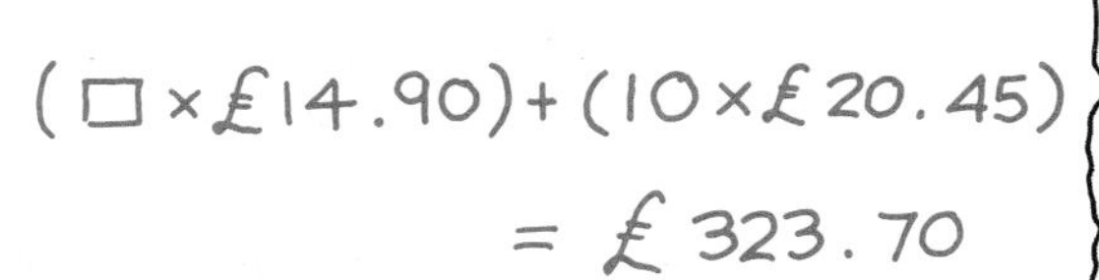

2 This is an expression for the area of rectangle ABGE.

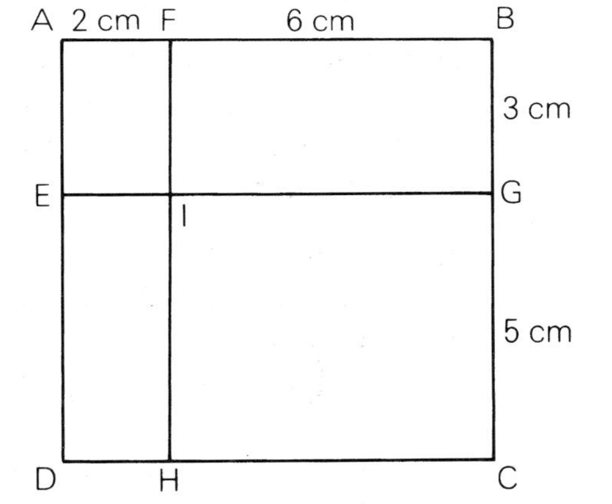

a) Here are some more area expressions. Which rectangles do they refer to?

A $2 \times (3+5)$ B $(2+6) \times (3+5)$

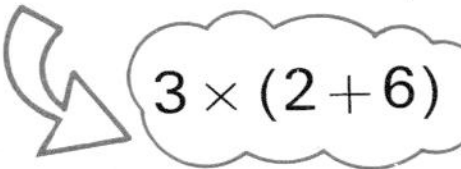

b) Write your own expression for the area of EGCD.

3 These are expressions for the areas of rectangles in the drawing.

a) $x + y \times p + q$ b) $p \times y + q \times y$

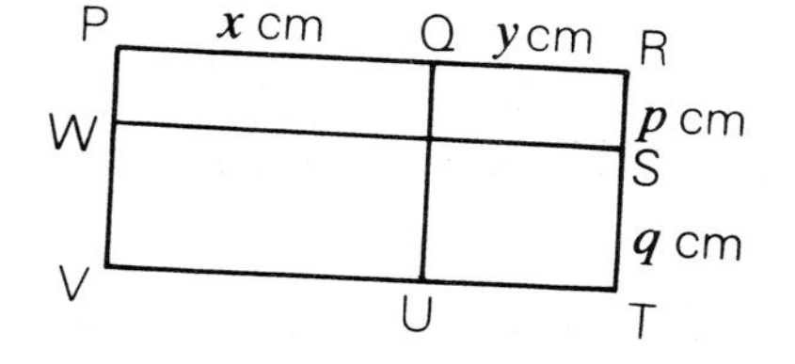

Put in brackets so that each one makes sense, and say which rectangle is involved.

Challenge

4 This expression gives the area of ABCD.

$c \times a + b + d - c \times a + d - c \times b$

Put in your own brackets so that the expression makes sense.

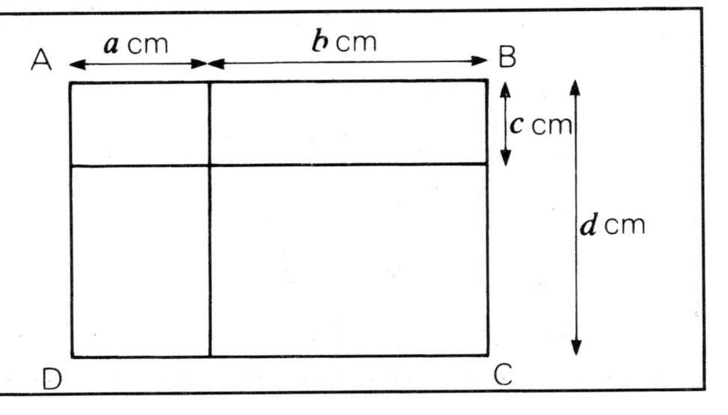

12

Take note

We use brackets to help show exactly what we mean.

$2 \times (4+3)$

means
'Add 4 and 3,
then multiply by 2'.
Result: 14

$(2 \times 4)+3$

means
'Multiply 2 by 4,
then add 3'.
Result: 11

Think it through

5 a) Try putting brackets in different positions for lots of expressions like these.

Example:
$(2+7)-4$
$2+(7-4)$

$2+7-4$ $9 \div 3-2$ $18-6 \times 2$

$3+7+8$ $8-3-2$

Try different kinds of operations.

$+, -, \div, \times$ are called 'operations.'

$+\,+$, $+\,-$, $\times\,\times$, $\div\,-$, etc.

Which pairs of operations always give the same result no matter where you place the brackets?

Example:
$(2+7)+3=12$
$2+(7+3)=12$

b) Which pairs of operations normally give different results?
Give an example for each pair of operations.

Example:
$(2 \times 3)+4=10$
$2 \times (3+4)=14$

6 Copy each number sentence.
Insert brackets on each side of the = sign so that the sentence is correct.

a) $12 \times 7-3=24 \div 2 \times 4$

b) $4+3 \times 2+5=9-6 \times 7-2$

one pair of brackets

two pairs of brackets

12

Take note

We don't need to use brackets in '+ +' or '× ×' expressions.
We get the same result no matter where the brackets go.

Examples: $(2+8)+6$ Result: 16
$2+(8+6)$ Result: 16

We just write $2+8+6$

$(3\times 4)\times 2$ Result: 24
$3\times(4\times 2)$ Result: 24

We just write $3\times 4\times 2$

7 This is a place setting for one person in a restaurant. There are seven pieces of cutlery for each person.

a) How many pieces of cutlery are there altogether on a table for eight?

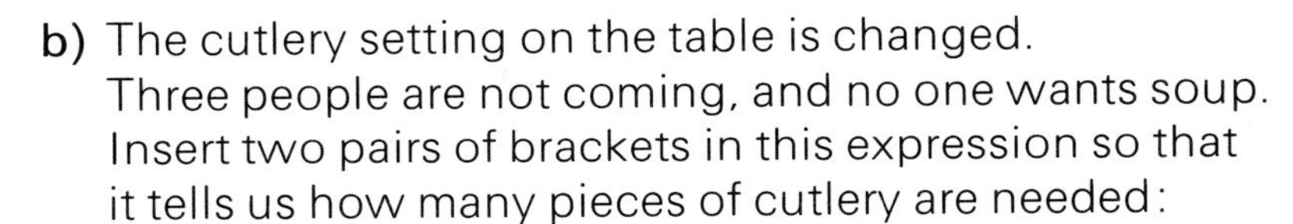

b) The cutlery setting on the table is changed. Three people are not coming, and no one wants soup. Insert two pairs of brackets in this expression so that it tells us how many pieces of cutlery are needed:

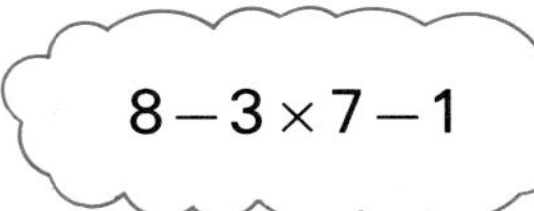

c) Another table is set for six people. Two extra people arrive, and nobody wants dessert. Write an expression, using two pairs of brackets, for the number of pieces of cutlery needed on the table.

12

Think it through

8 a) Insert brackets in this expression so that it is correct.

b) Check your result in **(a)** with a calculator.

$83+3\times 21 = 146$

c) Insert brackets to make this
(i) as large as possible.
(ii) as small as possible.

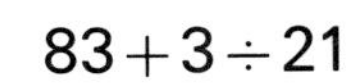

d) How many different ways are there of inserting brackets in this expression so that it is true?

$2+2\times 83-3\times 21>500$

Write down each one.

Brackets in your head

C 1 George needs four new tyres.
He works out the cost in his head.

a) Write down the cost.

b) Work out the cost of six tyres in your head.
Write down only the answer.

2 Here is George, working out 104×8:

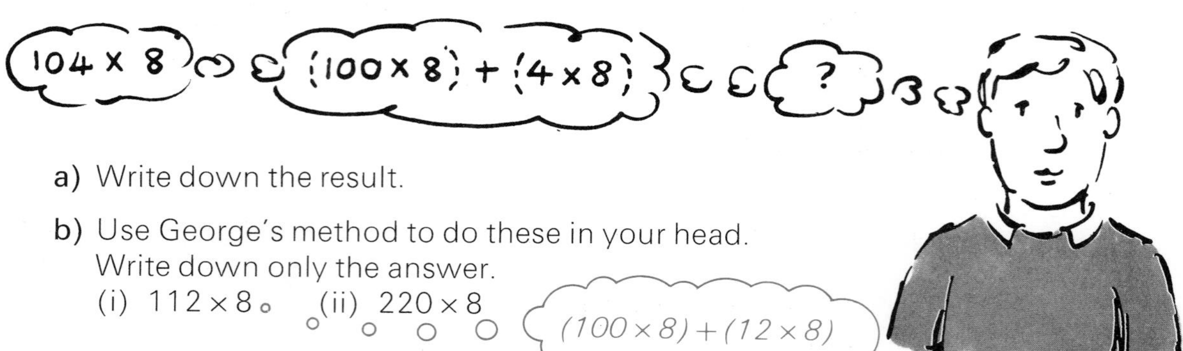

a) Write down the result.

b) Use George's method to do these in your head.
Write down only the answer.
(i) 112×8 (ii) 220×8

3 a) Do these in your head.
Write down only the answers.
(i) 101×12 (ii) 202×5 (iii) 725×3

b) Check your results with your calculator.

4 Now George is working out 99×9 in his head.

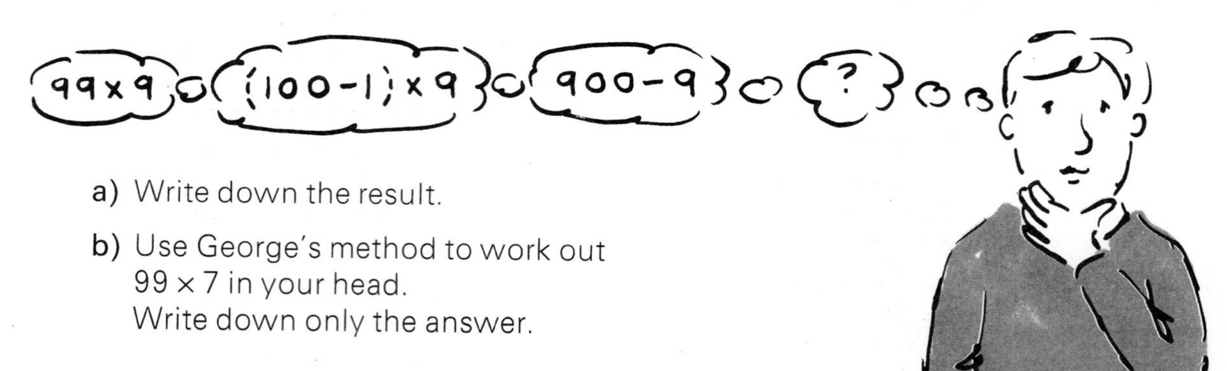

a) Write down the result.

b) Use George's method to work out 99×7 in your head.
Write down only the answer.

5 a) Do these in your head.
Write down only the answers.
(i) 98×4 (ii) 199×4 (iii) 999×7

b) Check your results with your calculator.

Let it flow

D 1

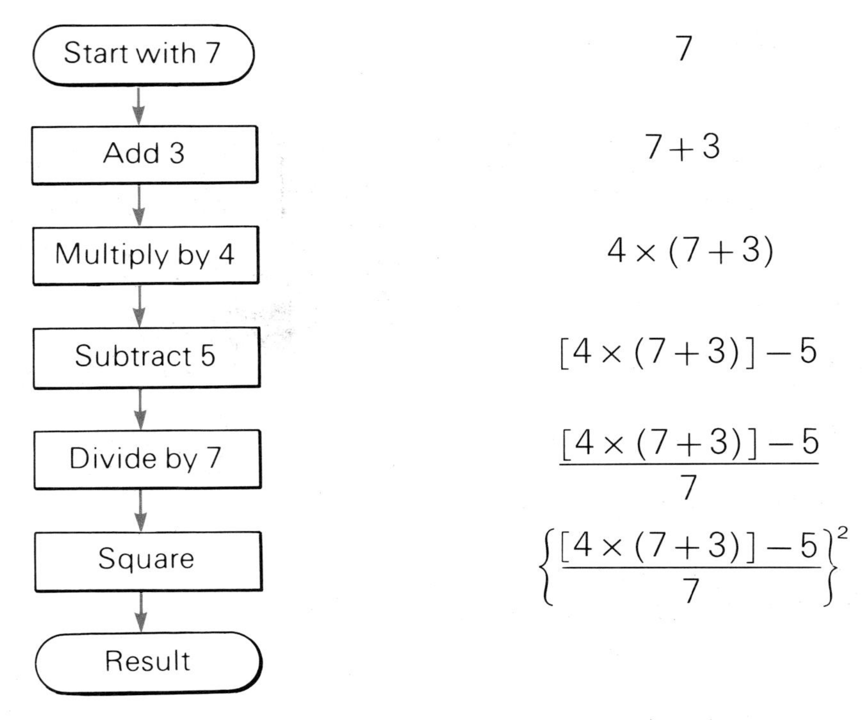

$$7$$

$$7+3$$

$$4\times(7+3)$$

$$[4\times(7+3)]-5$$

$$\frac{[4\times(7+3)]-5}{7}$$

$$\left\{\frac{[4\times(7+3)]-5}{7}\right\}^2$$

Work out the result
a) in your head. b) on your calculator.

2 a) Draw flow charts to show the stages in these.

(i) $\left[\frac{6\times(3-2)^2}{3}\right]+4$ (ii) $3\times\left\{\left[\frac{10}{(5+8)}\right]-6\right\}$

b) Work out each result.

3 a) Use brackets to show these.

b) Work out each result.

(i)

Multiply 5 by 3
↓
Subtract 7
↓
Divide by 4
↓
Square

(ii)

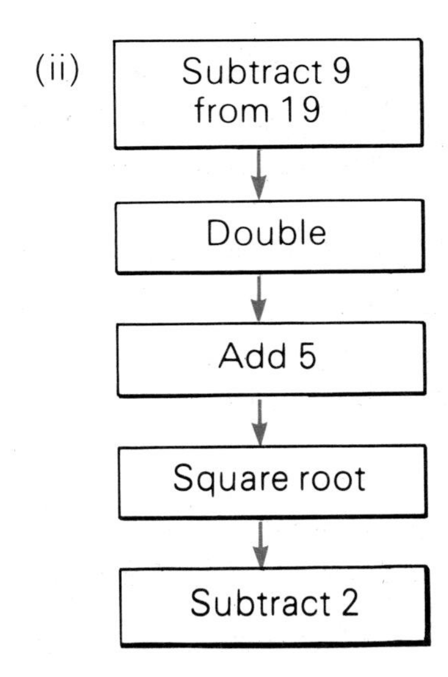

12

Know your calculator

E 1 Use your calculator.

a) Press C 2 + 7 × 8 =

Write down the result.

b) What does your calculator work out,
$(2+7)\times 8$ or $2+(7\times 8)$?

c) Use your calculator to find the other possible result.
Write down which keys you pressed.

2 a) Press C 1 5 − 9 ÷ 3 =

Write down the result.

b) What does your calculator work out,
$(15-9)\div 3$ or $15-(9\div 3)$?

c) Use your calculator to find the other possible result.
Write down which keys you pressed.

You may need to use the memory keys: MC, M+, MR, . . .

3 a) Press C 2 4 − 4 × 5 =

Write down the result.

b) What does your calculator work out,
$(24-4)\times 5$ or $24-(4\times 5)$?

c) Use your calculator to find the other possible result.
Write down which keys you pressed.

4 a) Press C 6 × 1 0 ÷ 5 =

Write down the result.

b) All calculators should give the result 12.
Write one or two sentences to explain why.

12

Think it through

5 Use your calculator to work out these.

a) $(2+3)(5+6)$

b) $2+(3\times 5)\div 6$

c) $(1.2^2+4)^2$

This means $(2+3)\times(5+6)$. We often leave out the × sign between pairs of brackets.

▲ 115
● Next chapter

F 1

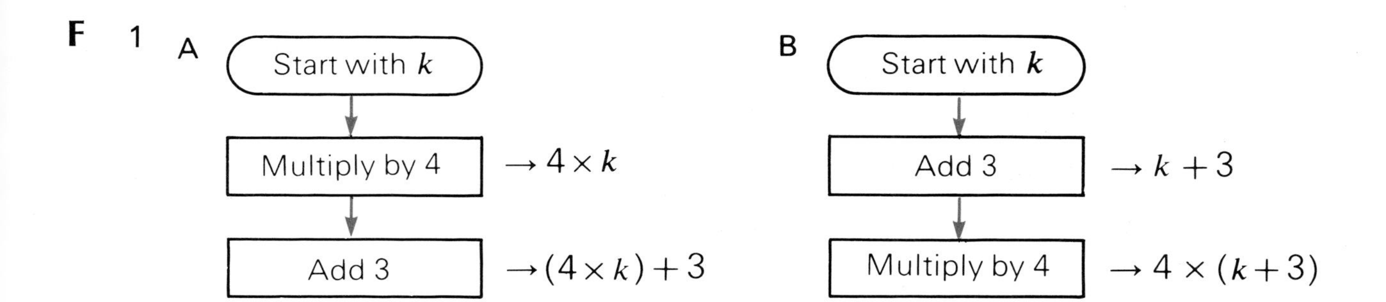

a) Do the two flow charts give the same result, no matter what k represents, or a different result?
Choose your own example for k to explain your answer.

b) Are there any replacements for k which will give the same result?
If you say **yes**, write down the replacements.
If you say **no**, explain how you know that there aren't any.

c) Solve this equation.

$2 \times (k + 2) = 3 + (k \times 4)$

Guess and test until you find the correct result.

2 a) Is this expression always true
sometimes true
or never true?

for all possible replacements for t

$(2 + t) \times 3 = (2 \times 3) + (t \times 3)$

b) Use the diagram to help you to explain what you decide.

A t B 2 C
$(t \times 3)$ cm²
(2×3) cm²
3
F E D

c) Write this expression using two brackets.

$(k + 4) \times 5$

Draw a diagram to help you to explain why your two expressions give identical results no matter which numbers you choose for k.

d) Write this expression using one bracket.

$(4 \times a) + (4 \times b)$

Draw a diagram to help you to explain why your two expressions give identical results no matter which numbers you choose for a and b.

12

3 a) Horace calculates this total cost.

$7 \times (£1.40 + £3.60)$

He gets:

$7 \times £1.40 = £9.80$
$+ £3.60$
$£13.40$

Is his calculation correct or incorrect?
If you say **incorrect**, say what he has done wrong, and write out the correct calculation.

b) Horace says this is true no matter what numbers we choose for a and b.

$$7 \times (a+b) = (7 \times a) + b$$

Is he correct?
If you say **no**, choose some numbers for a and b to help you to explain why.

c) Rupinder says this is true no matter what numbers we choose for a and b.

$$7 \times (a+b) = (7 \times a) + (7 \times b)$$

Is she correct?
If you say **no** choose some numbers for a and b to help you to explain why.

d) Investigate these expressions.
Write down the ones which are always true.

that is, the ones which are true no matter what numbers we choose for a and b

$$7 + (a+b) = (7+a) + (7+b)$$
$$7 \times (a-b) = (7 \times a) - (7 \times b)$$
$$7 \div (a+b) = (7 \div a) + (7 \div b)$$
$$7 \div (a \div b) = (7 \div a) \div (7 \div b)$$

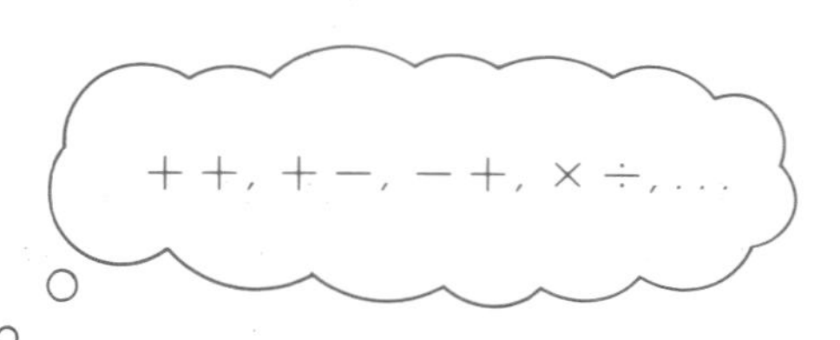

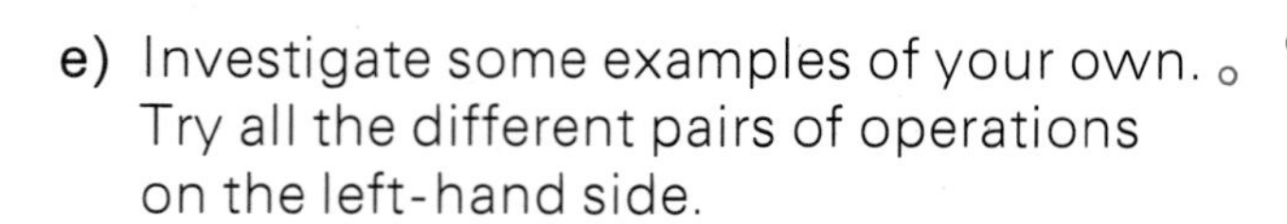

e) Investigate some examples of your own.
Try all the different pairs of operations on the left-hand side.

Write down each expression you find which is true no matter which numbers we choose for a and b.

12

13 Polygons

A 1 You need 1 cm squared paper.

a) Draw these five dots.

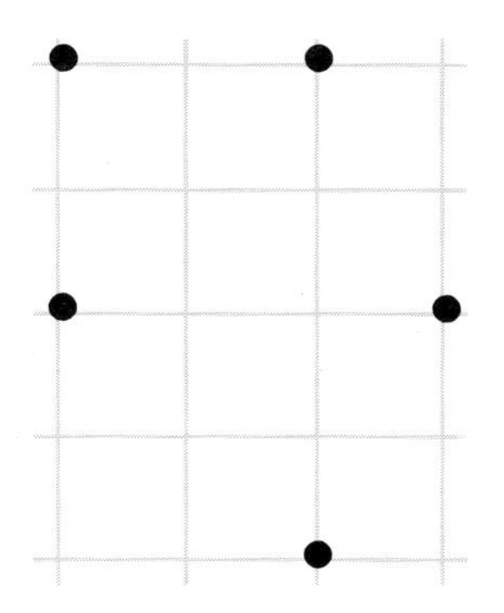

b) Some of the dots have been joined to make a trapezium.

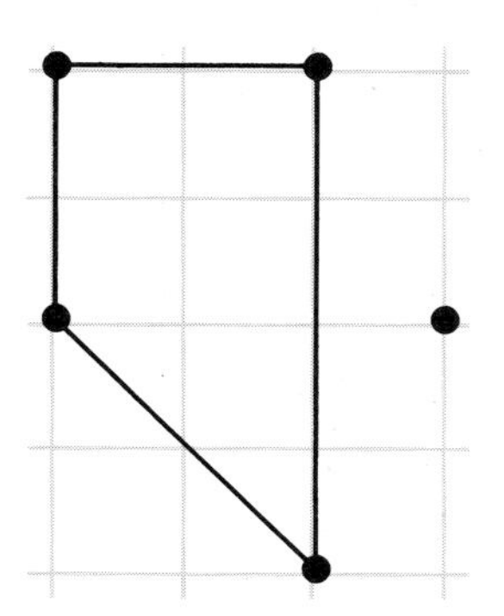

Draw **two** more different trapeziums on the five dots.

c) What is the area of each of your trapeziums?

2 The thick lines make an isosceles triangle.

a) Draw **two** more isosceles triangles on the five dots.

b) What is the area of each of your triangles?

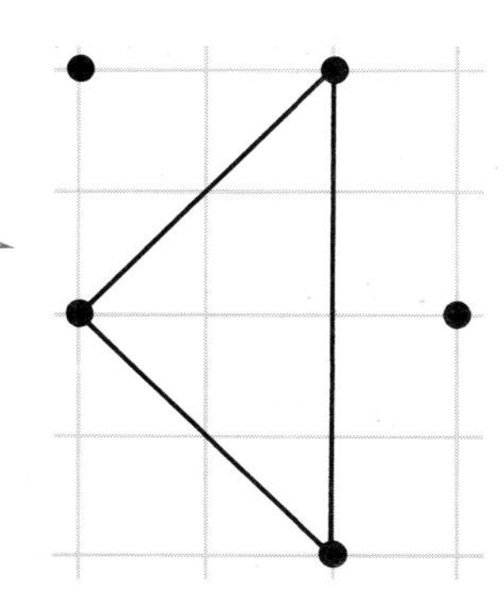

3 This is a **scalene** triangle.

a triangle with all its sides of different lengths

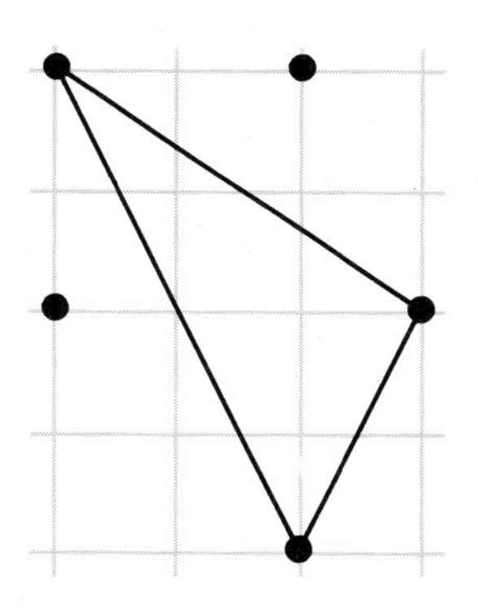

How many different scalene triangles can be drawn on the five dots?
Make a sketch of each one that you find.
On each of your sketches, mark with a **x** any angles which are greater than 90°.

4 Many triangles can be drawn on the five dots.
Draw the one with the smallest possible area.

5 You are to add one more dot to the five already drawn.
The dot must be on one of the 'crossing points' of the red lines.
Is it possible to position that dot so that an equilateral triangle can be drawn?
If you think so mark the dot and draw the triangle.

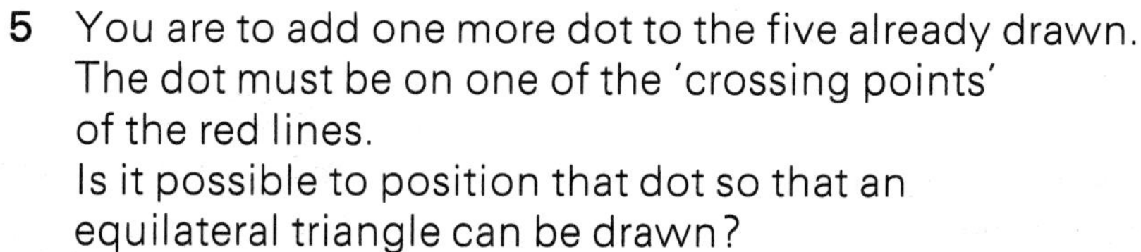

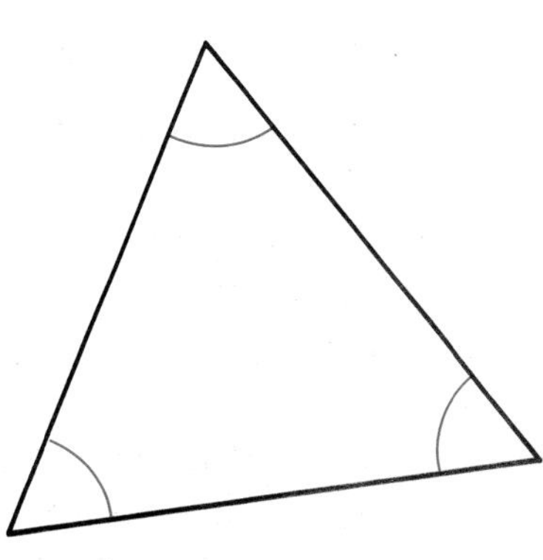

13

6 Here are some pentagons drawn on six dots.

A

B

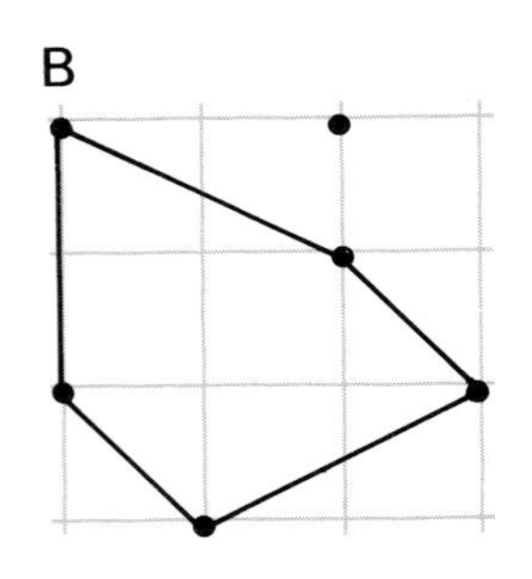

C

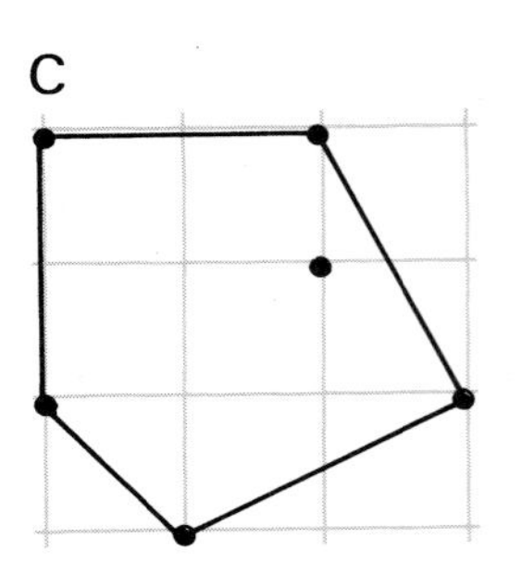

D

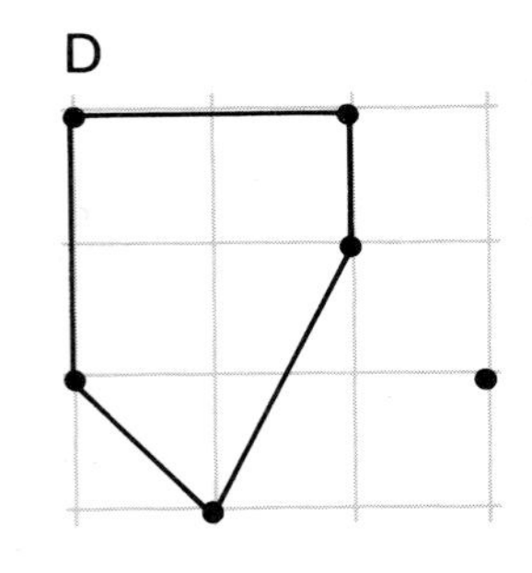

a) Sketch all the different pentagons which can be drawn on the six dots.

b) Pentagon **A** is **concave**.
The others are all **convex**.

All their interior angles are less than 180°.

Write **concave** or **convex** next to each of your pentagons.

Exploration

7 a) How many different rhombuses can be drawn on this 6×6 grid with the two dots as
(i) adjacent vertices?
(ii) opposite vertices?

b) ... What about for this pair of dots?

c) Investigate the situation for other pairs of dots.

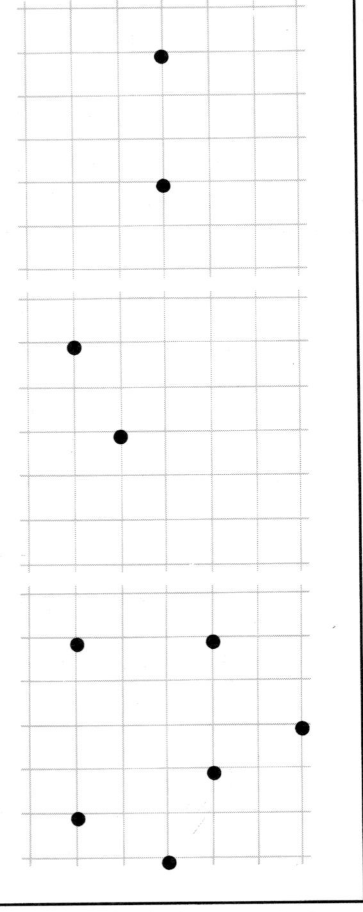

8 a) How many different concave quadrilaterals can be drawn on these six dots?

b) Investigate some more positions for the six dots.
What is the largest number of concave quadrilaterals that can be drawn?

13

Interior angles

B

With a friend

1 We can draw a hexagon with six equal angles

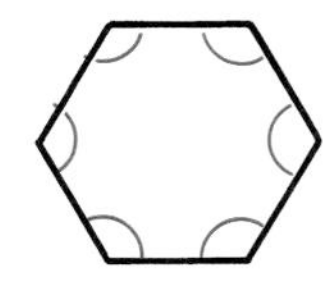

... and with five equal angles

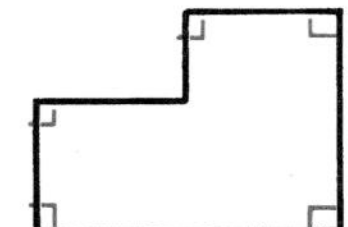

... and with three pairs of equal angles.

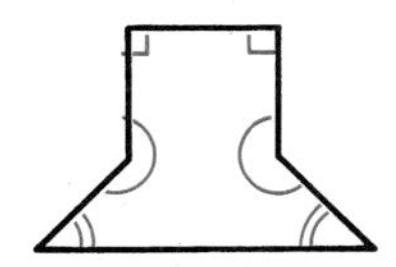

Draw hexagons for these other combinations:

4, 2 — four equal angles, two equal angles

3, 3 — three equal angles, three equal angles

3, 2, 1 — three equal angles, two equal angles, one different

Investigate the **quadrilaterals** you can draw

4 equal angles | 3 equal angles | 2 pairs of 2 equal angles

... and the **pentagons** you can draw

5 equal angles | 4 equal angles | 3 equal and 2 equal | 2 pairs of 2 equal

... and the **octagons**. *8 equal; 7 equal; 6, 2; 6, 1, 1; ...*

Each of you sketch the example you find for each polygon.
Are there any combinations for which a polygon cannot be drawn?

Think it through

2 Vince draws this quadrilateral. He cuts it into two triangles.

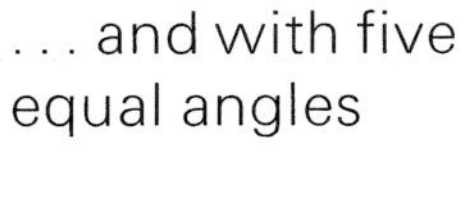

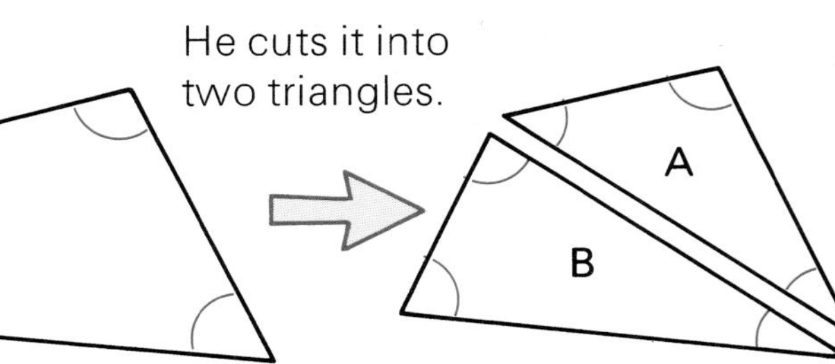

a) What is the angle sum of
(i) triangle **A**?
(ii) triangle **B**?

b) What does this tell you about the angle sum of the quadrilateral?

3 Repeat question **2** for a concave quadrilateral.

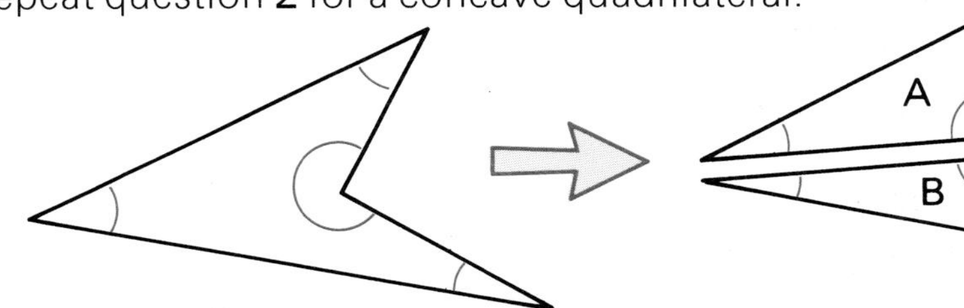

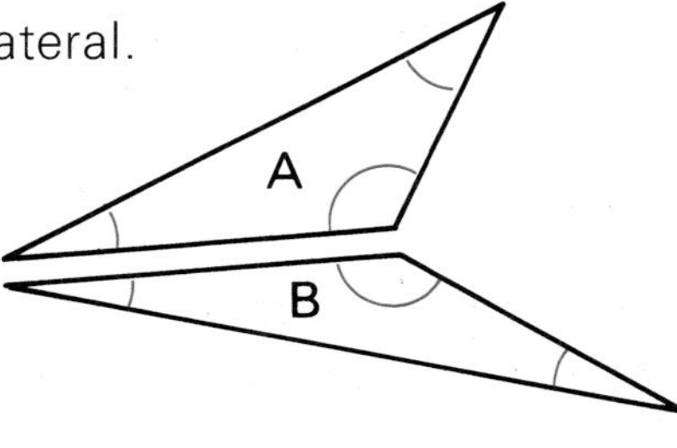

13

4 The interior angles of a triangle add up to 180°.
The interior angles of a quadrilateral add up to 360°.

Geoff thinks that the angle sum of a pentagon will be 720°.

Julia thinks it will be 540°.

What do **you** think?
Check your guess by drawing pentagons on dotted paper.

5 Guess the angle sum of a hexagon.
Check your guess by drawing hexagons on dotted paper.

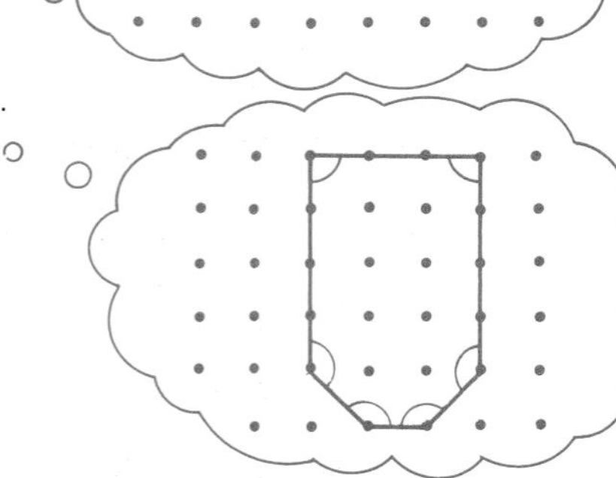

6 A hexagonal manhole cover has
two equal angles of 90°
two equal angles of 120°
and two more equal angles.
How many degrees is each of the angles?

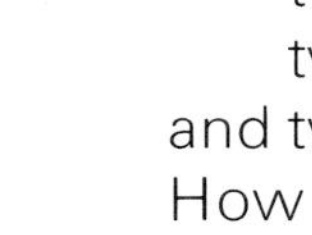

Think it through

7 The sum of the interior angles of this quadrilateral is 360°.

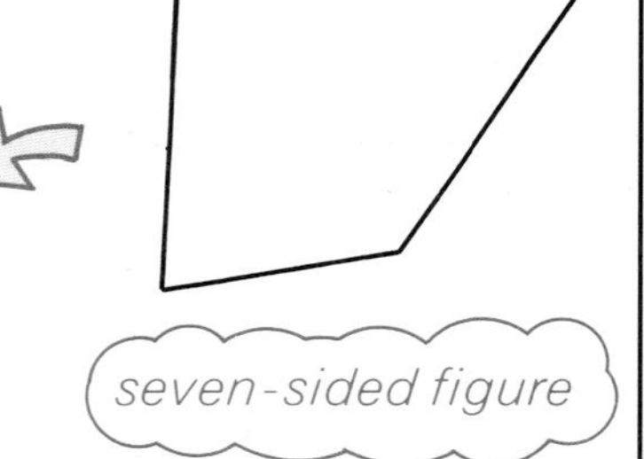

a) Study the drawing, then explain why the angle sum of this pentagon is 360° + 180°.

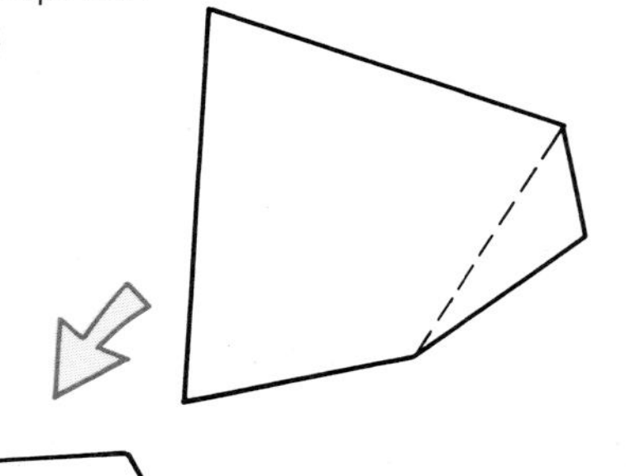

b) Write the angle sum like this:
□ × 180° = □°

c) What is the angle sum of this hexagon . . .?

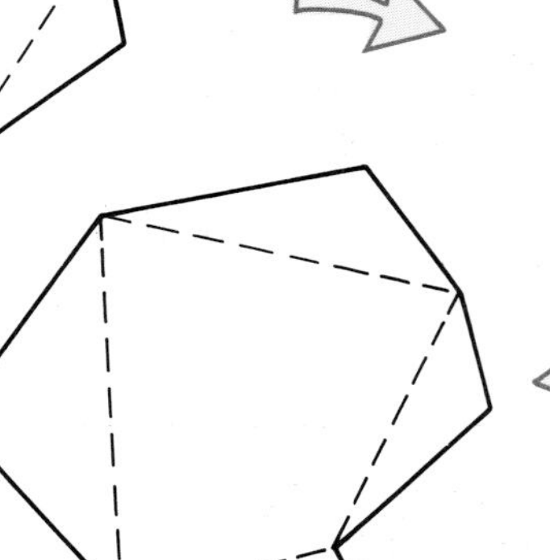

d) . . . of this heptagon . . .?

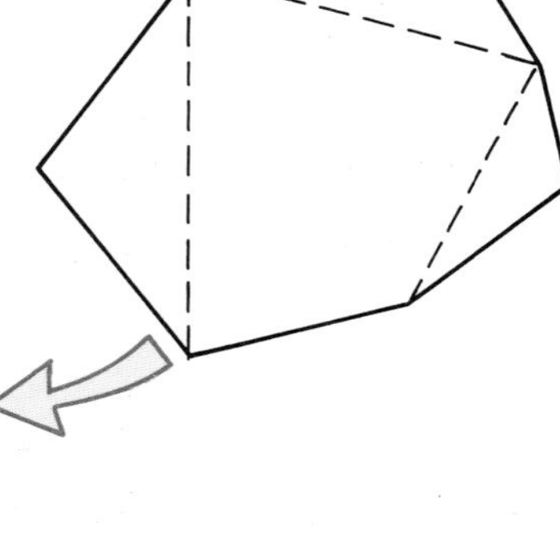

e) . . . and of this octagon?

f) What is the angle sum of a 12-sided figure?

g) Copy and complete this rule:
angle sum of any polygon is
(number of sides − □) × □

13

Regular polygons

C **Take note**

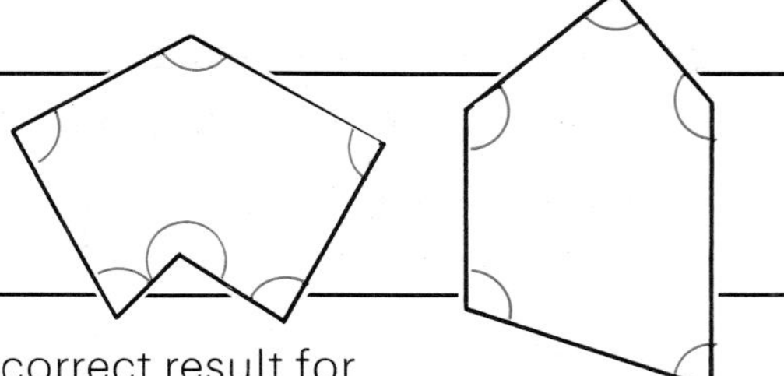

The angle sum of any polygon is
(number of sides − 2) × 180°.

1 a) Check that the rule in the **Take note** gives the correct result for
(i) squares. (ii) pentagons.

b) Use the rule to find the angle sum of a 20-sided figure.

c) What is the size of each angle in a 20-sided figure whose angles are all equal?

2 Use squared paper.

a) Draw a four-sided polygon which has
(i) all of its angles equal but not all of its sides equal.
(ii) all of its sides equal but not all of its angles equal.
(iii) all of its sides and all of its angles equal.

b) Use isometric paper.
Repeat **(a)** for polygons with six sides.

Do you remember . . .?

Polygons which have equal sides **and** equal angles are called **regular polygons**.

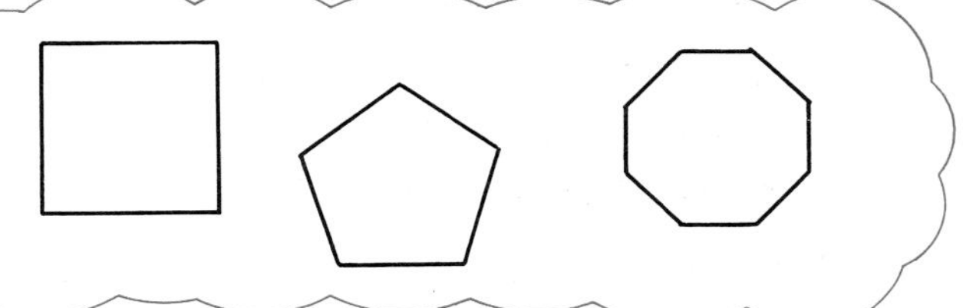

3 a) What is the angle sum of an octagon? *(eight sides)*

b) A carpenter is designing a table to seat eight people.
Its shape will be a regular octagon.
How many degrees is each of the eight angles of the table?

4 The shape of an old three-penny piece was a regular polygon with 12 sides.

What was the size of each interior angle?

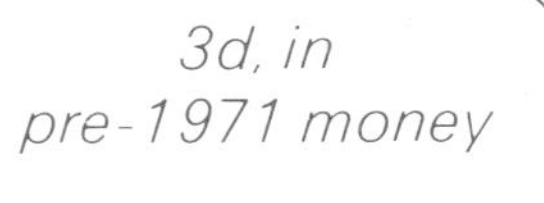

Challenge

5 A lily pond is in the shape of a regular polygon.
Each angle of the pond is 162°.
How many sides does the pond have?

13

6 These are all regular polygons.

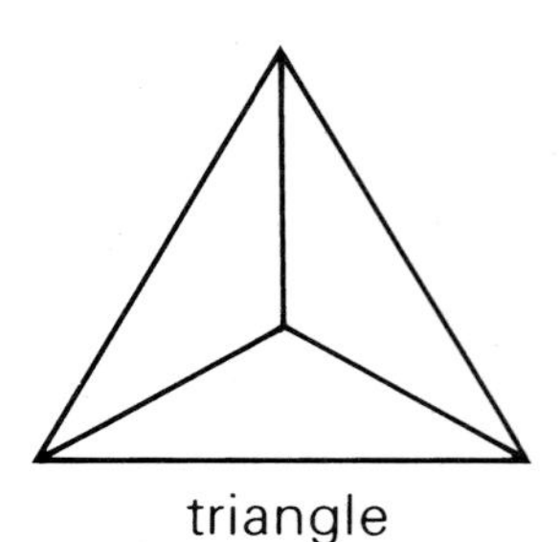

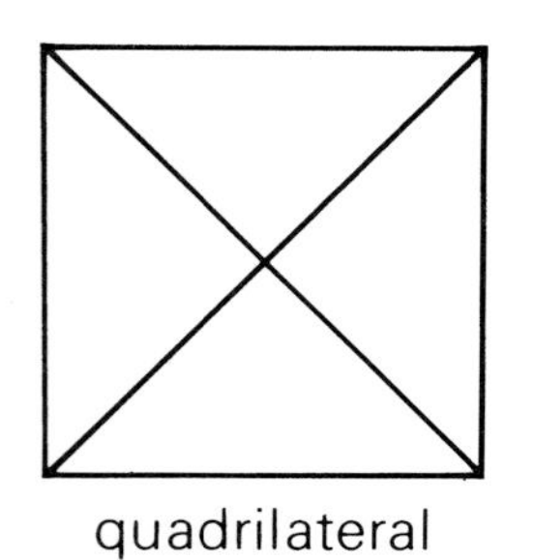

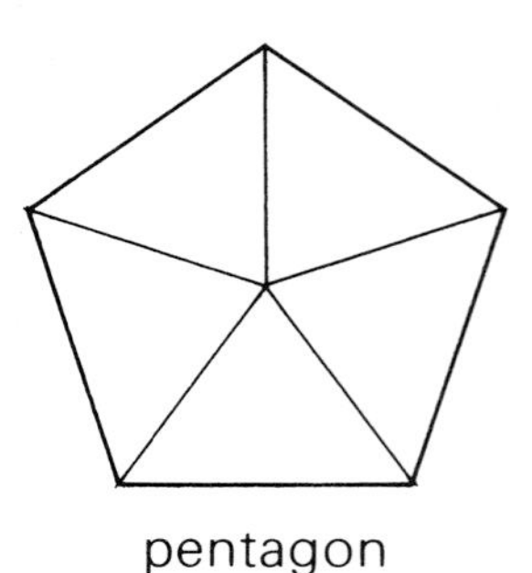

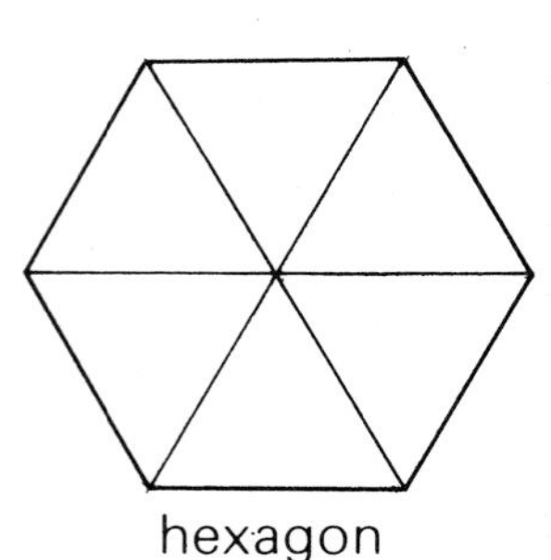

a) Each angle at the centre of a regular triangle is 120°.

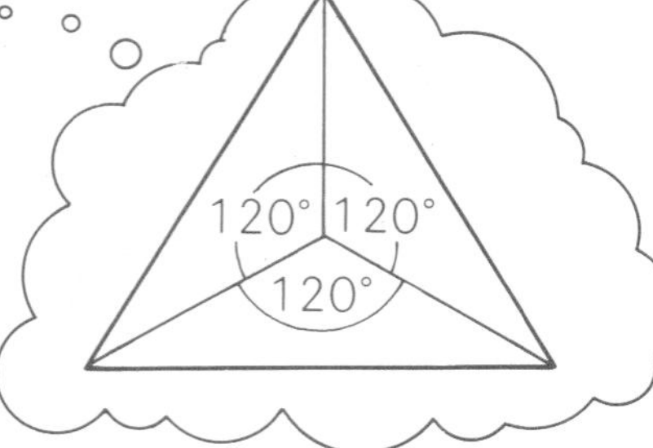

How many degrees is each angle at the centre of
(i) a regular quadrilateral?
(ii) a regular pentagon?
(iii) a regular hexagon?
(iv) a regular octagon?
(v) a regular decagon?
(vi) a regular 20-sided figure?

b) A regular polygon has n sides.
Write down an expression which tells us the size of each angle at its centre.

c) The angle at the centre of a regular polygon is 20°.
(i) How many sides does it have?
(ii) How many degrees is each interior angle of the polygon?

7 This is a regular pentagon.
Explain why each of these is true.

a) d is 72 b) k is 54

c) m is 108 d) n is 72

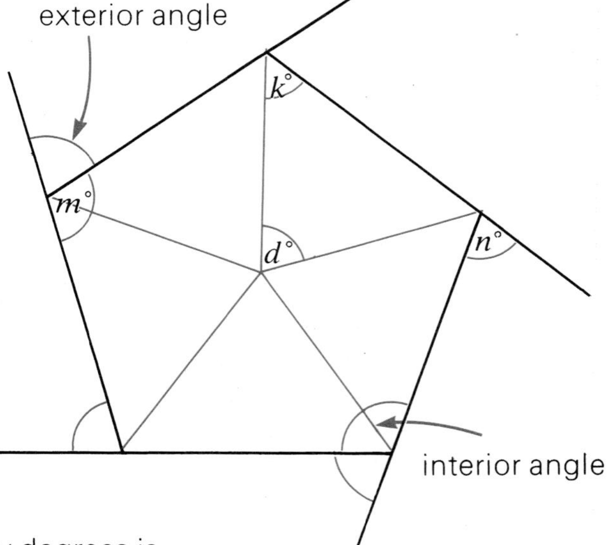

8 This is a regular octagon.

What is

a) a? b) b?

c) c? d) d?

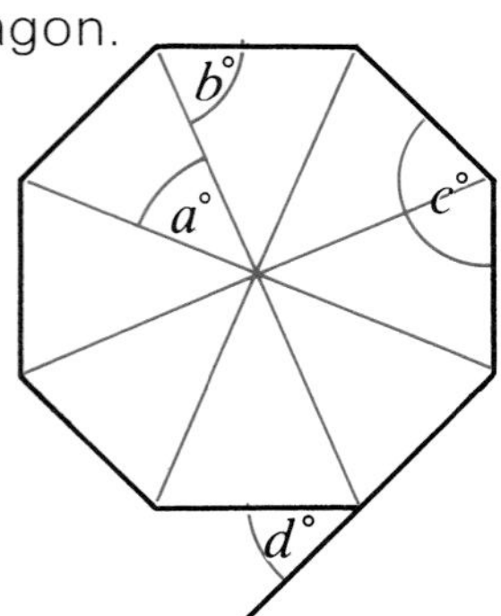

9 A regular polygon has 15 sides. How many degrees is

a) the interior angle? b) the exterior angle?

13

Exploration *Polygons and stars*

10 a) Follow these instructions.

START
40°
Move forward any distance.
Turn 40° clockwise.
Move forward the same distance.
Turn 40° clockwise.
Repeat until you return to START.

What kind of figure do you produce?

b) What kind of figure do you produce for these turns?

(i) 90° (ii) 60° (iii) 120° (iv) 45°

c) Sketch the figure produced by 144° turns.

START
144°
144°

d) Investigate some different turns.
Find out which turns produce regular polygons and which produce stars.

e) Find rules which tell you
   (i) how many sides a polygon will have, when you know the angle of turn.
   (ii) how many points a star will have, when you know the angle of turn.

f) Are there any turns which produce neither polygons nor complete stars?

▲ 124
● Next chapter

Changing shapes

D

1 Start with a square.
Mark the interior angles.
Each one measures 90° and they add up to 360°.

a) Redraw one of the sides at an angle.

b) What is the sum of the interior angles of this new shape (a trapezium)?

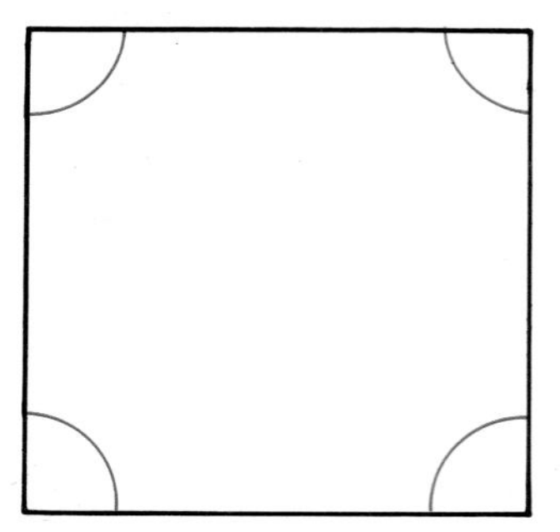

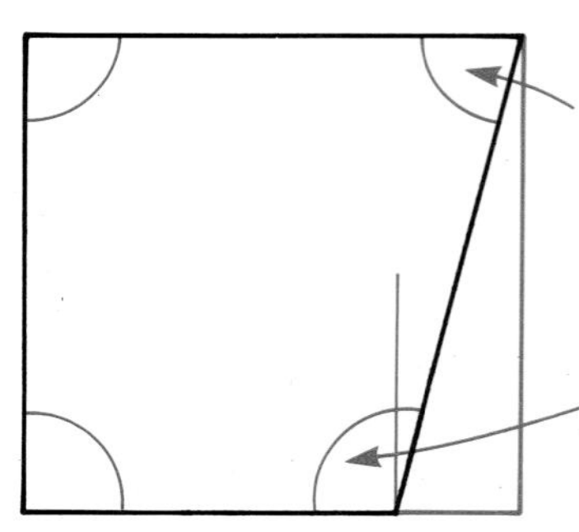

Notice:
The number of degrees 'lost' by this angle is equal to the number of degrees 'gained' by this angle.

Try to explain why.

c) 'Shrink' the trapezium into a thinner one.
What is the sum of the interior angles?
Why?

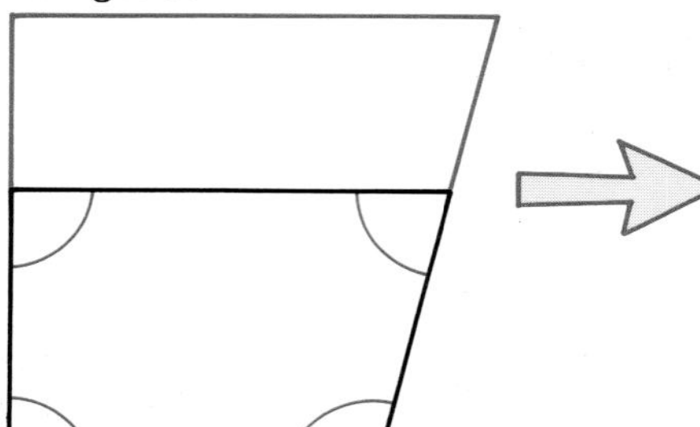

d) Redraw one of the sides at an angle.
What is the sum of the interior angles?
Why?

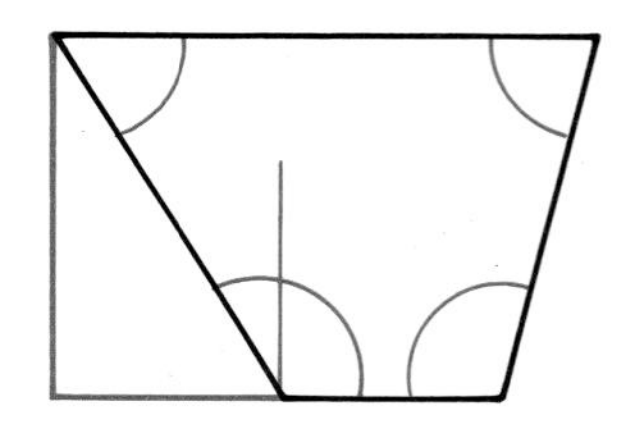

e) Stretch the trapezium into a taller one.
What is the sum of the interior angles?
Why?

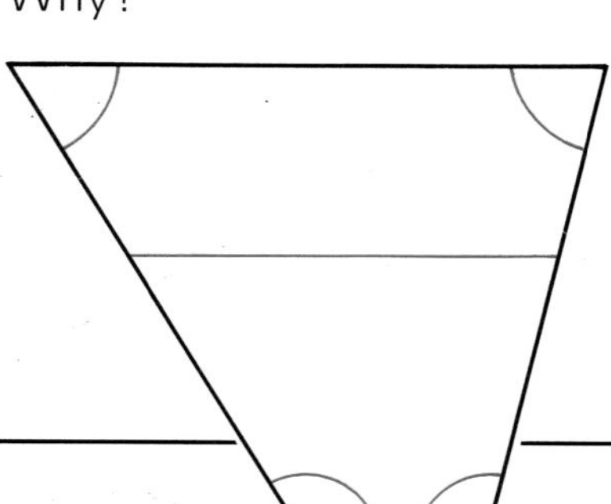

f) Redraw a third side at an angle.
Why does the sum of the interior angles remain the same throughout all these changes?

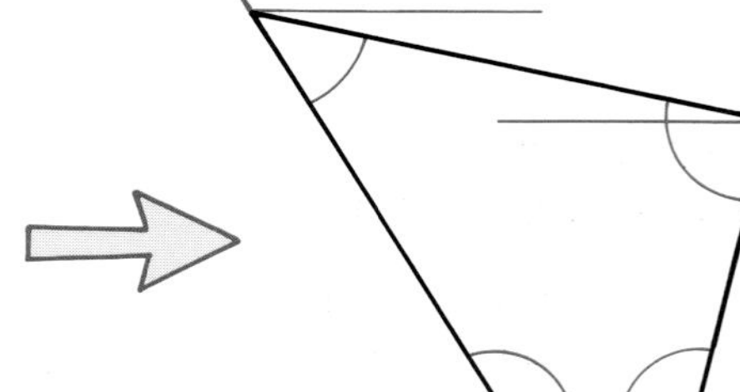

2 Carry out a similar experiment for a pentagon, starting with this one.

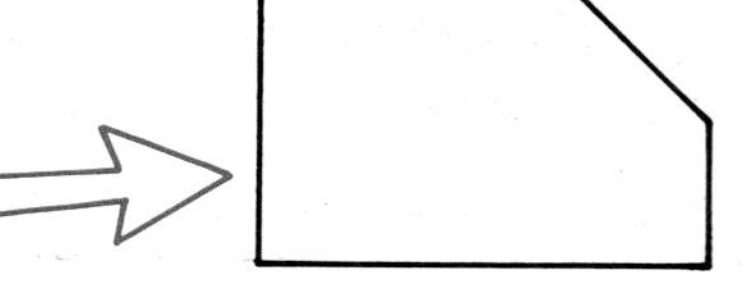

13

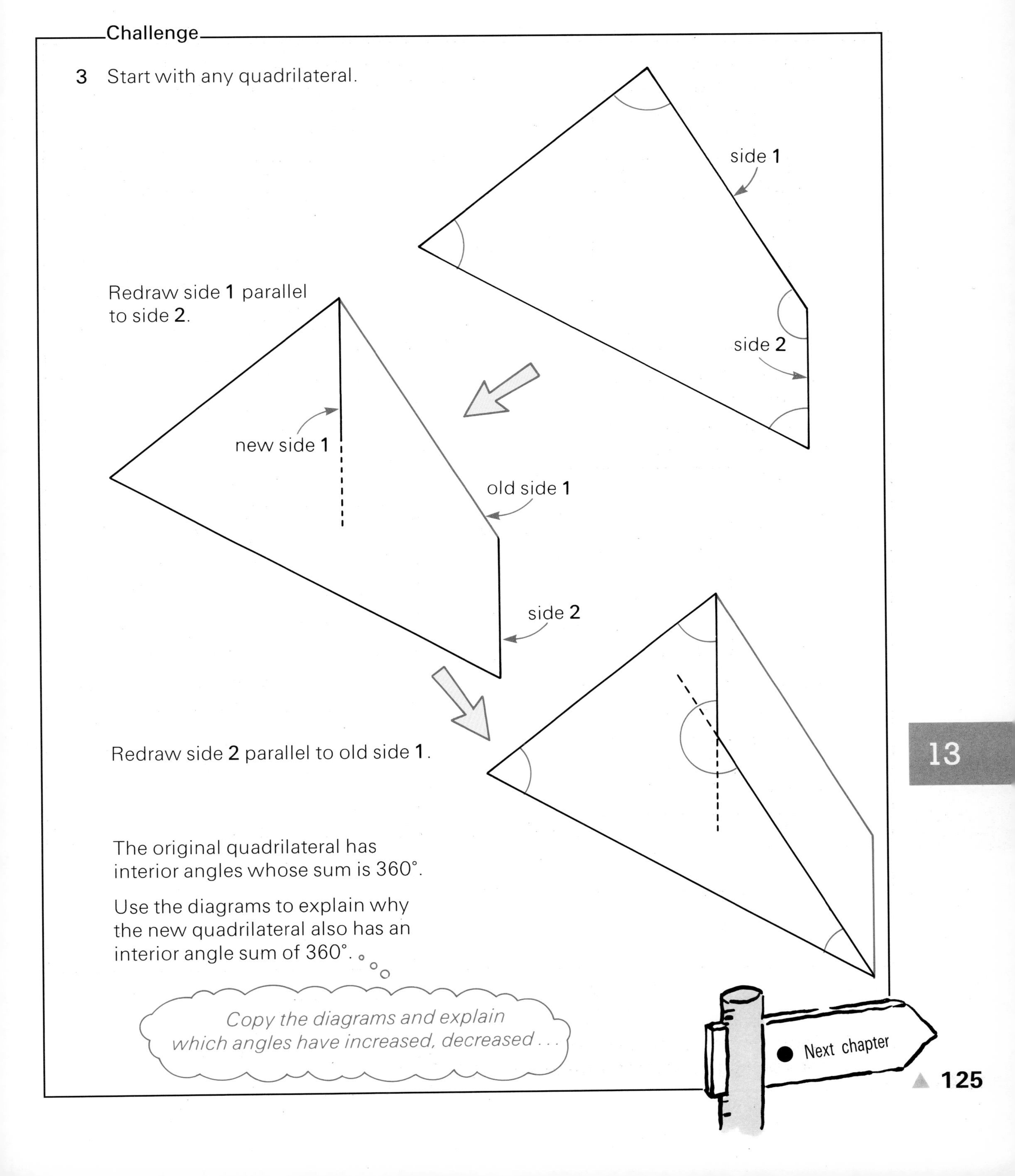

Challenge

3 Start with any quadrilateral.

Redraw side **1** parallel to side **2**.

Redraw side **2** parallel to old side **1**.

The original quadrilateral has interior angles whose sum is 360°.

Use the diagrams to explain why the new quadrilateral also has an interior angle sum of 360°.

14 Special numbers

A

Exploration

1 When Glenda tiles, she always starts in the middle of the floor, and works outwards in a spiral.

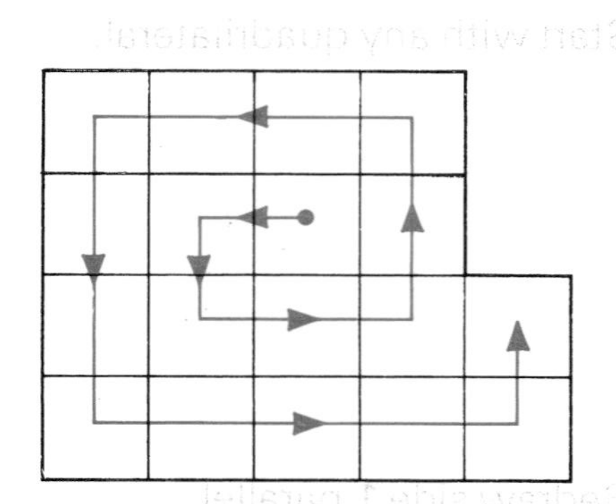

a) When she has laid four tiles, the pattern she has made is a square.

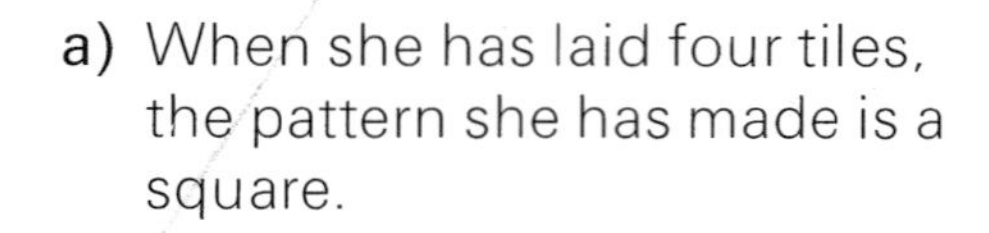

How many tiles has she laid the next time she gets a square?

When describing a floor pattern, Glenda refers to

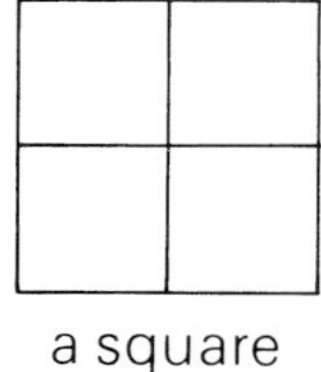

a square

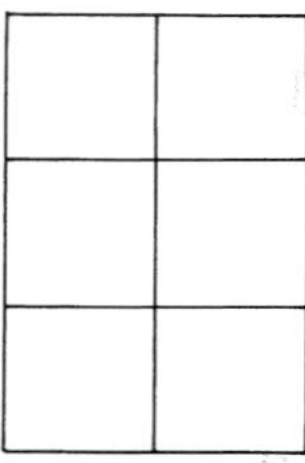

a rectangle

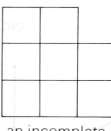

an incomplete square

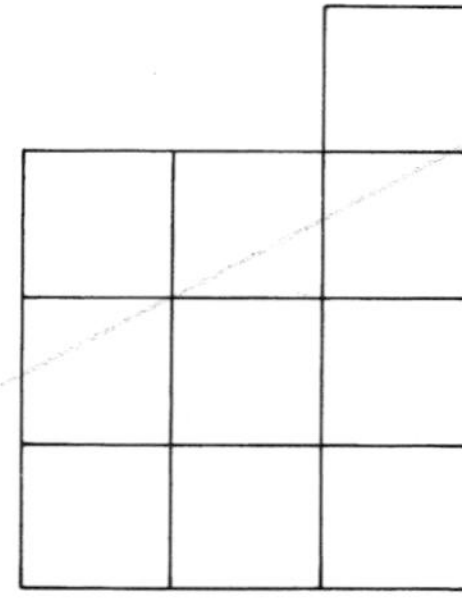

an incomplete rectangle

b) How would Glenda describe a floor pattern with

(i) 27 tiles? (ii) 64 tiles?

(iii) 105 tiles? (iv) 300 tiles?

c) Investigate the number of tiles that will give you a rectangle pattern.

Try to find a rule.

B 1 Starting with 1 and by adding consecutive odd numbers, we produce a well-known set of numbers.

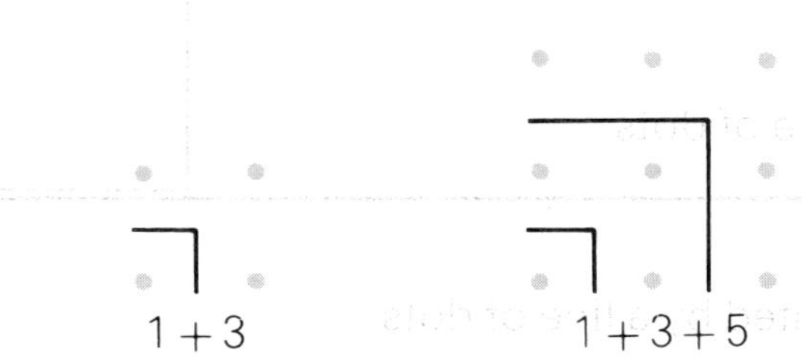

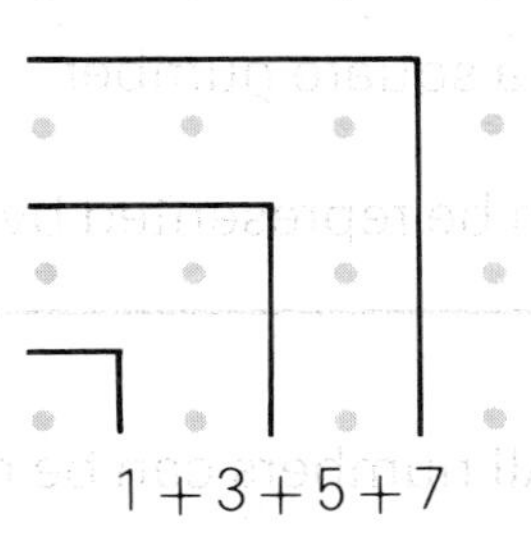

1   1 + 3   1 + 3 + 5   1 + 3 + 5 + 7

a) Draw the pattern for the next number in the set.

b) What special name is given to these numbers?

c) The first two square numbers are 1 and 4.
What is

(i) the tenth?   (ii) the hundredth?

d) Which square number does this sum represent?

1 + 3 + 5 + 7 + 9 + 11 + 13 + 15 + 17 + 19

e) The first six odd numbers are added.
Which square number does this give?

f) The first twenty-five odd numbers are added.
Which square number does this give?

g) Which consecutive odd numbers should be added to give these square numbers?
(i) 64   (ii) 225   (iii) 529

Challenge

2 **Do not use a calculator.**

The last odd number in this sum is 91.

1 + 3 + 5 + 7 + 9 + 11 + . . . + 91

Which square number does the sum represent?
Write one or two sentences to explain how you decided.

Take note

16 is a square number.

It can be represented by a square of dots.

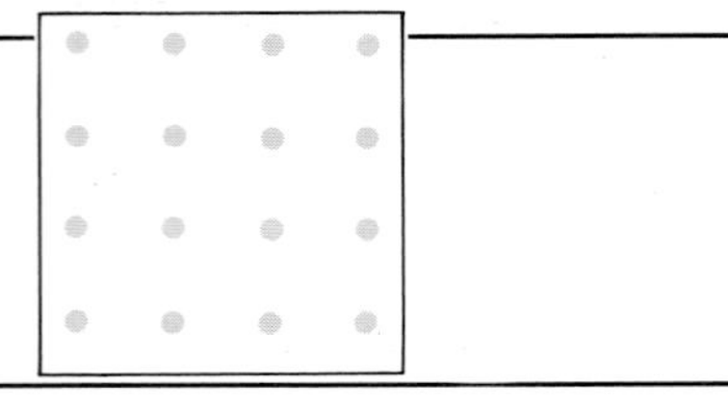

3 **All** numbers can be represented by a line of dots.

a) The first two numbers that can be represented by a rectangle or square pattern of dots are 4 and 6.

List the next five numbers of this kind.

b) Which of the first seven numbers that can be represented by a rectangle of dots can also be represented by a square of dots?

c) 16 can be represented by a square of dots **and** a rectangle of dots.

(i) Which of the first twelve square numbers **cannot** be represented by a rectangle of dots?

(ii) What have they in common?

d) List the first ten numbers that can only be represented by a line of dots.

14

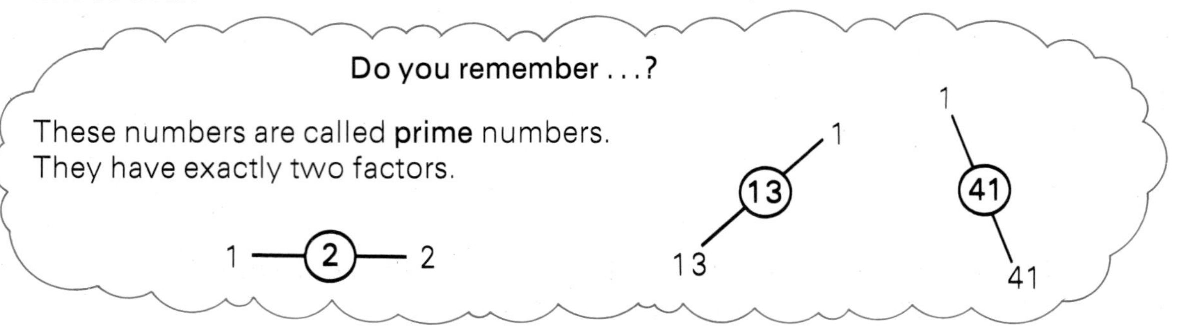

Do you remember . . .?
These numbers are called **prime** numbers.
They have exactly two factors.
1
2
2
1
13
13
1
41
41

e) Which number less than 100 can be represented by the greatest number of different rectangle patterns and square patterns?

C 1 Do you remember . . .?

1, 3, 6, 10 . . . are called **triangle numbers**.

1 3 6 10

a) Write down

(i) the fifth triangle number.

(ii) the tenth triangle number.

b) 190 is the nineteenth triangle number. What is

(i) the twentieth triangle number?

(ii) the sum of the nineteenth and twentieth triangle numbers?

c) What is the sum of the twentieth and twenty-first triangle numbers?

d) Investigate the sum of **consecutive** triangle numbers.
Write one or two sentences to explain what you find.

Challenge

2 36 is a very special number.

It can be represented by
a triangle of dots, a square of dots and a rectangle of dots.

Find the next example of a number which can be represented by
a triangle, a rectangle and a square of dots.

Use a calculator.
Be patient!

Exploration

3 In a large office block there are 100 offices, numbered 1 to 100.
There are 100 security guards.
At 6:00 pm each evening, the first security guard makes sure that all the office lights are **off**.
The second security guard comes along and switches every second light **on**.

The third security guard comes along and changes the state of every third light.

*So those that are **on** are switched **off**, and those that are **off** are switched **on**.*

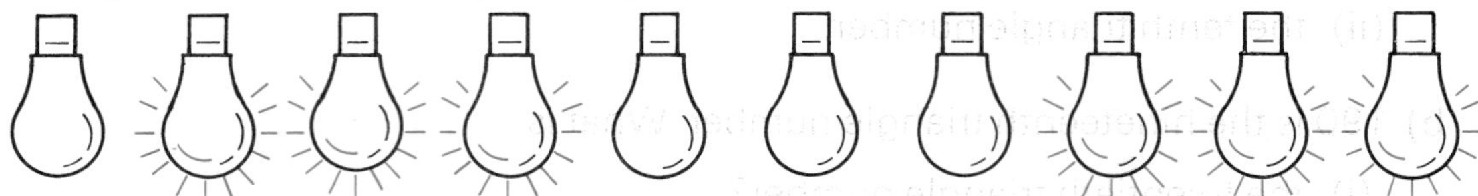

The fourth security guard comes along and changes the state of every fourth light
. . . and so on . . .
until all the guards have made their rounds.

a) When all the security guards have made their rounds,
  (i) how many lights are left on?
  (ii) which are they?

b) One evening, the second security guard fails to turn up for work.
Nobody fills in for her.
What effect does this have?

14

Product of primes

D 1 Study these 'prime factor' trees.

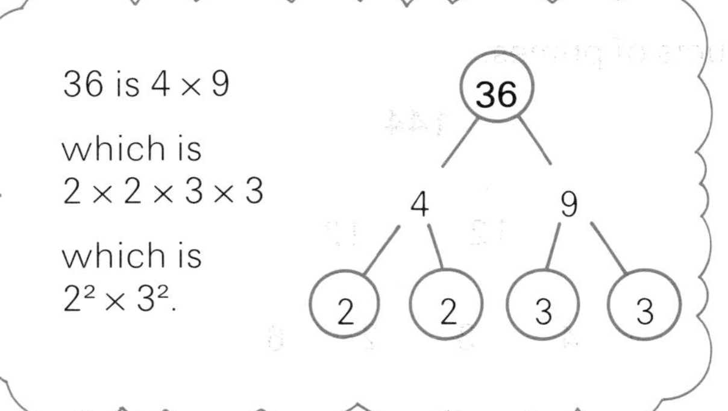

36 is 4×9
which is
$2 \times 2 \times 3 \times 3$
which is
$2^2 \times 3^2$.
36
4
9
2
2
3
3

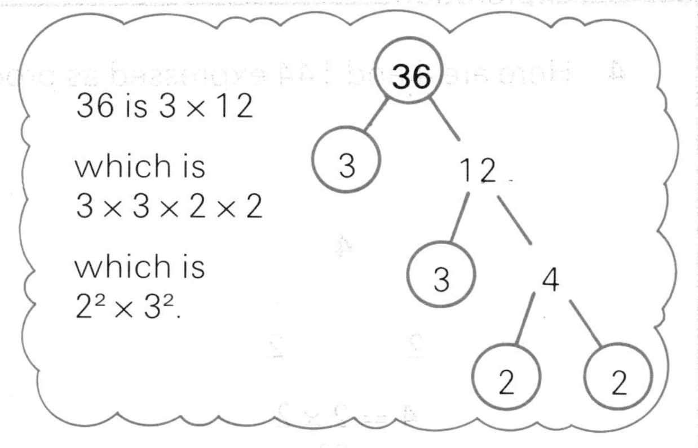

36 is 3×12
which is
$3 \times 3 \times 2 \times 2$
which is
$2^2 \times 3^2$.
36
3
12
3
4
2
2

a) Copy and continue trees (i) and (ii).
Stop when there is a prime number at the end of each 'branch'.

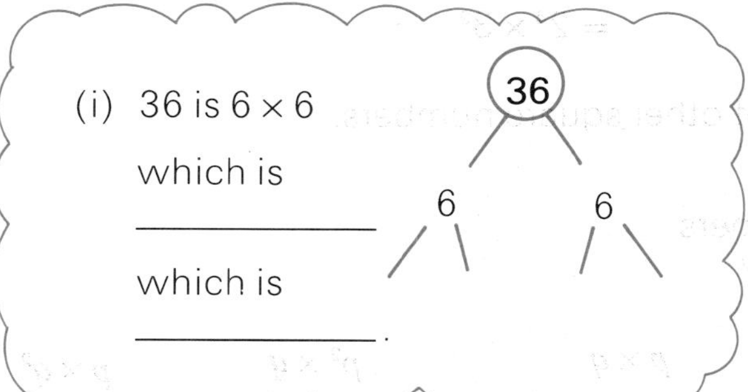

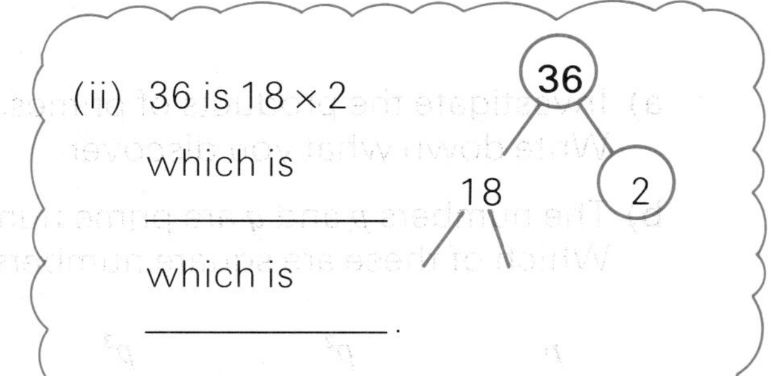

b) Draw all the prime factor trees for 24.
Write down what you notice about the results.

Take note

We say that, as a product of primes, 36 is $2 \times 2 \times 3 \times 3$
or $2^2 \times 3^2$.
No matter how we 'split' a number, we always finish up with the same product of primes.

2 Write each of these numbers as a product of primes.

a) 

b) 54

c) 121

d) 1225

Challenge *With a friend*

3 The numbers in this sequence are very special
6, 30, 210, 2310, 30030, 510510, . . .

a) Decide between you what is special about them.

b) What is the next number in the sequence?

14

Exploration

4 Here are 4 and 144 expressed as products of primes.

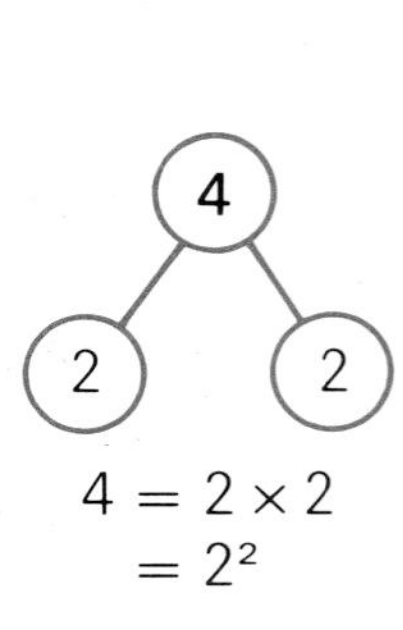

$4 = 2 \times 2$
$= 2^2$

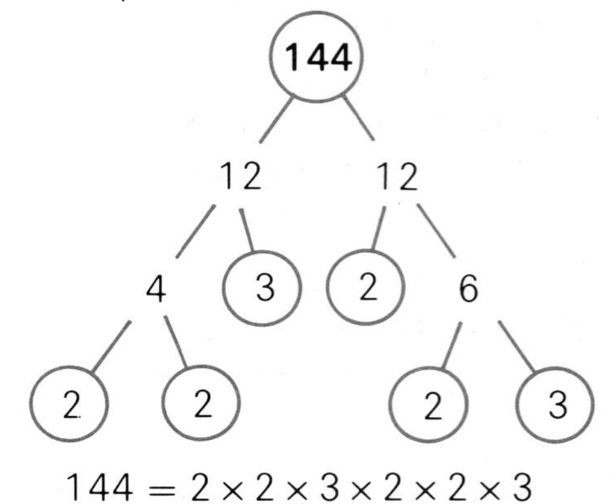

$144 = 2 \times 2 \times 3 \times 2 \times 2 \times 3$
$= 2^4 \times 3^2$

a) Investigate the products of primes of other square numbers. Write down what you discover.

b) The numbers p and q are prime numbers. Which of these are square numbers?

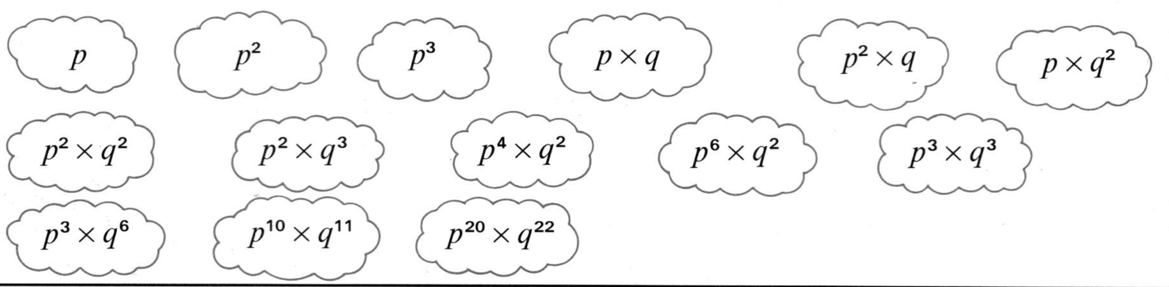

Challenge

5 This is a method which Meg uses to write a number as a product of primes.

2	60
2	30
3	15
5	5
	1

$60 = 2^2 \times 3 \times 5$

5	45
3	9
3	3
	1

$45 = 3^2 \times 5$

a) Try to explain how it works.

b) Use Meg's method to write these numbers as products of primes.

(i) 80 (ii) 216

(iii) 1225

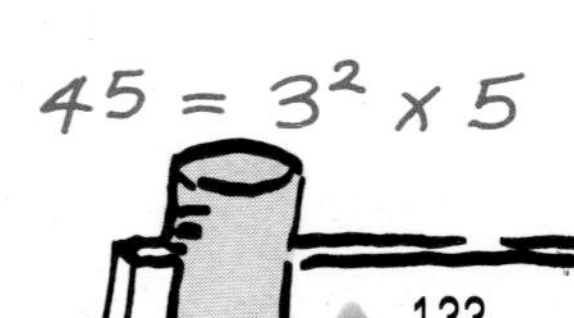

Diagrams and number patterns

E Explorations

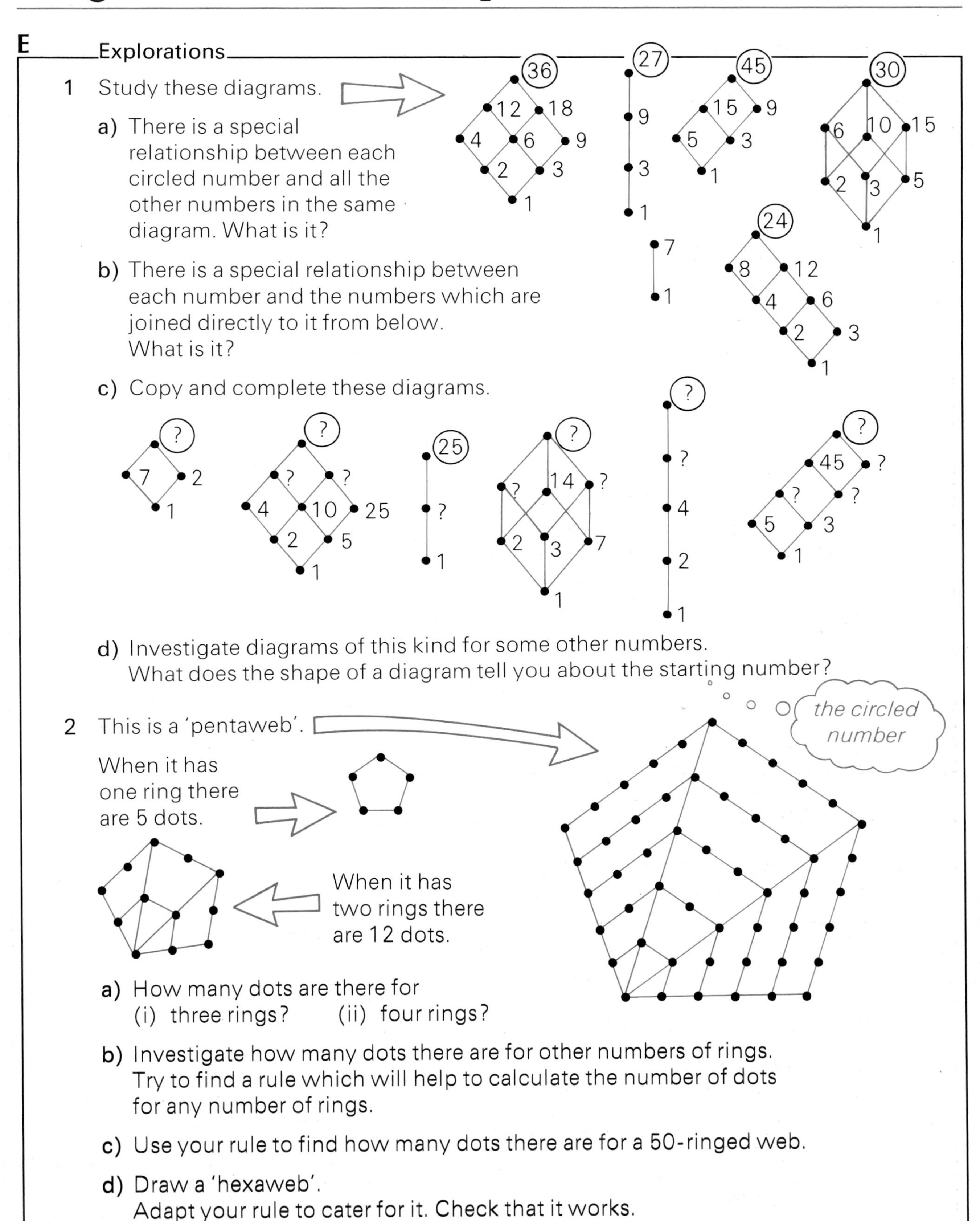

1 Study these diagrams.

a) There is a special relationship between each circled number and all the other numbers in the same diagram. What is it?

b) There is a special relationship between each number and the numbers which are joined directly to it from below. What is it?

c) Copy and complete these diagrams.

d) Investigate diagrams of this kind for some other numbers.
What does the shape of a diagram tell you about the starting number?

2 This is a 'pentaweb'.

When it has one ring there are 5 dots.

When it has two rings there are 12 dots.

a) How many dots are there for
(i) three rings? (ii) four rings?

b) Investigate how many dots there are for other numbers of rings.
Try to find a rule which will help to calculate the number of dots for any number of rings.

c) Use your rule to find how many dots there are for a 50-ringed web.

d) Draw a 'hexaweb'.
Adapt your rule to cater for it. Check that it works.

Challenges

3 Here are some sequences of numbers.
In each sequence, the numbers follow a simple pattern.

a) Find the next number in each sequence.

b) **Either** find the 50th number in the sequence
or describe how you would find the 50th number.

Try to describe the quickest possible method.

(i) 2, 4, 6, 8, 10, 12, . . .

(ii) 2, 4, 8, 16, 32, 64, . . .

(iii) 1, 4, 9, 16, 25, 36, . . .

(iv) 1, 1, 2, 3, 5, 8, . . .

(v) 1, 4, 7, 10, 13, 16, . . .

(vi) 1, 8, 27, 64, 125, 216, . . .

(vii) 1, 2, 4, 7, 11, 16, 22, . . .

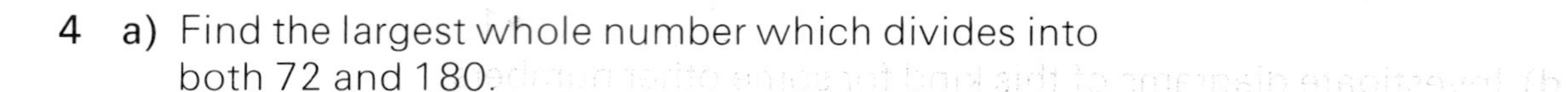

4 a) Find the largest whole number which divides into both 72 and 180.

b) Meg wants to find the largest whole number which divides into both 48 and 168,
. . . and the largest whole number which divides into both 150 and 400.

She writes:

$48 = 2^4 \times 3$
$168 = 2^3 \times 3 \times 7$
Largest number
$= 2^3 \times 3 = 24$

$150 = 2 \times 5^3$
$400 = 2^4 \times 5^2$
Largest number
$= 2 \times 5^2 = 50$

(i) Is she correct in each case?

(ii) Try to explain how Meg's method works.

(iii) Use Meg's method to find the largest whole number which divides into both 270 and 48.

15 Taking a chance

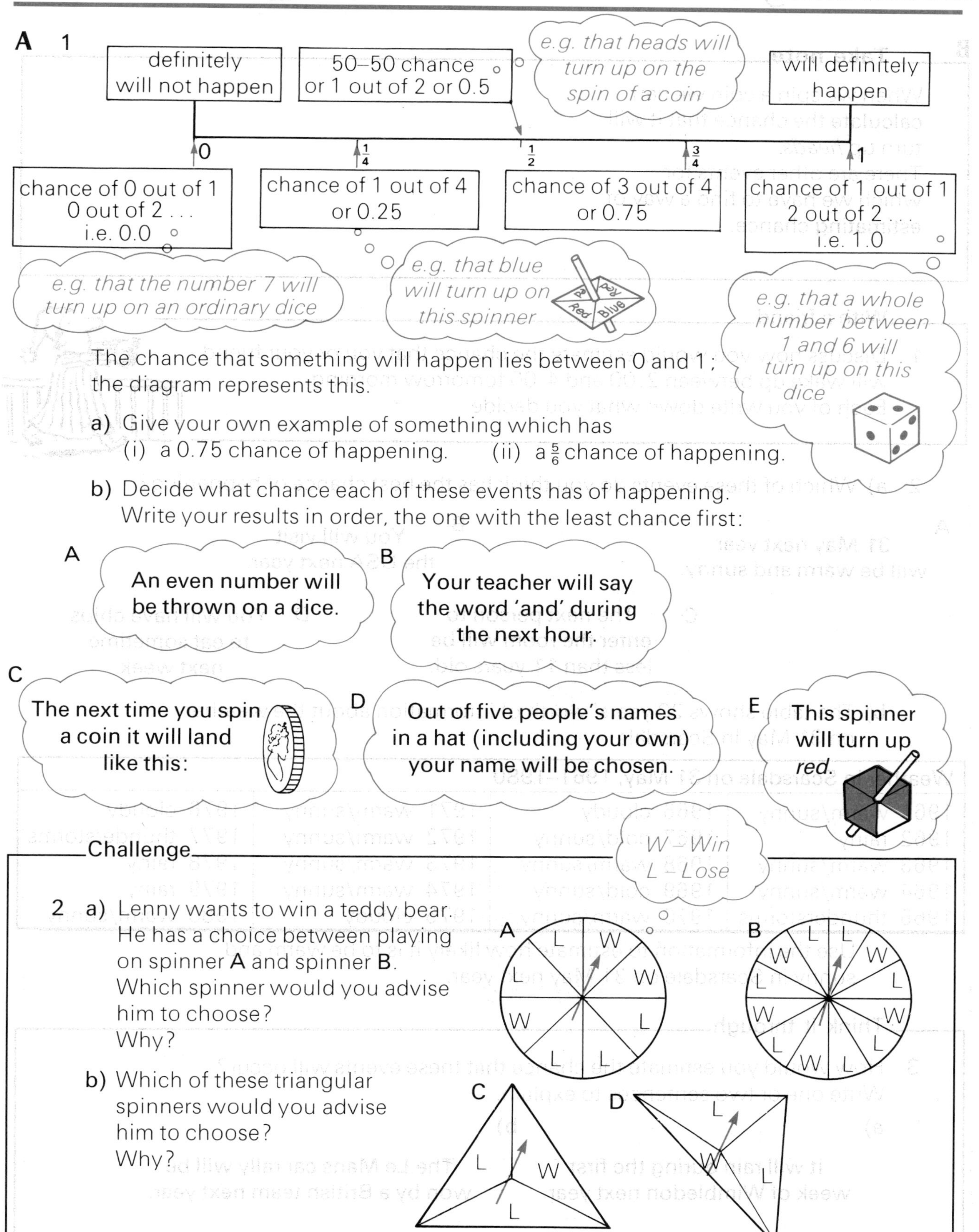
definitely will not happen
50–50 chance or 1 out of 2 or 0.5
e.g. that heads will turn up on the spin of a coin
will definitely happen
0
1/4
1/2
3/4
1
chance of 0 out of 1 0 out of 2 . . . i.e. 0.0
chance of 1 out of 4 or 0.25
chance of 3 out of 4 or 0.75
chance of 1 out of 1 2 out of 2 . . . i.e. 1.0
e.g. that the number 7 will turn up on an ordinary dice
e.g. that blue will turn up on this spinner
Red
Blue
e.g. that a whole number between 1 and 6 will turn up on this dice
A
An even number will be thrown on a dice.
B
Your teacher will say the word 'and' during the next hour.
C
The next time you spin a coin it will land like this:
D
Out of five people's names in a hat (including your own) your name will be chosen.
E
This spinner will turn up *red*.
W – Win
L – Lose
A
B
C
D
W
L

A 1

The chance that something will happen lies between 0 and 1; the diagram represents this.

a) Give your own example of something which has
(i) a 0.75 chance of happening. (ii) a $\frac{5}{6}$ chance of happening.

b) Decide what chance each of these events has of happening. Write your results in order, the one with the least chance first:

Challenge

2 a) Lenny wants to win a teddy bear. He has a choice between playing on spinner **A** and spinner **B**. Which spinner would you advise him to choose? Why?

b) Which of these triangular spinners would you advise him to choose? Why?

Estimating chance

B

Take note

When we spin a coin we can **calculate** the chance that it will turn up *heads*.
There are other events for which we have to find a way of **estimating** chance.

For example, we cannot calculate the chance that you will wake up between 2:00 and 4:00 tomorrow morning . . . but we might be able to make an estimate.

With a friend

1 Discuss how you would estimate the chance that you or your friend will wake up between 2:00 and 4:00 tomorrow morning.
Each of you write down what you decide.

2 a) Which of these events do you think has the best chance of happening?

A 31 May next year will be warm and sunny.

B You will visit the USA next year.

C The next person to enter the room will be less than 11 years old.

D You will have chips to eat sometime next week.

b) The table shows 20 years' worth of information about the weather on 31 May in Scarsdale.

Weather in Scarsdale on 31 May, 1961–1980			
1961 warm/sunny	1966 cloudy	1971 warm/sunny	1976 cloudy
1962 rainy	1967 cold/sunny	1972 warm/sunny	1977 thunderstorms
1963 warm/sunny	1968 warm/sunny	1973 warm/sunny	1978 rainy
1964 warm/sunny	1969 cold/sunny	1974 warm/sunny	1979 rainy
1965 thunderstorms	1970 warm/sunny	1975 cloudy	1980 warm/sunny

Use the information to estimate how likely it is to be warm and sunny in Scarsdale on 31 May next year.

15

Think it through

3 How would you estimate the chance that these events will occur?
Write one or two sentences to explain.

a) It will rain during the first week of Wimbledon next year.

b) The Le Mans car rally will be won by a British team next year.

Take note

Sometimes we can estimate the chance that something will happen by collecting information about what has happened in the past.

For example, between 1951 and 1980 it snowed on Christmas Day in London on two occasions. We might use this to estimate the chance that it will snow next year on Christmas Day in London as $\frac{2}{30}$ or $\frac{1}{15}$

Activity *Dropping pins*

4 Pepper drops a drawing pin onto the desk. 'I'll bet it lands point up', she says to Lenny.

a) Do you think she has made the correct choice? Write down what you think.

b) Do this experiment.

Drop a drawing pin onto this page 100 times. Count how many times it lands point down:

Use your results to estimate Pepper's chance of winning the bet.

Take note

Sometimes we can estimate the chance that something will happen by carrying out an experiment.

as in the drawing pin experiment

Activity *Dice game*

5 Play the game with a friend. You need two dice.

RULES
Take turns to throw the dice a total of 72 times.
The one who has the first throw is called Blue.
The other is called Red.
Blue scores a point if the difference of the scores on the two dice is zero or even.
Otherwise Red scores a point.
The one with most points wins.

a) Play the game and find out who wins.

b) Use your results to estimate the chance that
 (i) Blue will win on the next throw you make.
 (ii) Red will win on the next throw you make.

Take note

Another word for 'chance' is **probability**.
We say that the probability that a coin will turn up heads is $\frac{1}{2}$ or 0.5.

Calculating probabilities

C 1 a) Check that the probability you will get red (R) on this spinner is $\frac{5}{8}$ or 0.625.

b) What is the probability that you will score blue (B)?

There are 8 equally likely outcomes. 5 of them are red.

2 The probability that you will throw an even number on this octahedral dice is $\frac{4}{8}$ or 0.5.

a) Explain why. *8-sided*

b) What is the probability that you will throw a number equal to or greater than 2?

c) What is the probability that you **will not** throw a 3?

3 This is a very simple game of chance. You win if the ball drops into the WIN cup.

a) **Guess** the probability that it will do this.

b) Here is one route the ball can take to reach the WIN cup. How many routes are there altogether into the WIN cup?

c) How many different routes are there altogether that the ball can take?

d) ... so what is the probability that you win?

Was your guess correct in (a)?

4 This is the game of chance in question **3**. It costs 10p for each try. You win 10p plus your stake money if the ball lands in the middle cup. The probability that the ball goes into the middle cup is $\frac{2}{4}$ or 0.5.

Did you get this in question 3?

a) In 50 goes, how many times would you expect to win?

b) How many times would you expect to lose?

c) Would you expect to win, lose or break even over the whole 50 tries?

5 Here is the game of chance again.

This time the payout has been changed. In 50 tries would you expect to win, lose or break even? If you say 'win' or 'lose', write down how much you would expect to win or lose.

15

Think it through

6 a) Calculate the probability that the ball will fall into
   (i) cup A. (ii) cup B.
   (iii) cup C. (iv) cup D.

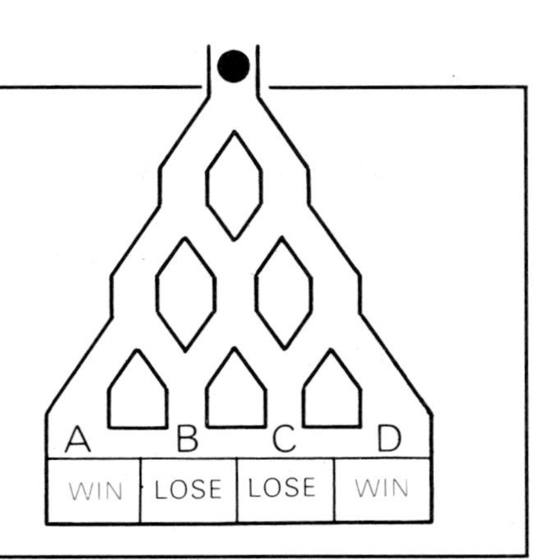

b) It costs you 10p a go to play.
You win 10p plus your stake money.
You have 40 goes.
Would you expect to win or lose, and how much?

7 In a coin game you spin three coins.
You win if you get three *heads* or three *tails*.

a) Do you think the probability of doing this is more than 0.25, less than 0.25 or exactly 0.25?
Have a guess.

b) List all the possible outcomes for one throw, like this.

H h **H**
H h **T**
H t **H**
H t **T**
T t ...
⋮

c) What is the probability that you will get
   (i) *H h* ***H***? (ii) *T t* ***T***?

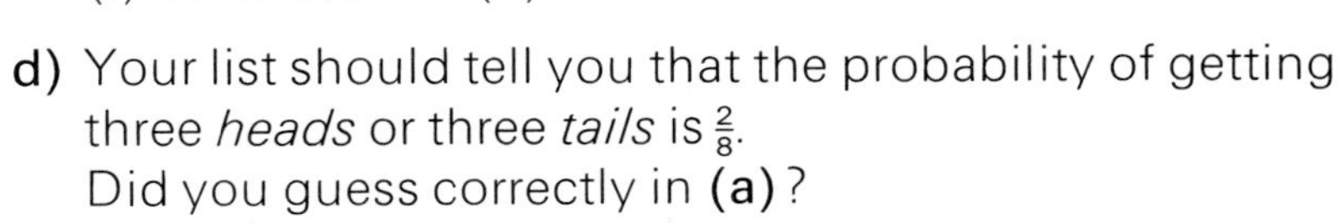

d) Your list should tell you that the probability of getting three *heads* or three *tails* is $\frac{2}{8}$.
Did you guess correctly in **(a)**?

e) In another coin game you spin four coins together.
You win if you get three or more *heads* or three or more *tails*.
What is the probability that you will win on any one throw?

Challenge

8 The Italian national flag is red, green and white.
Winston has to draw it for an exhibition
... but he can't remember which stripe is which.

Finally, in frustration, he guesses.
What is the probability that he is correct?

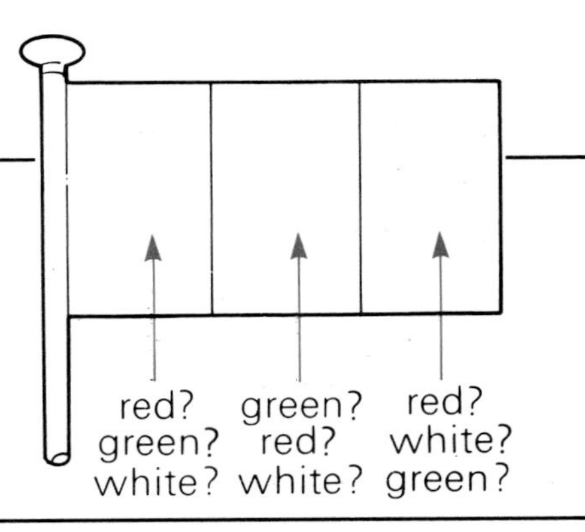

Take note

To calculate a probability,
divide the number of equally likely outcomes which are acceptable
by the total number of equally likely outcomes.

15

Challenge

9 In a game, two dice are thrown.
The difference between the two scores is recorded.
If this is odd, player **1** wins.
If it is zero or even, player **2** wins.

gives '4'

a) One possible outcome is ⚅ ⚂
How many possible outcomes are there altogether?

b) How many of the possible outcomes give the result 'odd'?

c) What is the probability that player **1** wins on any throw?

d) Two players, Red and Blue, play the game 72 times.
Red wins if the difference is odd.
Blue wins if the difference is even or zero.
How many times would you expect (i) Red to win? (ii) Blue to win?

e) Compare your results in **(d)** with the results of the game you played in question **5** on page 138.
Does your calculation give a *good*, *OK* or *poor* prediction of the result of your game?

Take note

Calculations of probabilities can be used to predict what will happen.

10 Two dice are rolled together.

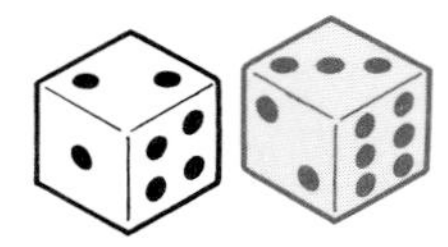

The circles on the diagram represent all the possible outcomes.

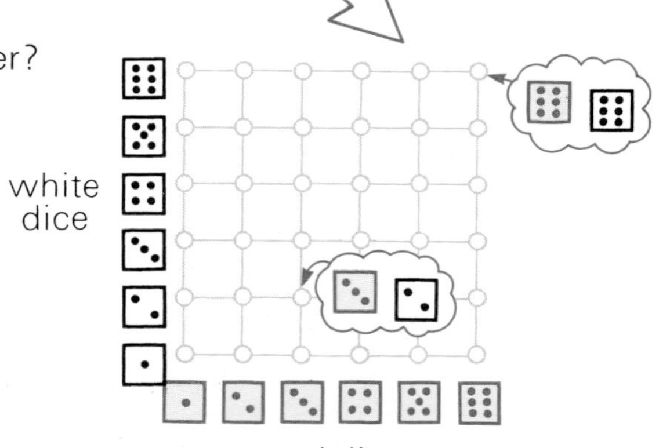

a) How many outcomes are there altogether?

b) How many of the outcomes give a total score of 4?

c) What is the probability that you will throw a total of 4 with two dice?

d) Alan rolls two dice 360 times.
Use your result in **(c)** to predict how many times he will score a total of 4.

e) How many times would you expect a total of
(i) 2 (ii) 9 to be scored in the 360 throws?

f) In a game you have to predict the total score on two dice.
Which number would you choose? Why?

Equally likely

D 1 a) Three of these tetrahedral dice are thrown together. The score is the number which lands 'face-down'. Which total score is most likely to result? Explain why.

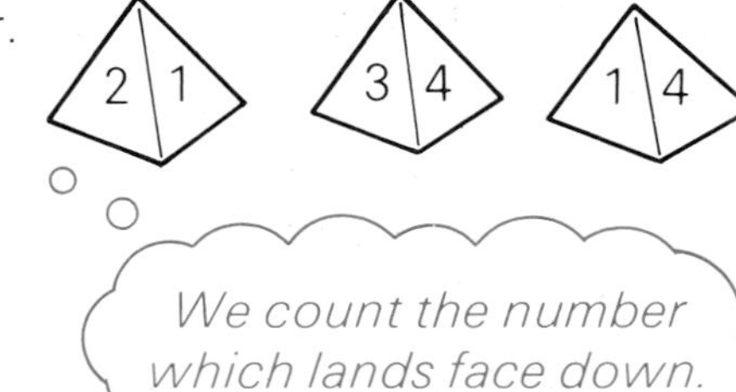

b) What is the probability that these totals will be scored?
(i) 3 (ii) 8

2 When the dice is thrown, [1] and [3] are equally likely outcomes.

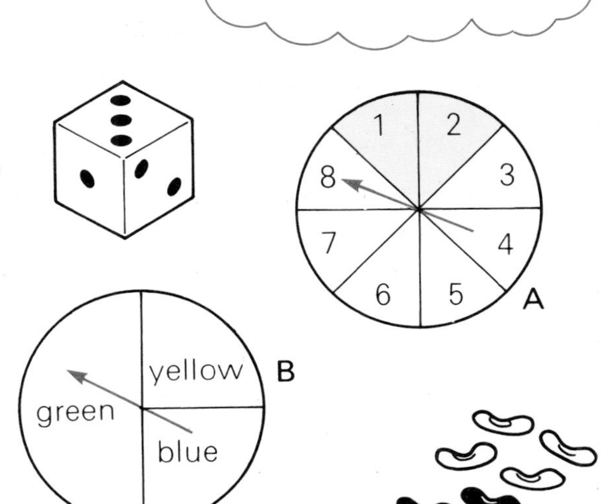

Which of these are equally likely outcomes?
Write **yes** or **no** for each one.
If you say **no**, explain why.

a) throwing [1] and an even number

b) spinning a *1* and a *3* on spinner **A**

c) spinning *red* and *white* on spinner **A**

d) spinning *yellow* and *green* on spinner **B**

e) spinning *yellow* and *blue* on spinner **B**

f) choosing a black bean from pile **C** and choosing a black bean from pile **D**

g) Choosing a black bean from pile **C** and choosing a black bean from pile **E**

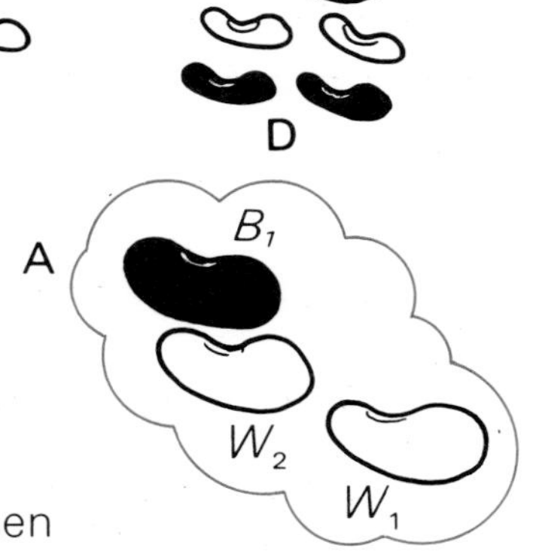

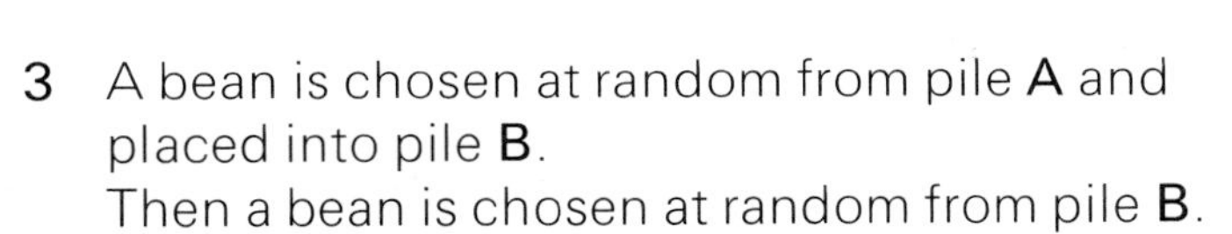

3 A bean is chosen at random from pile **A** and placed into pile **B**.
Then a bean is chosen at random from pile **B**.

a) Which of these are equally likely outcomes?
If you say **no**, explain why.
(i) B_1 is the final bean chosen; B_2 is the final bean chosen
(ii) W_3 is the final bean chosen; B_2 is the final bean chosen
(iii) B_1 is the final bean chosen; W_2 is the final bean chosen
(iv) a white (W) bean is finally chosen; a black (B) bean is finally chosen

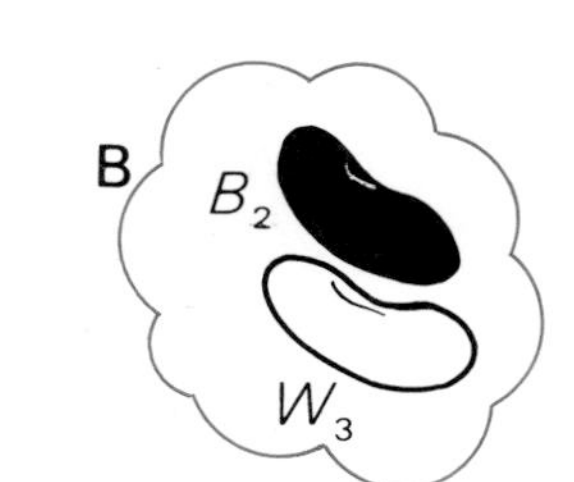

b) How many equally likely outcomes are there? List them.

c) What is the probability that
(i) B_1 is finally chosen? (ii) B_2 is finally chosen?
(iii) a black bean is finally chosen?

d) In a game of chance you have to predict the colour of the bean finally chosen.
Which colour would you choose? Why?

15

4 In this game of chance you have to
- choose one of the two shapes at random from a bag
- roll the shape.

These are the faces of the tetrahedron: 

These are the faces of the cube: 

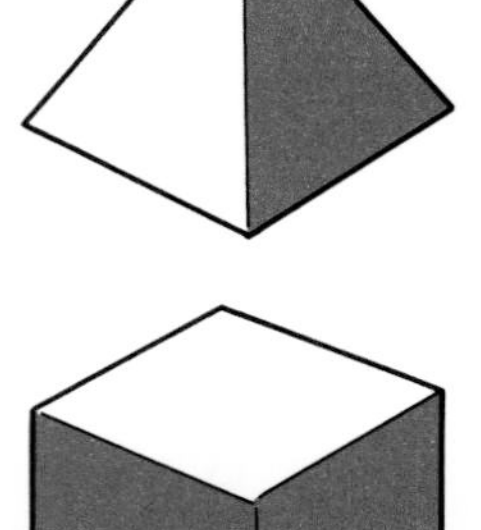

a) Which colour would you predict, white or red? Why?

b) How many times in 120 goes would you expect to get

(i) a △? (ii) a ■? (iii) a white face?

Activity

c) Make the two solids, and check your predictions in **(a)** and **(b)**. Are your experimental results what you would expect? Why?

Assignment

5 a) Make up a game of chance of your own, something like the ones in questions **3** and **4**. Decide how much you should charge for each go, which outcomes are 'win', which outcomes are 'lose', and how much should be paid for each separate 'win' possibility. Make up a tariff board showing the winning combinations and the payments. *as on a fruit machine*

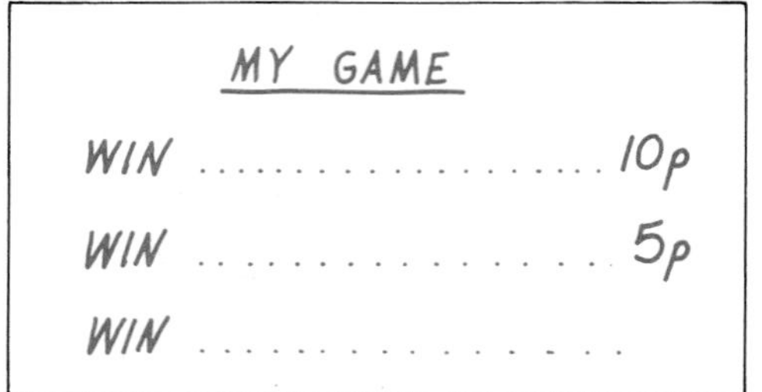

b) Test your game. *not with real money!* Modify it, if you need to, until you are sure that it
- gives a reasonable return to those who play, so that they keep on playing
- wins for you in the long run.

c) Write a report in which you describe your game and explain how you tested it. Explain how you made sure it met the conditions in **(b)**.

16 On reflection

A 1 Glenda visits Grimsby in the school bus.
She fixes this sticker to the rear window.

Sketch what the sticker looks like from inside the bus.

2 Someone has written a message on one of the windows.
What does it say?

3 This sticker is also on the rear window.
Sketch what it looks like from inside the bus.

4 Glenda sits in the driving seat.
In the rear-view mirror she looks like this.

a) What does it say on her headband?

b) This is the Grimsby sticker.
Sketch what the writing looks like to Glenda, looking in the mirror.

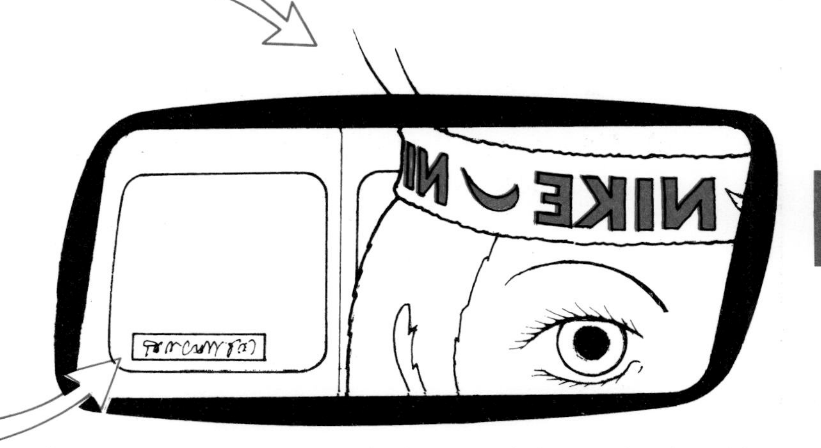

Mirror-writing

B With a friend

You will each need a sheet of carbon paper, some plain paper and a mirror.

1 a) Lay your sheet of carbon paper **ink-side-up**. Cover it with a blank sheet of paper. Write a word in mirror-writing. Check your writing by turning over the paper.

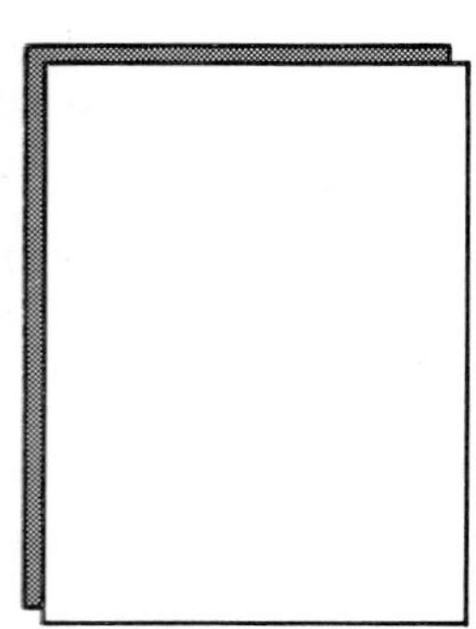

b) Practise mirror-writing until you can do it fairly easily.
Use the carbon paper to check your writing.

c) Write a message in mirror-writing without the carbon paper.
Check it with a mirror.
Pass the message to your friend.
Who is the better at writing, and reading, mirror-messages?

2 a) Lay your carbon paper ink-side-up again.

Cover it with two **half** sheets of paper.

Copy these lists of words.
Use CAPITAL letters.

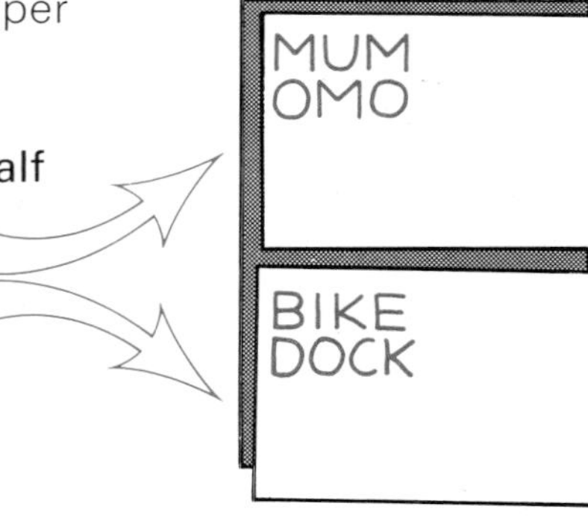

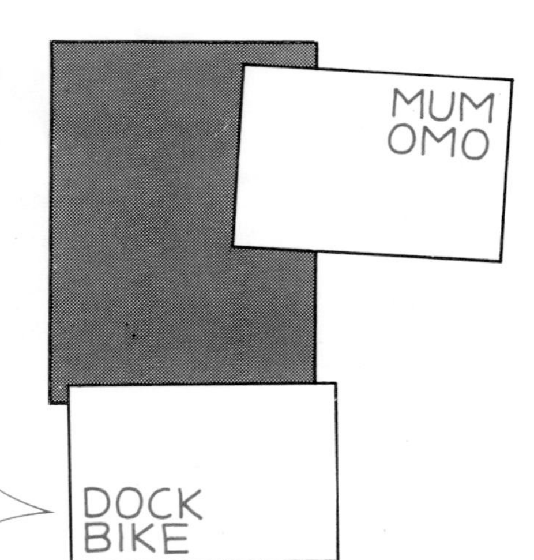

b) Turn over the half sheets, like this.
What do you notice?

c) Continue the lists with the same sorts of words.
See who can get more words.

Folding patterns

C 1 Glenda writes this message on a piece of card.

What does it say?

2 Glenda folds the card.

She wants to wrap it in this tissue paper. (The paper is transparent; the pattern shows through on both sides.)

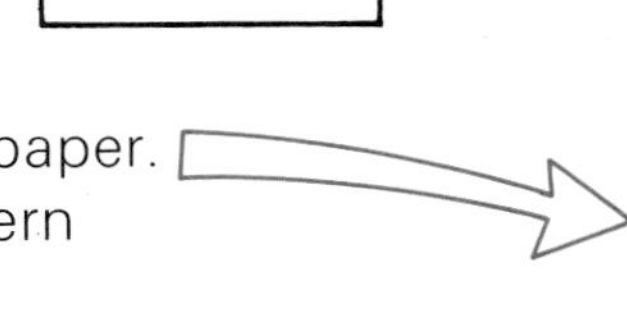

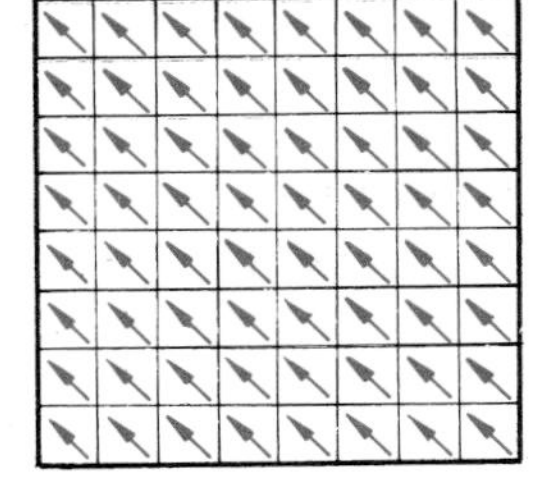

This is how she wraps the card.

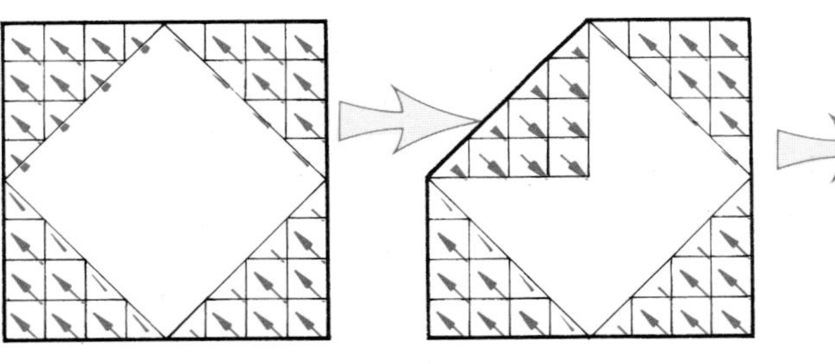

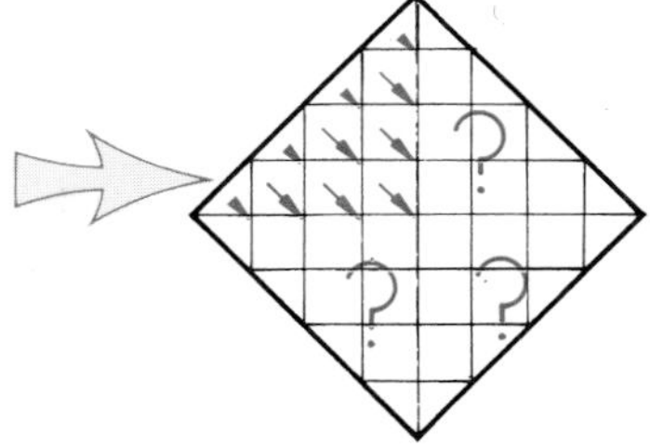

a) Copy the wrapped card onto 1 cm squared paper.

b) Complete the pattern.

3 Glenda wraps some more cards in tissue paper.

On squared paper, draw what the wrapped cards look like.

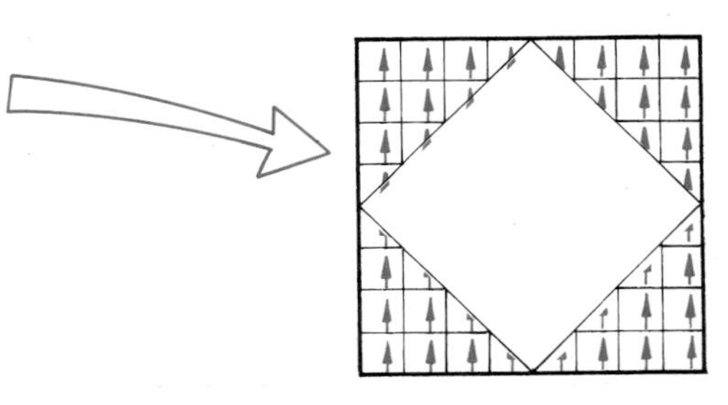

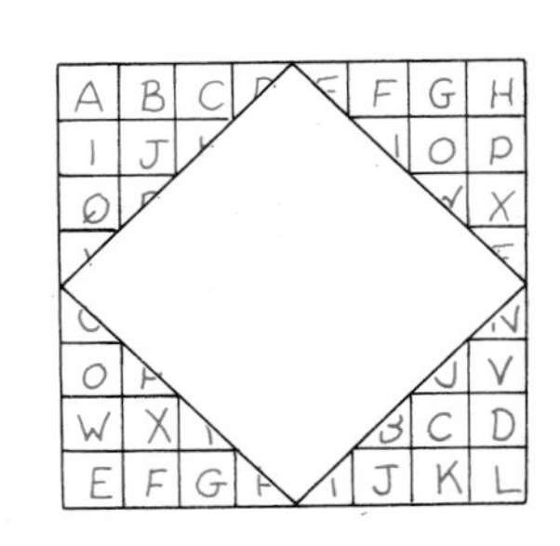

4 Glenda's school tie looks like this.

Sketch what the stripes look like on the back.

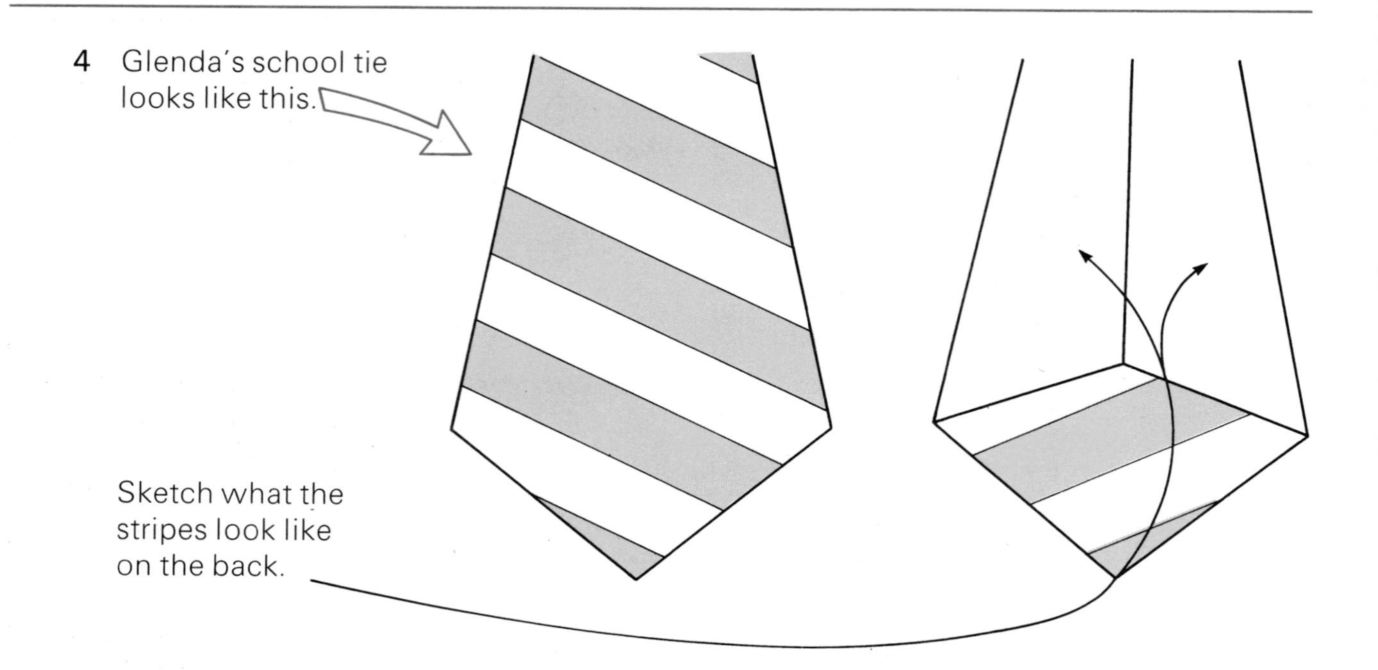

5 You need some squared paper.
Check your answer to question **4** like this.

a) Draw this.

Use the dots to help you.

b) Cut around the broken line.

c) Fold along lines **A** and **B**.

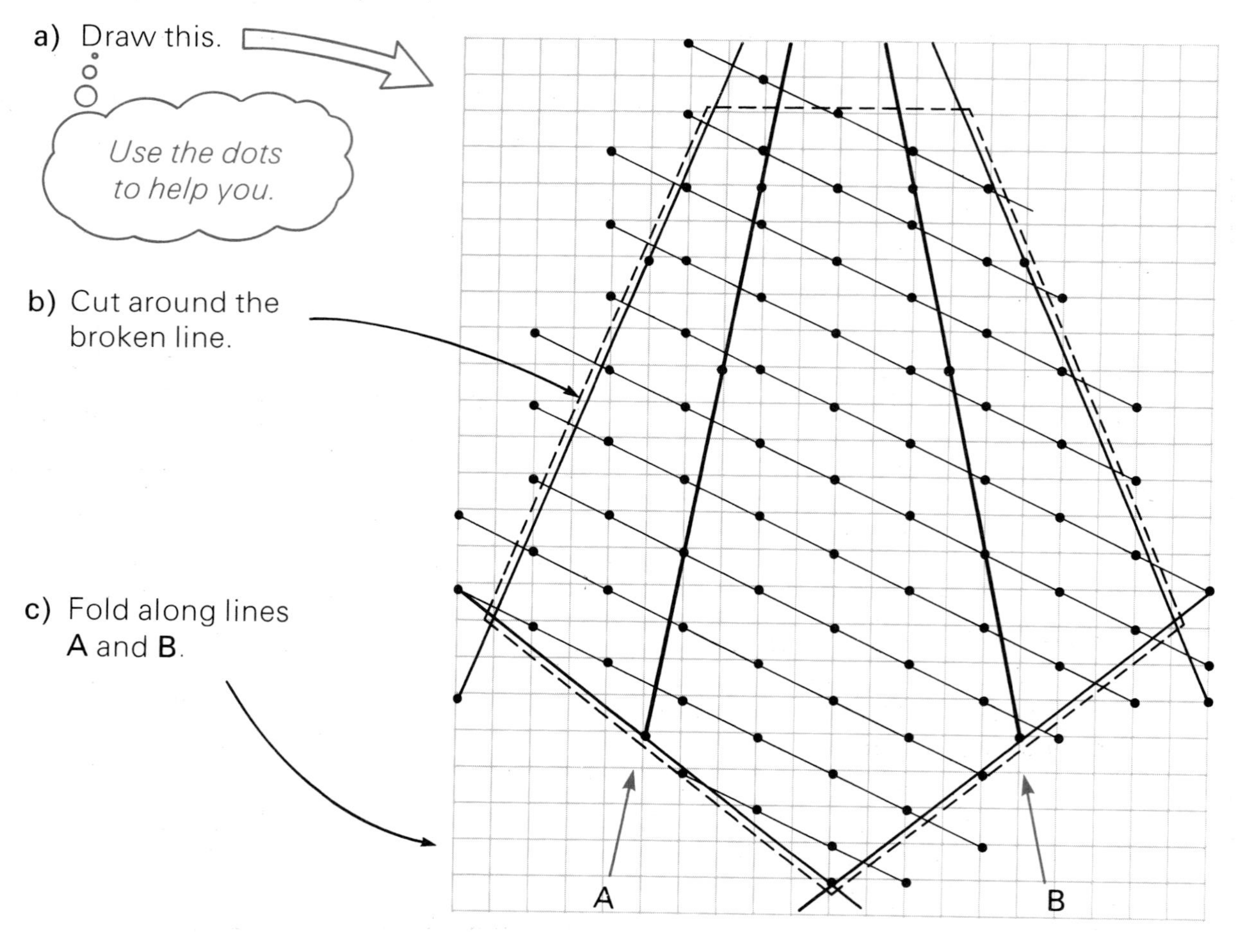

Folding

D

Activity

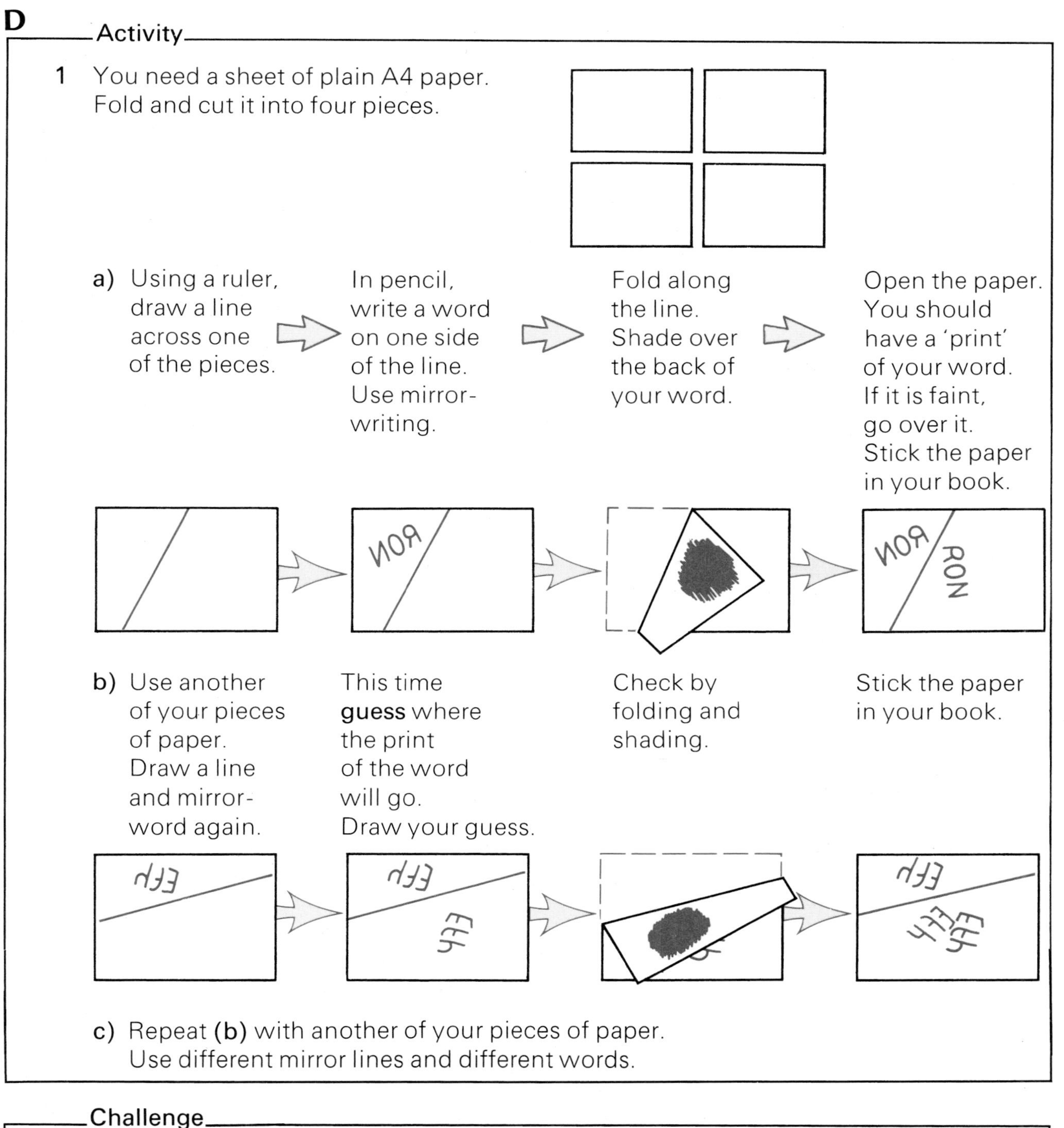

1 You need a sheet of plain A4 paper.
Fold and cut it into four pieces.

a) Using a ruler, draw a line across one of the pieces. → In pencil, write a word on one side of the line. Use mirror-writing. → Fold along the line. Shade over the back of your word. → Open the paper. You should have a 'print' of your word. If it is faint, go over it. Stick the paper in your book.

b) Use another of your pieces of paper. Draw a line and mirror-word again. → This time **guess** where the print of the word will go. Draw your guess. → Check by folding and shading. → Stick the paper in your book.

c) Repeat **(b)** with another of your pieces of paper.
Use different mirror lines and different words.

Challenge

2 Use your last piece of paper.
Draw two dots at least 8 cm apart anywhere on the paper.
Imagine one dot is the print of the other.
Find the fold line.
Draw it on the piece of paper.
Stick the paper in your book.

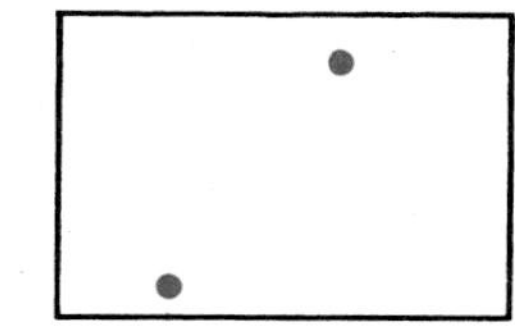

3 In these drawings the thick black line is a fold line.

Is the arrowed figure a print of the other figure?
Write **yes** or **no** for each drawing.
(Use a mirror to help you.)

a)

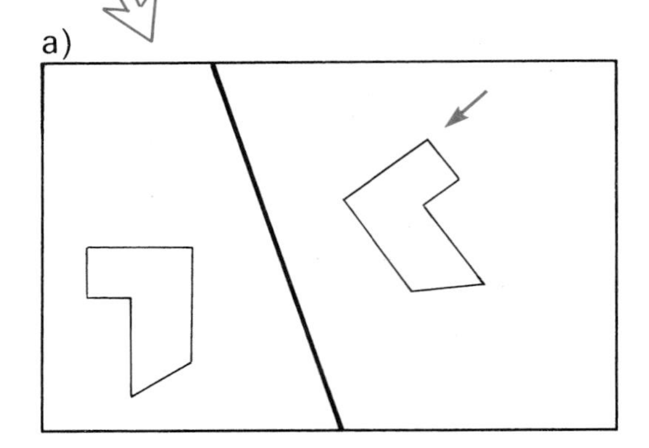

b)

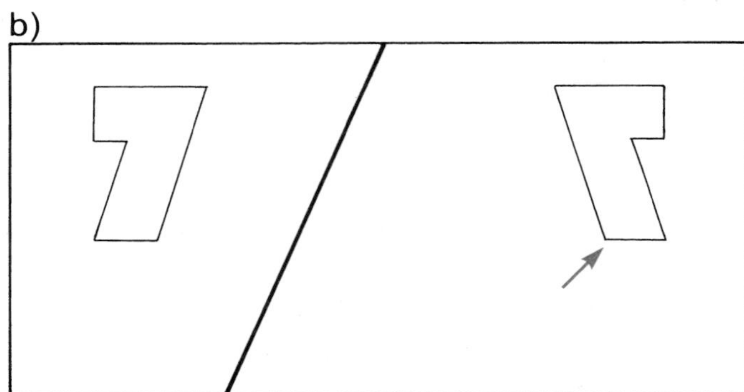

c)

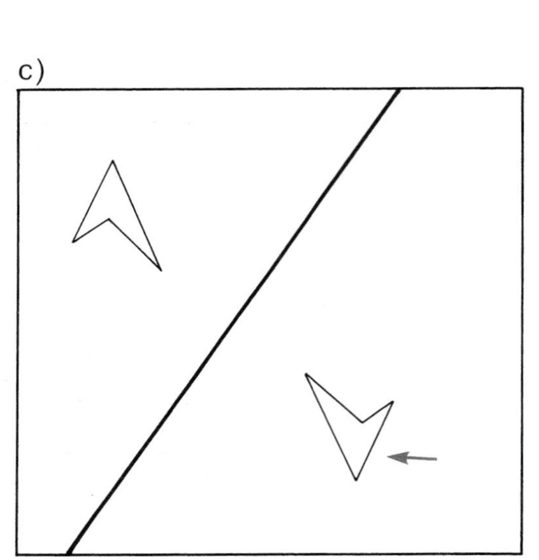

d)

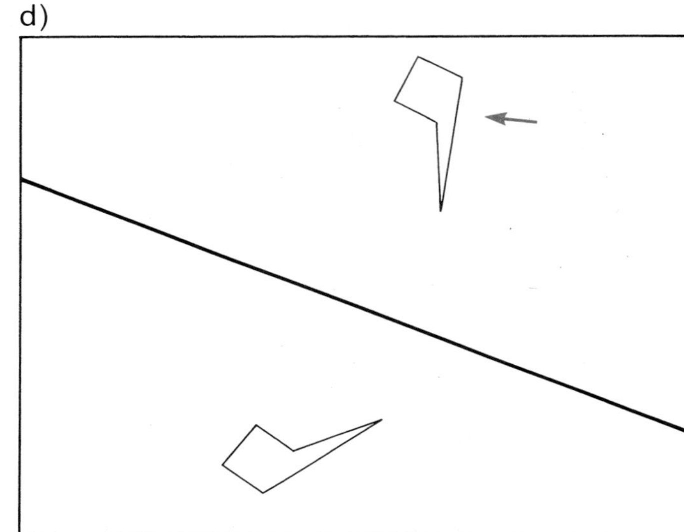

4 a) Copy this drawing onto 1 cm squared paper.

b) Imagine folding along the slanting line.
Draw the print of the '1'.
(Use a mirror to help you.)

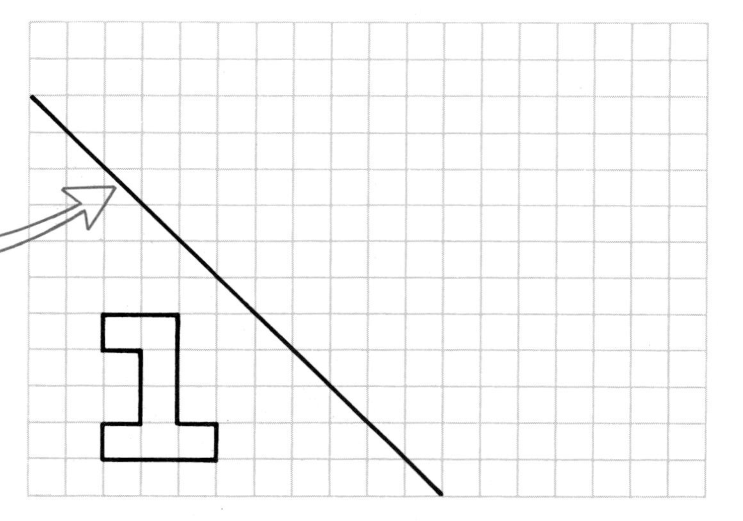

16

Reflection

E Take note

In this drawing the upper figure folds onto the lower figure.

We call this line the **mirror line**.

We say the lower figure is the **reflection** of the upper figure.

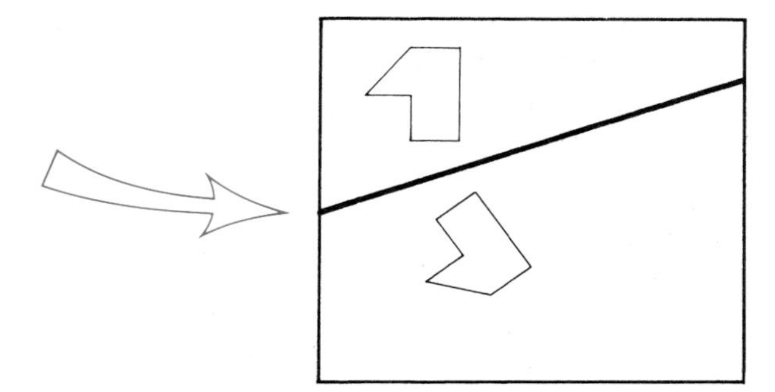

1 Look at these drawings.
The line m is the mirror line.
Is the arrowed figure the reflection of the other figure?
Write **yes** or **no** for each drawing.
(Use a mirror to help you.)

a)

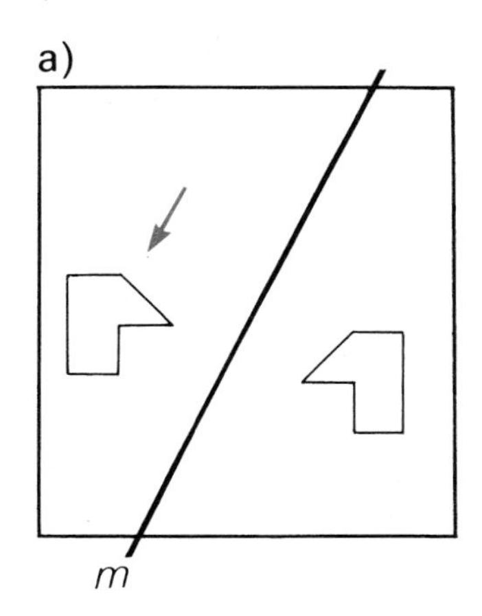

b)

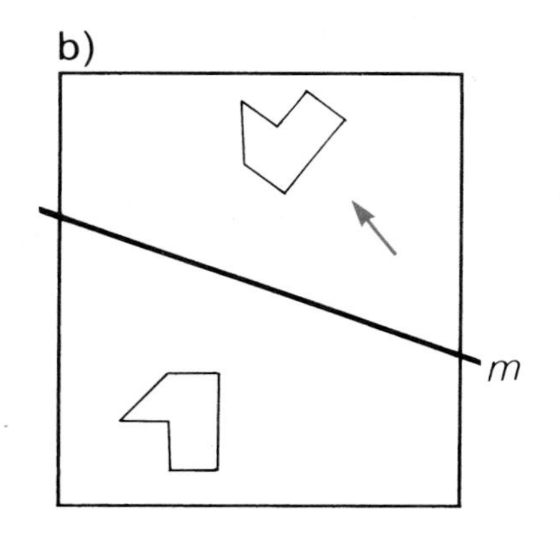

c)

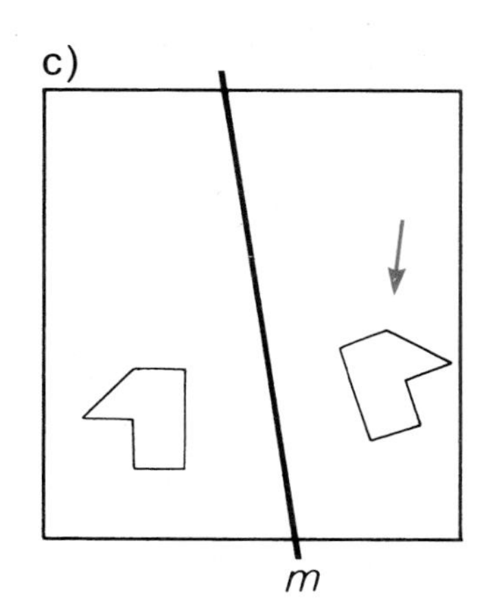

2 Copy these drawings onto squared paper.
Draw the reflection of each flag.

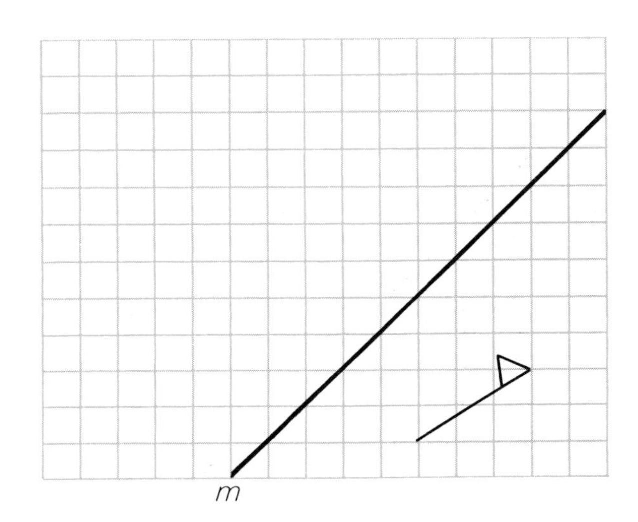

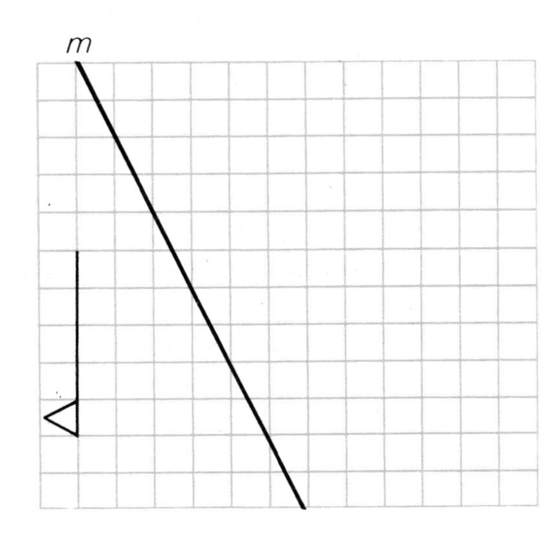

16

3 The point X is to be reflected in line m.
Here are some students' answers. 

a) Mark the answers out of 10.

b) Explain your method of marking.

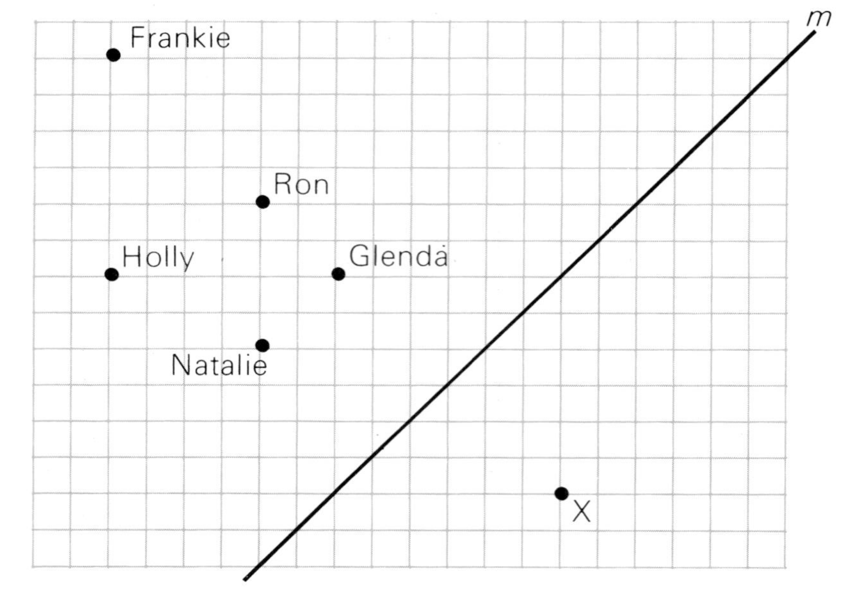

4 a) The two triangles are to be reflected in line m.
Write down the new coordinates of points A, B, C, D, E and F.

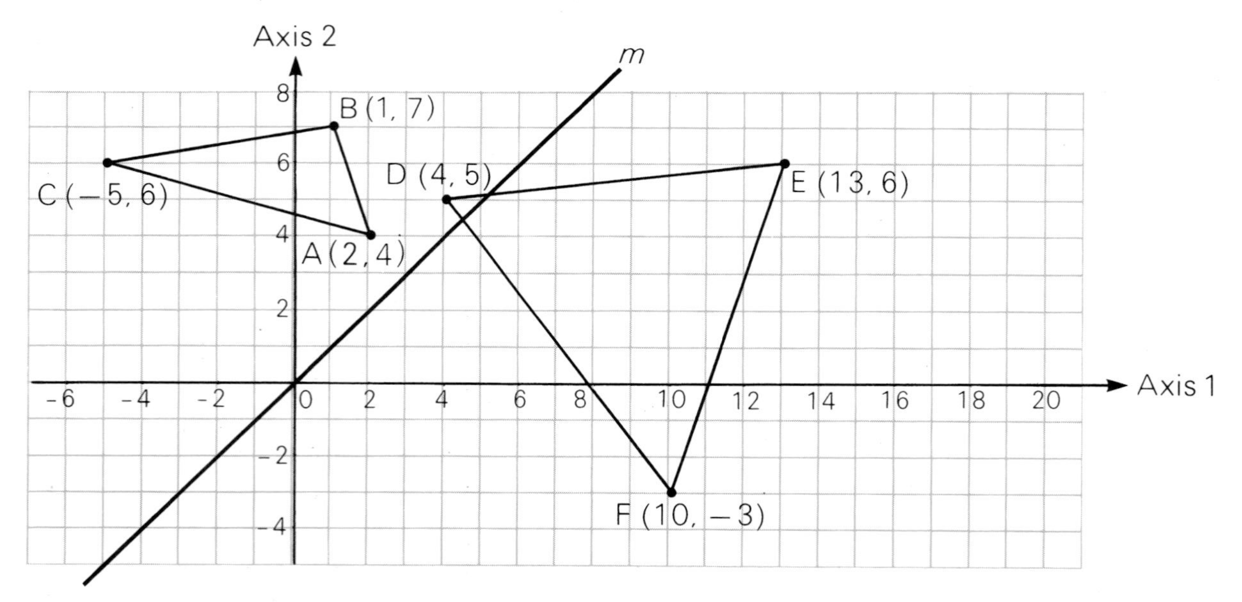

b) These points are also to be reflected in line m.

G (2, 6) H (1, 20) I (−1, 4) J (7, 7)  K (39, 82)

Write down their new coordinates.

5 These points are to be reflected in line m.

A (8, 2) B (8, 7) C(12, 1) D(18, 3) E (1, 4) F (−5, 8)

Write down their new coordinates.

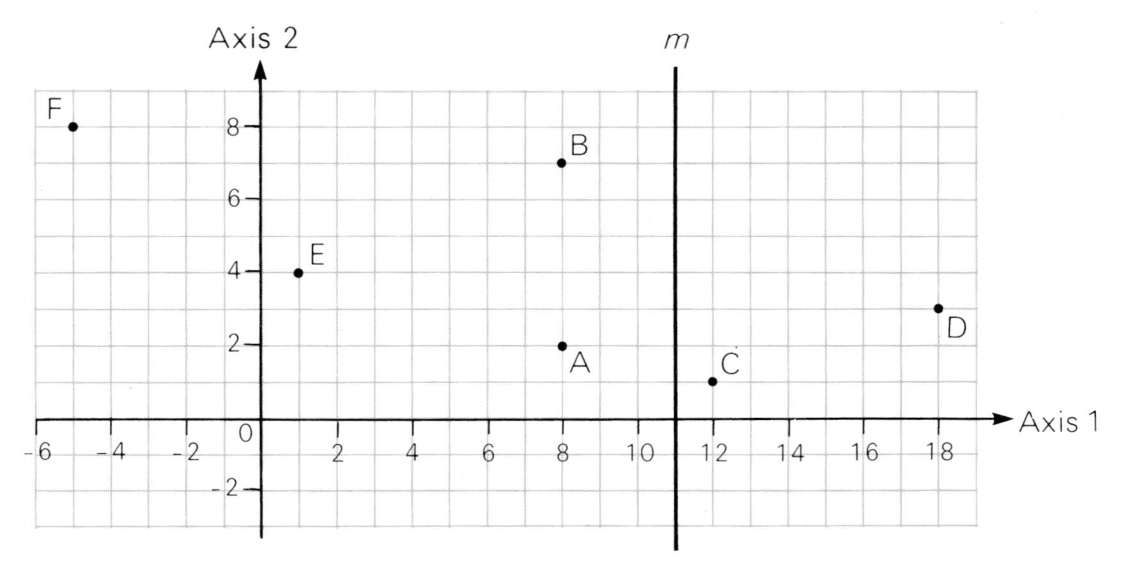

6 The three triangles are to be reflected in line m.
Then, the answers are to be reflected in line n.

Copy the diagram on squared paper.
Draw the **final** position of each triangle.
What do you notice?

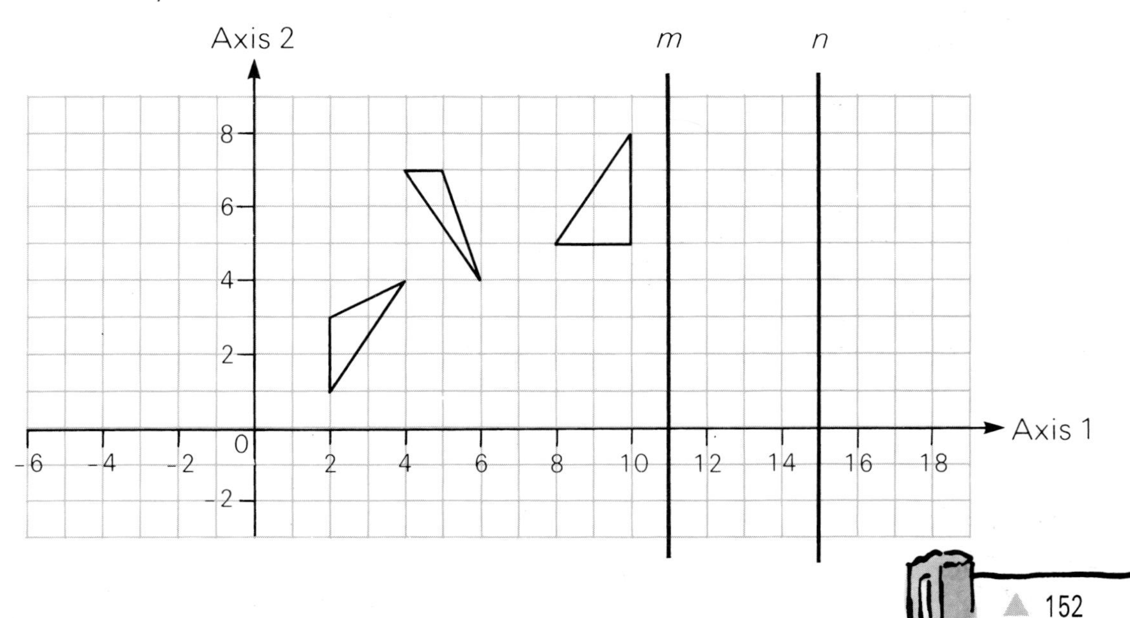

7 Repeat question **6**, but reflect in n first, then in m.
What do you notice this time?

▲ 152
● Next chapter

Reflections of solids

F 1 This object is made from five 'alphabet cubes', glued together.

Here it is again, in front of two vertical mirrors. Copy the object and its two reflections. (Use isometric paper.) Complete the lettering.

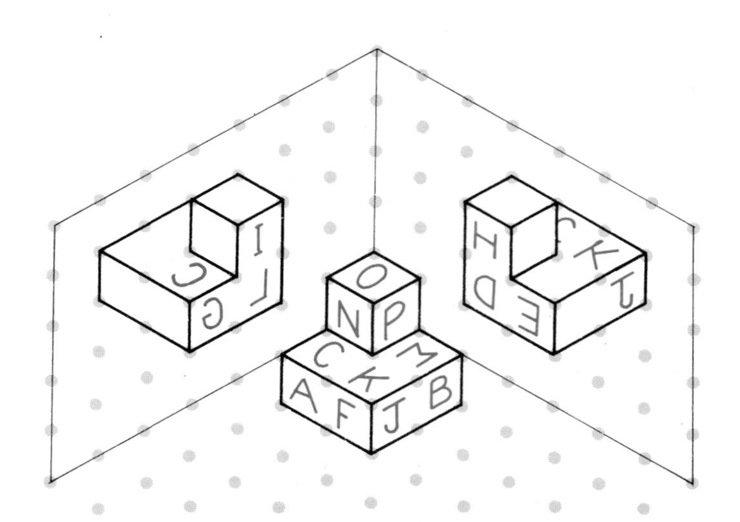

2 These drawings show the same object. Again, it is in front of two vertical mirrors. Copy and complete each drawing.

a) b)

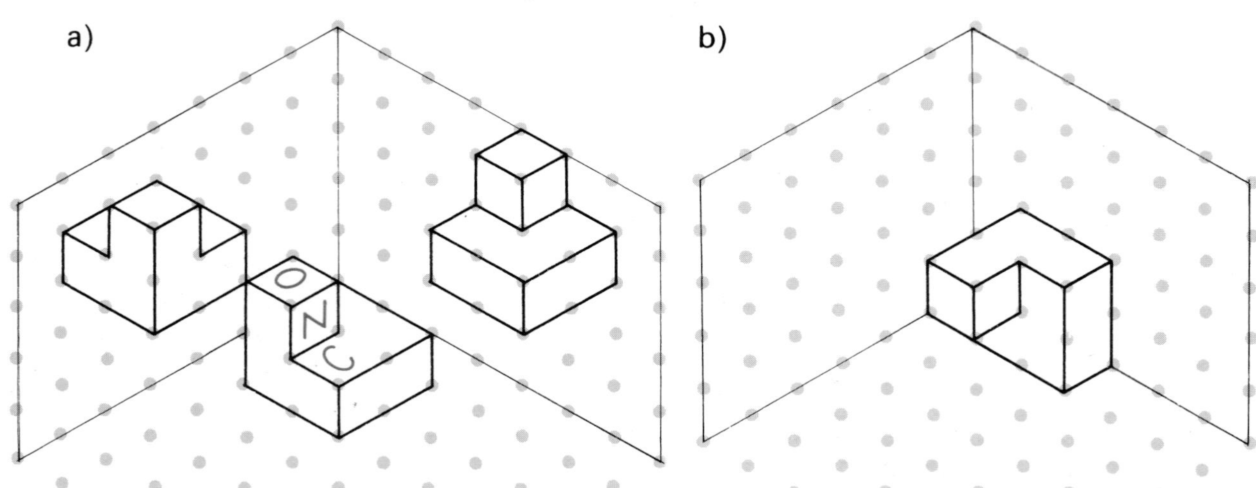

16

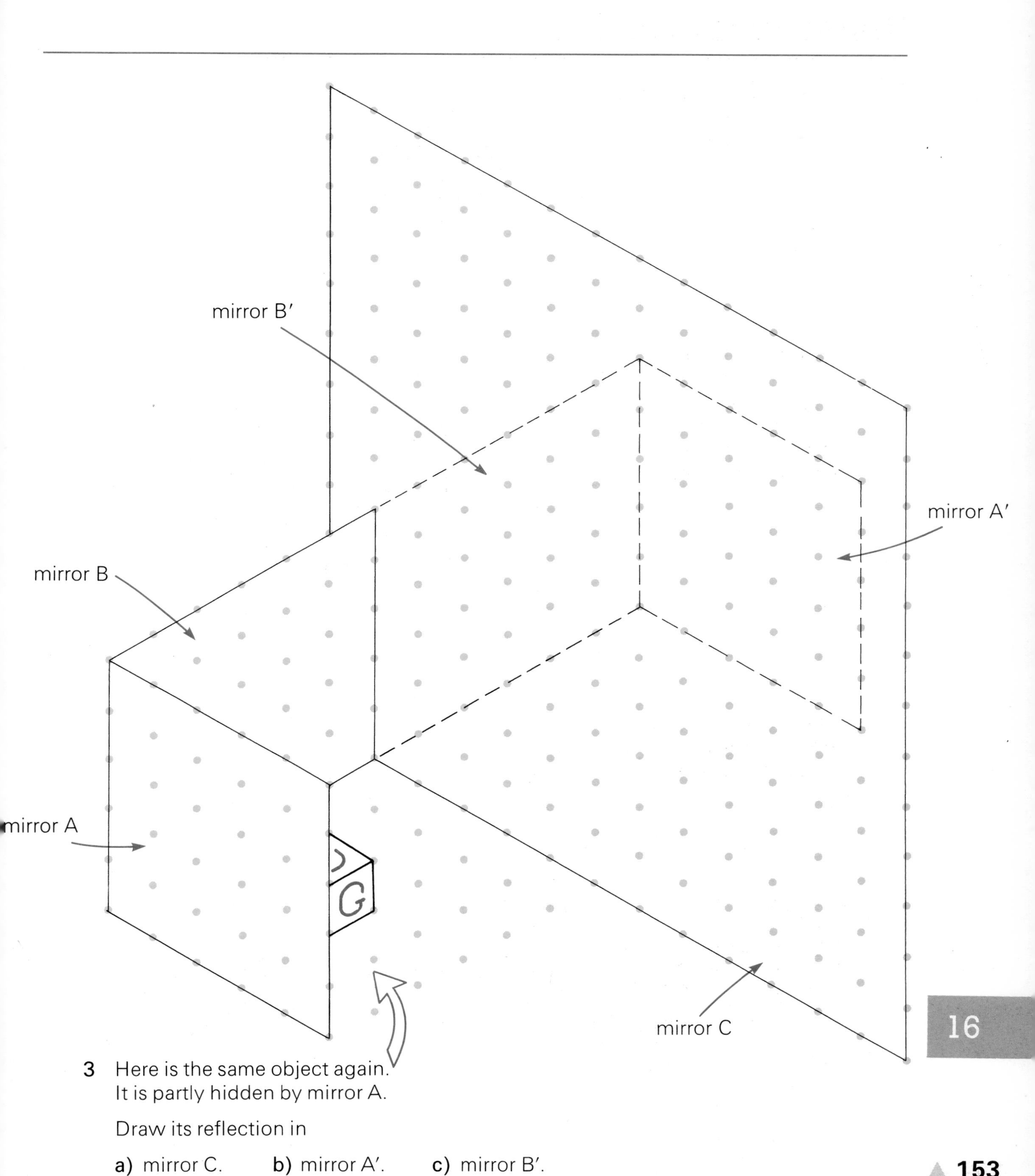

3 Here is the same object again.
It is partly hidden by mirror A.

Draw its reflection in

a) mirror C. b) mirror A′. c) mirror B′.

How many bounces?

G Exploration

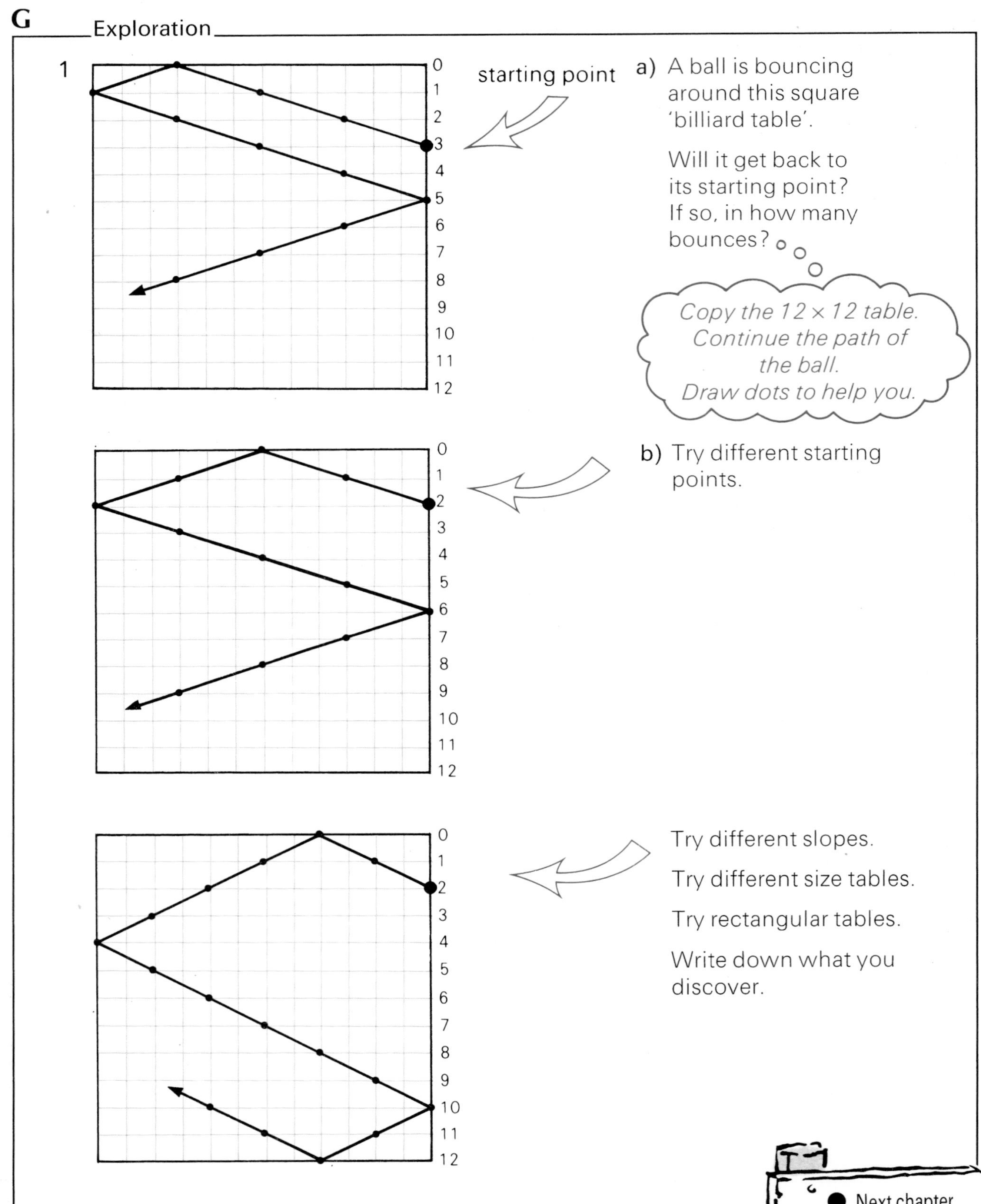

a) A ball is bouncing around this square 'billiard table'.

Will it get back to its starting point? If so, in how many bounces?

Copy the 12 × 12 table. Continue the path of the ball. Draw dots to help you.

b) Try different starting points.

Try different slopes.

Try different size tables.

Try rectangular tables.

Write down what you discover.

17 Similar shapes

A

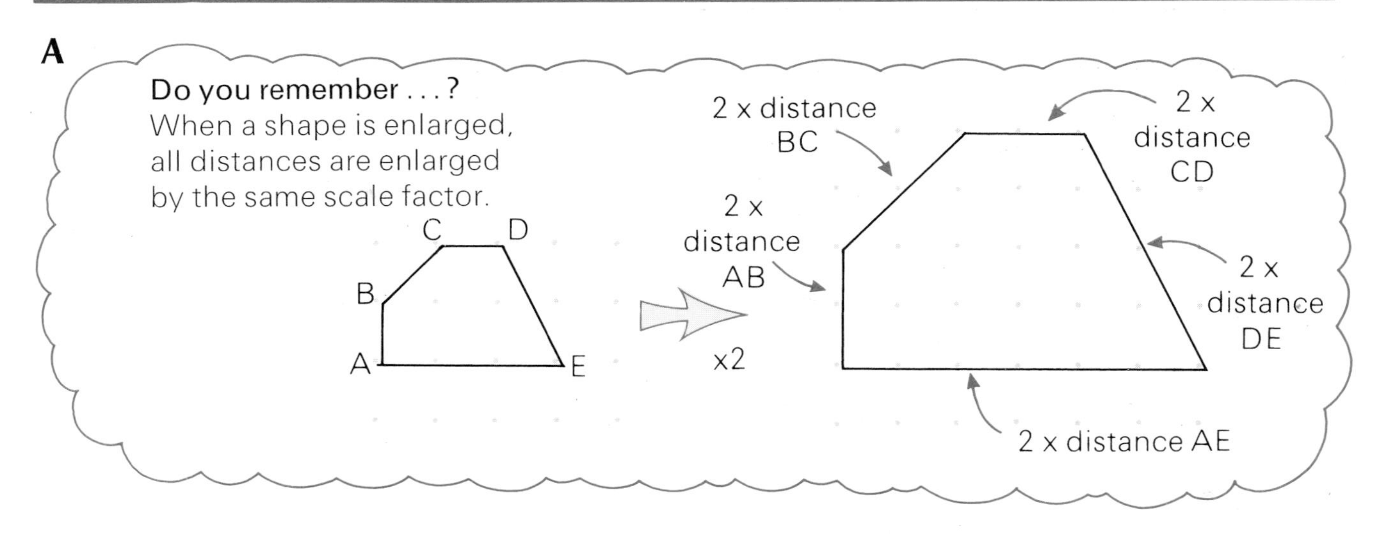

1 What happens to the angles of a shape when it is enlarged;
do they
- increase in size
- decrease in size
- stay the same size

or
- do some increase and some decrease?

Activity

2 You need 1 cm squared dotted paper.
The black lines are part of an enlarged shape.
The red lines are part of the original shape.
Copy and complete **both** shapes.

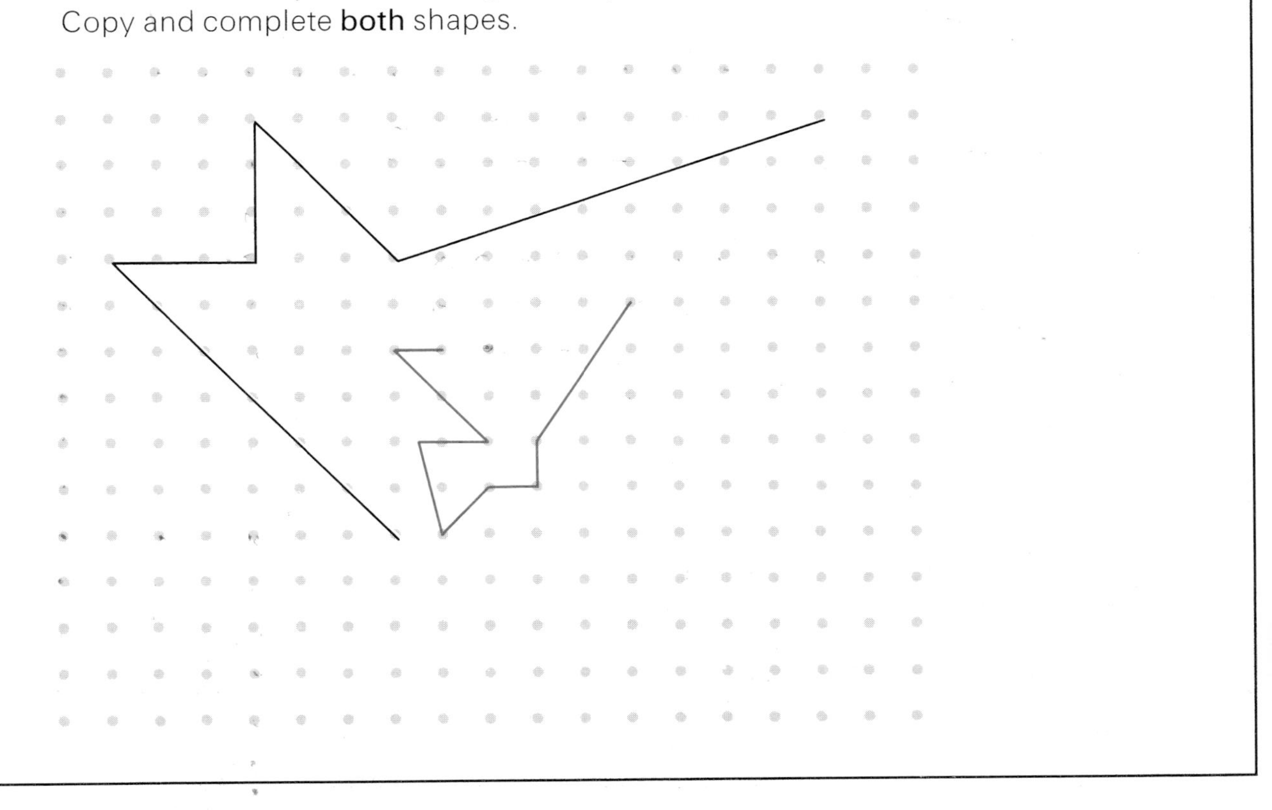

Take note

When a shape is enlarged, its angles stay the same size.

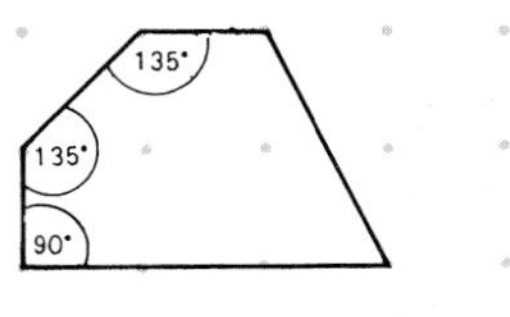

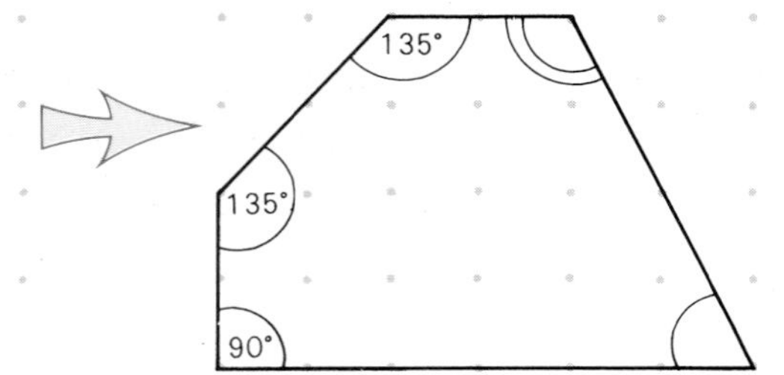

3 a) Copy and continue this shape.

Use 1 cm squared dotted paper.

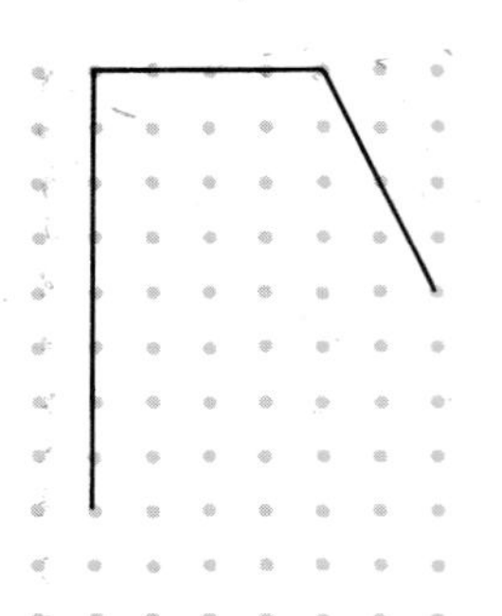

Make it have the same number of sides as this shape. Make the angles the same size as the corresponding angles of the smaller shape. Draw your shape so that it is **not** an enlargement of the smaller shape.

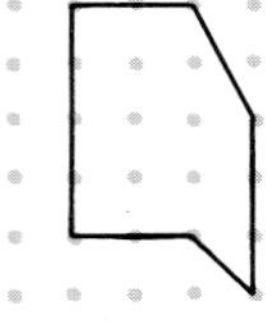

b) Explain why your shape is not an enlargement of the smaller shape.

Take note

same size angles, but not an enlargement

Two shapes with the same size angles are not always connected by an enlargement.

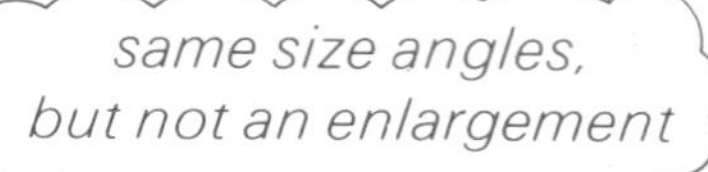

Shape **B** is an enlargement of shape **A**. We say that **A** and **B** are similar. Shape **D** is not an enlargement of shape **C**. Shapes **C** and **D** are **not** similar.

even though they have the same size angles

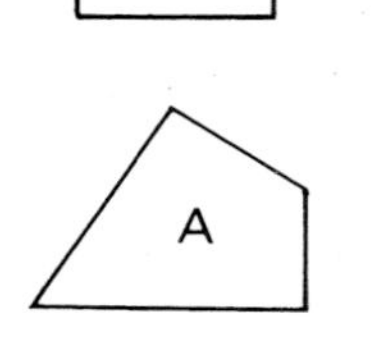

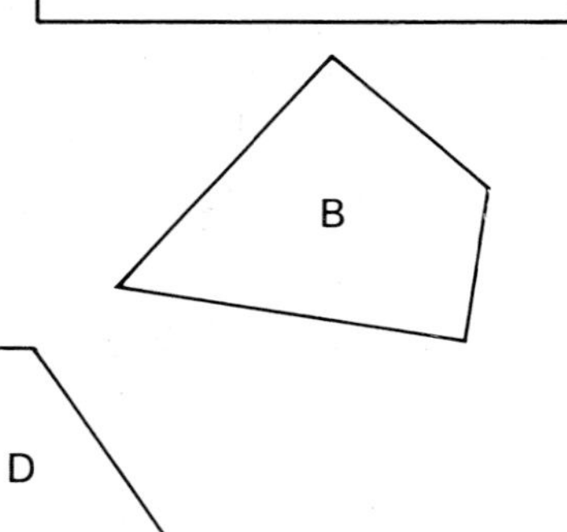

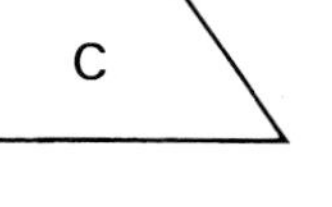

D

4 Start with this shape.

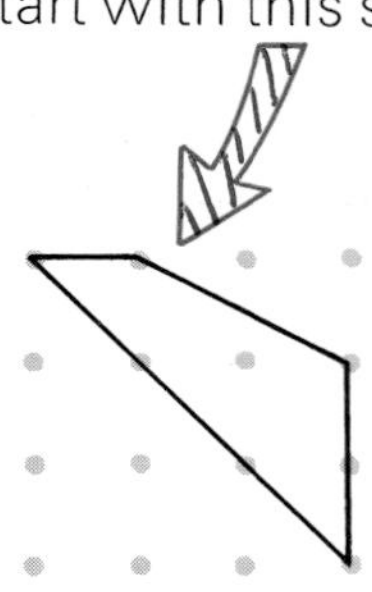

a) Draw a set of at least three similar shapes.

b) Sketch a set of at least three shapes which are **not** similar, but where the corresponding angles are the same size.

c) How many different sets could be drawn in **(b)**? Sketch three members of each set.

Similarity, lengths and angles

B

Exploration

1 You need 1 cm squared dotted paper.
These two pentagons have corresponding angles of the same size.

a) Explain why they are not similar.

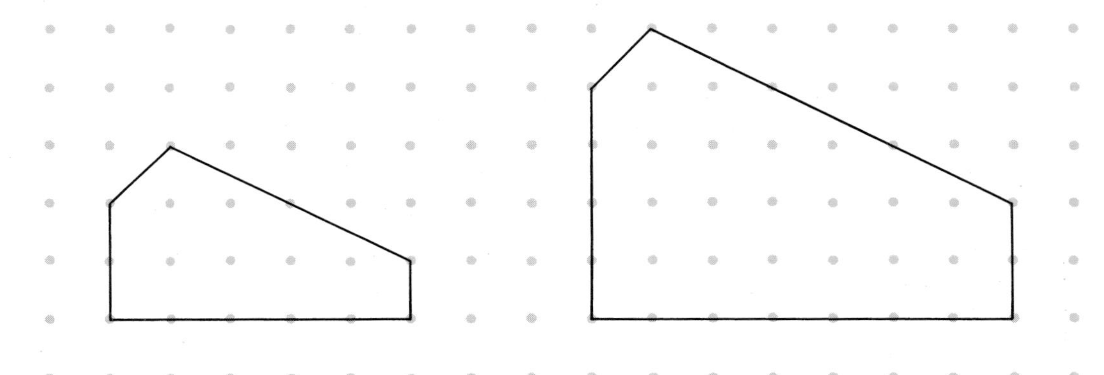

b) Draw two hexagons (*six sides*)
which have corresponding angles of the same size . . .
but which are not similar.

c) Repeat **(b)** for two rectangles.

d) Investigate some triangles.
Can you find two triangles which have corresponding angles of the same size but which are not similar?
If so, draw them.

e) What about two squares?

f) . . . two kites?

g) . . . two trapeziums?

h) . . . two parallelograms?

i) . . . two rhombuses?

Take note

Triangles which have the same size angles are similar.
One is an enlargement of the other.
We **cannot** draw two triangles with the same size angles which are not similar!

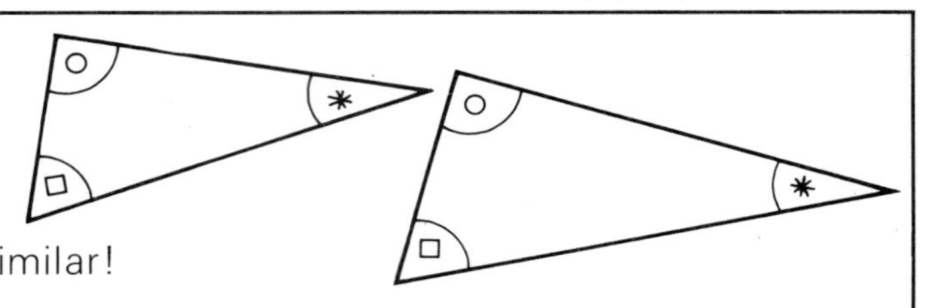

In your head

2 Do not draw anything.
Look and think.
Triangle DEF doesn't quite fit . . . sorry!
Are triangles ABC and DEF similar?
Explain how you know.

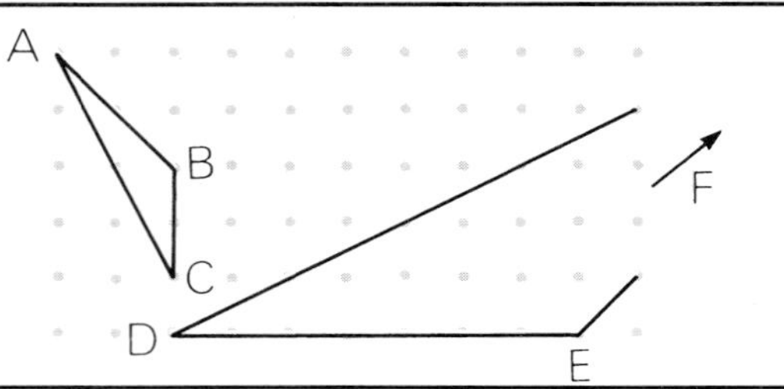

3 Which of these triangles are similar to the red triangle?

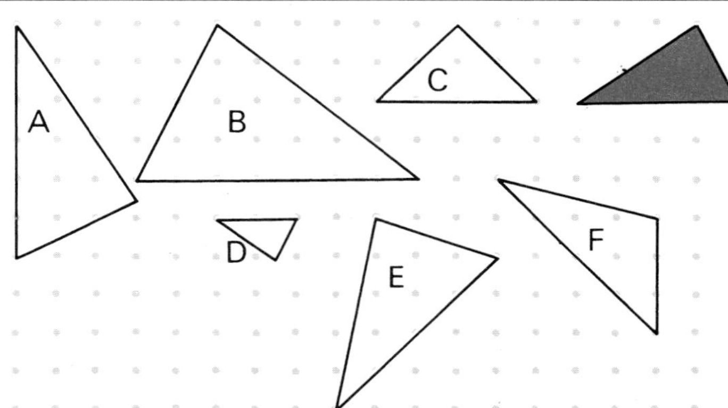

Activity

4 You need squared dotted paper.
Draw a triangle similar to this one, with this line as its base.

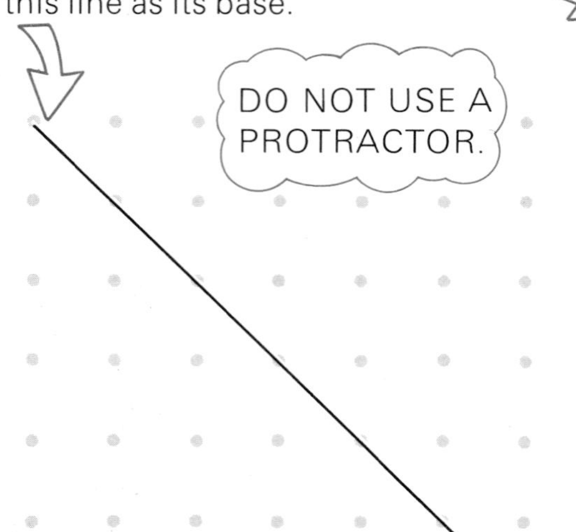

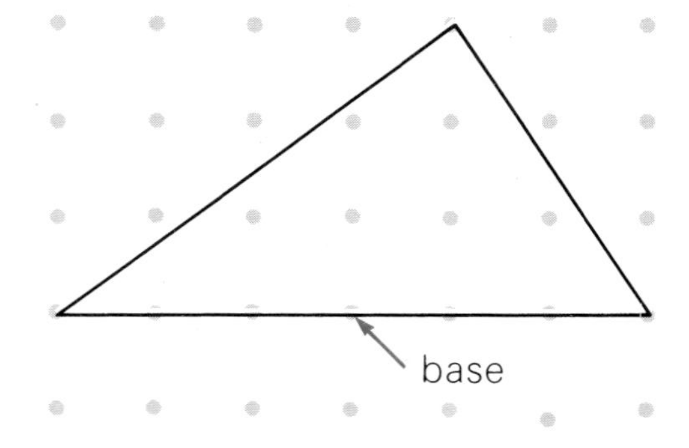

Challenge

5 You want to draw a triangle A′B′C′ which is similar to triangle ABC.

If this dot is A′, and if B′ and C′ are at dots on the grid, which dot(s) could be C′?

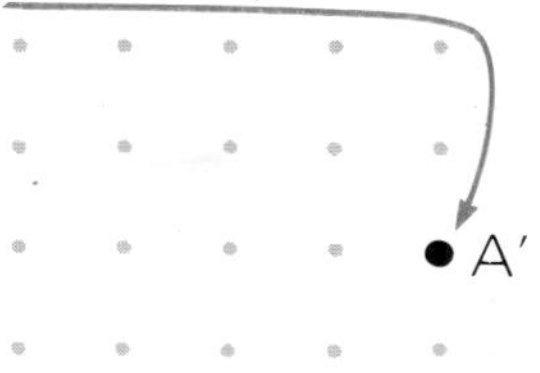

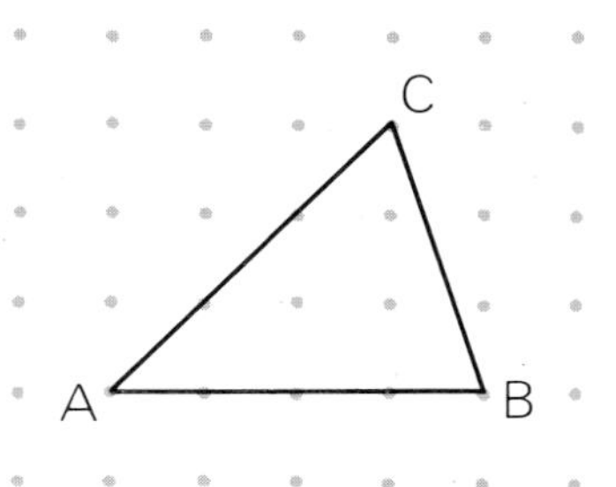

6 Which of these triangles are similar to the red triangle?

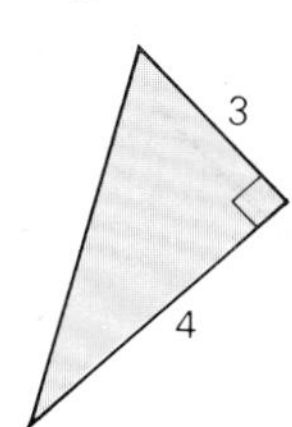

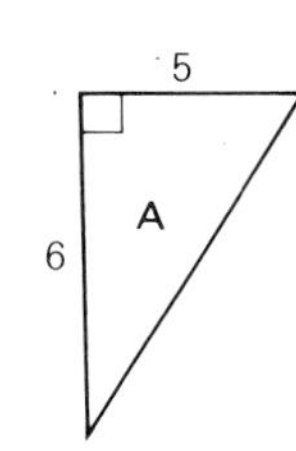

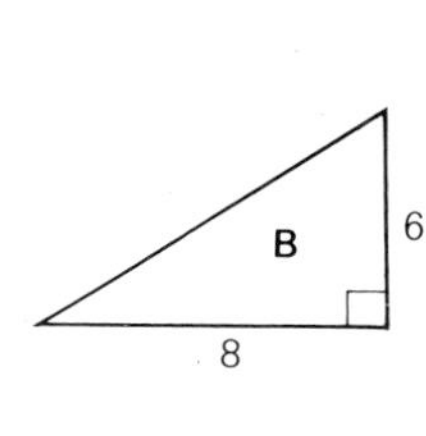

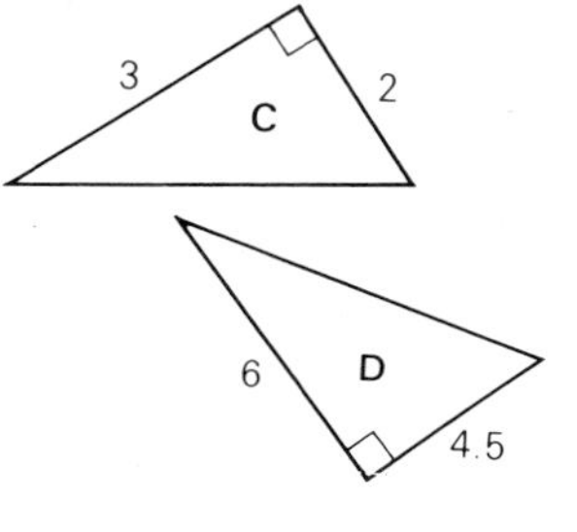

For those which are not, explain why not.

7 a) How many similar triangles can you see in each of these diagrams?
For each diagram, name the triangles you find.

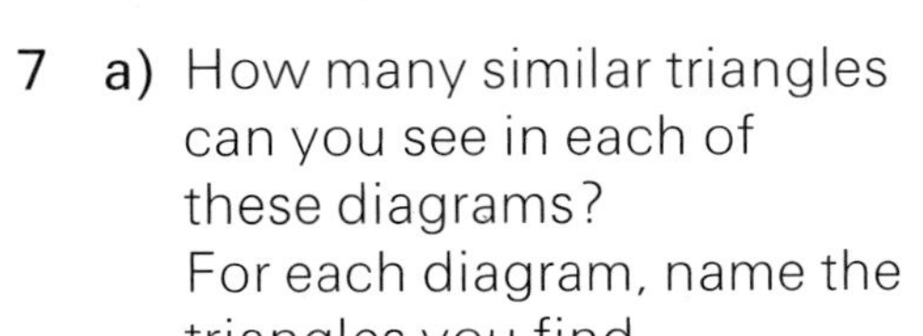

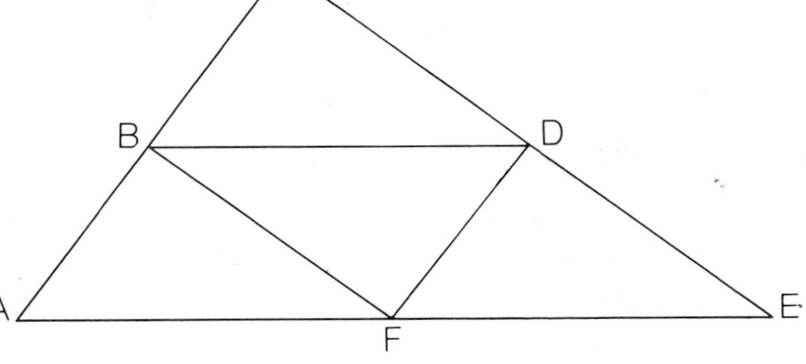

b) In X, BD = 5 cm
BF = 4 cm
FD = 3 cm.
Find as many of the other distances as possible.

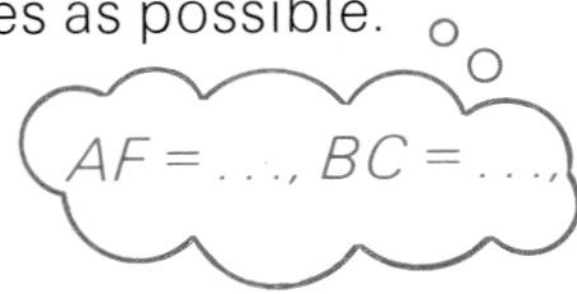

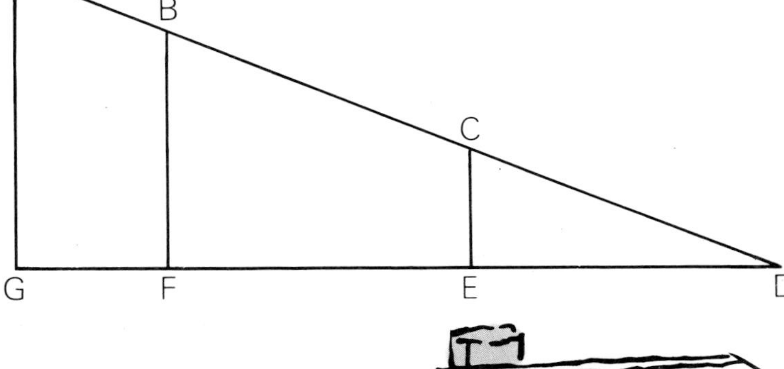

c) In Y, CE = 2 cm
AG = 5 cm
BF = 4 cm
and ED = 5 cm.
Calculate
(i) GD. (ii) FE.

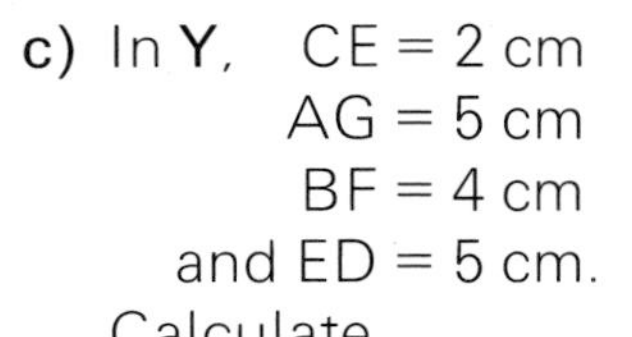

▲ 160
● Next chapter

Similarity, area and capacity

C

Challenges

1 You need 1 cm squared dotted paper.

a) Copy and complete this set of similar shapes.

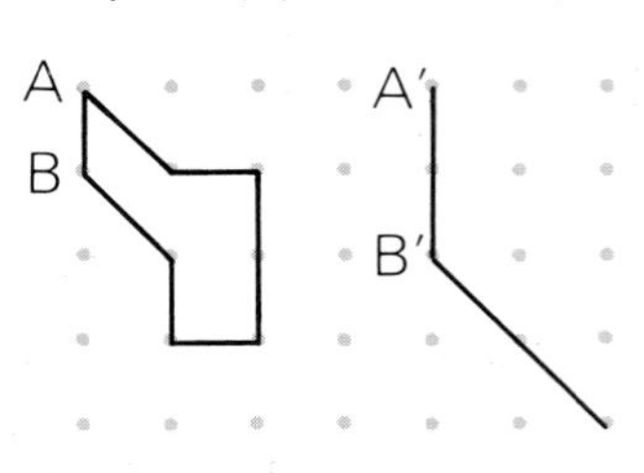

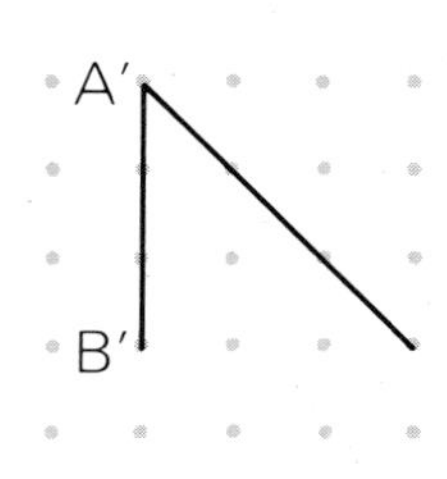

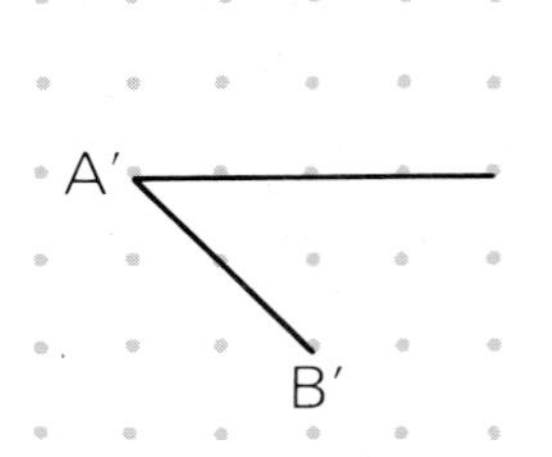

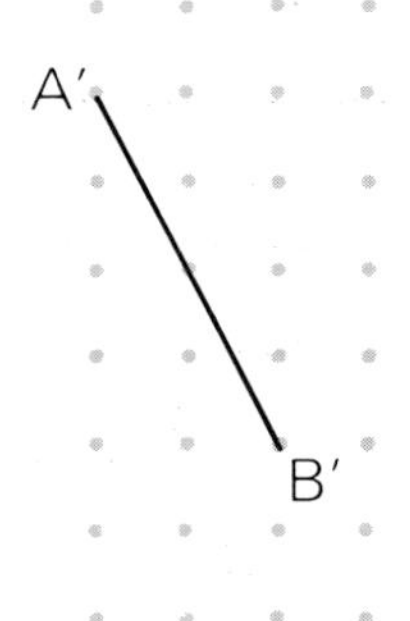

b) The area of the original shape is 3 cm².
How many times bigger is the area of each of the similar shapes?
How are your answers related to the distance A′B′?

2 a) Two of these boxes are similar.
Which two?

b) Investigate the capacities of similar boxes.
How are the capacities related?

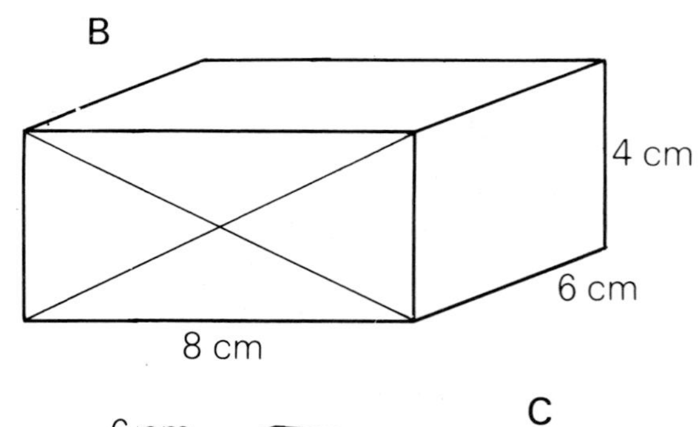

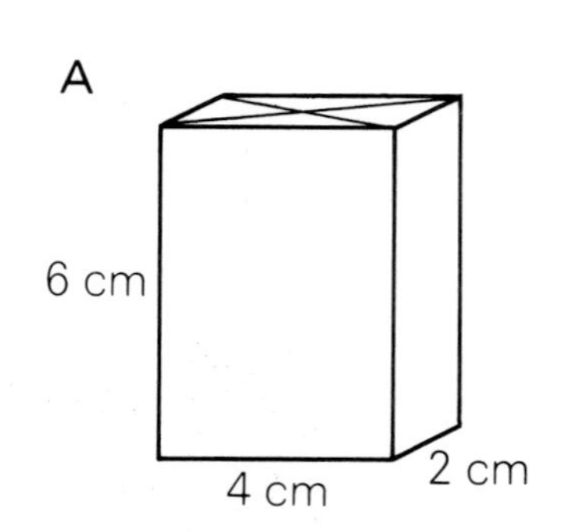

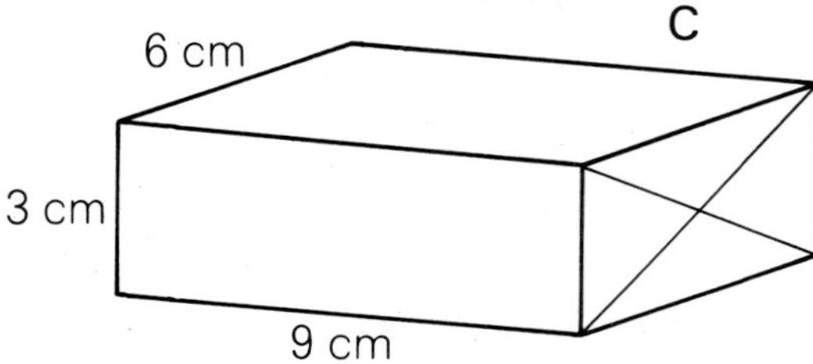

Looking for similarity

D

Assignment

1 Collect some pairs of containers.

for example, different milk bottles, different breakfast cereal packets, different soup tins

Collect an example of a pair which are similar and an example of a pair which are not similar. For the two containers which are not similar explain how you could change one so that they are similar, without changing its capacity.

Exploration

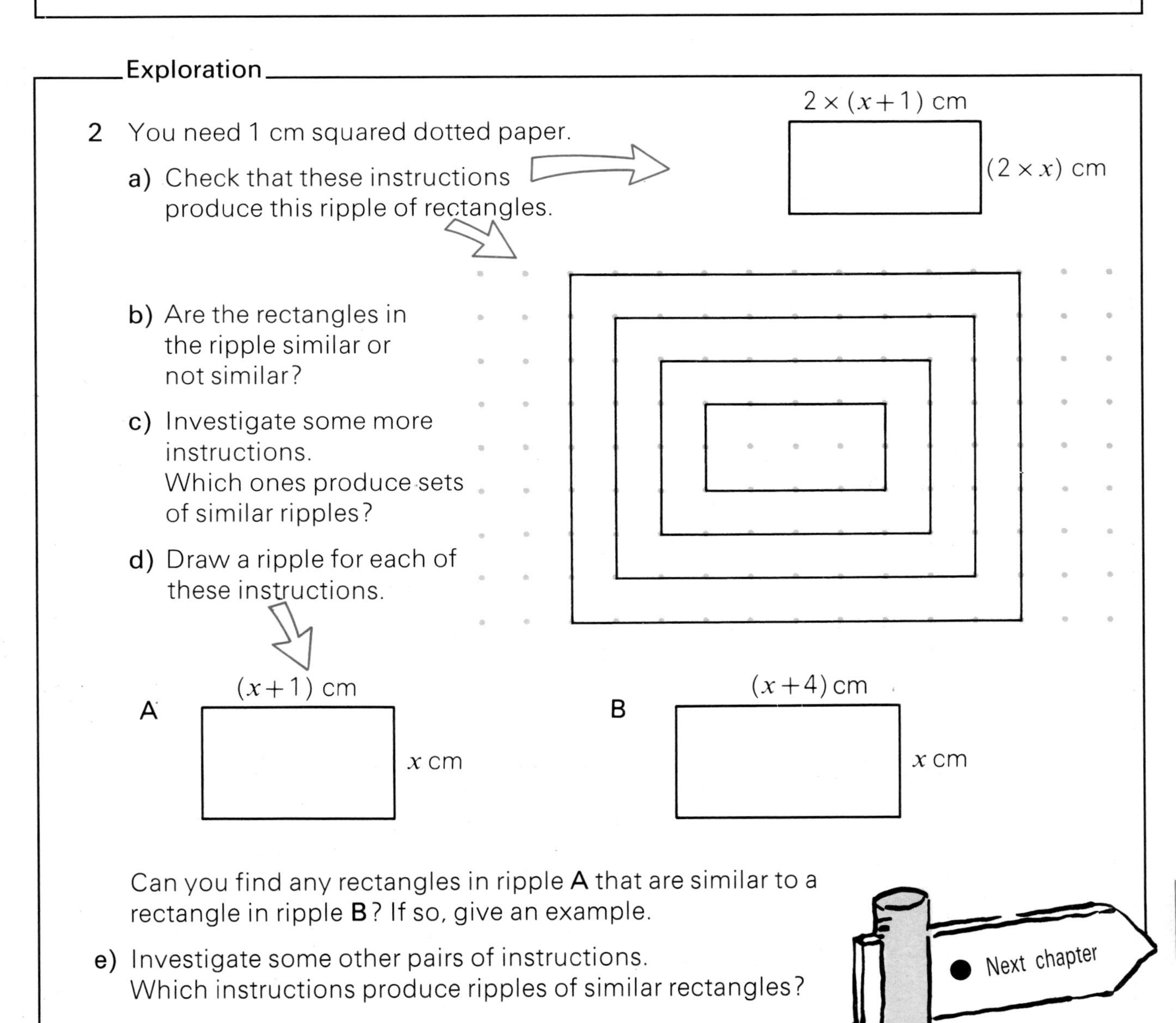

2 You need 1 cm squared dotted paper.

a) Check that these instructions produce this ripple of rectangles.

b) Are the rectangles in the ripple similar or not similar?

c) Investigate some more instructions. Which ones produce sets of similar ripples?

d) Draw a ripple for each of these instructions.

Can you find any rectangles in ripple **A** that are similar to a rectangle in ripple **B**? If so, give an example.

e) Investigate some other pairs of instructions. Which instructions produce ripples of similar rectangles?

18 Dealing with numbers

A In your head

1 **Do not use a calculator.**
Write down only the results.

a)

The flour is tipped into five equal piles.
(i) How many kilograms is this in each pile?
(ii) Copy and complete:

$1\text{ kg} \div 5 = \square.\square\text{ kg}$
$5 \times \square.\square\text{ kg} = 1\text{ kg}$

b)

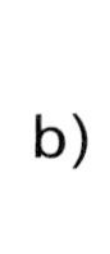

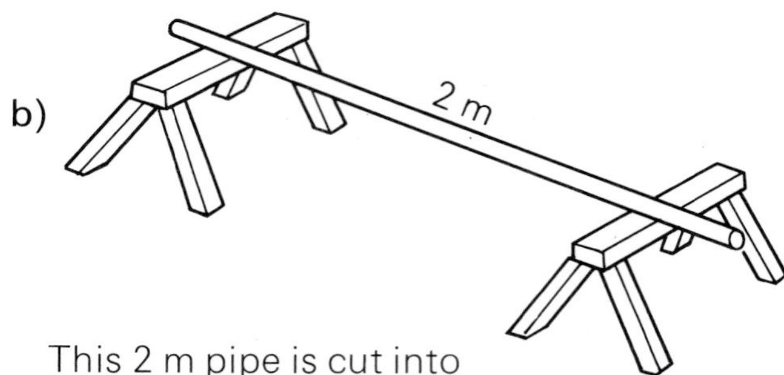

This 2 m pipe is cut into lengths of 0.4 m.
(i) How many pieces does this give?
(ii) Copy and complete:

$2 \div 0.4 = \square$
$\square \times 0.4 = 2$

c) Each section of this tower is 0.3 m tall.
(i) What is its height, in metres?
(ii) Copy and complete:

$0.3\text{ m} \times 8 = \square.\square\text{ m}$
$\square.\square \div 8 = 0.3$
$\square.\square \div 0.3 = 8$

d) Altogether, these four identical cushions need 4.8 m² of material.
(i) How many square metres each is this?
(ii) Copy and complete:

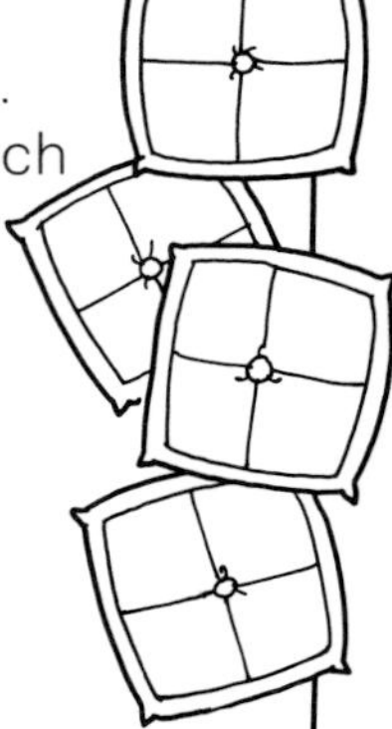

$4.8\text{ m}^2 \div 4 = \square.\square\text{ m}^2$
$4.8 \div \square.\square = 4$
$\square.\square\text{ m}^2 \times 4 = 4.8\text{ m}^2$

2 Check your results in question **1** with your calculator.

*Think about piles of flour … or stacking cubes … as in question **1**.*

3 **Do not use a calculator.**
Copy and complete each of these.

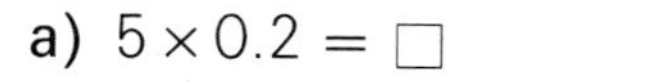

a) $5 \times 0.2 = \square$ b) $0.1 \times 6 = \square$ c) $8 \times 0.3 = \square$
d) $10 \times 0.4 = \square$ e) $4 \div 0.1 = \square$ f) $5 \div 0.2 = \square$
g) $8 \div 0.5 = \square$ h) $0.9 \div 3 = \square$ i) $1.2 \div 2 = \square$
j) $2.8 \div 7 = \square$ k) $3.6 \div 4 = \square$ l) $10.8 \div 0.9 = \square$

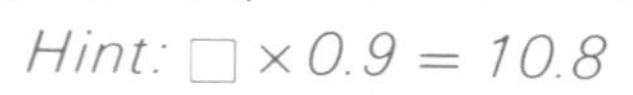

4 Check your results in question **3** with your calculator.

5 Do these in your head.
Write down only the results.

a) 5×0.2 b) 5×0.3 c) 10×0.1 d) 0.4×5
e) $4 \div 0.1$ f) $2 \div 0.2$ g) 4×0.5 h) $6 \div 0.3$
i) $0.8 \div 0.2$ j) $0.9 \div 0.1$ k) $4 \div 0.4$ l) $0.3 \div 0.6$

6 Check your results in question **5** with your calculator.

7 Do this in your head.

a) Multiply 0.6 by each whole number, in turn, up to 10.

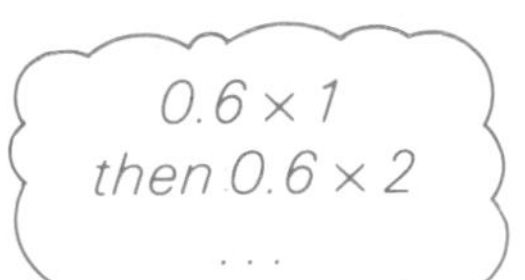

b) Divide these by 0.6:
(i) 6 (ii) 5.4 (iii) 4.8 (iv) 4.2
(v) 1.8 (vi) 1.2 (vii) 0.6 (viii) 0.3

8 **Do not use a calculator.**
These cling-wrapped wedges of cheese each weigh 0.8 kg.

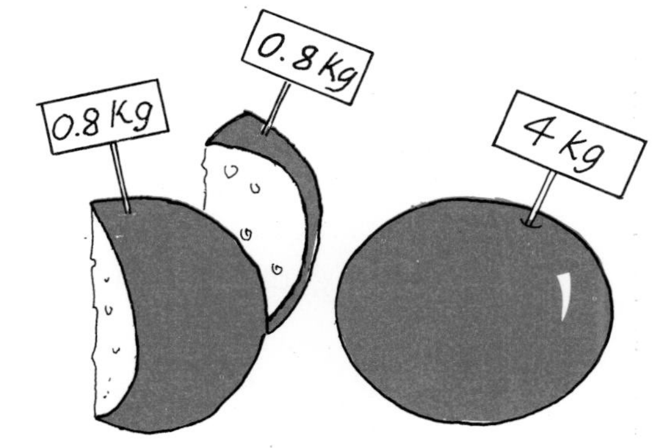

a) How many cheese wedges can be cut from the 4 kg ball of cheese?

b) Copy and complete:
(i) $4 \div 0.8 = \square$ (ii) $0.8 \times \square = 4$
(iii) $6 \div 0.4 = \square$ (iv) $\square \times 0.4 = 6$

9 Check your results in question **8** with your calculator.

Think it through

10 **Do not use a calculator.**

a) For each of these, write **smaller than 1** or **larger than 1**.

You do not have to make an exact calculation.

(i) $1 \div 0.7$ (ii) $2.41 \div 3$
(iii) 4×0.29 (iv) 0.09×8

b) Copy and complete each of these in two ways:
(1) to give a result greater than 1.
(2) to give a result smaller than 1.

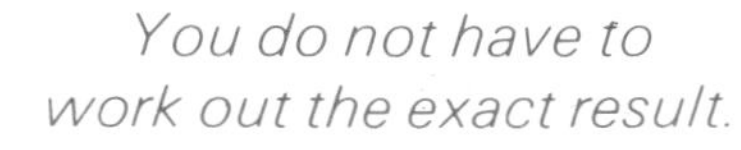

You do not have to work out the exact result.

(i) $0.85 \div \square$ (ii) $\square \times 0.32$

11 Use your calculator to check your results in question **10**.

Dividing

B 1 Meg is doing this division.

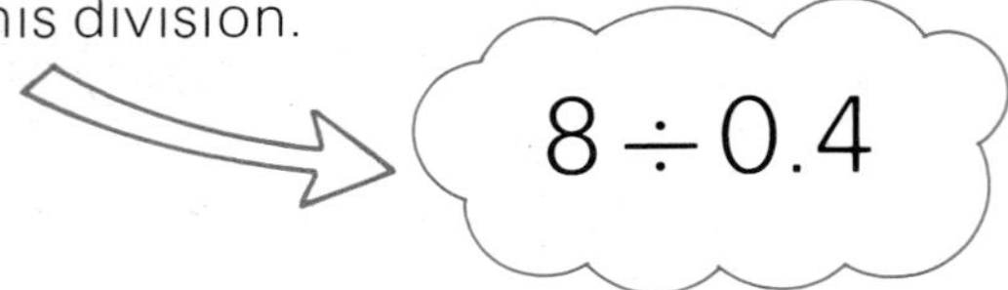

a) **Do not use a calculator.**

Should her result be larger than 16
or smaller than 16?

Why?

b) Here is her working.
Use your calculator to check
that her result is correct.

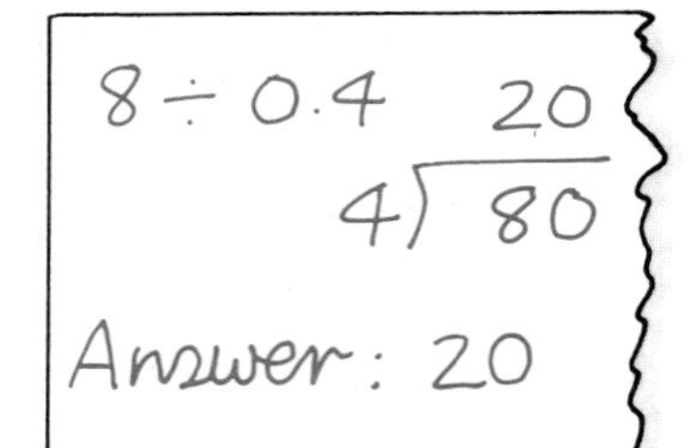

c) Why does 80÷4 give the same result as 8÷0.4?
Write one or two sentences to explain.

d) Use Meg's method to do this division:

$4.8 \div 0.6$

e) Check your result in **(d)** with your calculator.

2 a) **Do not use a calculator.**

Here are some of Meg's and Ron's divisions.

Copy and complete each one.

(i)
2·4 ÷ 0·4
= 24 ÷ □
= □

(ii)
0·56 ÷ 0.08
= 56 ÷ □
= □

(iii)
0·48 ÷ 0·8
?
?) 4·8
Answer: ?

(iv)
0.84 ÷ 1.2
?
12) ?
Answer: ?

b) Use your calculator to check your results in **(a)**.

3 The bolt from the crossbow travelled 9.4 m in 0.08 s.

a) **Do not use a calculator.**
Would you say the speed of the crossbow bolt is

A 

B 

or C 

?

b) Here is how Meg makes the estimate:

9 ÷ 0.1

90 ÷ 1

ABOUT 90

Check the result with your calculator.

c) Copy and complete Meg's accurate calculation:

?
8 ⟌ ?
Answer: ?

4 Use Meg's method to estimate

a) $2.34 \div 0.26$. b) $6.8 \div 0.03$.

Check your estimates with your calculator.

Think it through

5 **Do not use a calculator.**

a) Glenda needs to cut this copper pipe into 0.3 m lengths.
How many pieces will she get?

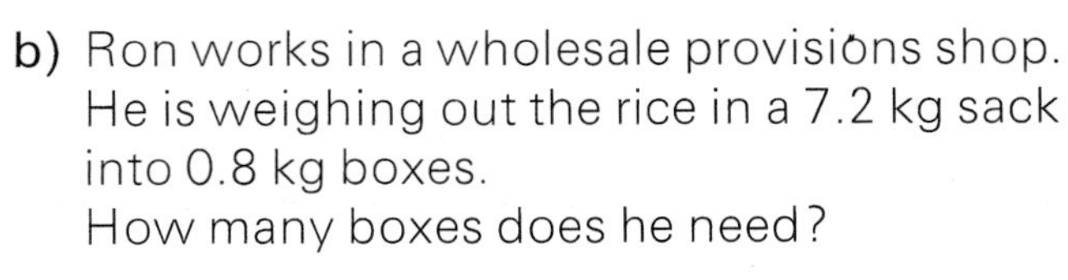

5.7 m

b) Ron works in a wholesale provisions shop.
He is weighing out the rice in a 7.2 kg sack into 0.8 kg boxes.
How many boxes does he need?

c) Each charm on this bracelet uses 0.04 kg of silver.
How many charms can be made from a 0.6 kg block?

6 Check your results in question **5** with your calculator.

7 a) **Do not use a calculator.**
Two of these divisions give the same result as division **A**.
Which two are they?

A $2.3 \div 0.7$ B $23 \div 70$ C $0.23 \div 7$ D $3.3 \div 1.7$

E $23 \div 7$ F $0.23 \div 0.7$ G $0.23 \div 0.07$

b) Check your results with your calculator.

8 a) **Do not use a calculator.**
Each of these divisions is supposed to give the same result as division **A**.
Copy and complete each one.

A $19.6 \div 1.4$ B $196 \div \square$ C $0.196 \div \square$ D $1.96 \div \square$

E $0.0196 \div \square$ F $\square \div 140$ G $\square \div 0.014$ H $\square \div 14$

b) Check your results with your calculator.

Take note

$2.6 \div 0.41$ gives the same result as

⋮ ⋮
$0.026 \div 0.0041$
$0.26 \div 0.041$
$26 \div 4.1$
$260 \div 41$
⋮ ⋮

multiply or divide by the same number

Multiply or divide by 10, 100, 1000 ... or any other number (except 0!)

Challenge

9 When we divide two numbers we can get a result which is larger than the dividend or smaller than the dividend or equal to the dividend.

Take note

12.0	÷	0.6	=	20.0
↑		↑		↑
dividend		divisor		result

Investigate which divisions give which results.

Write down what you discover.

Multiplying

C 1 The suitcase weighs 16.5 kg.

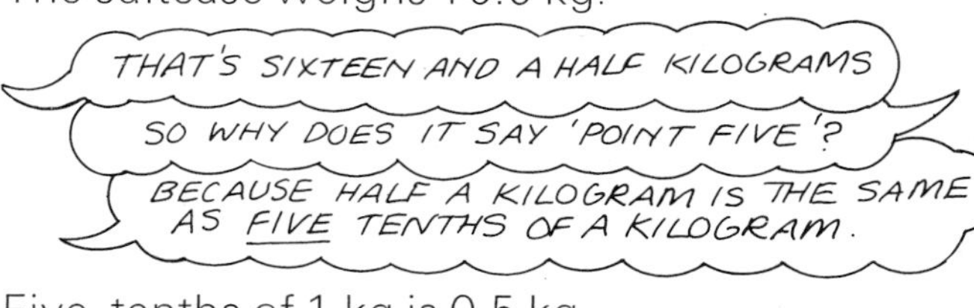

THAT'S SIXTEEN AND A HALF KILOGRAMS
SO WHY DOES IT SAY 'POINT FIVE'?
BECAUSE HALF A KILOGRAM IS THE SAME AS FIVE TENTHS OF A KILOGRAM.

Five-tenths of 1 kg is 0.5 kg.

What is a) one-tenth of 1 kg? b) eight-tenths of 1 kg?
c) three-tenths of 1 kg? d) six-tenths of 1 kg?

2 'One-tenth' is 0.1. So 0.1 of 1 kg is 0.1 kg.

What is a) 0.2 of 1 kg? b) 0.3 of 1 kg?
c) 0.6 of 1 kg? d) 0.8 of 1 kg?

Take note

3×4 kg means 3 'lots of' 4 kg = 12 kg.
0.1×1 kg means 0.1 lots of 1 kg or one-tenth lots of 1 kg = 0.1 kg.

3 **Do not use a calculator.**

What is a) 0.1×1 l? b) 0.2×1 l? c) 0.4×1 l?
d) 0.8×1 l? e) 0.9×1 l? f) 0.6×1 l?

4 **Do you remember . . . ?**

100 ml = 0.1 l 10 ml = 0.01 l 1 ml = 0.001 l

Do not use a calculator.

a) What is 0.1×100 l (i) in millilitres? (ii) in litres?

b) What is 0.1×10 ml (i) in millilitres? (ii) in litres?

5 Check your results in question **4** with your calculator.

6 **Do not use a calculator.**
Copy and complete:

a) 0.1×0.1 l = □ l b) 0.1×0.01 l = □ l c) 0.2×0.1 l = □ l

d) 0.4×0.1 l = □ l e) 0.2×0.01 l = □ l f) 0.8×0.01 l = □ l

g) 0.01×1 l = □ l h) 0.01×0.1 l = □ l i) 0.02×1 l = □ l

one-hundredth

j) 0.07×0.1 kg = □ kg k) 0.02×2 kg = □ kg l) 0.07×0.2 kg = □ kg

7 Check your results in question **6** with your calculator.

8 Meg is helping to buy supplies for a large, expensive dinner party.
The speciality food shop has 8.3 kg of smoked salmon costing £4.20 per kilogram.
Meg has £35; she wonders if that will be enough to buy all the smoked salmon.

a) She forgot to bring her calculator, so she starts to calculate 8.3×4.2. Why?

b) Meg writes:

4·2
× 3·8
32·0
1·2
1·6
0·6
34·4
4 × 8
4 × 0.3
0.2 × 8
0.2 × 0.3
You may use your calculator to check.

Is Meg correct?

c) If you say **no** in **(b)**, rewrite her working correctly.

9 a) **Do not use a calculator.**
Use Meg's method to do these calculations.
(i) 3.1×2.1 (ii) 4.6×1.8

... but don't make mistakes.

b) Check your calculations in **(a)** with your calculator.

Challenge

10 Use your calculator if you wish.
When we multiply two numbers the result can be

larger than each number	or	larger than just one of the numbers	or	equal to just one of the numbers

or smaller than each number or equal to both numbers

Investigate which multiplications give which results.
Write down what you discover.

D With a friend *The Close-that-gap Game*

1 Use **one** calculator between you.

Game 1

Player 1: Enter 1.2 into the calculator.

Player 2: Multiply 1.2 by a number.
Try to get a result as close to 100 as possible.
You have only one try.

Player 1: Multiply the new result by a number.
Try to get a result as close to 100 as possible.
You have only one try.

⋮

and so on.

for example, 99.91 or 100.08

99.9 100 100.1

The first player to get **within** 0.1 of 100 wins a point.

Try the game with starting numbers of your own.

The first player to 5 points wins.

Game 2

Player 1: Enter 77.8 into the calculator.

Player 2: Divide 77.8 by a number.
Try to get a result as close to 0.3 as possible.

Player 1: Divide the new result by a number.
Try to get a result as close to 0.3 as possible.

⋮

and so on.

You have only one try.

for example, 0.294 or 0.309

0.29 0.3 0.31

The first player to get **within** 0.01 of 0.3 wins a point.

Try the game with some starting numbers of your own.

The first player to 5 points wins.

2 **Do not use a calculator.**
Choose the best estimate for each calculation.

a) $\frac{2.7 \times 0.9}{0.8}$ A 30 B 3 C 0.3 D 300

b) $(0.8 \div 0.03) \times 1.74$ A 5 B 50 C 100 D 25

3 Use your calculator to check your results in question **2**.

With a friend *The Close-call Game*

4 Player 1: Choose one of these two kinds of calculation.

$$\frac{\square \times \square}{\square} \qquad (\square \div \square) \times \square$$

Player 2: Choose a number for one of the boxes.
Player 1: Choose a number for a second box.
Player 2: Choose a number for the last box.
Player 1: Estimate the result of the calculation.
Player 2: Estimate the result of the calculation.

Be daring! Use decimals!

Together,
- use a calculator to find out what the result actually is
- agree on whose estimate was closer to the true result
- award 1 point to the player who made the closer estimate.

Play again, but switch roles this time.
The first player to 5 points wins.

5 If you drop a stone down a well, the depth of the well may be calculated from this formula:

$$\text{depth (in m)} = \frac{9.81 \times \text{time before splash (in s)} \times \text{time before splash (in s)}}{2}$$

Find the depths of these wells, first by estimating and then by using a calculator.

a) time before splash = 5.2 s *measured with a stopwatch*

b) time before splash = 2.7 s

Write down both your estimate and the calculator result.

6 The time it takes money to earn a certain amount of interest when it is invested at simple interest may be calculated from this formula:

$$\text{time (in years)} = \frac{\text{interest (in £)} \times 100}{\text{amount invested (in £)} \times \text{interest rate (in \% per year)}}$$

Find the time it takes for £24 to earn these amounts of interest at these interest rates.

Estimate first, then use a calculator.

a) £5 at 9.75% per year b) £18 at 8.83% per year

Calculating with decimals

E 1 Different sponges soak up different amounts of water. They can be compared by using an 'absorption factor', which is calculated from this formula:

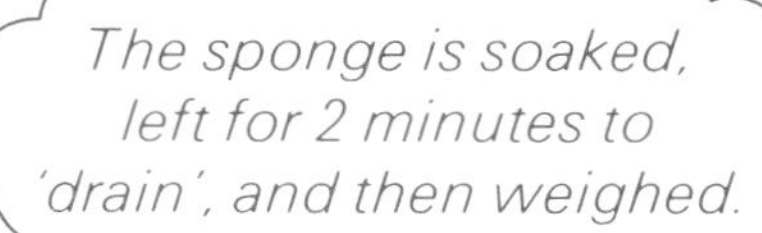

The sponge is soaked, left for 2 minutes to 'drain', and then weighed.

$$\text{absorption factor} = \frac{\text{weight soaked} - \text{weight dry}}{\text{weight dry}}$$

A

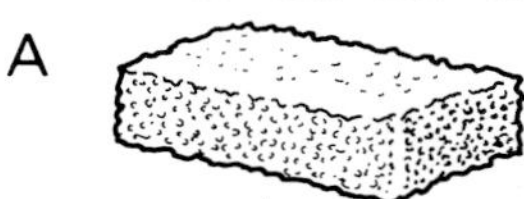

weight dry 0.03 kg
weight soaked 0.2 kg

B

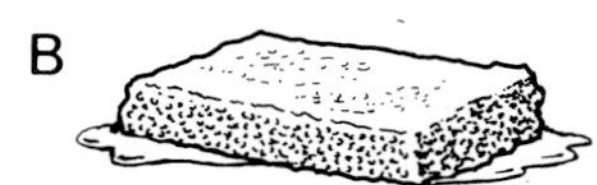

weight dry 0.048 kg
weight soaked 0.28 kg

a) **Estimate** the absorption factors for these sponges:

b) **Calculate** the absorption factors.
Which sponge do you think is more efficient at mopping up water?

c) Copy and complete this statement:
The larger is the absorption factor the . . .

d) The perfect sponge for the bathroom has an absorption factor of 8.
What should a 0.05 kg bathroom sponge weigh when it is soaked?

Test and check until you find the weight, or use another method.

2 Speeds in miles per hour can be changed to metres per second using this formula:

$$\text{speed (in m/s)} = \frac{1609}{3600} \times \text{speed (in mph)}$$

mph

A car advert claims 0–60 in 8.4 seconds

a) How fast is the car travelling, in m/s, after 8.4 s?

b) By how many m/s does it increase its speed each second, if its speed increases steadily during this time?

*that is, what is its **acceleration** in m/s per second*

3 Jane Huxton visits the USA on business once or twice a year.
One November she goes when £1 is worth \$1.75. *(1.75 dollars)*
She sees a handbag she likes and works out its cost in sterling as £12.84 – but she decides not to buy it. *(British money)*

The next March she sees the same handbag at the same dollar price, but by then £1 is only worth \$1.63.

a) Write the new cost of the handbag in sterling like this: $£\dfrac{\square \times \square}{\square}$,

using numbers you have been given in place of the $\square$s.

b) Estimate the new cost in sterling.

c) Roughly how much more, measured in sterling, would you have to pay for the bag?

d) Use a calculator to find out exactly how much the sterling price has gone up.

4 Glenda is going to France and Spain for her holidays.
She buys some foreign currency in advance from her bank.
The bank charges a fee of 1% for each currency transaction.

a) Including the fees, she pays £40 for the French francs and £60 for the Spanish pesetas.

£1 =
French francs 9.89 FF
Spanish pesetas 203.7 pts

(i) Explain why she only receives

$$\frac{40 \times 9.89}{1.01} \text{ FF.}$$

(ii) The bank rounds the calculation in (i) to the next 1 FF.
How many French francs does Glenda receive?

(iii) How many Spanish pesetas does Glenda receive?

b) Glenda spends 8565 pesetas.
When she gets home, she changes her remaining pesetas back into sterling.
By now, £1 = 206.2 pts. How much does she receive in sterling?

5 The volume of a pyramid is given by the formula

$$\text{volume} = \tfrac{1}{3} \times \text{area of base} \times \text{height}$$

This is the Great Pyramid of Giza, finished about 2580 B.C.
Originally it was 146.5 m high, and its square base measured 230 m along each side.
It has been estimated that a permanent work force of 4000 men spent 30 years building it from 2.3 million limestone blocks.
Its total weight was originally about 584 000 tonnes.

a) What is the average weight of the blocks used to build the pyramid?

b) Roughly, what area of land does the pyramid stand upon?

in hectares

c) What was the average time taken to put each block in place?

d) Roughly, what is the volume, in m³, of the pyramid?

Assignment

6 Calculate the 'absorption factor' of two different types of sand.

Use a method like that in question 1 (page 171).

Which sand would be best for use in sandbags to hold back floods?
Why?

19 Using letters for rules

A

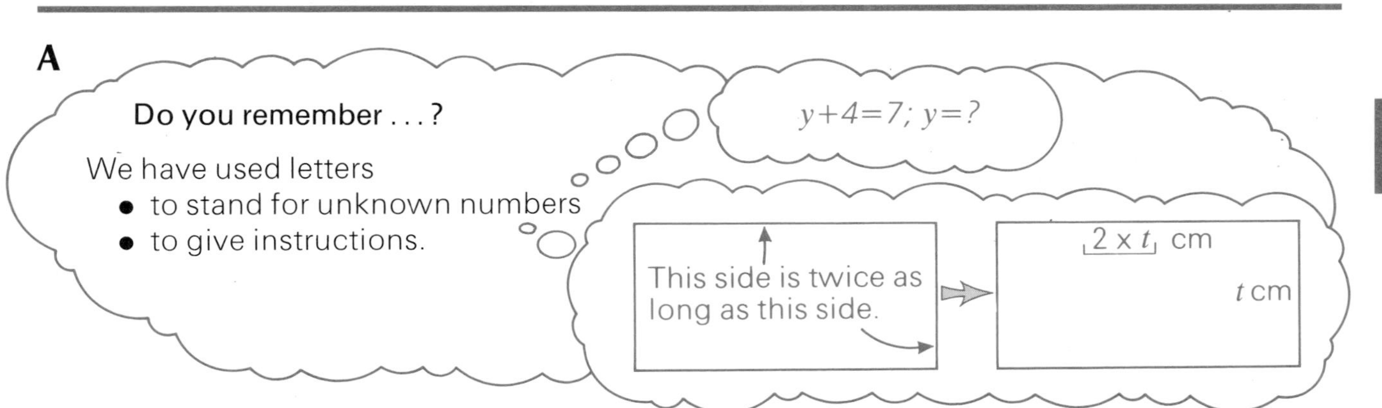

Do you remember . . . ?
We have used letters
• to stand for unknown numbers
• to give instructions.
y+4=7; y=?
This side is twice as long as this side.
2 x t cm
t cm

1 Find the number which y replaces in each sentence.

a) $y+4=7$ b) $y \times 7 = 21$ c) $\frac{y}{7} = 3$ d) $72-(3 \times y) = 27$

2 Here are some instructions for drawing rectangles.

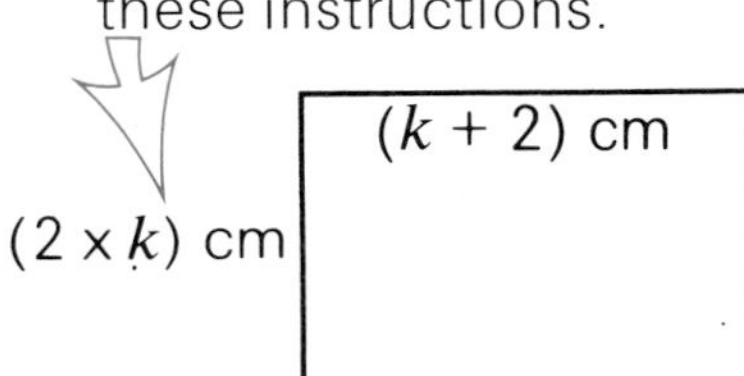

This side is 2 cm less than three times the length of this side.

Copy and complete this version of the instructions.

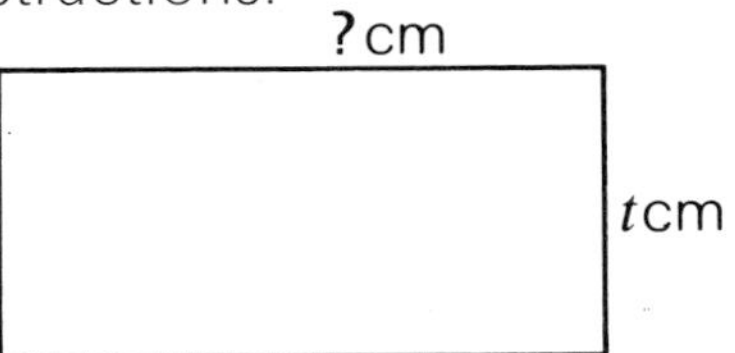

3 The ripple of rectangles was drawn using these instructions.

(2 x k) cm

(k + 2) cm

k was chosen to be 1 for the smallest rectangle. What value was chosen for k for each of the other rectangles?

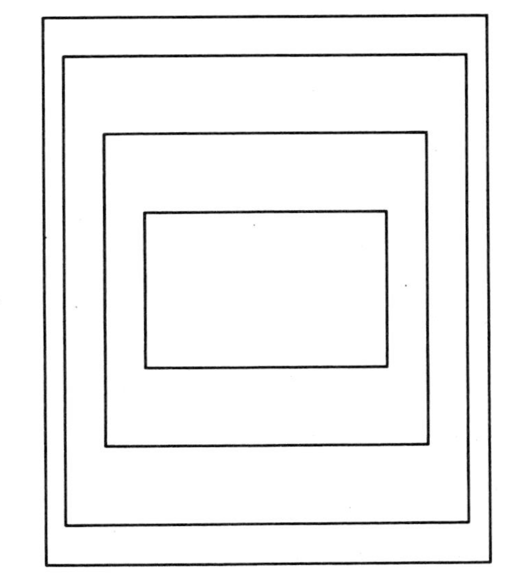

Think it through

4 These instructions give this ripple of triangles. Copy and complete the instructions.

? cm

t cm

B 1 a) Meg has collected these piles of sticks and pebbles.
The numbers of sticks and pebbles in each pile obey a special rule.
Try to decide what the rule is.
Write down what you decide.

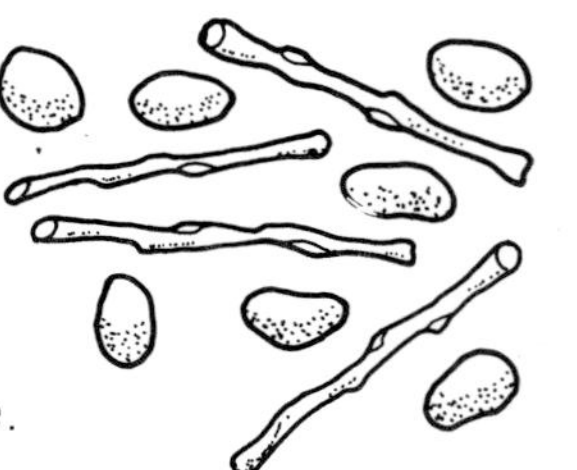

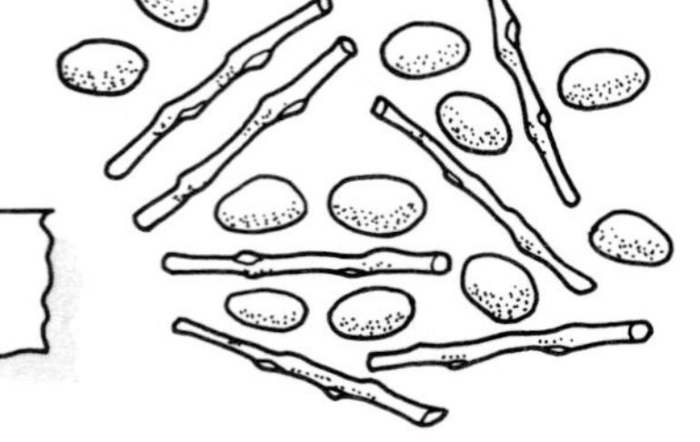

b) Ben decides he has found the rule.
He writes:

> *Number of pebbles in each pile = (2 × Number of sticks in each pile) − 1*

Do you agree with him?

c) Write Ben's rule in another way, like this:

> *Number of sticks in each pile =* $\dfrac{\text{Number of pebbles in each pile} + \square}{\square}$

d) Horace says he can write Ben's rule more simply . . . using letters.

He writes:

$$p = (2 \times s) - 1$$

'p' means 'number of pebbles in each pile'

's' means 'number of sticks in the pile'

Use letters to write the other version of the rule, like this:

$$s = \ldots$$

e) Here is another way of writing the same rule.
Copy and complete it.

> *(2 × Number of sticks in each pile) − Number of pebbles in each pile =* $\square$

f) Now write this version of the rule using letters.

> $(2 \times s) - \ldots = \ldots$

Think it through

2 Write this rule in two more ways. ⇨ $(3 \times s) - 2 = p$

C

Take note

There are several ways of writing the same rule:

$(2 \times p) - 1 = s \qquad 2 \times p = s + 1 \qquad (2 \times p) - s = 1 \qquad \dots$

1 a) This rule gives a connection between the number of posts and rails needed for different lengths of this type of fencing.

$k = (t - \square) \times \square$

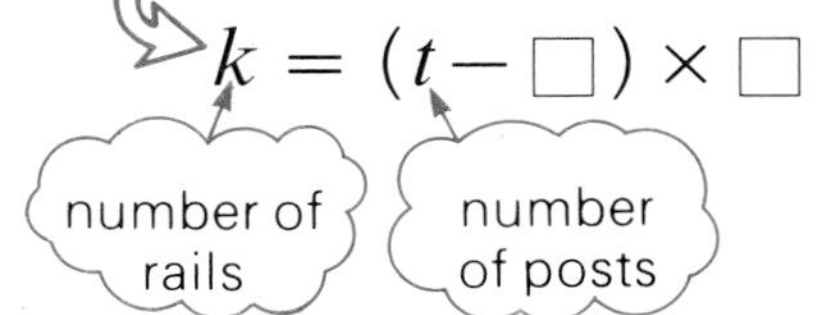

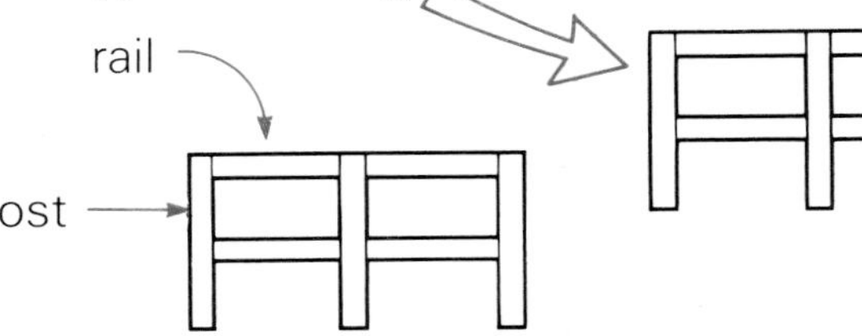

Copy and complete it.

b) Here is another way of writing the same rule:

$t = (k \div \square) + \square$

Copy and complete it.

c) ... and here is a third way:

$(2 \times t) - k = \square$

Copy and complete it.

2 These rules give a connection between the number of red buttons and white buttons on each strip.

$(2 \times p) + \square = n$

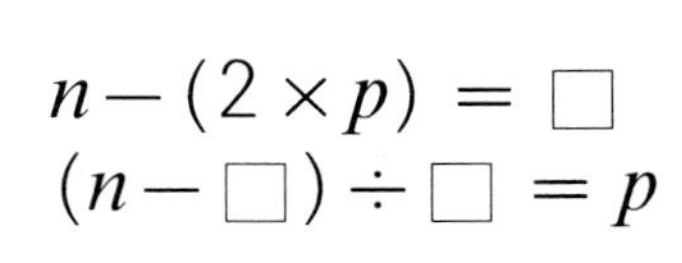

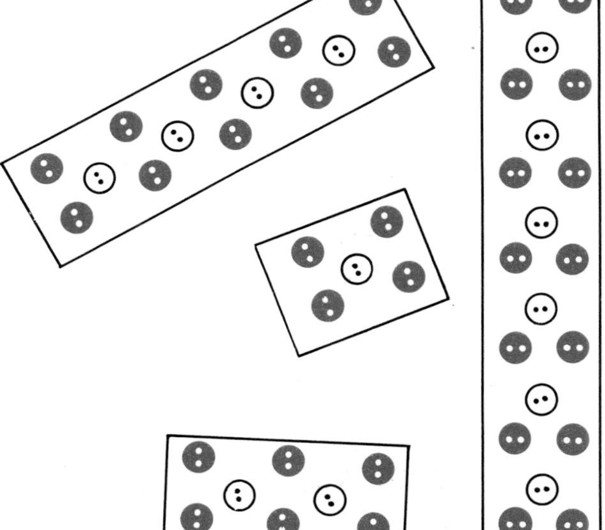

$n - (2 \times p) = \square$

$(n - \square) \div \square = p$

Copy and complete each rule.

Think it through

3 Find a rule which connects the number of posts and rails needed for these fences.
Write your rule in three different ways.

Choose your own letters.

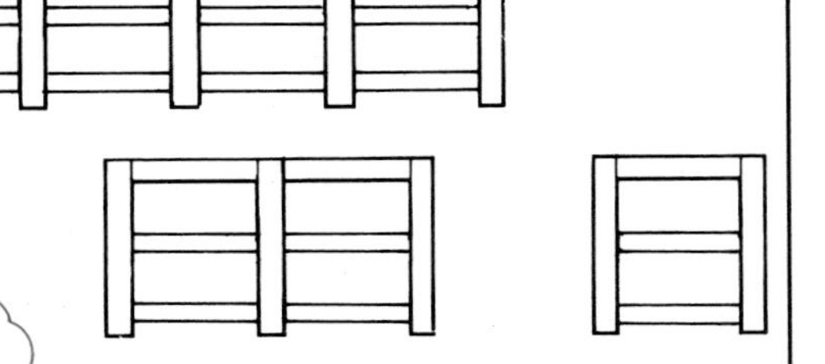

4 This is Meg's 'pins and buttons' box.

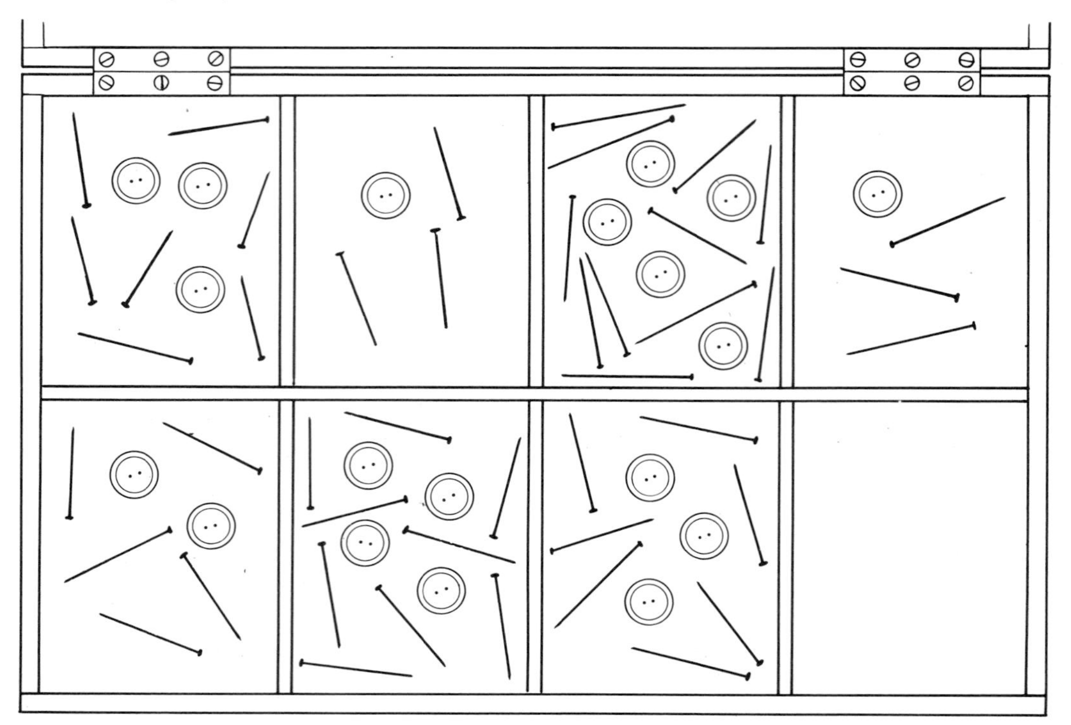

Meg always makes sure that the numbers of pins and buttons in each tray obey a special rule.

a) What is the rule for the box? *Ignore the empty tray.*
Write it like this:

number of pins in each tray = number of buttons in each tray . . .

. . . and like this:

number of buttons in each tray = . . .

. . . and like this:

number of pins in each tray − . . . = 1

b) Meg puts 25 pins in the empty tray.
How many buttons should she put into it?

c) Meg writes the rule she is using on the lid of the box.

Copy and complete each version of the rule.

number of pins in each tray

number of buttons in the tray

$p = (\square \times b) + \square$

$b = \ldots$

$p - \ldots = 1$

d) Sketch a box like Meg's with empty trays.
Draw pins and buttons in each tray which obey this rule:

$2 \times p = 20 - b$

D 1 a) Only one of these rules is correct for the connection between days and weeks. Which one is it?

A $d = w \div 7$ B $w = d \div 7$ C $w \div d = 7$

(number of days) (number of weeks)

b) Here is another way of writing the rule: $d = w \times 7$.
Check that when w is chosen to be 3, d is 21.

c) What is d when w is chosen to be 4?
Is this correct for the number of days and weeks?

d) Are there any replacements for w which do not give the correct number of days? If you say **yes**, write one down.

e) Find another way of writing the same rule.

2 No matter how many triangles we draw there will be three times as many sides as triangles.

(*no two of them with a side in common*)

a) Which of these is a correct rule?

A $n = 3 \times c$ B $3 \times n = c$

(number of sides) (number of triangles)

b) Choose c to be 7 in **A**. What value does this give for n?
Is this correct for the number of sides?

c) Are there any replacements for c in **A** which do not give the correct number of sides? If you say **yes**, write one down.

d) Find two more ways of writing the same rule.

3 One of these is the correct rule for the area of a square.

A $k = t + t$ B $k = t^2$

(area (cm²)) (length of side (cm))

a) Which one is it?

b) Choose t to be 2 in **A**. Does this give the correct value for k?

c) Are there any replacements for t in **A** which do not give the correct area? If you say **yes**, write one down.

d) Are there any replacements for t in **B** which do not give the correct area? If you say **yes**, write one down.

Take note

Don't be satisfied with checking a rule by using just one or two replacements.
Always ask yourself if the rule works for all the numbers you want it to work for.
For example,

$k = t + t$ (*question 3*)

works for the area of a square when t is 2
but it isn't the correct rule; it doesn't work, for example, when t is 3.

Challenge

4 Winston says that this rule gives a prime number for each whole-number replacement for n:

($n \times n$)

$$p = (n^2 - n) + 41$$

For example, choose n to be 0.
Then p is 41 ... which is a prime number.

a) Find what happens when n is chosen to be (i) 1. (ii) 2.

b) Is Winston correct? Why or why not?

5 Here are some rules written using letters. Some are correct and some are incorrect. For each part, write down the ones which are correct.

Test and check each rule with numbers until you are sure it is correct or incorrect.

(number of days) (number of hours)

a) (i) $24 \times d = h$ (ii) $24 \div h = d$ (iii) $h \div d = 24$

(iv) $d \div h = 24$ (v) $24 \div d = h$ (vi) $h \div 24 = d$

(number of centimetres) (number of millimetres)

b) (i) $10 \times k = n$ (ii) $k \times n = 10$ (iii) $n \div 10 = k$

(iv) $10 \div k = n$ (v) $n \times 10 = k$ (vi) $\frac{n}{k} = 10$

(area of a square) (length of a side)

c) (i) $A = l^2$ (ii) $l^2 \div A = 1$ (iii) $l \div A = l$

(iv) $\frac{1}{A} = l^2$ (v) $\frac{A}{l} = l$ (vi) $\frac{A}{l^2} = 1$

(number of lenses) (number of pairs of spectacles)

d) (i) $2 \times l = s$ (ii) $l \div 2 = s$ (iii) $s \div l = 2$

(iv) $\frac{l}{s} = 2$ (v) $\frac{s}{2} = l$ (vi) $\frac{2}{l} = s$

6 This rule is called Ohm's Law. ⇨ $V = I \times R$

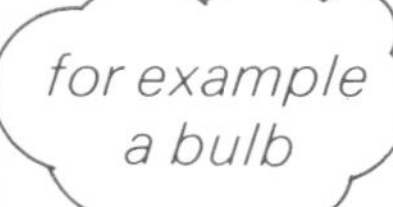

It tells us that the voltage (V) across an electrical component is the current (I) in the component multiplied by the resistance (R) of the component.

Write the rule in four more ways.

Test and check each of your rules with numbers until you are sure that it is correct.

7 One of these rules connects the angle sum of polygons with the number of sides they have:

$$s = (180 \times n) - 180$$

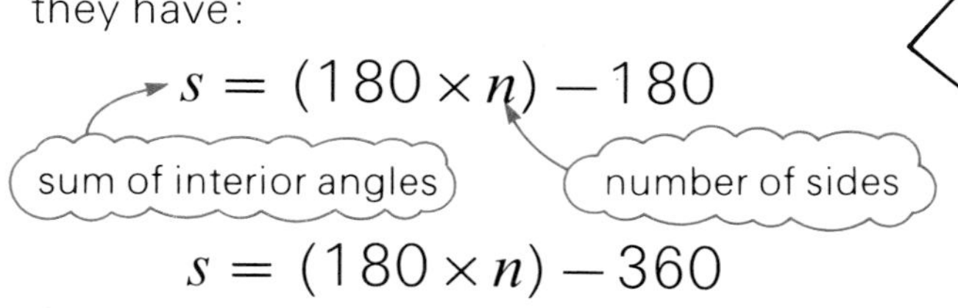

$$s = (180 \times n) - 360$$

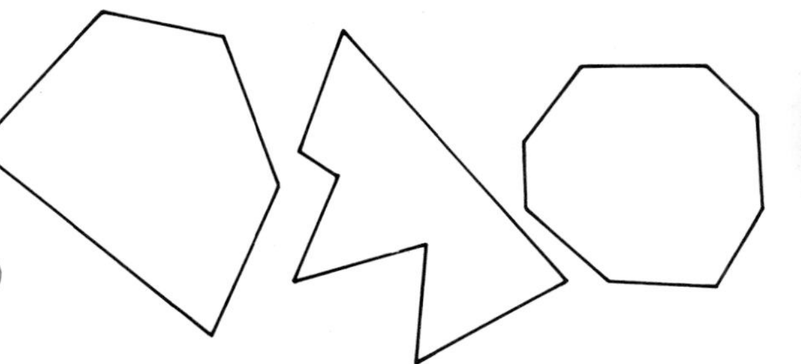

a) Think about triangles.
What is their angle sum?
Choose n to be 3 in each rule, and find out what this gives for s each time.

b) Which of the two rules gives the correct result for triangles?

c) Check that the second rule gives the correct result for quadrilaterals.

d) Use the second rule to find the sum of the interior angles of a polygon with nine sides.
Check that your result is correct by finding the angle sum from this diagram.

e) Does the second rule work for all polygons?
If you say **no**, give an example.

8 Below is a famous rule which was discovered by a mathematician called Euler.

$$F + V - E = 2$$

It gives a relationship between the number of faces (F), vertices (V) and edges (E) of solids like these:

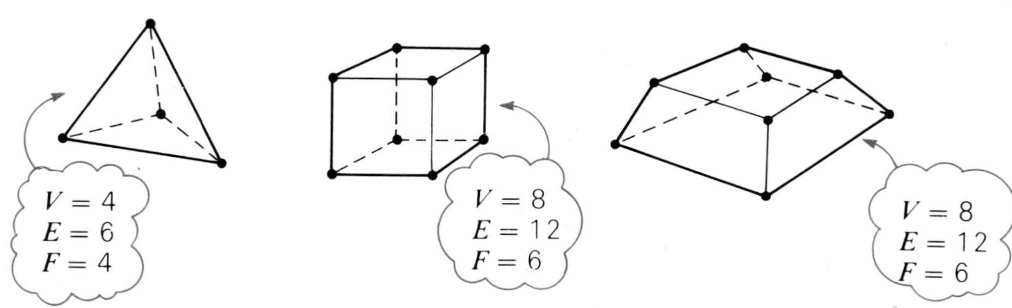

a) Check that the rule works for each of the three solids.

b) Draw a solid of your own which obeys the rule.

c) Horace says he has drawn a solid for which F is 4, V is 6 and E is 7.
Do you believe him?
Why?

9 This rule connects °C (degrees Celsius) and °F (degrees Fahrenheit).

$$F = (C \times \square \div \square) + \square$$

Here are some corresponding temperatures:

(0°C, 32°F)   (100°C, 212°F)   (20°C, 68°F)

Use them to help you to find the missing numbers in the rule.

*For example, choose **C** to be 0 and **F** to be 32. What does this tell you about this number? $F = (C \times \square \div \square) + \square$*

10 This rule gives us a rough idea of how far the stone has fallen at various times:

$$d \approx 5 \times t^2$$

time in seconds since stone was released

distance fallen in metres

a) How far does the stone fall during the first three seconds?

b) How far does it fall in the next three seconds?

c) The well is 60 metres deep.
Roughly how long will the stone take to reach the water?

Challenge

11 The drawing shows a circle with three lines drawn across it.
The lines cross each other in two places.
They divide the circle into six regions.

marked with dots

They are numbered to help you.

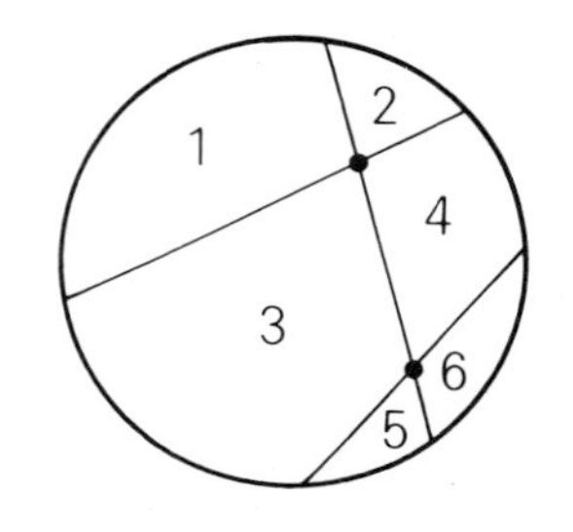

a) How many regions are there when
   (i) three lines meet in one point?
   (ii) four lines meet in three points?

b) Investigate some more examples of your own.
Find a rule which connects the number of regions (R), the number of lines (L) and the number of crossing points (C).

c) Use your rule to find out
   (i) how many regions are made by six lines which cross in four points.
   (ii) how many crossing points there must be if five lines make thirteen regions.

d) Draw an example diagram for each situation in **(c)**.

e) Find an arrangement of lines which makes fifteen regions.

This may take some time!

f) Meg says, 'With four lines, the maximum possible number of regions is eleven.'
Do you agree?
Why?

Think it through — *Trinket box*

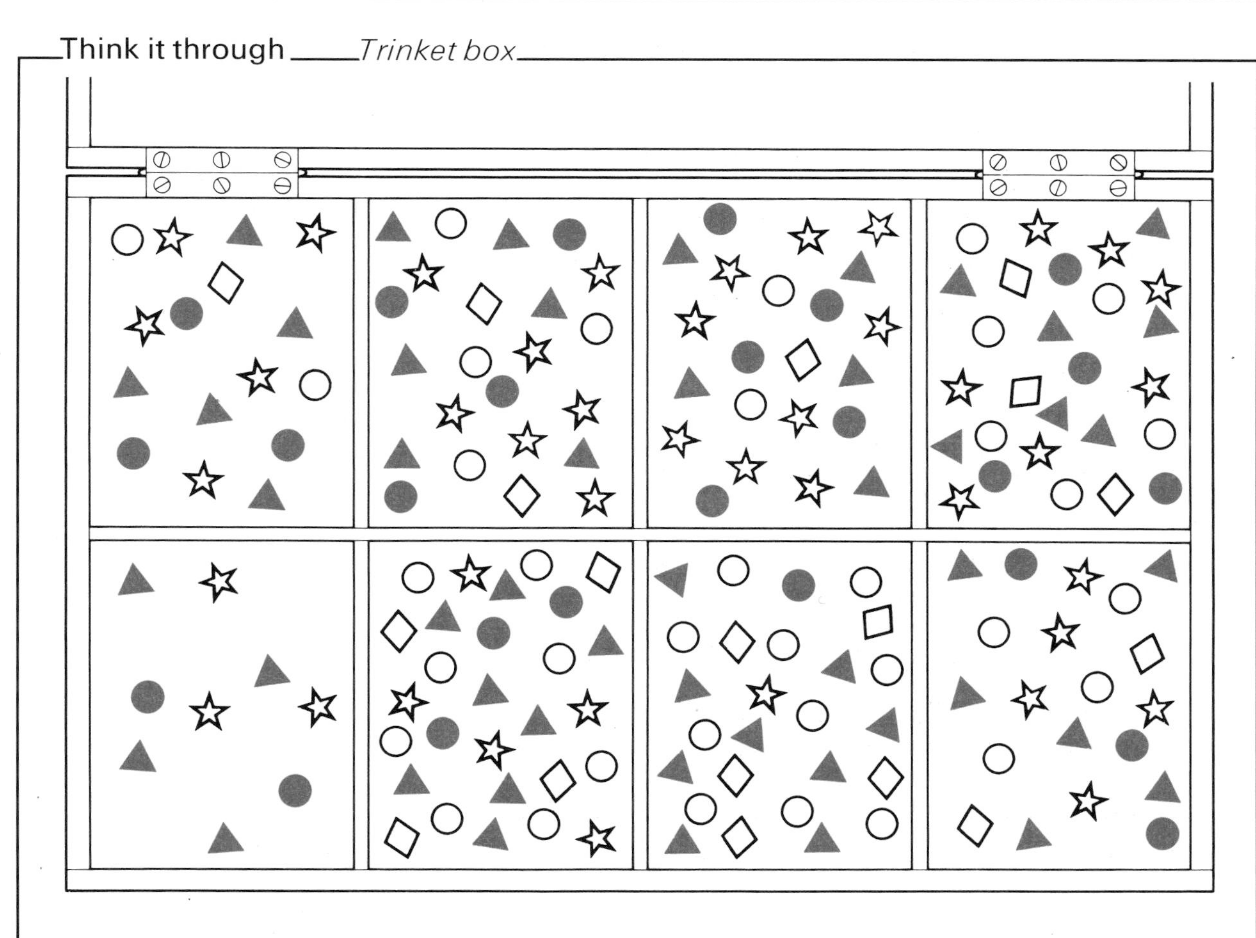

12 The trinket box contains rings (○), stars (☆), triangles (▲), blobs (●) and diamonds (◇).

The numbers of trinkets in each tray obey various rules.

For example, there is a rule which connects the number of rings and the number of diamonds in each tray.

a) Find all the rules that you can which connect two kinds of trinkets.

There are three.

b) Write down each rule you find in as many different ways as you can.
Use 'r' for 'number of rings'
's' for 'number of stars'
't' for 'number of triangles'
'b' for 'number of blobs'
'd' for 'number of diamonds'.

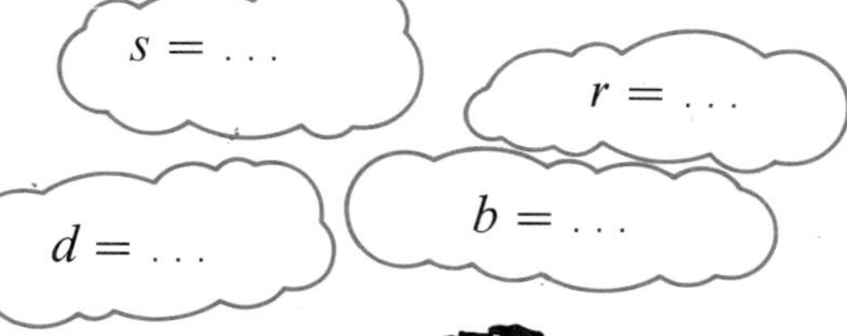

▲ 182
● Next chapter

Rules, rules, rules!

E 1 Henry Macintosh and Beth Morris have studied many thunderstorms.
They recorded the time between seeing lightning and hearing thunder.
They also recorded how far away each storm was.
Here are some of their results.

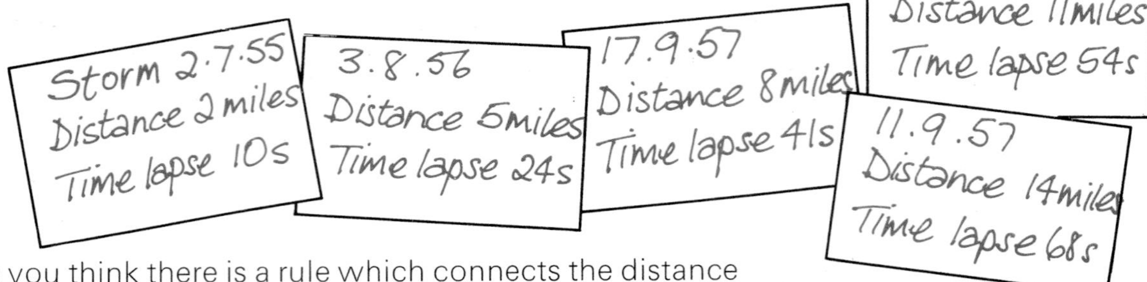

Storm 2.7.55
Distance 2 miles
Time lapse 10s
3.8.56
Distance 5miles
Time lapse 24s
17.9.57
Distance 8miles
Time lapse 41s
9.8.56
Distance 11miles
Time lapse 54s
11.9.57
Distance 14miles
Time lapse 68s

Do you think there is a rule which connects the distance from a storm and the time between seeing lightning and hearing thunder?
If you think there is a rule, write it down in words and letters.

2 In parts of the USA, speeding fines are calculated like this:

$50 plus $2 for each mile per hour over the speed limit

a) What would the fine be if a motorist was caught speeding at v mph in an area where the speed limit was L mph?
Express the fine using letters.

b) Just about every speed limit in the USA is one of these: (in mph)

15 25 30 35 45 55 65

(i) A motorist travelling at 50 mph is ticketed for speeding.
What might the fine be?
List each possibility.

(ii) Describe three different situations in which a speeder would be fined $100.

c) Do you agree with this statement?

> Emily Parry was driving at 70 mph and was fined for speeding.
>
> Gary Heal was driving at 60 mph and was fined for speeding.
>
> Emily Parry must have had the heavier fine.

Why?

Challenge

d) Helen and Joseph were caught in the same speed trap within half an hour of each other.
Joseph was fined twice as much as Helen.
What can you say about the relationship between the speeds at which Joseph and Helen were driving?

3 The distance from a lens to the point at which it focuses the sun's rays is called the **focal length** of the lens. Midge has been doing some science experiments with lenses.

such as a magnifying glass

the object

She lined up a lens, an illuminated crosswire and a screen, like this.

Then she moved the screen towards and away from the lens until the image of the crosswire was in focus on the screen. Here are her results for various lenses:

Focal length (cm)	10	10	15	15	20	20	25
Object distance (cm)	20	30	20	40	30	45	45
Image distance (cm)	20	15	60	24	60	36	56.25

a) There is a relationship between the focal length, the object distance and the image distance. Write it down in words and in letters.

b) Check your relationship in **(a)** with the help of this information:

Focal length 18 cm
Object distance 45 cm
Image distance 30 cm

c) When the object distance is equal to the focal length, what can you say about the image distance?

d) Can the image distance ever be equal to the focal length? If you say **yes**, explain how. If you say **no**, give as much information as you can about the possibilities for the image distance.

Challenge

Hint: You may find your answer to ***(c)*** *useful.*

e) The crosswire and its image are 30 cm apart. Find two possibilities for the focal length and position of the lens.

4 A lorry is travelling along a straight road at u km/h.
The lorry is l metres long.
Mary O'Keefe drives up behind the lorry at v km/h and overtakes it.

She starts to overtake when her car is 25 m behind the lorry, and she completes her manoeuvre when her car is 25 m in front of it.

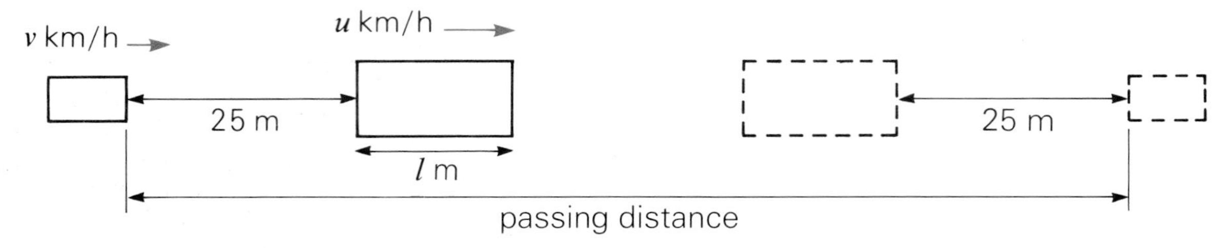

Mary's passing distance may be calculated like this:

$$\text{passing distance (in m)} = \frac{(50+l) \times v}{v-u}$$

a) What is her passing distance if she overtakes a 5 m long coach going at 60 km/h when she is doing 80 km/h?

b) Mary took 300 m to overtake a car going at 82 km/h. She was travelling at 100 km/h.

How long is the car she overtook?

c) In Spain, the law requires that all overtaking manoeuvres must be completed within 400 m. What is the minimum speed at which Mary could legally overtake a 10 m long juggernaut lorry doing 85 km/h in Spain?

d) Mary is planning to drive to Spain for her next summer holiday. Make up an information table for her, showing the minimum speeds at which she could overtake vehicles of various lengths travelling at various speeds.

20 Thinking about circles

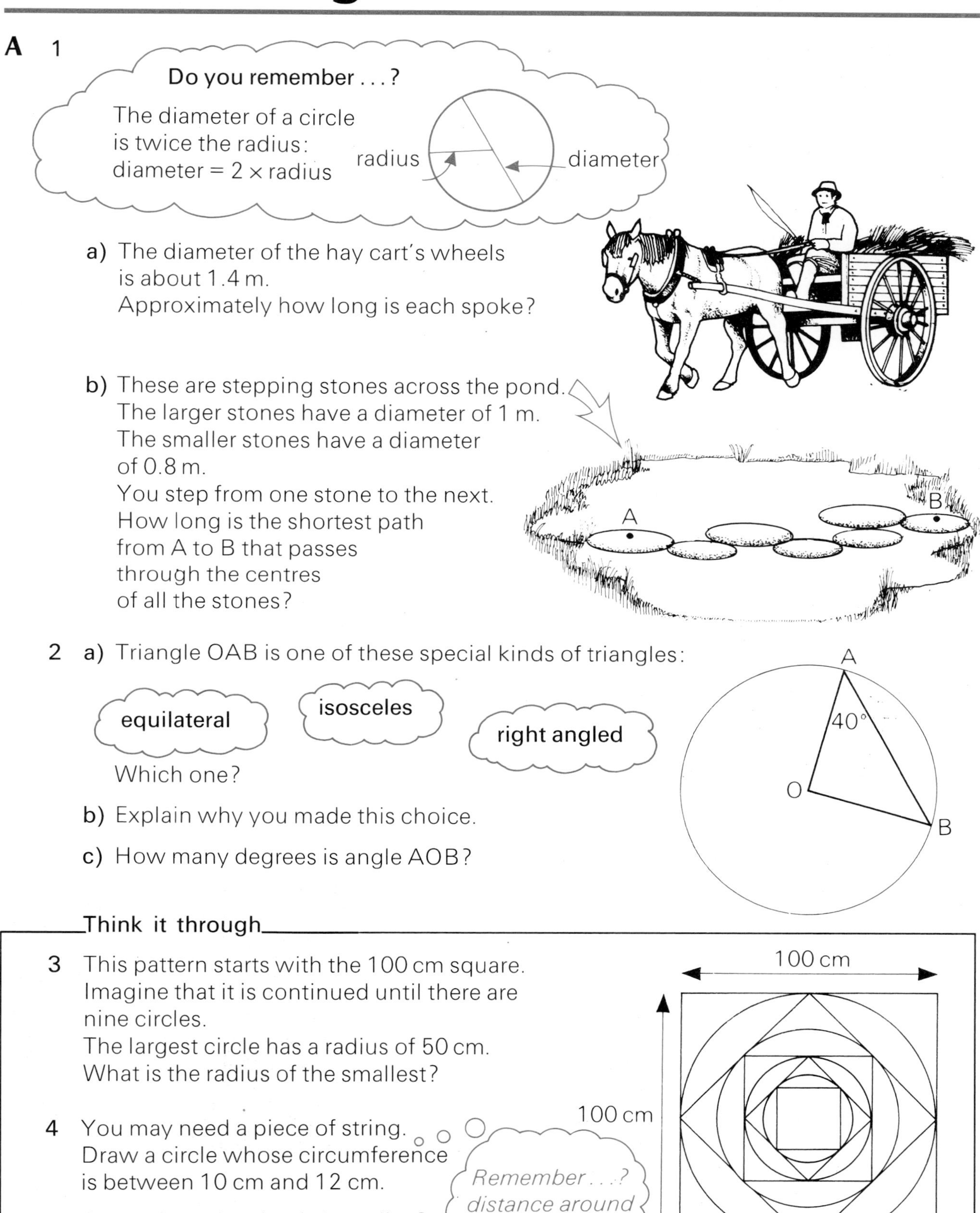

A 1

Do you remember . . . ?

The diameter of a circle is twice the radius:
diameter = 2 × radius

a) The diameter of the hay cart's wheels is about 1.4 m.
Approximately how long is each spoke?

b) These are stepping stones across the pond.
The larger stones have a diameter of 1 m.
The smaller stones have a diameter of 0.8 m.
You step from one stone to the next.
How long is the shortest path from A to B that passes through the centres of all the stones?

2 a) Triangle OAB is one of these special kinds of triangles:

equilateral　isosceles　right angled

Which one?

b) Explain why you made this choice.

c) How many degrees is angle AOB?

Think it through

3 This pattern starts with the 100 cm square.
Imagine that it is continued until there are nine circles.
The largest circle has a radius of 50 cm.
What is the radius of the smallest?

4 You may need a piece of string.
Draw a circle whose circumference is between 10 cm and 12 cm.

Approximately what is its radius?

Remember . . . ? distance around the circle

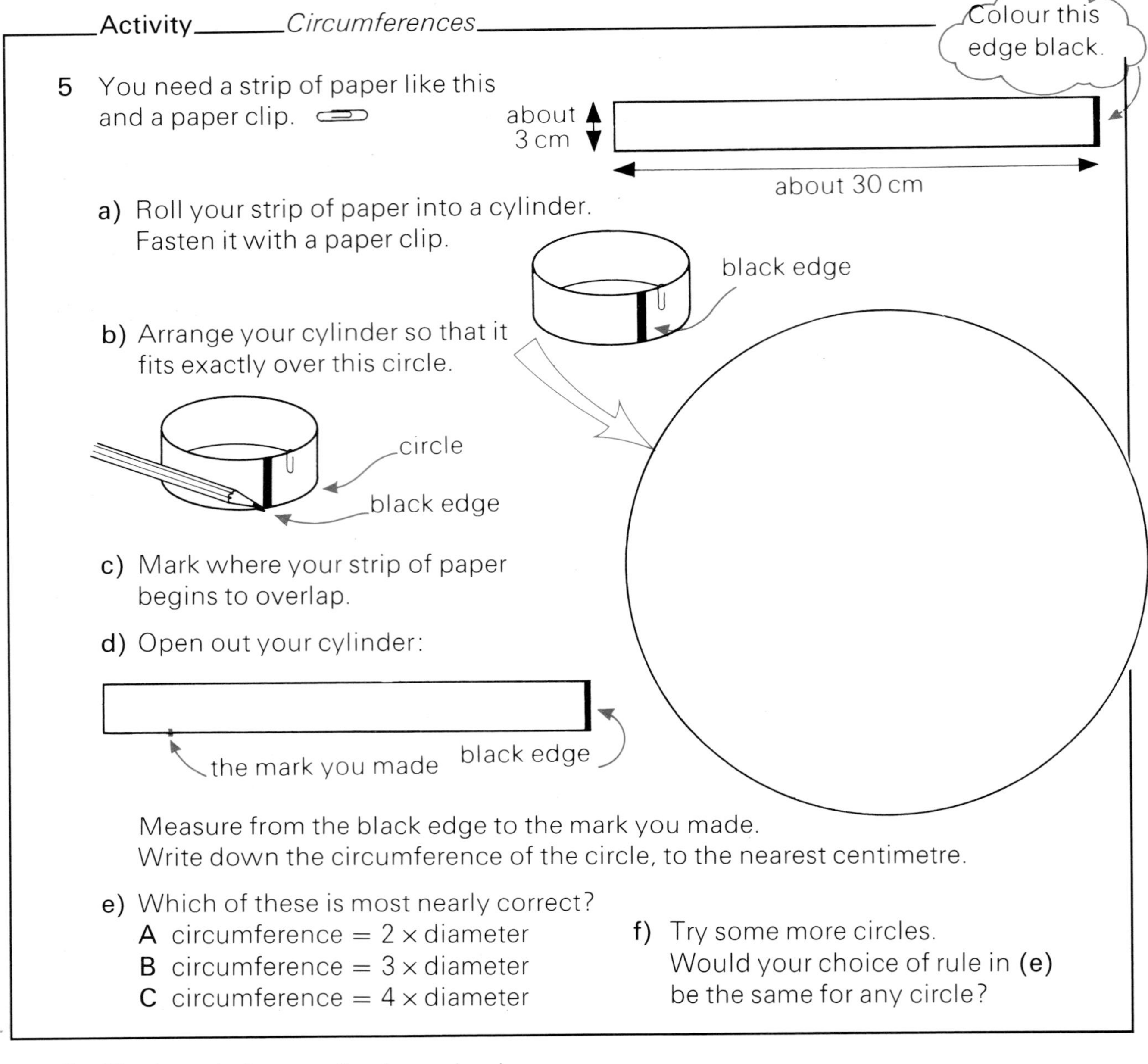

Activity *Circumferences*

5 You need a strip of paper like this and a paper clip.

a) Roll your strip of paper into a cylinder. Fasten it with a paper clip.

b) Arrange your cylinder so that it fits exactly over this circle.

c) Mark where your strip of paper begins to overlap.

d) Open out your cylinder:

Measure from the black edge to the mark you made.
Write down the circumference of the circle, to the nearest centimetre.

e) Which of these is most nearly correct?
A circumference = $2 \times$ diameter
B circumference = $3 \times$ diameter
C circumference = $4 \times$ diameter

f) Try some more circles.
Would your choice of rule in **(e)** be the same for any circle?

6 The hospital tea trolley has wheels the same size as the circle in question **5**.
Every day it is pushed about 1.5 km around the wards.
About how many times does each wheel turn?

7 The hour hand of Big Ben is 2.7 m long.
The minute hand is 4.3 m long.
Roughly, how far does the tip of each hand travel in 1 year (365 days)?

8 The circumference of the earth is about 40 000 km.
Roughly what is the straight-line distance between the North and South poles?

Circles and their circumferences

B Challenges

1 This is a swing-ball game.
As the players hit the ball, it swings around the pole in circles.
The string is 2 m long.

a) Roughly, what is the **greatest** distance that the ball travels in one complete circuit?
Explain how you arrived at your estimate.

b) About how far does the ball travel in one circuit

A scale drawing on 1 cm squared paper will help.

(i) on this track? (ii) on this track?

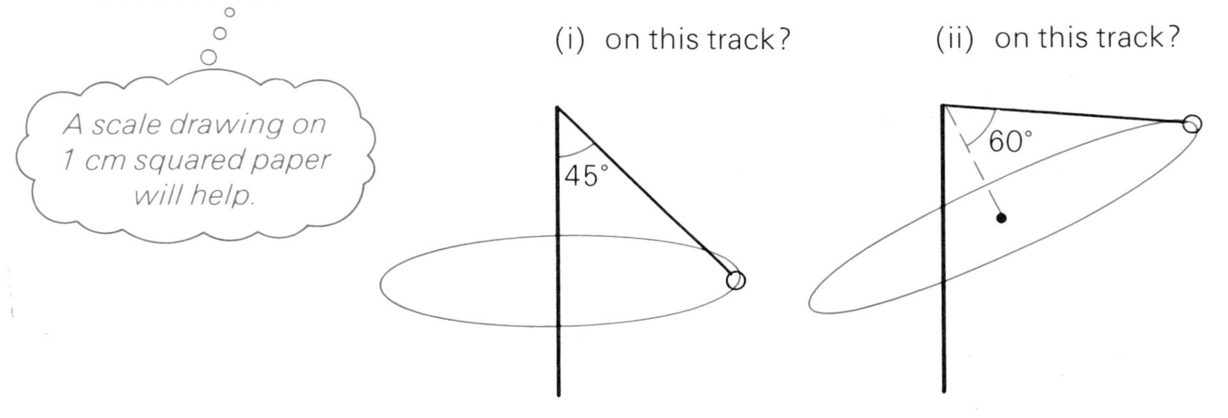

2 a) Imagine that you walk in the largest possible circle inside your classroom.

Roughly, how far would you walk in one circuit?

b) Roughly, what is the greatest distance that a fly can travel in one circuit of a circle in your classroom?

c) The photograph shows Silbury Hill in Wiltshire, the largest artificial mound in Europe.
It involved the moving of 681 000 tonnes of chalk to make a cone 39 m high with a base area of 2 hectares.
Horace walked around its base.
About how far did he walk?
Write one or two sentences to explain how you made your estimate.

Squared paper will help.

3 This 'trackometer' is used to measure map distances very accurately. It can be used to measure the circumferences of circles.

trackometer
12.7 cm
measurement
wheel

Here are some results:

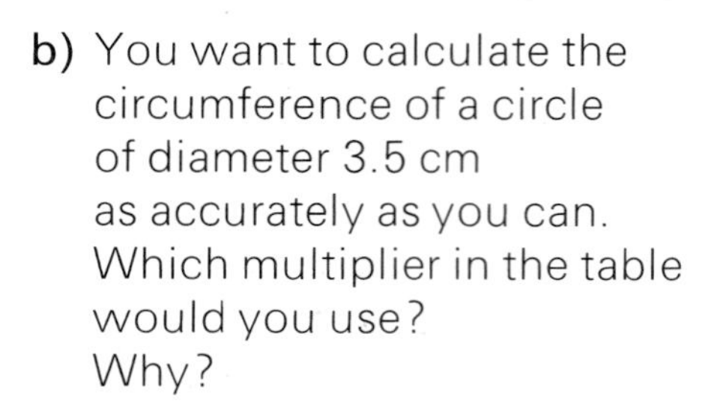

a) Calculate the missing multipliers.

b) You want to calculate the circumference of a circle of diameter 3.5 cm as accurately as you can. Which multiplier in the table would you use? Why?

Circle diameter (cm)	Approximate circumference (cm)	Multiplier to give circumference from diameter
1	3.1	3.1
2	6.3	3.15
3	9.4	
4	12.6	
5	15.7	
6	18.8	

4 The largest dome in Great Britain is the one forming the roof of the Bell Sports Centre, Perth, Scotland. Its diameter is 67 m and its circumference is 210.5 m.

a) What is the ratio of the circumference of the dome to the diameter?

b) Copy and complete:

circumference of Sports Centre dome $\approx$ □ $\times$ diameter

Take note

The circumference of any circle is approximately its diameter multiplied by 3:

circumference $\approx 3 \times$ diameter

We get a better approximation if we multiply by 3.1:

circumference $\approx 3.1 \times$ diameter

For an even better approximation, we can multiply by 3.14:

circumference $\approx 3.14 \times$ diameter

The **exact** multiplier cannot be written using digits because the string of digits after the decimal point

- never ends, and
- has no repeating pattern.

π

It starts like this:

3.141 592 653 589 793 238 462 643 383 279 502 884 197 169 399...

When writing the **exact** circumference of a circle we use the Greek letter π ('pi') to stand for the exact multiplier:

circumference $= \pi \times$ diameter

5 Check whether or not your calculator has a key (or a 'function key') labelled 'π'.
If it does, what value does your calculator display for π?

Do you remember . . . ?

An **approximation** for the circumference of this circle is 3.1×2 cm.
A better approximation is 3.14×2 cm.
The calculator in the photograph gives the approximation 3.1415927×2 cm.
The **exact** circumference is $\pi \times 2$ cm.
We cannot write π using only digits.
We would be writing forever!

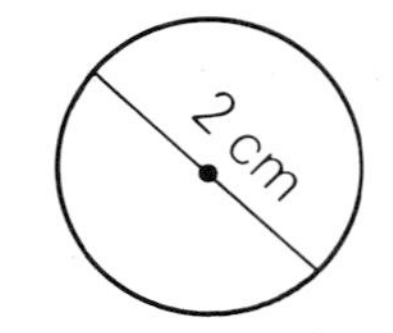

We usually write this as $2 \times \pi$ cm or 2π cm.

6 A circular sewer inspection cover has **radius** 60 cm.

a) Find its approximate circumference (i) in centimetres. (ii) in metres.
Use $\pi \approx 3.14$.

b) Write down its exact circumference, using π.

7 Old haycarts had metal strips around their wheels.
In an original design, the metal strip was to measure 3.3 m.

a) Roughly, what was the **radius** of the wheel? Use $\pi \approx 3.1$.
b) Write down its **exact** radius, by using π in your result.

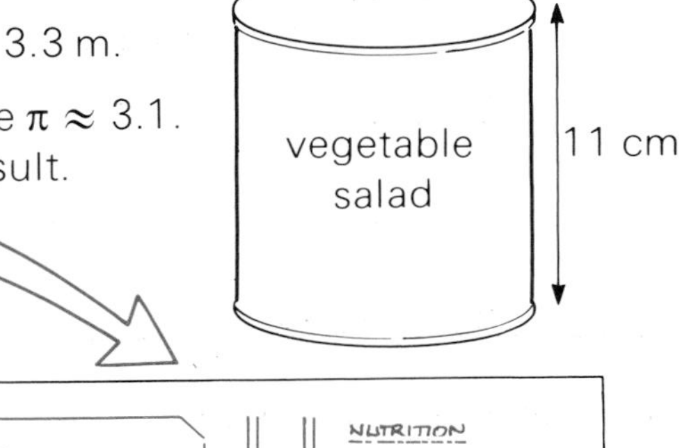

8 The tin of vegetable salad needs a designer label.
Its height is 11 cm, and its diameter is 9 cm.
What size do you think the label should be?

9 The diameter of the moon is about 3475 km.
Roughly what is the greatest circular distance you can travel in a moon buggy on the moon?

10 The circumference of a circle is 24π cm.
What is a) its radius? b) its diameter?

Think it through

11 Each of these are parts of circular discs.

Use $\pi \approx 3.1$.

a) Approximately how long is the curved part of each one?

(i)

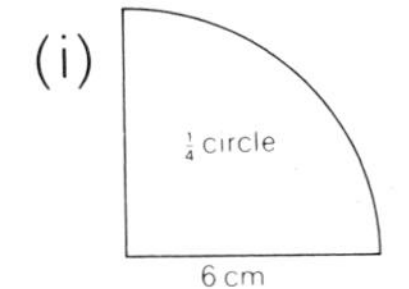

(ii)

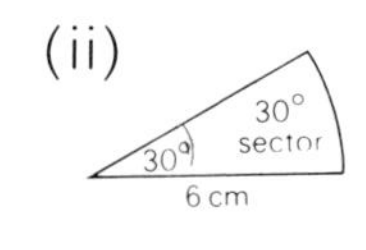

(iii)

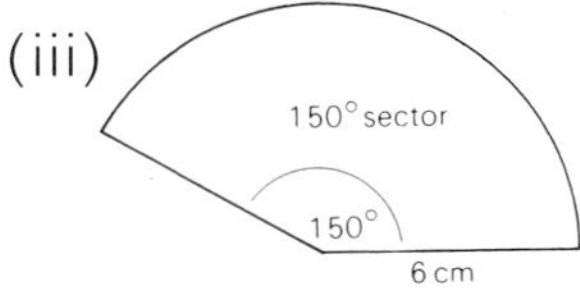

b) What is the **exact** length of the curved part of each one?

Circles and their areas

20

C 1 a) These expressions give the areas of the large square, each of the four smaller squares, and the 'dotted' square.
Copy and complete each one.

$5^2 \times \square$ cm², $\square^2$ cm², $5^2 \times \square$ cm²

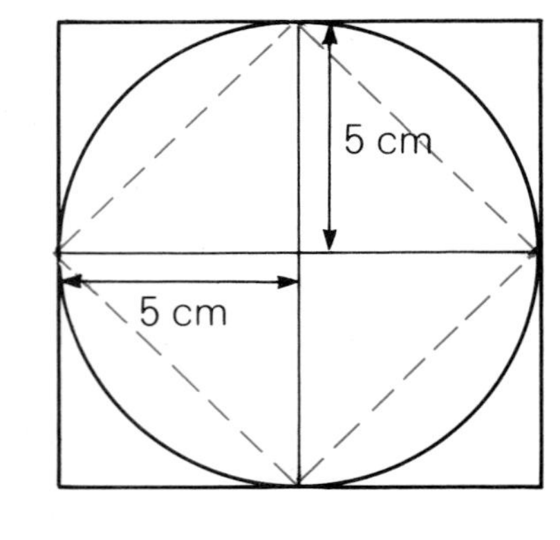

b) Which of these do you think is most nearly correct?
The area of the circle is

A $5^2 \times 4$ cm²
B $5^2 \times 3$ cm²
C $5^2 \times 5$ cm²
D $5^2 \times 2$ cm²

c) Write one or two sentences to explain why you made this choice.

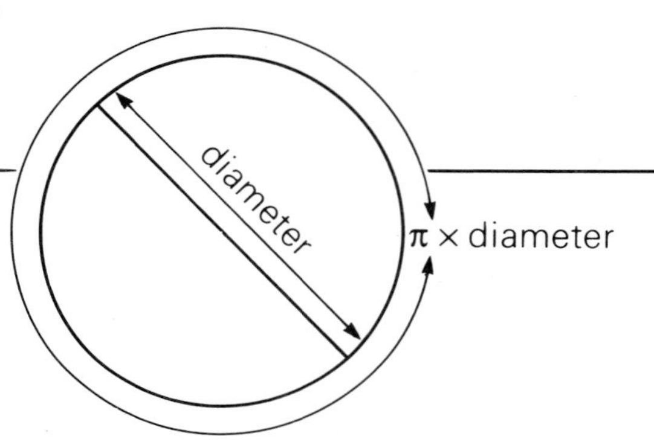

2 One of these is the correct rule (or 'formula') for the area of a circle.

A area of a circle
$= \frac{1}{2} \times \pi \times$ diameter

B area of a circle
$= \frac{1}{8} \times \pi \times$ diameter $\times$ diameter

C area of a circle
$= \frac{1}{2} \times \pi \times$ radius $\times$ radius

D area of a circle
$= \pi \times$ radius $\times$ radius

From the drawings, decide between you which one is most likely.
Each of you write down why you made this choice.

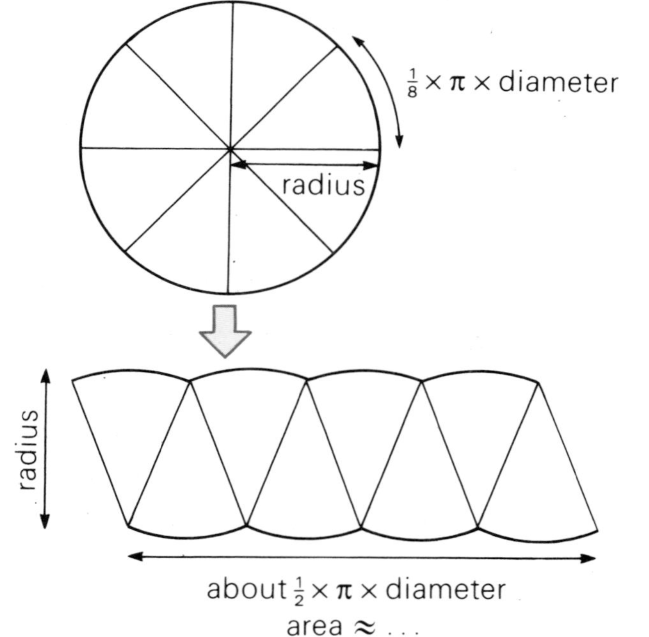

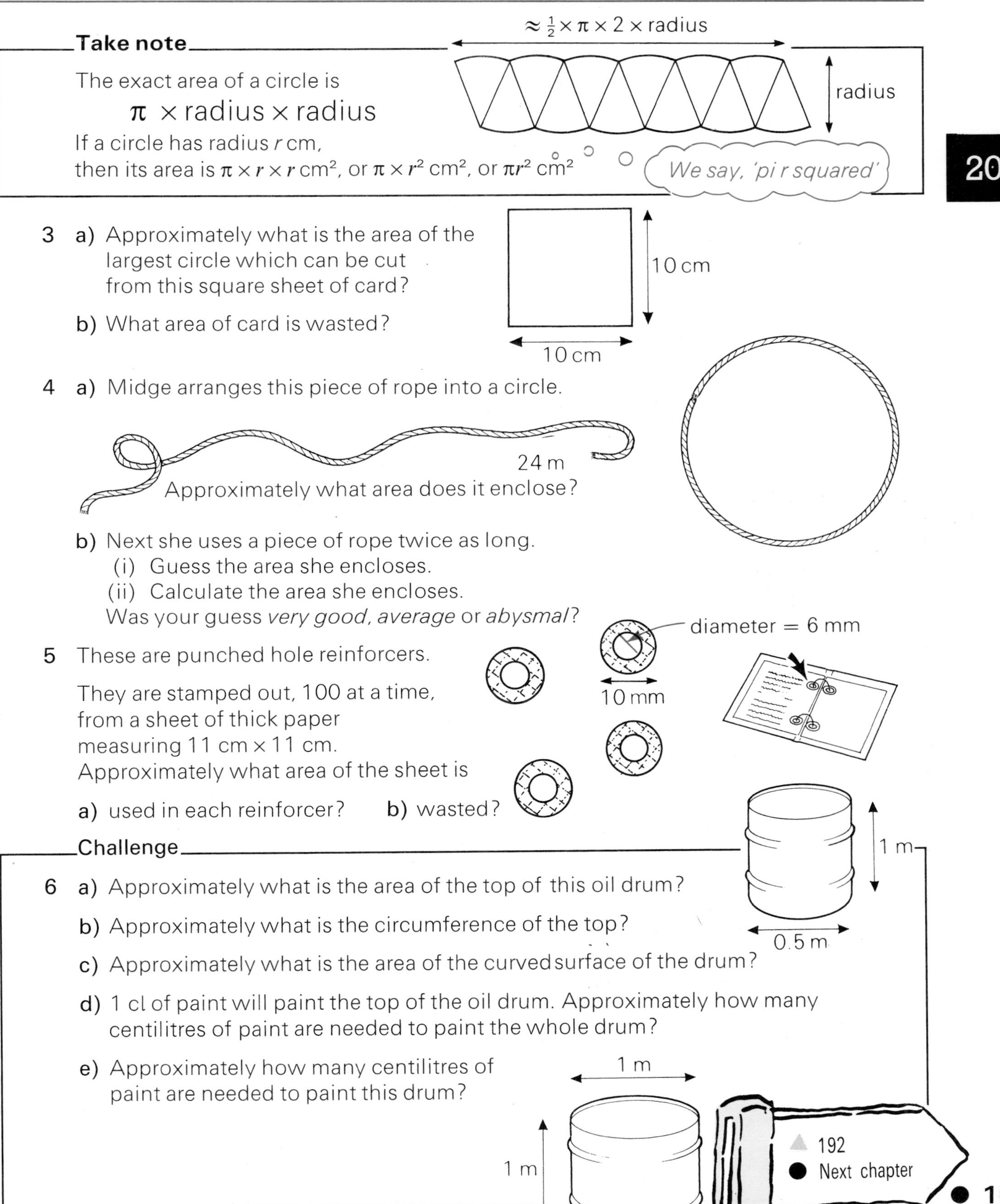

Take note

The exact area of a circle is

$$\pi \times \text{radius} \times \text{radius}$$

If a circle has radius r cm,
then its area is $\pi \times r \times r$ cm², or $\pi \times r^2$ cm², or πr^2 cm²

3 a) Approximately what is the area of the largest circle which can be cut from this square sheet of card?

b) What area of card is wasted?

4 a) Midge arranges this piece of rope into a circle.

Approximately what area does it enclose?

b) Next she uses a piece of rope twice as long.
   (i) Guess the area she encloses.
   (ii) Calculate the area she encloses.
   Was your guess *very good*, *average* or *abysmal*?

5 These are punched hole reinforcers.

They are stamped out, 100 at a time, from a sheet of thick paper measuring 11 cm × 11 cm.
Approximately what area of the sheet is

a) used in each reinforcer? b) wasted?

Challenge

6 a) Approximately what is the area of the top of this oil drum?

b) Approximately what is the circumference of the top?

c) Approximately what is the area of the curved surface of the drum?

d) 1 cl of paint will paint the top of the oil drum. Approximately how many centilitres of paint are needed to paint the whole drum?

e) Approximately how many centilitres of paint are needed to paint this drum?

▲ 192
● Next chapter

Working with circles

D 1 This is a Big Wheel by the river Rhein in West Germany.
Its diameter is 63.5 m.
Approximately how far apart (measured along the circumference of the wheel) are the chairs?

2 This is a penny-farthing bicycle.
The diameter of the large wheel is about 2 m.
The diameter of the small wheel is about 0.5 m.

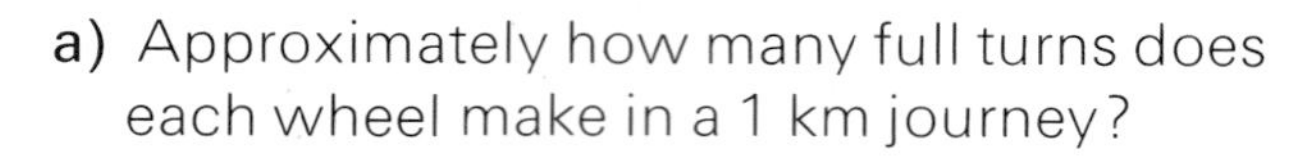

a) Approximately how many full turns does each wheel make in a 1 km journey?

b) How much more quickly would you expect the back wheel to wear out than the front wheel?

3 The drawing shows Billy the goat at the end of his tether.

He can just reach 20 m away from the corner of his owner's house.
Approximately what area of grass can he eat?
(Use $\pi \approx 3.1$.)

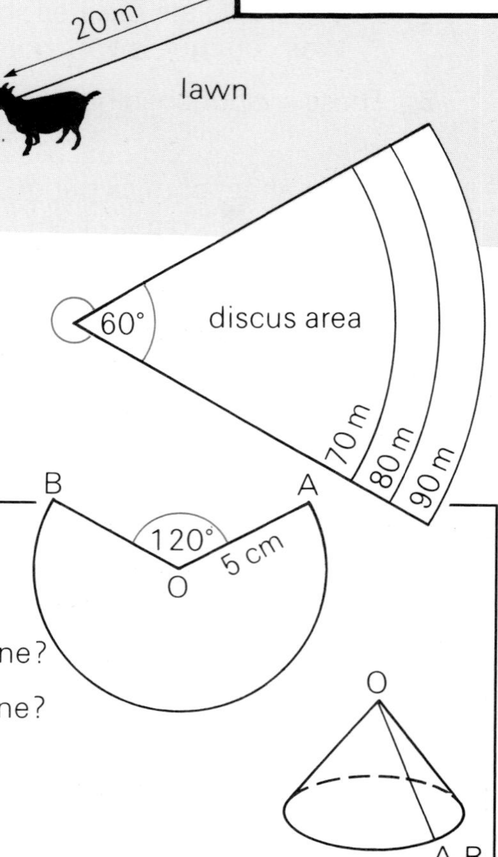

4 This is the discus area on a sports field.
Calculate

a) the **exact** length of the curved 80 m line, in metres.
b) the **exact** area of field needed for the discus area, up to a 100 m throw, in square metres.
c) the **exact** area enclosed between the 70 m and 80 m lines, in square metres.

Activity *Making a cone*

5 This sector of a circle will make the cone shown below it.

a) What is the circumference of the base of the cone?

b) What is the area of the curved surface of the cone?

c) What is the radius of the base of the cone?

d) Make a cone whose base radius is π cm.

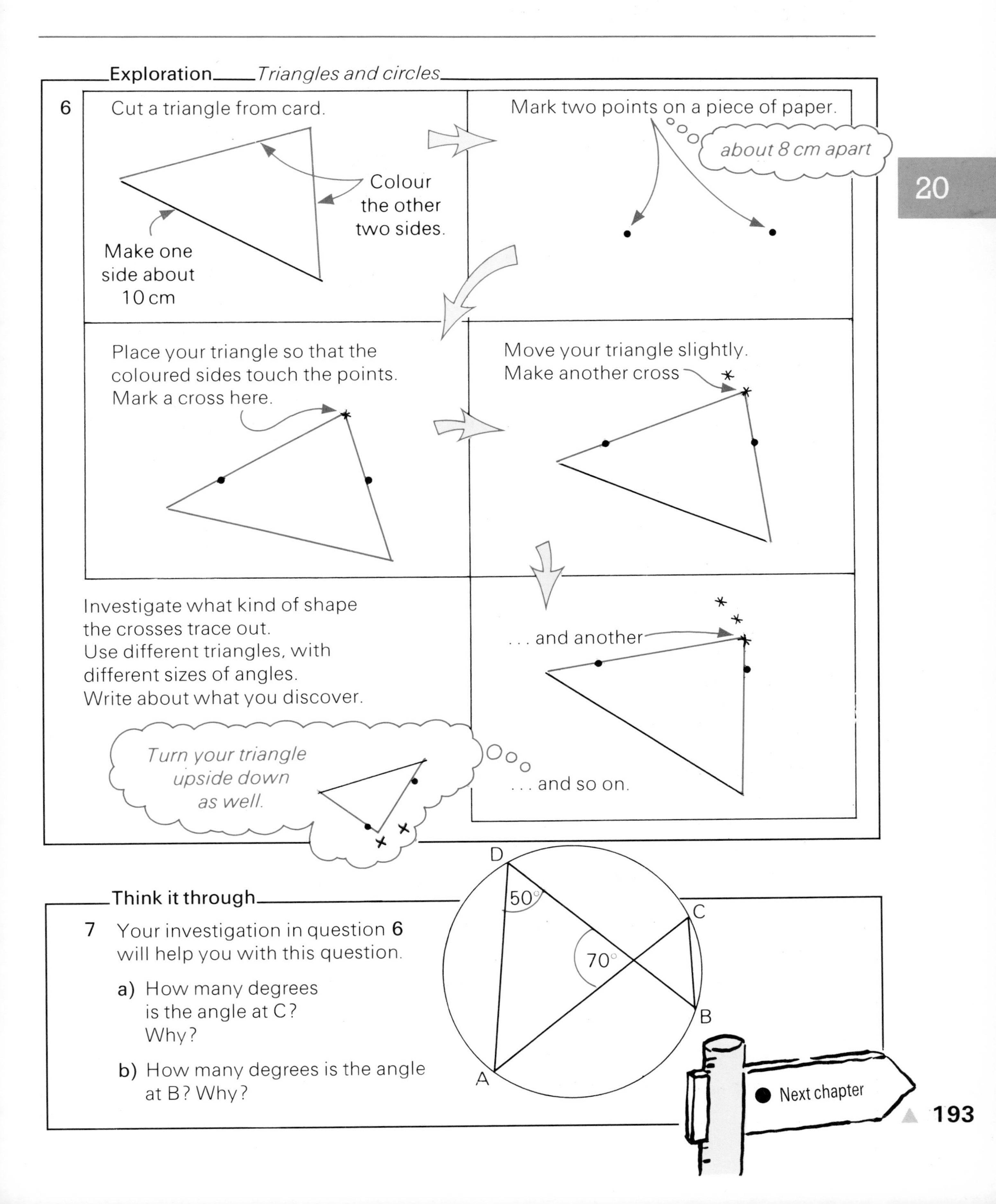
Exploration
Triangles and circles
6
Cut a triangle from card.
Colour the other two sides.
Make one side about 10 cm
Mark two points on a piece of paper.
about 8 cm apart
Place your triangle so that the coloured sides touch the points.
Mark a cross here.
Move your triangle slightly.
Make another cross
Investigate what kind of shape the crosses trace out.
Use different triangles, with different sizes of angles.
Write about what you discover.
. . . and another
Turn your triangle upside down as well.
. . . and so on.
Think it through
7 Your investigation in question 6 will help you with this question.
a) How many degrees is the angle at C? Why?
b) How many degrees is the angle at B? Why?
D
50°
C
70°
B
A
Next chapter
20

21 Working with fractions

21

A 1 Do these in your head.
Write down only the answers.

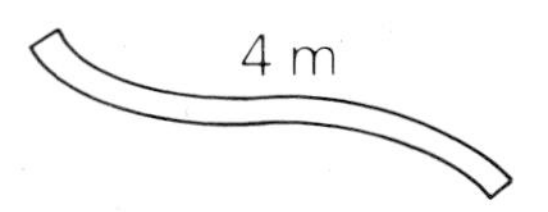

a) The flour is separated into $\frac{1}{2}$ kg piles. How many piles is this?

b) The flour is separated into 15 equal piles. How many kilograms is this in each pile?

c) The plastic strip is cut into $\frac{1}{3}$ m lengths. How many pieces is this?

d) The canoe travels 10 km at $2\frac{1}{2}$ km per hour. How long does the journey take?

e) Every half hour Greg walks $\frac{2}{3}$ km. How far does he walk in 3 hours?

f) A robot shuffles $\frac{1}{2}$ km at $\frac{1}{4}$ km per hour. How many hours does this take?

2 Write down the correct sign (+, ÷, × or −) in each of these sentences.
Your results in question **1** will help you.

a) $3 \ ? \ \frac{1}{2} = 6$ b) $5 \ ? \ 15 = \frac{1}{3}$ c) $4 \ ? \ \frac{1}{3} = 12$

d) $10 \ ? \ 2\frac{1}{2} = 4$ e) $\frac{2}{3} \ ? \ 6 = 4$ f) $\frac{1}{2} \ ? \ \frac{1}{4} = 2$

3 Copy and complete:

a) $12 \times \frac{1}{2} = \square$ b) $4 \div \frac{1}{2} = \square$ c) $7 \div \frac{1}{4} = \square$ d) $8 \times \frac{1}{4} = \square$

e) $8 \div \frac{1}{4} = \square$ f) $5 \times \frac{1}{4} = \square$ g) $2 \div \frac{1}{5} = \square$ h) $2\frac{1}{5} \div \frac{1}{5} = \square$

4 This diagram shows that $2 \div \frac{2}{3} = 3$, and $3 \times \frac{2}{3} = 2$.

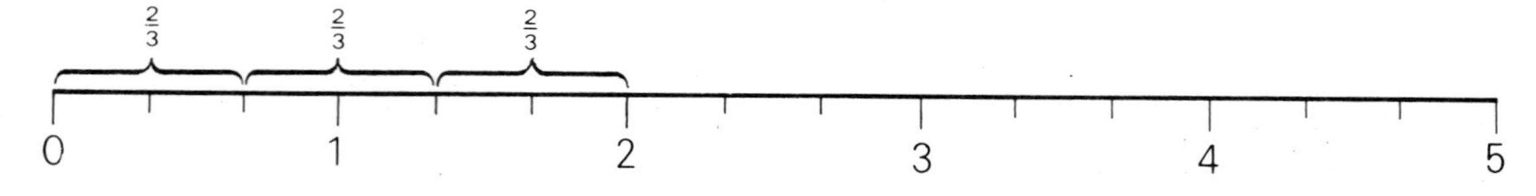

Use it to help you do these calculations:

a) $4 \div \frac{2}{3} = \square$ b) $\frac{2}{3} \times 6 = \square$ c) $3\frac{1}{3} \div \frac{2}{3} = \square$

Think it through

5 Draw a diagram to help you to do these calculations:

a) $4 \div \frac{4}{5} = \square$ b) $6 \div \frac{3}{5} = \square$ c) $3\frac{3}{5} \div 9 = \square$

6 Pepper is doing this division:

$5\frac{3}{5} \div 7$

This is her working. 

$5\frac{3}{5}$ $5 \times 5 \rightarrow 25$
$+ \ 3$
28
$28 \div 7 = 4$
Answer $\frac{4}{5}$

She is not very neat!
And she doesn't make things very clear.
Study carefully what Pepper has done, then write an explanation for each of these.

a) Why did she multiply 5×5, then add 3?
What does the result tell her?

b) Why did she divide 28 by 7 to get 4, and then write $\frac{4}{5}$?

7 Use Pepper's method to do this:

$8\frac{4}{5} \div 11$

8 Use any method you like to do this:

$7\frac{1}{12} \div 17$

9 This is another example of Pepper's divisions:

$4\frac{4}{9} \div 5$
$4 \times 4 \rightarrow 16$
$+ \ 9$
25
$25 \div 5 = 5$
Answer $5/9$

This time she has made mistakes.
Find out what they are.
Rewrite her working correctly.

21

10 Here are some different ways, using $\div$, of getting $\frac{3}{5}$l

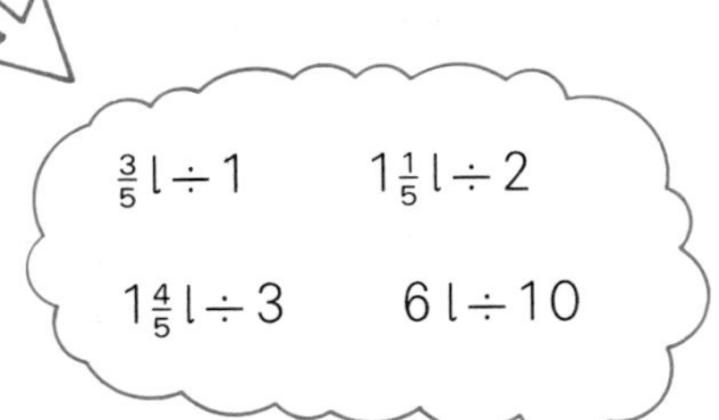

Write four different ways, using $\div$, of getting this amount.

$\frac{2}{3}$ kg

11 Here are some different ways, using $\times$, of getting $\frac{3}{8}$ km.

$\frac{3}{8}\text{km} \times 1$ $\frac{3}{40}\text{km} \times 5$
$\frac{3}{16}\text{km} \times 2$ $\frac{1}{2} \times \frac{3}{4}\text{km}$

Write four different ways, using $\times$, of getting this amount.

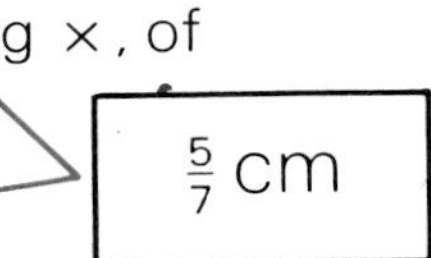

Multiplying fractions

B 1

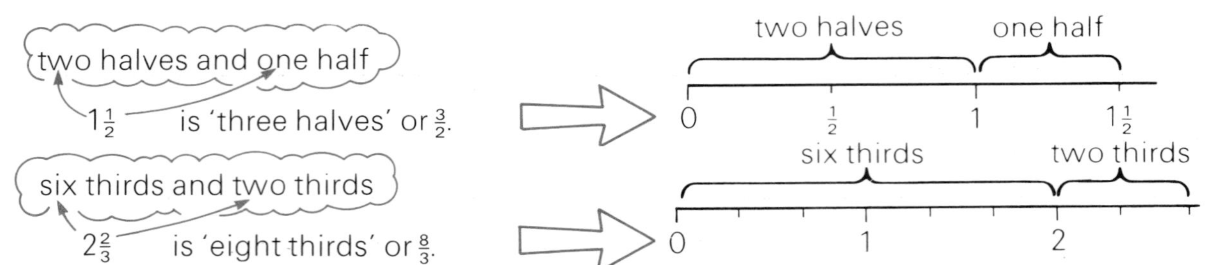

a) Write each of these in the form $\frac{\square}{\square}$.

(i) $2\frac{1}{4}$ (ii) $3\frac{2}{3}$ (iii) $4\frac{7}{8}$ (iv) $5\frac{9}{10}$ (v) $11\frac{5}{9}$

b) Write each of these in the form $\square\frac{\square}{\square}$.

(i) $\frac{7}{3}$ (ii) $\frac{9}{5}$ (iii) $\frac{13}{7}$ (iv) $\frac{63}{13}$ (v) $\frac{100}{19}$

2 a) Do these calculations:

(i) $(4 \div 2) \times (6 \div 3)$ (ii) $(12 \div 6) \times (9 \div 3)$ (iii) $(25 \div 5) \times (8 \div 4)$

b) Write down a similar calculation of your own, and find the result.

c) Calculate

(i) $(4 \times 6) \div (2 \times 3)$ (ii) $(12 \times 9) \div (6 \times 3)$ (iii) $(25 \times 8) \div (5 \times 4)$

d) Compare your results in **(a)** and **(c)**. Write down what you notice.

e) Copy and complete:

$$(8 \div 4) \times (9 \div 3) = (\square \times \square) \div (\square \times \square)$$

f) Try some more examples of your own like the divisions and multiplications in **(a)** and **(c)**.

Is what you found in **(d)** always true?
If you say **no**, give an example where it is not true.

g) (i) Write this multiplication as two divisions multiplied together.

$\frac{2}{3} \times \frac{5}{7}$

as in ***(c)***

(ii) Now write it as one multiplication divided by another multiplication.

(iii) Copy and complete:

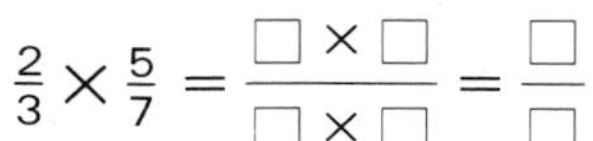

$$\frac{2}{3} \times \frac{5}{7} = \frac{\square \times \square}{\square \times \square} = \frac{\square}{\square}$$

h) Do these calculations:

$\frac{3}{2}$

(i) $\frac{4}{5} \times \frac{2}{3}$ (ii) $\frac{3}{8} \times \frac{5}{9}$ (iii) $1\frac{1}{2} \times \frac{3}{5}$ (iv) $2\frac{1}{2} \times \frac{7}{8}$

(v) $2\frac{1}{3} \times 3\frac{1}{3}$ (vi) $\frac{5}{8} \times 3\frac{1}{2}$ (vii) $1\frac{1}{2} \times \frac{2}{3} \times \frac{5}{8}$

Take note

$\frac{2}{3} \times \frac{4}{5} = \frac{2 \times 4}{3 \times 5} = \frac{8}{15}$ $\qquad$ $\frac{7}{3} \times 4 = \frac{7}{3} \times \frac{4}{1} = \frac{7 \times 4}{3 \times 1} = \frac{28}{3}$

3 a) After how many multiplications is $\frac{2}{3} \times \frac{2}{3} \times \frac{2}{3} \times \frac{2}{3} \ldots$ less than $\frac{1}{10}$?
   b) After how many multiplications is $1\frac{1}{3} \times 1\frac{1}{3} \times 1\frac{1}{3} \times 1\frac{1}{3} \times \ldots$ greater than 10?

4 A rectangular lawn measures $8\frac{1}{2}$ m by $3\frac{1}{3}$ m.
What is its area?

the result when multiplied together

5 Find three different pairs of fractions whose product is $\frac{24}{35}$.

6 $\frac{2}{3} \times \frac{a}{b} = \frac{8}{9}$ $\qquad$ What is $\frac{a}{b}$?

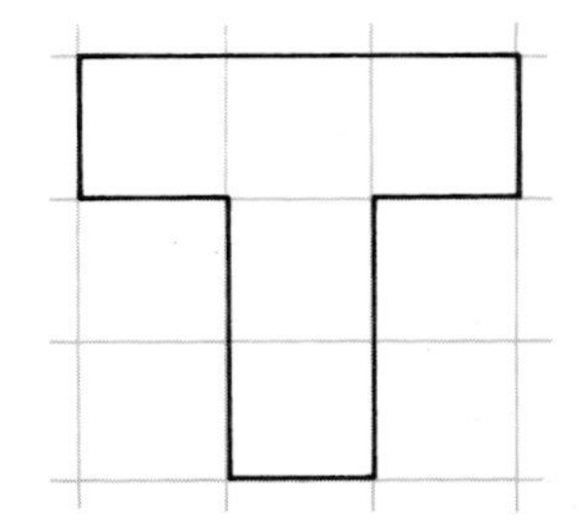

7 The letter **T** shown here is enlarged by a photocopier using a scale factor $\times 1\frac{1}{2}$.
The resulting **T** is enlarged again using a scale factor $\times 2\frac{1}{3}$.
How tall and how wide is the final **T**?

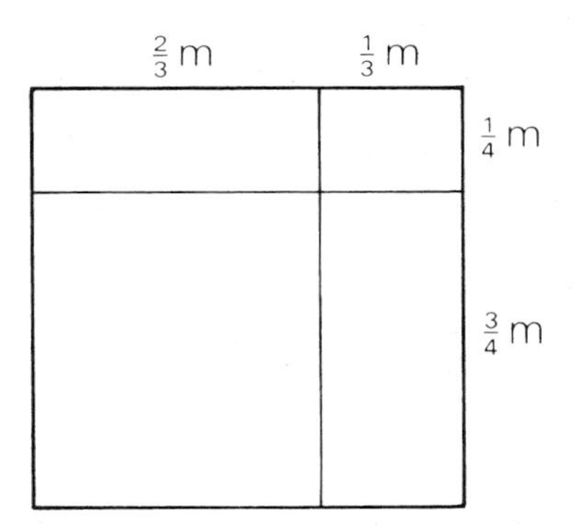

8 Calculate the area of glass needed for each section of the window.
Check that your four areas amount to 1 m² of glass.

For example, 6 ÷ 12 is the same as 1 ÷ 2, so $\frac{6}{12} = \frac{1}{2}$.

9 a) Copy and complete the pattern.
   ⟶ means 'multiply by $\frac{2}{3}$'.
   ↓ means 'multiply by $\frac{3}{4}$'.
   Write each fraction as simply as you can.

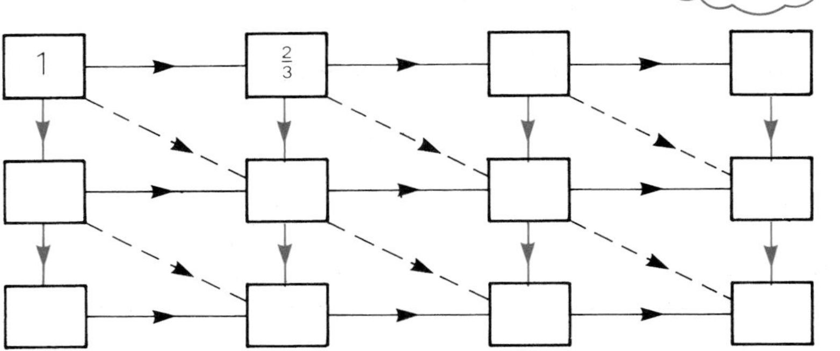

   b) What does - -▸- - mean?

Dividing by a fraction

C 1 a) Do these calculations:
(i) $(8 \div 4) \div (12 \div 6)$ (ii) $(24 \div 4) \div (4 \div 2)$ (iii) $(30 \div 6) \div (4 \div 8)$

b) Write down a similar calculation of your own, and find the result.

c) Calculate
(i) $(8 \div 4) \times (6 \div 12)$ (ii) $(24 \div 4) \times (2 \div 4)$ (iii) $(30 \div 6) \times (8 \div 4)$

d) Compare your results in **(a)** and **(c)**.
Write down what you notice.

e) Copy and complete:
$(8 \div 5) \div (9 \div 4) = (\square \div \square) \times (\square \div \square)$.

f) Try some more examples of your own like the divisions and multiplications in **(a)** and **(c)**.

Is what you found in **(d)** always true?
If you say **no**, give an example where it is not true.

g) (i) Write this division as the division of two whole-number divisions.
$\frac{2}{3} \div \frac{3}{5}$

(ii) Now write the division as a multiplication of two whole-number divisions.
(iii) Copy and complete:

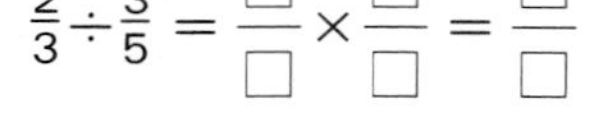

$$\frac{2}{3} \div \frac{3}{5} = \frac{\square}{\square} \times \frac{\square}{\square} = \frac{\square}{\square}$$

h) Copy and complete:

(i) $\frac{3}{4} \div \frac{2}{5} = \frac{3}{4} \times \frac{\square}{\square}$
$= \frac{3 \times \square}{4 \times \square}$
$= \frac{\square}{\square}$

(ii) $1\frac{1}{2} \div 2\frac{1}{4} = \frac{\square}{\square} \div \frac{\square}{\square}$
$= \frac{\square}{\square} \times \frac{\square}{\square}$
$= \frac{2}{\square}$

i) Do these calculations:
(i) $\frac{5}{6} \div \frac{8}{9}$ (ii) $1\frac{1}{3} \div 2\frac{1}{3}$ (iii) $4\frac{1}{2} \div \frac{3}{4}$

j) Write down a division of fractions of your own which gives the result
(i) $\frac{2}{3}$ (ii) $\frac{4}{7}$

Take note

$\frac{2}{3} \div \frac{4}{7}$ gives the same result as $\frac{2}{3} \times \frac{7}{4}$, which is $\frac{14}{12}$ (or $\frac{7}{6}$).

2 Find three different pairs of fractions whose quotient is $\frac{2}{3}$.

the result when one is divided by the other

3 $\frac{2}{3} \div \frac{c}{d} = \frac{6}{21}$. What is $\frac{c}{d}$?

4 Calculate (i) $\frac{2}{3} \div (\frac{4}{7} \times \frac{1}{2})$. (ii) $(\frac{2}{3} \div \frac{4}{7}) \times \frac{1}{2}$.

5 a) How many divisions are needed before the result is greater than 10?

$\frac{2}{3} \div \frac{2}{3} \div \frac{2}{3} \div \frac{2}{3} \div \frac{2}{3} \ldots$

b) How many divisions are needed before the result is less than 0.1?

$1\frac{1}{2} \div 1\frac{1}{2} \div 1\frac{1}{2} \div 1\frac{1}{2} \div 1\frac{1}{2} \ldots$

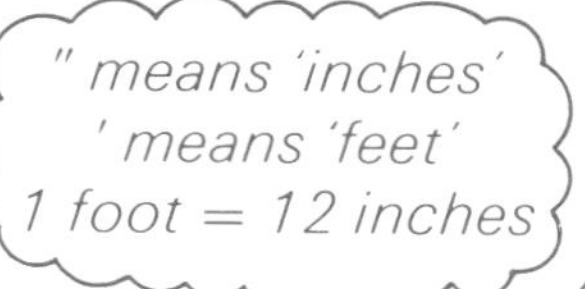

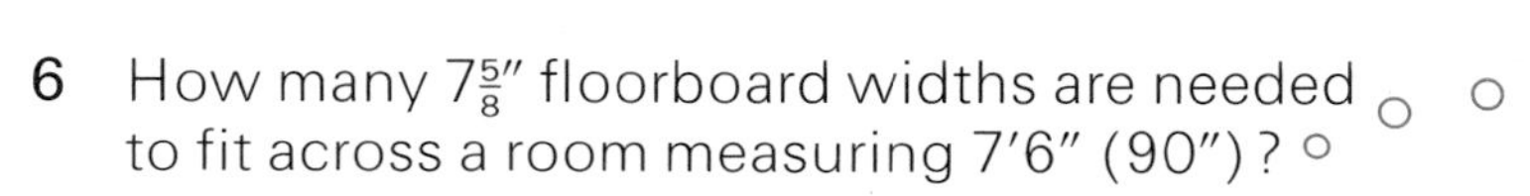

6 How many $7\frac{5}{8}$" floorboard widths are needed to fit across a room measuring 7'6" (90")?

7 Continue dividing up the 1 m square like this.

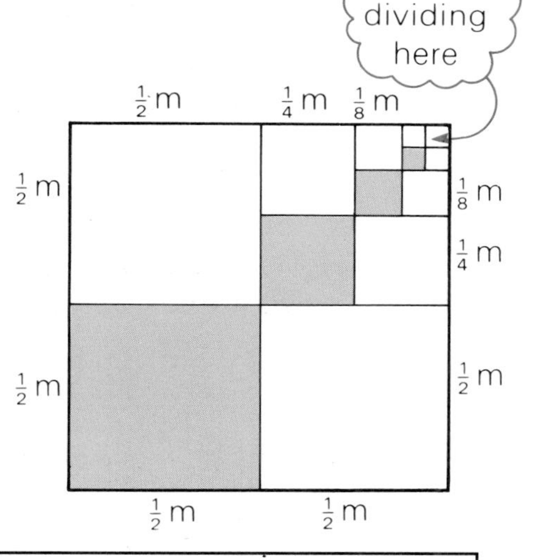

a) What is the area of (i) the largest red square?
(ii) the next largest red square?
(iii) the fifth largest red square?

b) If the pattern is continued, what fraction of the whole square will eventually be red?

c) What result does this 'infinite' addition give?

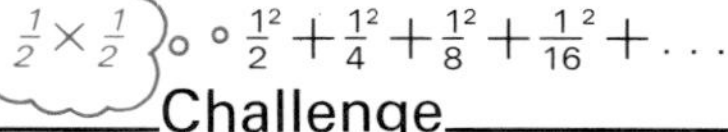

$\frac{1}{2}^2 + \frac{1}{4}^2 + \frac{1}{8}^2 + \frac{1}{16}^2 + \ldots$

Challenge

8 An approximation which is often used for π is $3\frac{1}{7}$.

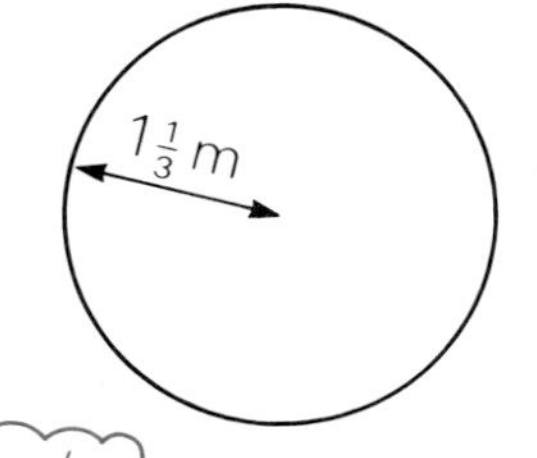

a) Roughly what is
(i) the circumference ($C = \pi \times D$)
(ii) the area ($A = \pi \times r^2$)
of a circle whose radius is $1\frac{1}{3}$ m?
Write your results using mixed numbers.

for example, $7\frac{1}{3}$, $19\frac{1}{4}$ ...

b) The circumference of a circle is $12\frac{1}{2}$ cm.
Roughly what is its diameter?
Write your result using a fraction.
Use $\pi \approx 3\frac{1}{7}$

Calculating with fractions

D 1 $\frac{k}{3} \times \frac{4}{m} = \frac{5}{6}$.

Find k and m.

2 $\frac{p}{q}$ and $\frac{r}{s}$ are two fractions.

Copy and complete each of these statements:

a) $\frac{p}{q} \times \frac{r}{s} = \frac{\square \times \square}{\square \times \square}$ b) $\frac{p}{q} \div \frac{r}{s} = \frac{\square \times \square}{\square \times \square}$

3 $\frac{a}{b}$ and $\frac{c}{d}$ are two fractions.

a) Use $a = 4,\ b = 2,\ c = 6,\ d = 3,$ to decide which of these are definitely **not** correct statements:

(i) $\frac{a}{b} + \frac{c}{d} = \frac{a+c}{b+d}$ (ii) $\frac{a}{b} + \frac{c}{d} = \frac{a+c}{b \times d}$

(iii) $\frac{a}{b} + \frac{c}{d} = \frac{a \times c}{b+d}$ (iv) $\frac{a}{b} + \frac{c}{d} = \frac{(a \times d) + (c \times b)}{b \times d}$

b) For the statement you did not reject in **(a)**,
try these values for a, b, c and d:
$a = 8,\ b = 2,\ c = 12,\ d = 3$.
Does this also give a correct statement?

c) Try some more values of a, b, c and d in **(a)** (iv).
Do you always get a correct statement?

d) Use what you found in **(c)** to write down the result of

this addition: $\frac{2}{7} + \frac{1}{4}$

e) (i) Explain why $\frac{a}{b} \times \frac{d}{d} = \frac{a}{b}$. (ii) Explain why $\frac{c}{d} \times \frac{b}{b} = \frac{c}{d}$.

(iii) Use your results in (i) and (ii) to explain why

$$\frac{a}{b} + \frac{c}{d} = \frac{a \times d}{b \times d} + \frac{c \times b}{d \times b}$$

f) Use the expression in **(e)** (iii) to write these additions as additions of two fractions with the same denominators:

(i) $\frac{2}{7} + \frac{1}{4}$ (ii) $\frac{3}{8} + \frac{2}{9}$

22 Rules and graphs

A 1 Meg is collecting sticks and pebbles into piles.
She is using this rule to make each pile:

number of sticks in each pile = (number of pebbles in the pile − 1) × 2

Here are three of her collections:

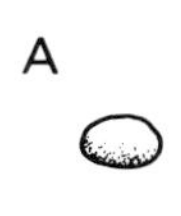

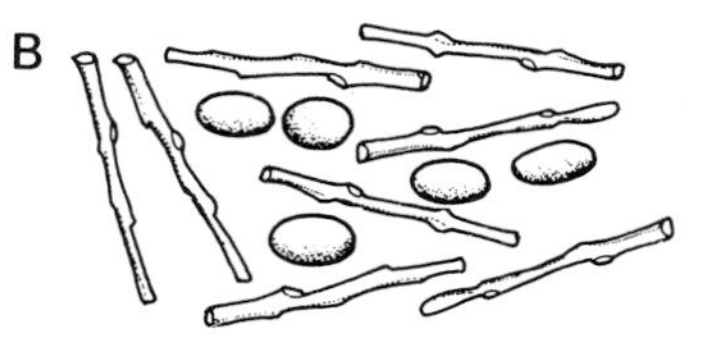

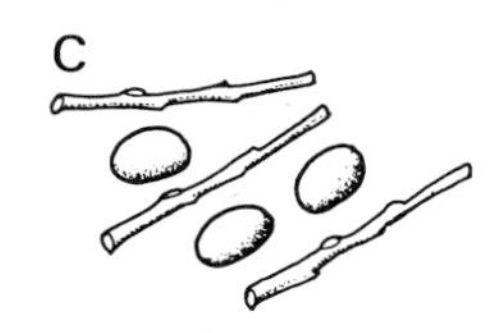

a) One of the piles doesn't fit the rule. Which one?

b) It is not possible for one of Meg's piles to have the number of sticks in the 'odd pile out'. Why?

c) How should the number of sticks in the 'odd pile out' be adjusted so that the pile obeys Meg's rule?

d) Meg removes a stick from pile **C**.
How many pebbles should she remove so that the result obeys her rule?

e) Copy and complete this table to show how many sticks and pebbles are needed for different piles.

Number of pebbles in a pile	1	2	3	5		10
Number of sticks in the pile	0			8	14	

f) **Do you remember . . .?**

We can use letters to help us write rules more simply.
For example,

numbers of rings = number of stars + 2

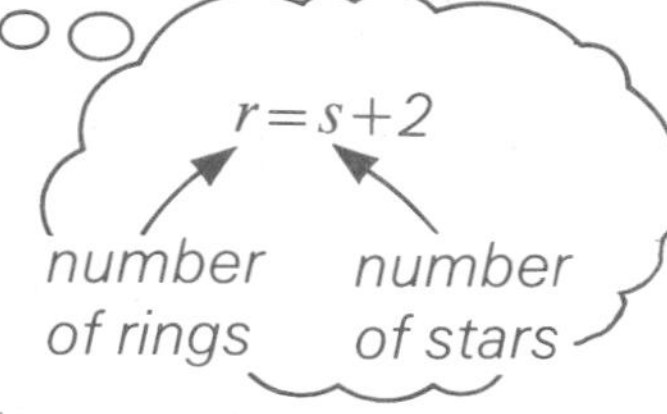

Write Meg's 'sticks and pebbles' rule using letters.

2 p is the number of pins; n is the number of needles.
Write down what this rule tells us.
$p = (4 \times n) - 3$
Give three examples for the number of pins and the corresponding number of needles.

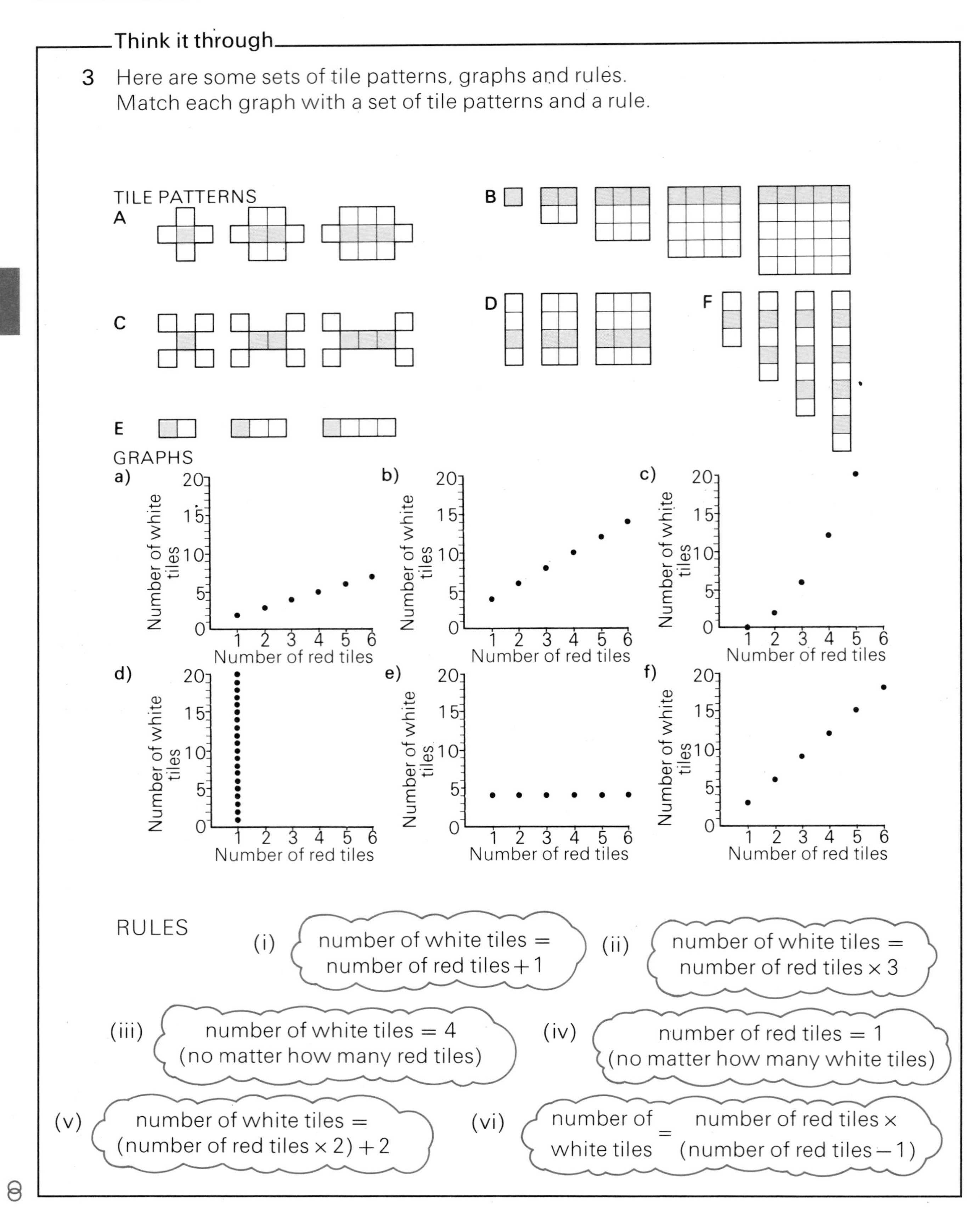
Think it through
3 Here are some sets of tile patterns, graphs and rules.
Match each graph with a set of tile patterns and a rule.
TILE PATTERNS
A
B
C
D
E
F
GRAPHS
a)
b)
c)
d)
e)
f)
Number of white tiles
Number of red tiles
0 5 10 15 20
1 2 3 4 5 6
RULES
(i) number of white tiles = number of red tiles + 1
(ii) number of white tiles = number of red tiles × 3
(iii) number of white tiles = 4 (no matter how many red tiles)
(iv) number of red tiles = 1 (no matter how many white tiles)
(v) number of white tiles = (number of red tiles × 2) + 2
(vi) number of white tiles = number of red tiles × (number of red tiles − 1)

From rules to graphs . . . and back

B **Take note**

We can 'picture' rules using graphs.

Challenge

1 This collection of tiled rectangles may just look like a mess! But there is a rule which connects the numbers of red tiles and white tiles in each rectangle.

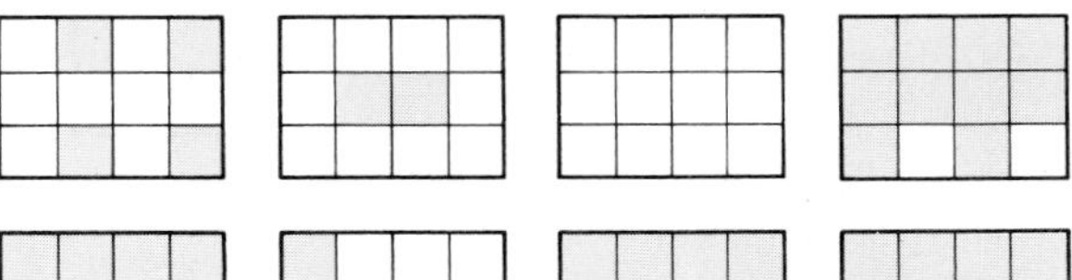

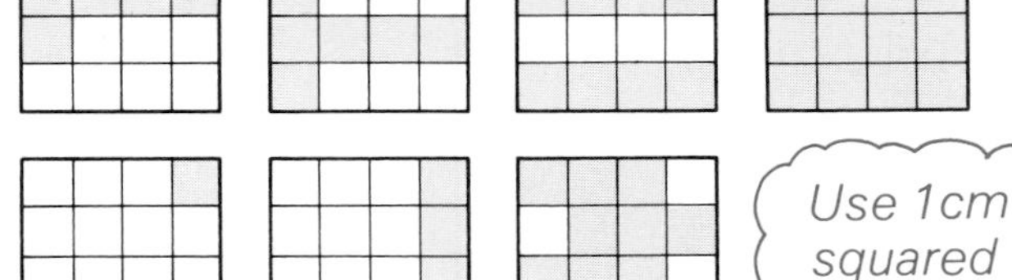

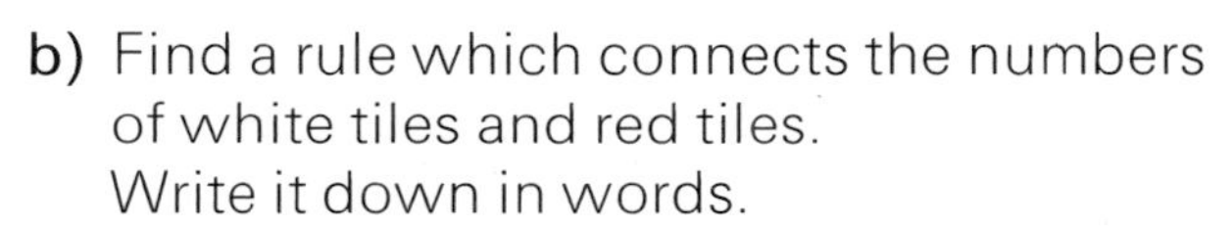

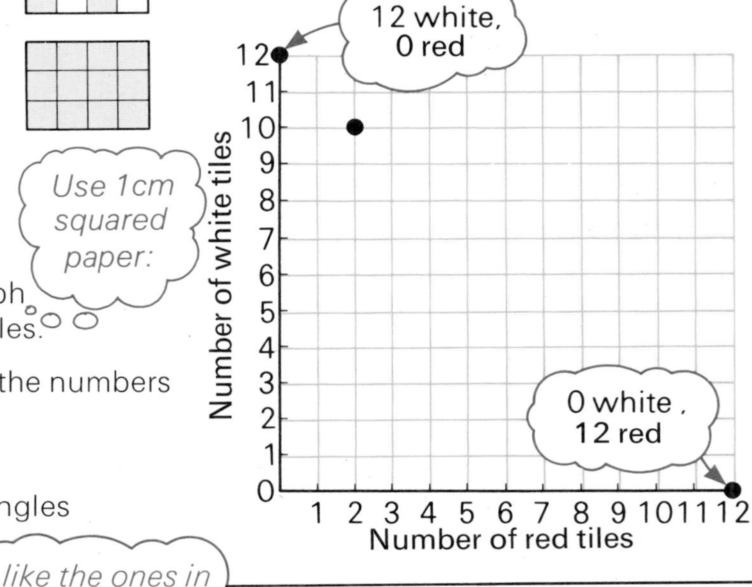

Use 1cm squared paper:

a) Copy and complete the graph for the collection of rectangles.

b) Find a rule which connects the numbers of white tiles and red tiles. Write it down in words.

c) Sketch two more tiled rectangles which obey the same rule.

like the ones in question **1**

2 Here is a shorthand rule for another set of tiled rectangles.

$$w + r = 10$$

w	0	1	2	3	4	5	6	7	8	9	10
r	10				6						0

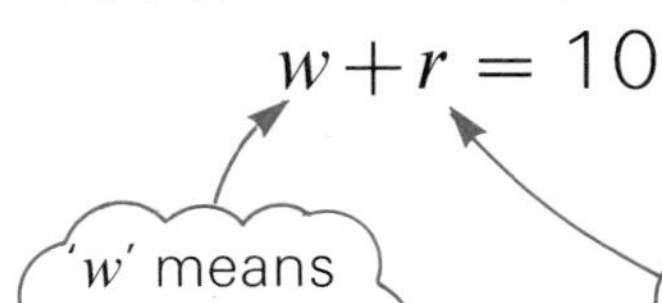

'r' means 'number of red tiles'

a) Write the rule in words.

b) Draw two different tiled rectangles which obey the rule.

c) Copy and complete the table.

d) Copy and complete the graph.

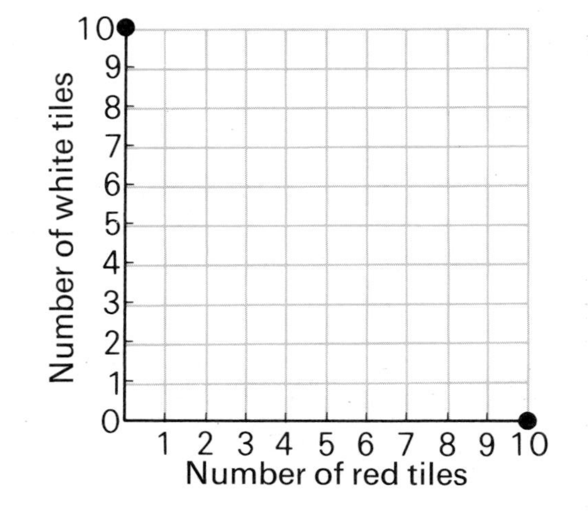

C 3 These letter **L**s are as tall as they are long.
They are all 1 cm wide.

a) Copy and complete this table for **L**s like them.

Height (cm)	2	3	4	5	6	7	8
Area (cm²)							

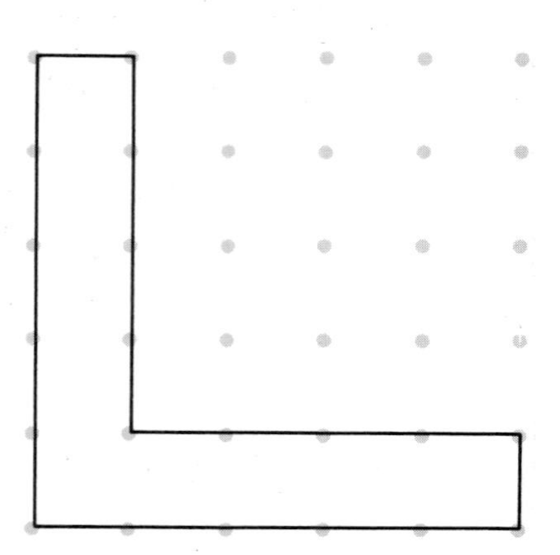

b) From your table, predict the area of an **L** which is 18 cm tall.

c) This rule connects the area and the height.

$$A = (\square \times h) - \square$$

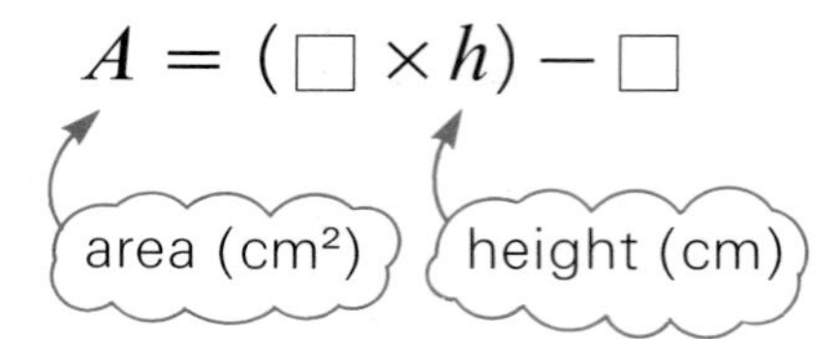

Copy and complete it.

d) The area of an **L** 40 cm tall is 79 cm².
Check that your rule agrees with this result.

e) This is a graph of the rule.
Use it to check your results in **(a)**.

f) Use the graph to predict the area of an **L** which is $3\frac{1}{2}$ cm tall.
Check your prediction by drawing an **L**.

g) Does the rule work for an **L** which is $\frac{1}{2}$ cm tall?
Draw an **L** to explain your answer.

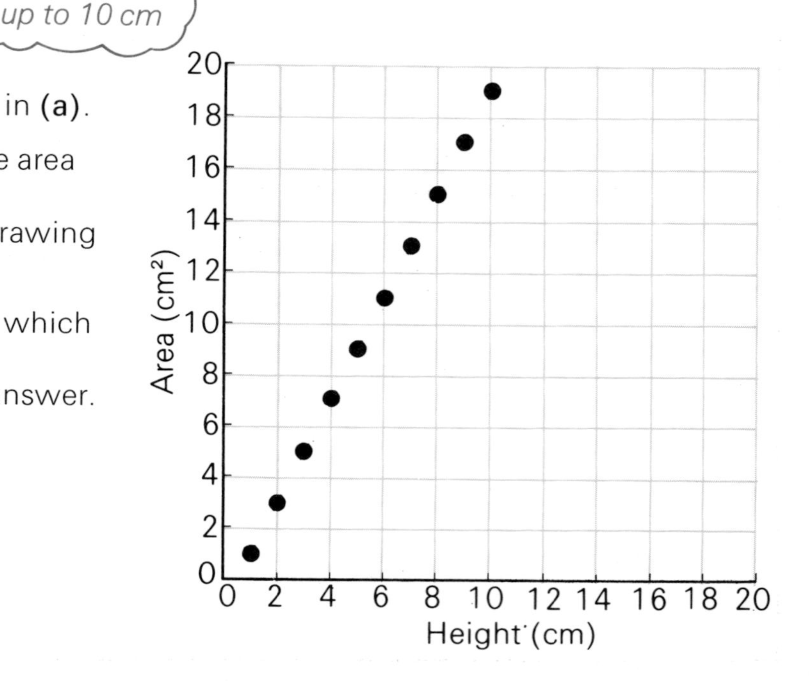

Assignment

4 These letter **C**s are all as wide as they are tall.

a) Copy and complete this table for **C**s up to 8 cm tall.

Height (cm)	3	4	5	6	7	8
Area (cm²)						

b) Copy and complete this rule:

$$A = (\square \times h) - \square$$

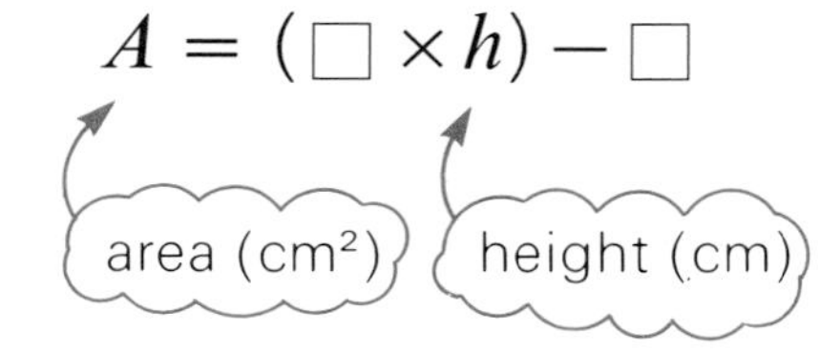

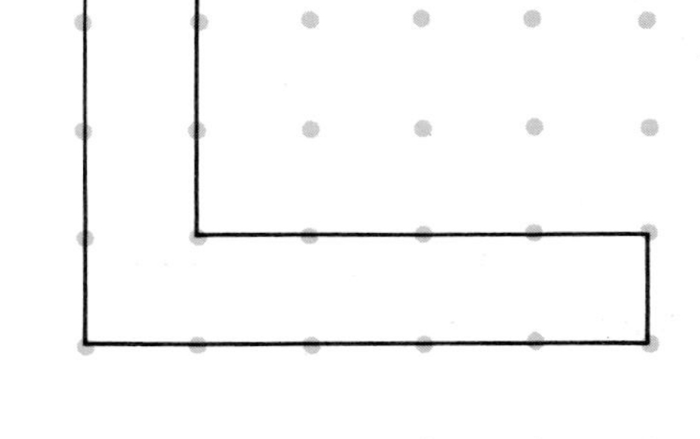

c) Draw a graph to show the rule.

d) Investigate some more letters. Look for some more rules.

Draw graphs for the rules you find.

e) Are there rules which connect the heights and perimeters of the letters?

Draw graphs for any rules you find.

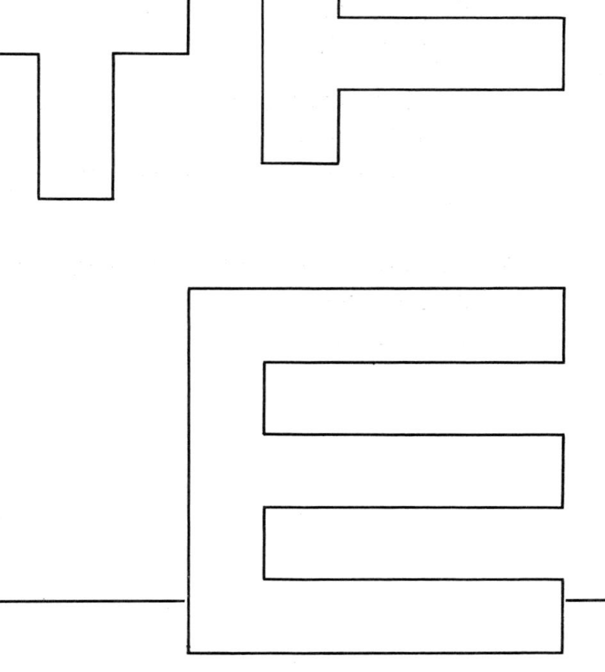

22

Activity *Balancing triangles*

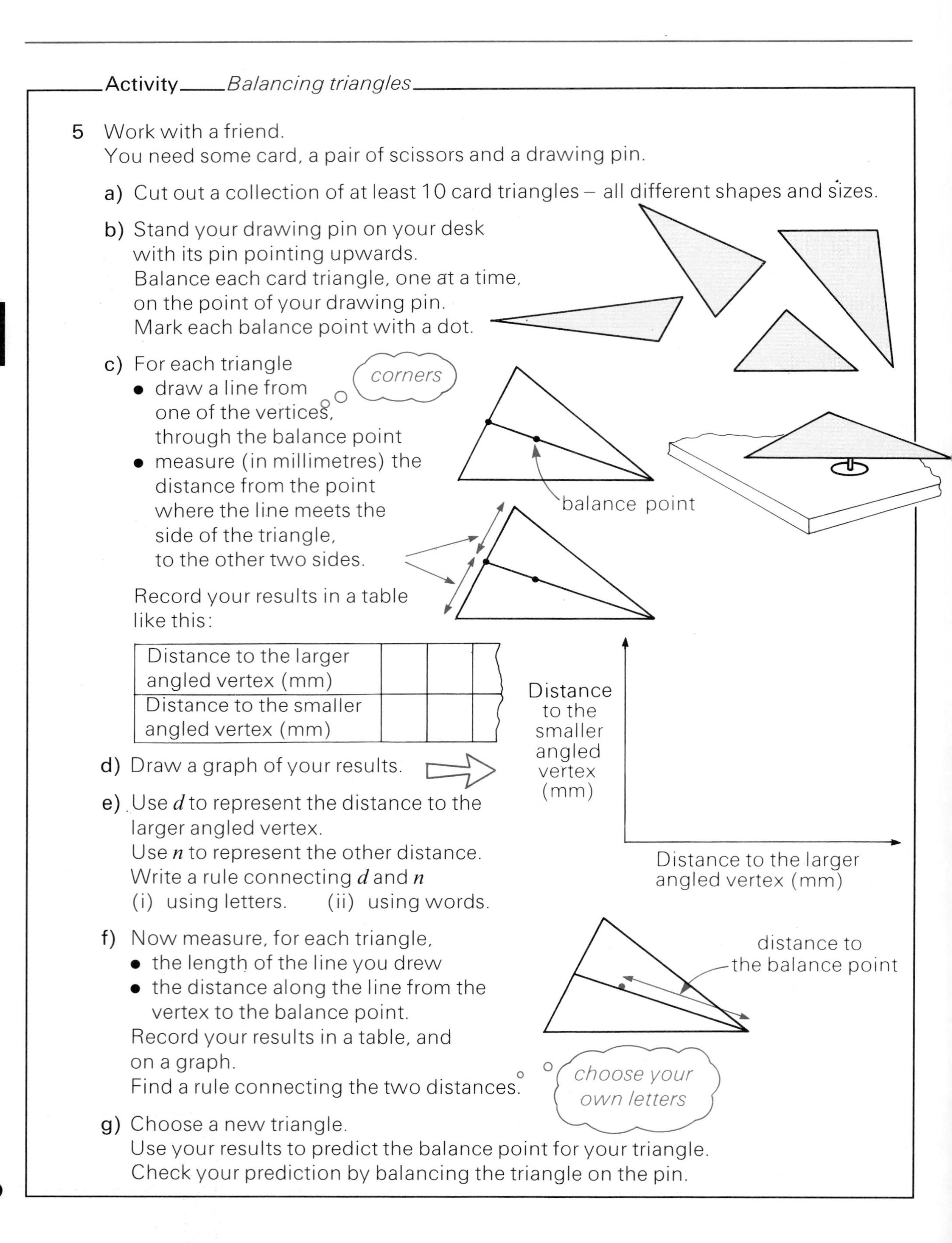

5 Work with a friend.
You need some card, a pair of scissors and a drawing pin.

a) Cut out a collection of at least 10 card triangles – all different shapes and sizes.

b) Stand your drawing pin on your desk with its pin pointing upwards.
Balance each card triangle, one at a time, on the point of your drawing pin.
Mark each balance point with a dot.

c) For each triangle
- draw a line from one of the vertices, through the balance point
- measure (in millimetres) the distance from the point where the line meets the side of the triangle, to the other two sides.

Record your results in a table like this:

Distance to the larger angled vertex (mm)			
Distance to the smaller angled vertex (mm)			

d) Draw a graph of your results.

e) Use d to represent the distance to the larger angled vertex.
Use n to represent the other distance.
Write a rule connecting d and n
(i) using letters. (ii) using words.

f) Now measure, for each triangle,
- the length of the line you drew
- the distance along the line from the vertex to the balance point.

Record your results in a table, and on a graph.
Find a rule connecting the two distances.

g) Choose a new triangle.
Use your results to predict the balance point for your triangle.
Check your prediction by balancing the triangle on the pin.

Challenges

6 This table gives the lengths of shoes for various British shoe sizes.

Shoe size	2	$3\frac{1}{2}$	5	$6\frac{1}{2}$	8	$9\frac{1}{2}$	11
Length of shoe (inches)	9	$9\frac{1}{2}$	10	$10\frac{1}{2}$	11	$11\frac{1}{2}$	12

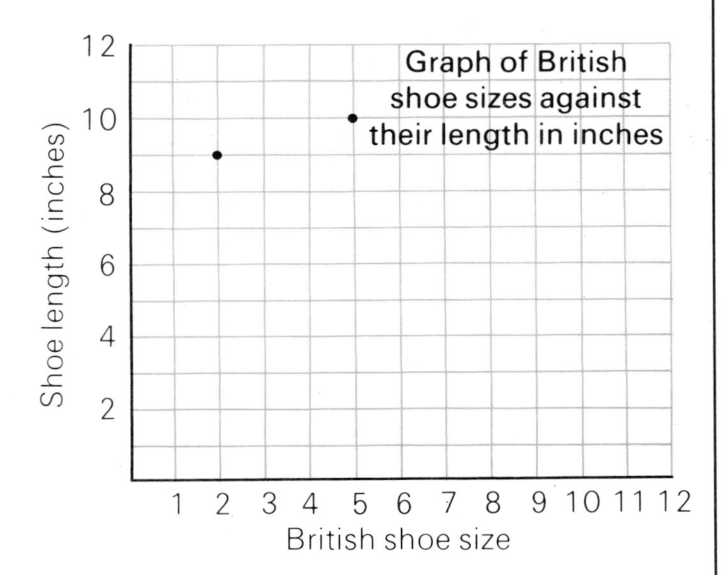

a) Copy and complete this graph.

b) Use your graph to predict the length of a size $12\frac{1}{2}$ shoe.

c) Use your graph to estimate the length of
 (i) a size $2\frac{1}{2}$ shoe.
 (ii) a size 4 shoe.
 (iii) your own size of shoe.

d) There is a rule which connects the length of shoe and British shoe size.
Find it, and write it in words and in letters.

e) Use your rule to check that the length of a size 4 shoe is $9\frac{2}{3}$ inches.

If your rule does not give this, it is not correct; go back to **(d)** *and try again.*

f) Use your rule to calculate the length of each shoe in **(c)**.
How close were your estimates?

7 This table gives some Continental shoe sizes and lengths.

Shoe size	33	36	39	42	45
Length of shoe (cm)	22	24	26	28	30

a) Measure your shoe in centimetres.
What do you think its size is, using the Continental system?

b) Draw a graph of Continental shoe size against shoe length (in centimetres).
There is a rule connecting the length of shoe and Continental shoe size.
Find it, and write it in words and in letters.

c) Use your rule to find the length of a size 40 shoe.
If anyone in your class takes a size 40 shoe, check your result by measuring.

Working with graphs

C 1 One Saturday morning at 10:00 am, Tina sets out on foot from her house in Tadcaster for Scarsdale, 16 km away. She walks steadily at 6.4 km/h.

a) Find a rule connecting her journey time (in hours) and her distance from home (in kilometres).

b) Draw a graph of Tina's journey time (in hours) against her distance from home (in kilometres).

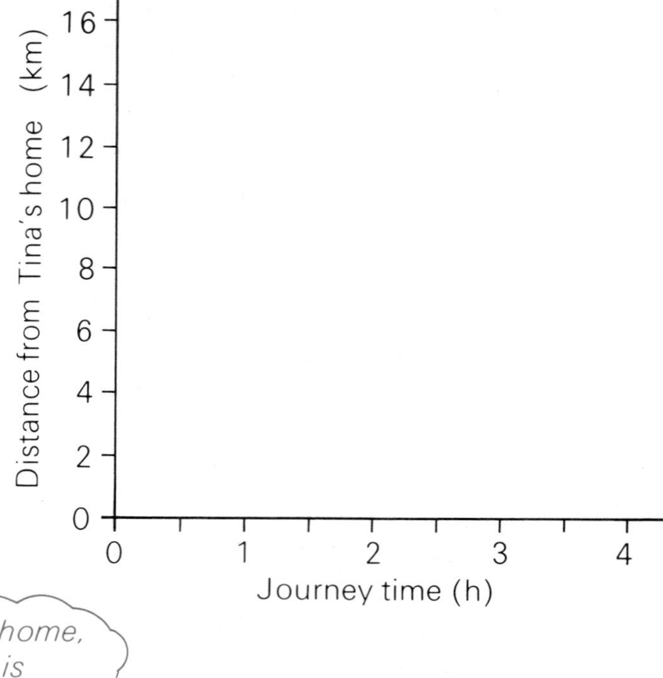

c) On the same Saturday and also at 10:00 am, Bernie sets out from his home in Scarsdale to cycle to Tadcaster.
He cycles steadily at 12.8 km/h along the same road Tina is using.
Find a rule connecting his journey time (in hours) and his distance from home (in kilometres).

his own home, that is

d) Now find a rule connecting Bernie's journey time (in hours) and his distance (in kilometres) from **Tina's** home.

e) Draw a graph of your rule in **(d)** on top of the graph you drew in **(b)**.

f) Your two graphs should cross one another. What do you think this means?

g) From your graphs, **estimate** the distance from Tina's house and the journey time represented by the point where the graphs meet.

h) Using d km for distance from Tina's house and t hours for journey time, write your rules for Tina's and Bernie's journeys.

i) By thinking and testing, find a value for d and a value for t which satisfy **both** rules at once.

__Warning:__ Both the values you are looking for are fractions. It may help you to know that $6.4 = 6\frac{2}{5} = \frac{32}{5}$ and $12.8 = \frac{64}{5}$.

j) What do these values for d and t have to do with the point where your graphs meet?

k) Write a short paragraph describing the events along the road between Tadcastle and Scarsdale on this Saturday morning.

2 The power output (in watts) of an electric component, its resistance (in ohms), and the electric current (in amperes) are related by this rule:

such as a light bulb, or a hair dryer

or amps, for short

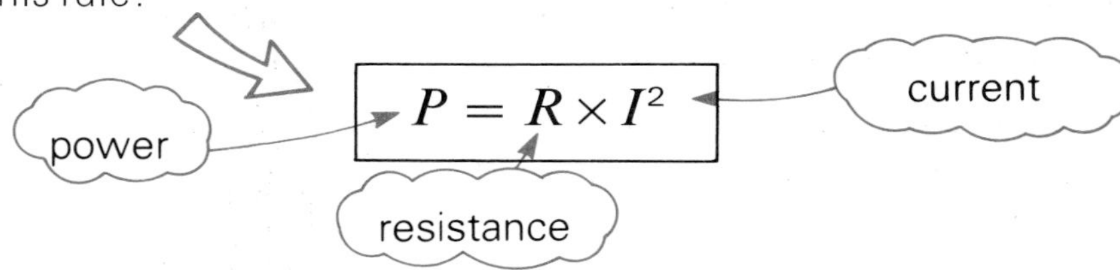

a) Which of these do you think shows a graph of the power output of a 576 ohm light bulb with various strengths of electric current passing through it?

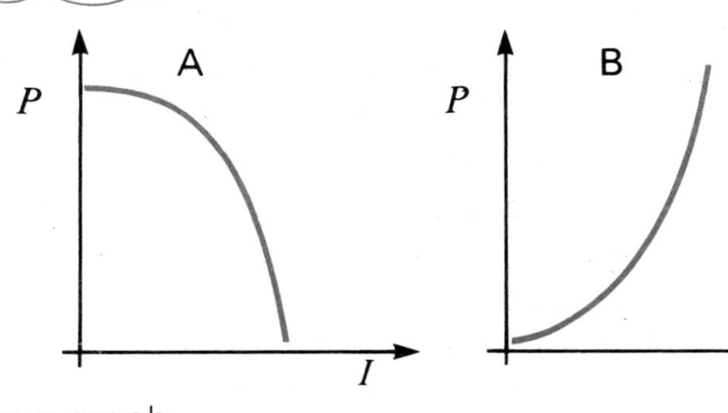

b) Check your answer in **(a)** by making a table of values for P corresponding to various values for I from 1 to 10, and then drawing a graph.

c) Would it be wise to put a 5 amp fuse in the plug of a 1500 watt hair dryer, which has a resistance of 38.4 ohms? Why?

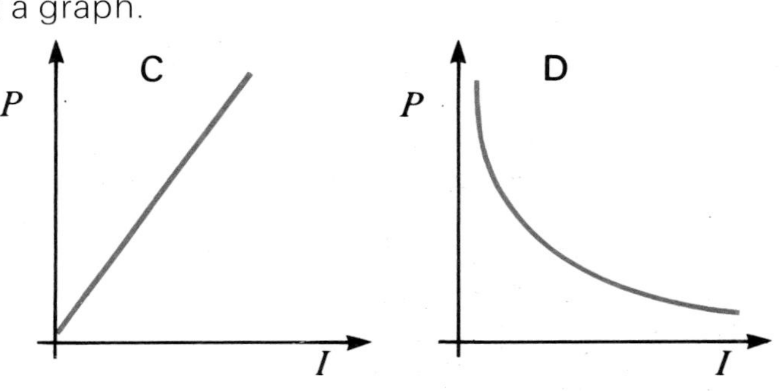

Challenge

3 A new car is being tested on the manufacturer's private test track.
Here are some measurements of the car's speed during one ten-second time interval.

When the stopwatch was started, the car was already going at $12\frac{1}{2}$ m/s.

Time (in s)	0	2	5	8	10
Speed (in m/s)	12.5	17.5	25	32.5	37.5

During the ten seconds the car accelerated steadily.
Find a rule connecting time and speed during the ten-second time interval.

Write your rule in words and in letters.
From your rule, find out what the car's acceleration was in m/s per second.

23 Rotation

A 1 Glenda puts this record on her turntable.
She spins it through a quarter-turn clockwise.

Copy and complete this sketch of the label.
Use plain paper.

2 Glenda writes her initial on this record label.

a) You need some squared paper.
Copy just the initial and the centre of the record.

b) The record spins through a quarter-turn clockwise.
Add the new position of the initial to your drawing.

3 Glenda opens a pickle jar.
Here is a filmstrip of the lid opening.
Copy and complete the last two frames.

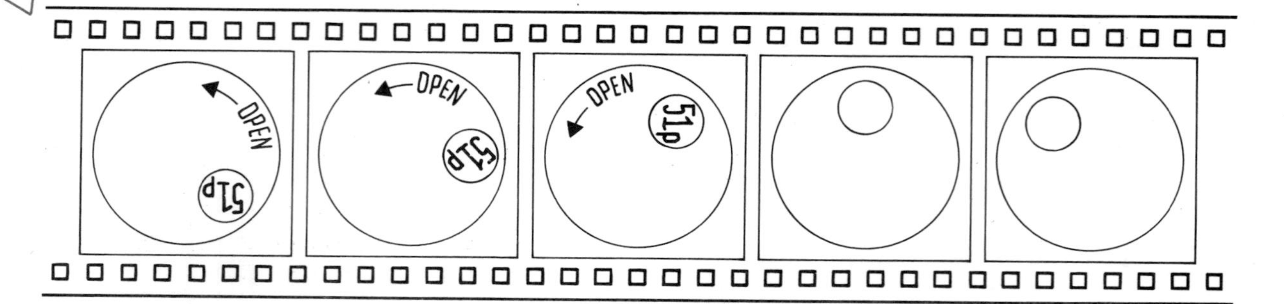

4 Glenda has stuck her name on this door handle.

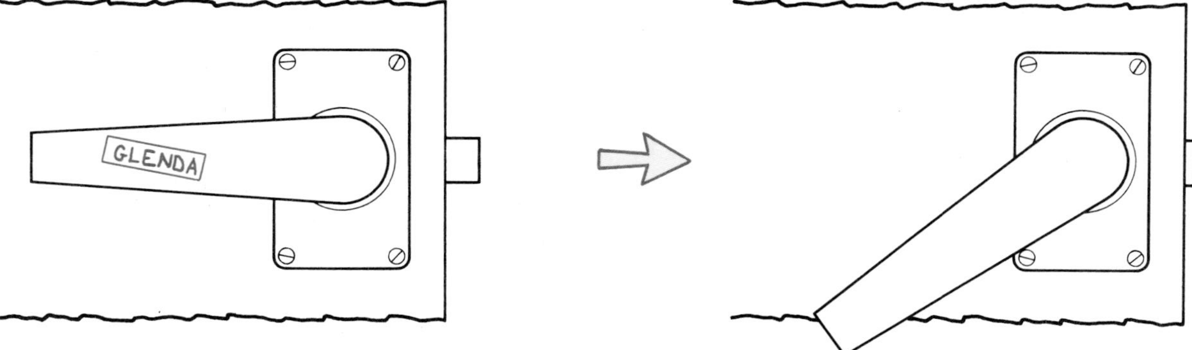

Sketch the handle in the new position. Include Glenda's name.

5 Horace has made his own handle.

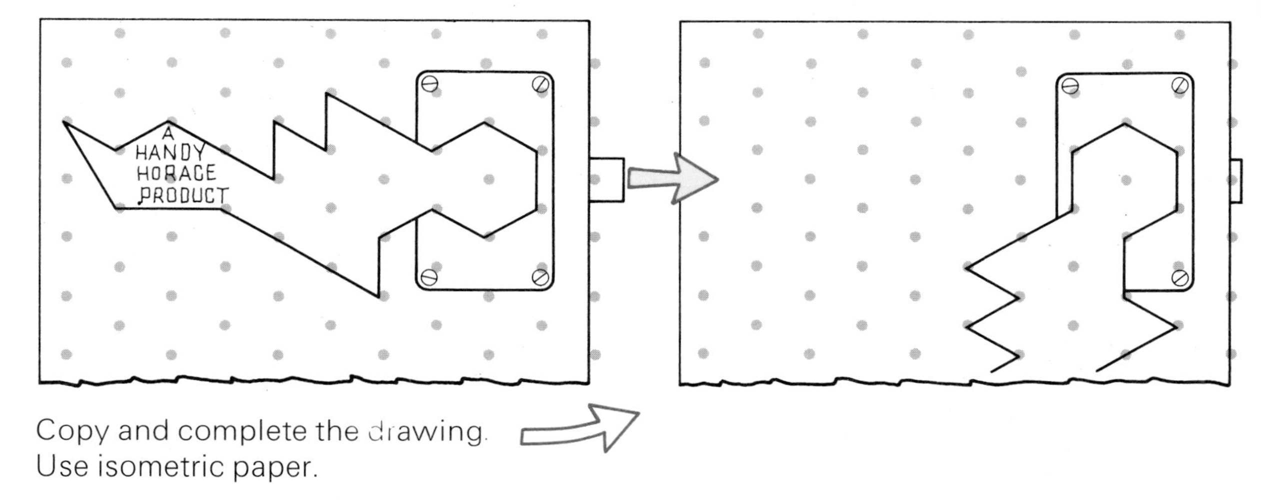

Copy and complete the drawing.
Use isometric paper.

Think it through

6 Here is another of Horace's door handles.

a) Copy it on isometric paper.

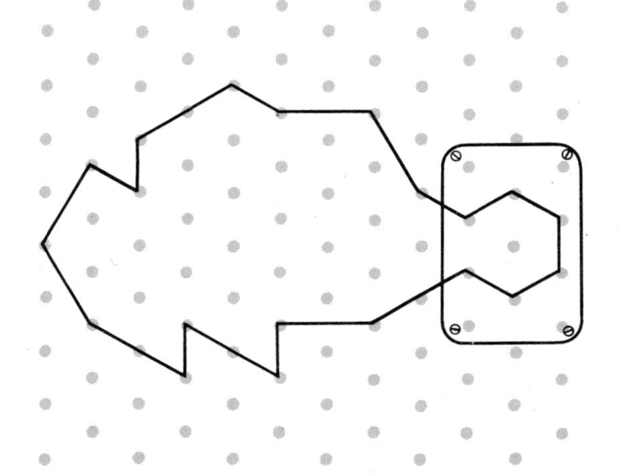

b) Horace writes his initial on the handle. When he turns the handle the initial moves like this:

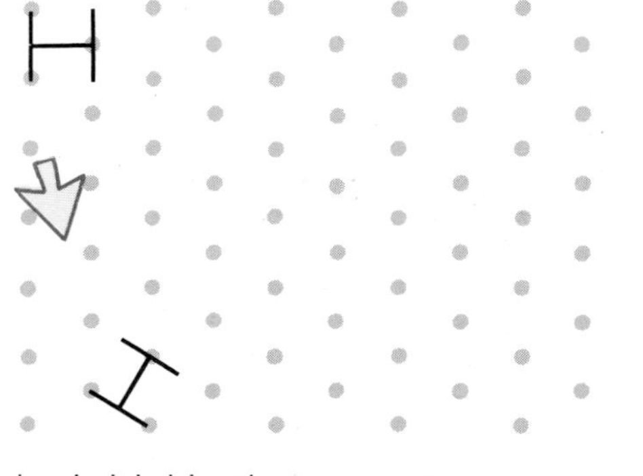

Draw the initial in the correct position on the handle.

B 1 Glenda is in class 3U9. She pins these cards to the class notice board.

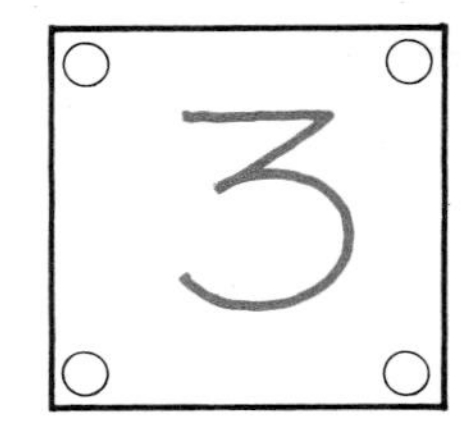

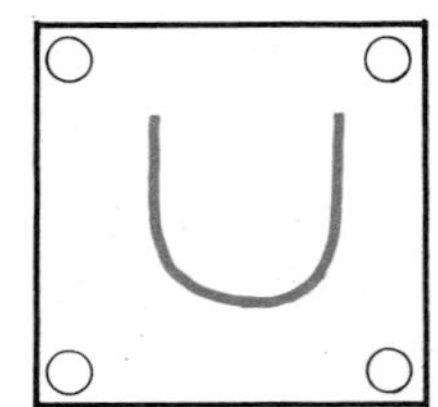

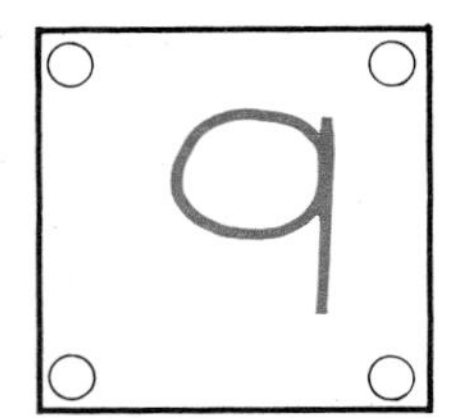

Someone borrows some of the drawing pins.

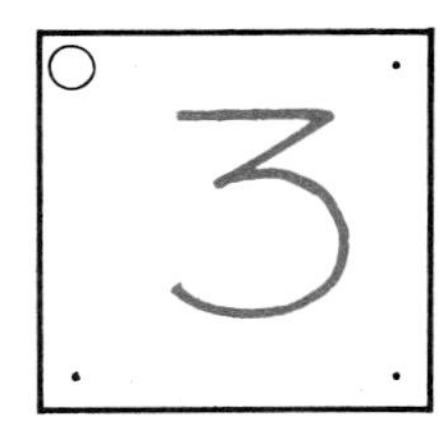

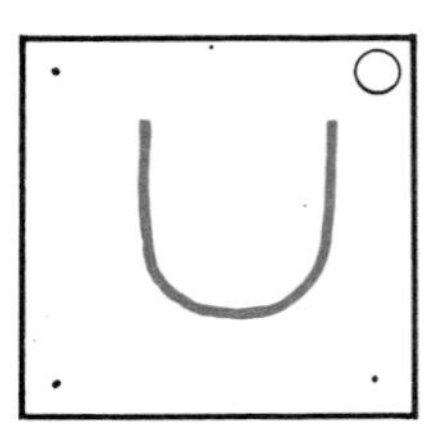

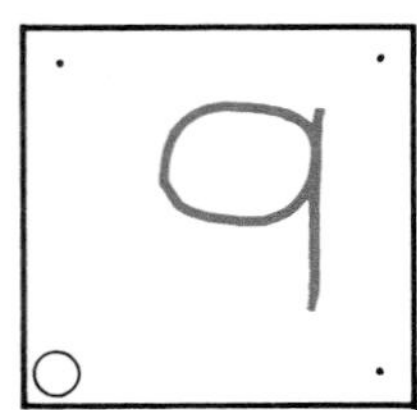

23

The cards swing down. Copy and complete this drawing of the cards.

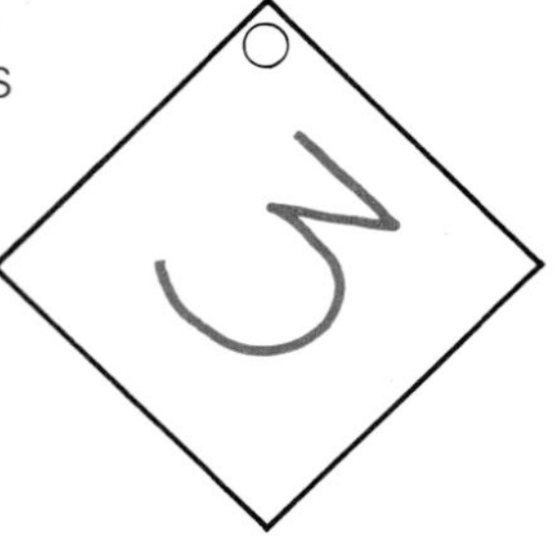

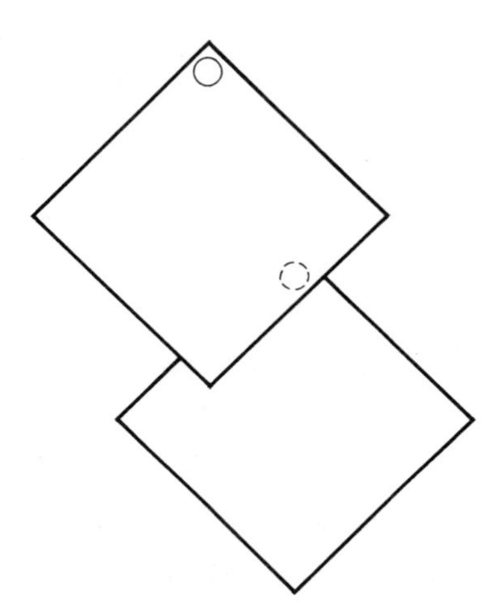

2 Glenda pins up these cards.

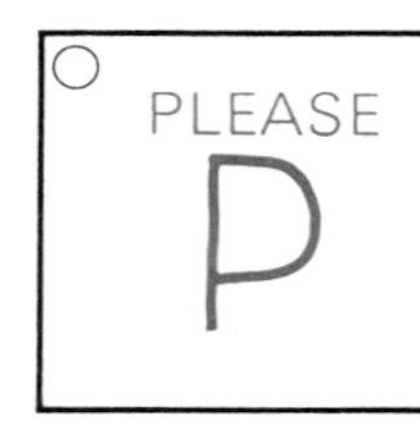

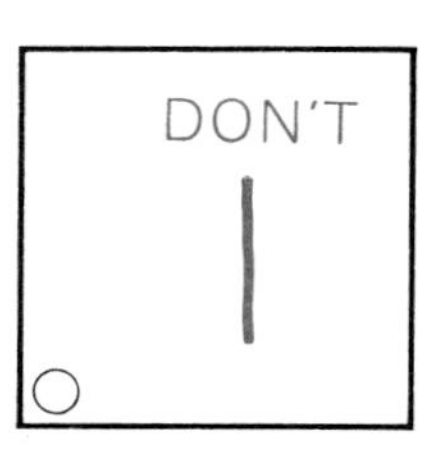

a) Copy the cards.

b) Underneath, draw the cards after they have swung down.

c) Repeat **(a)** and **(b)** for these cards.

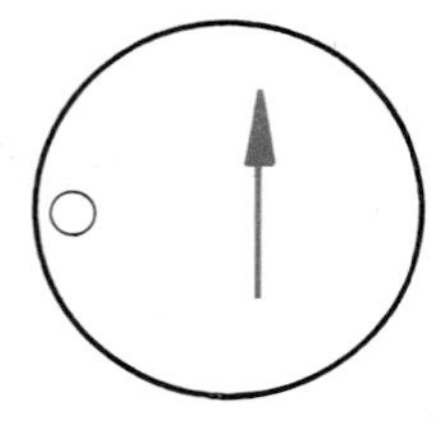

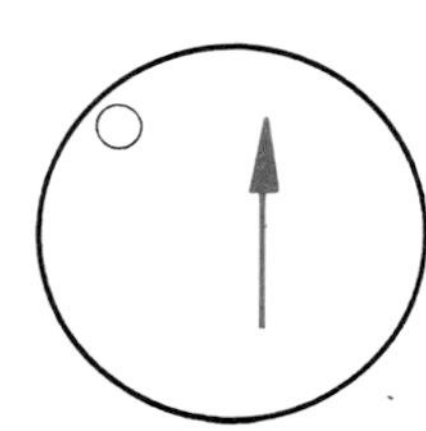

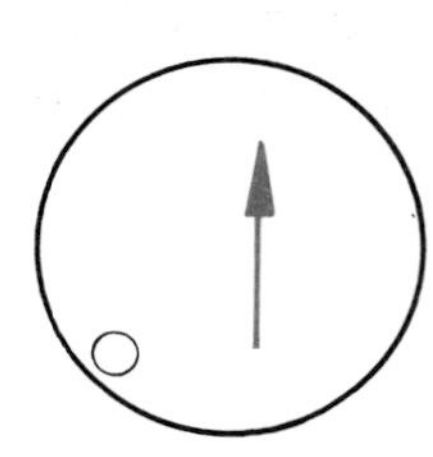

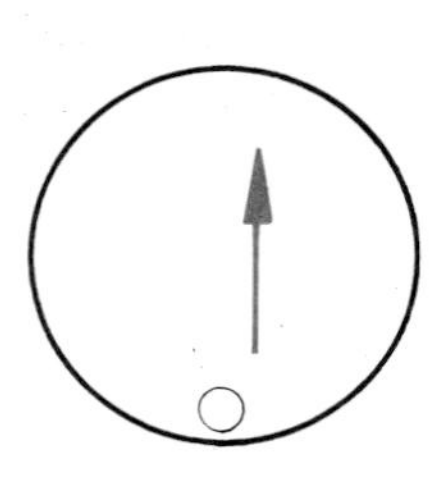

Centres of rotation; images

C 1 Glenda is undoing a nut with a spanner.

a) She loosens it one-sixth of a turn, by turning it anticlockwise. Draw the new position of the spanner on isometric paper. Include the trademark.

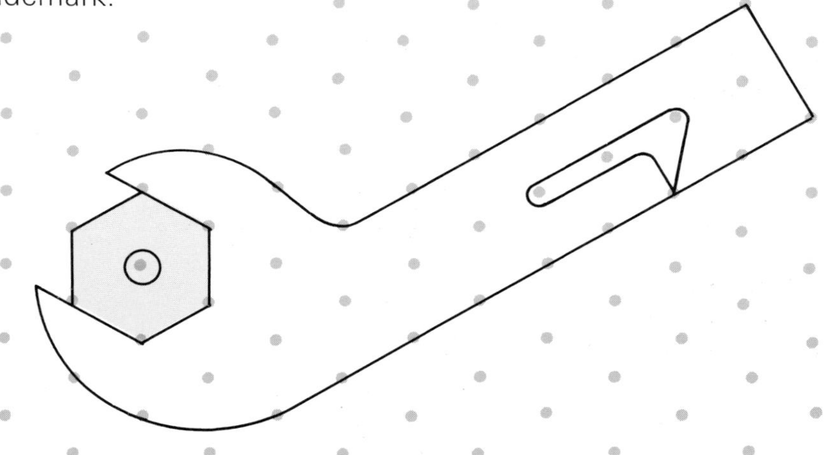

b) She loosens the nut another one-sixth of a turn. On your drawing, add the new position **of the trademark**. (Do not draw the whole spanner.)

c) Get some tracing paper and trace your original drawing. Press a pencil point on the centre of the nut. Turn the tracing paper one-sixth of a turn. Check your answer to **(b)**.

2 Horace is also undoing a nut.

a) Copy the nut and the trademark. Do not bother with the spanner. Use isometric paper.

b) Horace loosens the nut one-sixth of a turn. Draw the new position of the trademark.

c) Trace the nut and the trademark. Check your answer to **(b)**.

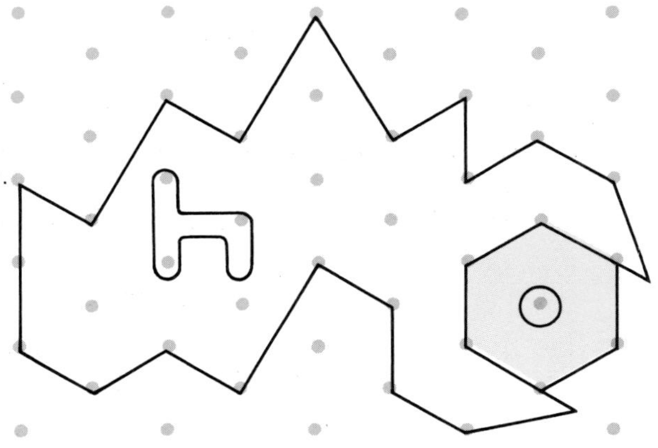

3 These drawings show different positions of trademarks on some spanners.
Each spanner is undoing a nut.

a) Copy the drawings onto isometric paper.

b) Trace the first trademark.
Hold the tracing paper down with a pencil point.
Turn the tracing paper.
Find the centre of the first nut.

c) Repeat **(b)** for the other trademarks.

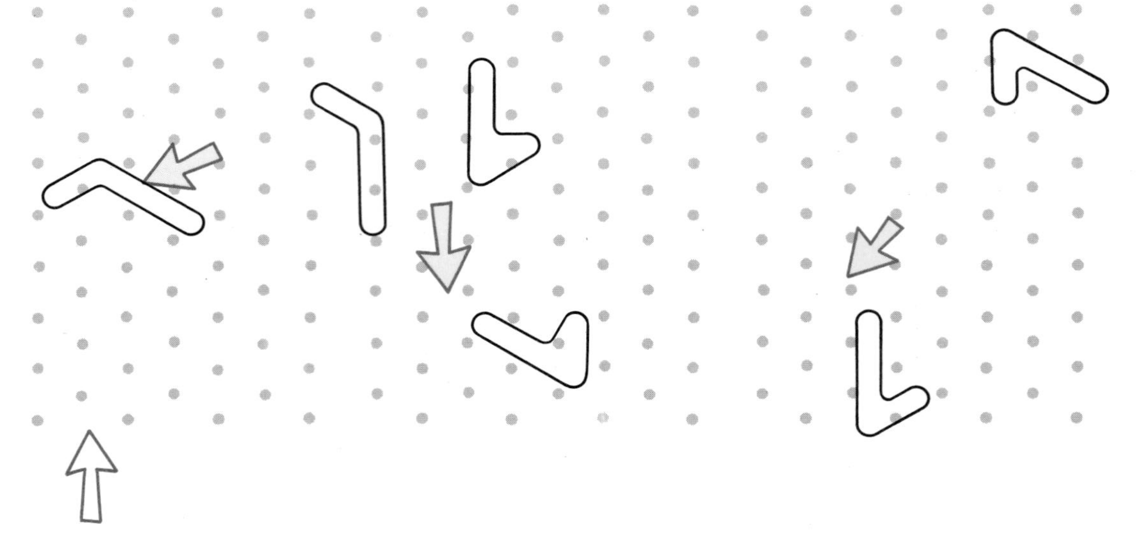

For this drawing, the centre of the nut is at point C (see the **Take note**).

23

Take note

We say:
the point C is the **centre of rotation**;
object **A′** is the **image** of object **A**.

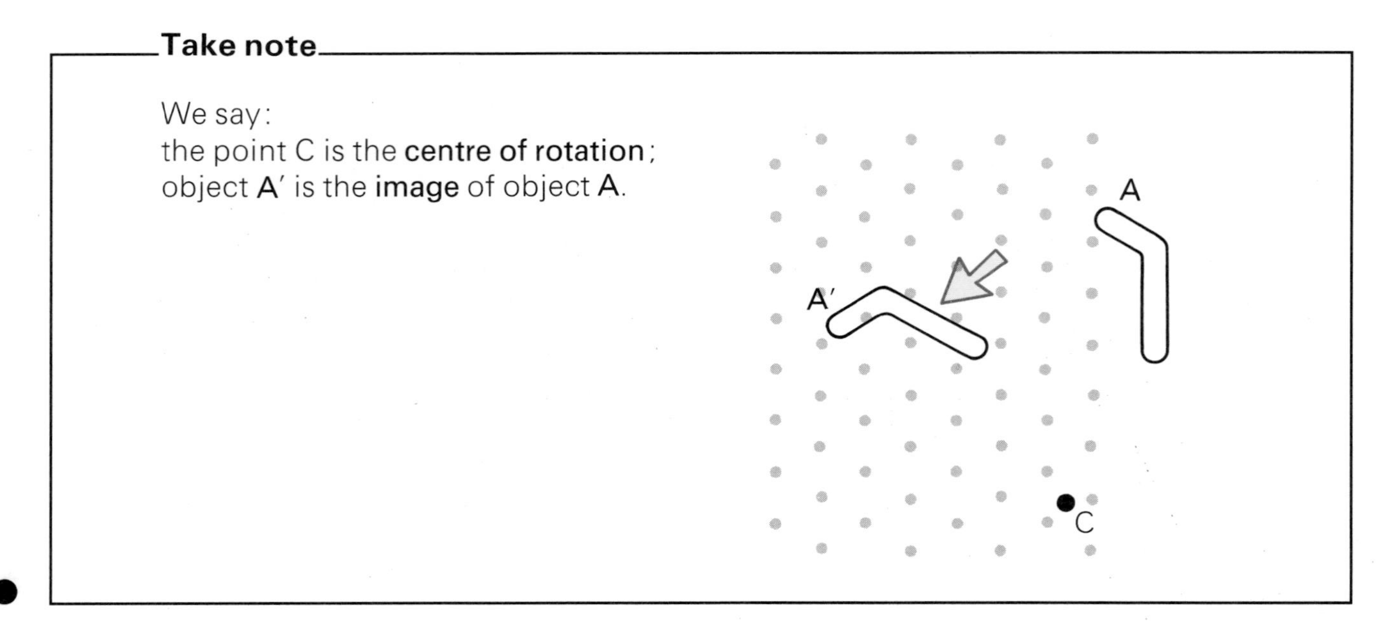

4 a) Copy this drawing.

b) The flag is to be rotated $\frac{1}{6}$-turn anticlockwise. Draw its **image** when the **centre of rotation** is at
   (i) point A.
   (ii) point B.
   (iii) point C.
   Use tracing paper to help you.

c) Write down what you notice about the three images.

5 Here are some drawings.
Each shows a flag and its image after rotation.
Write down the size of each rotation as a fraction of a turn.

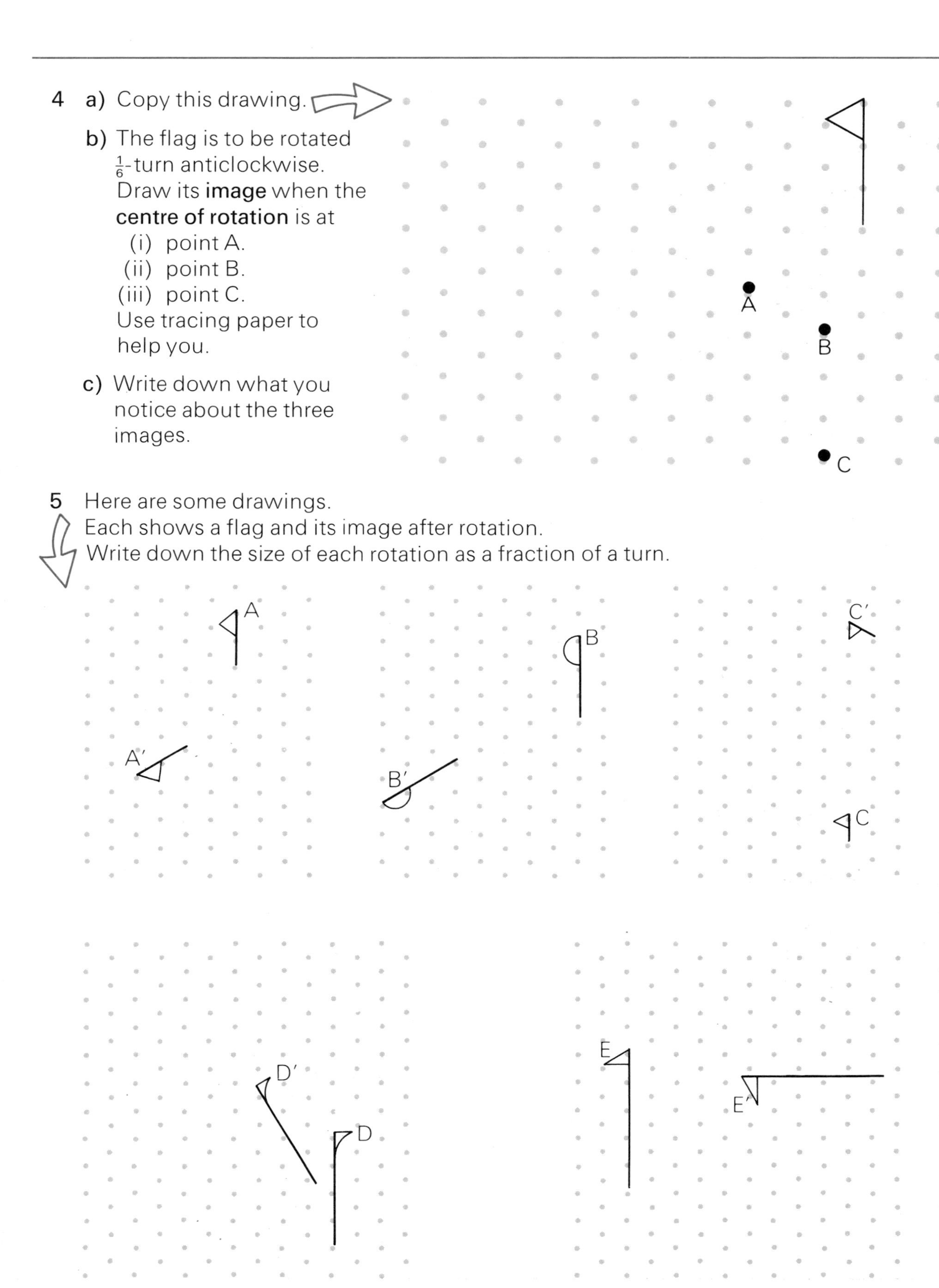

With a friend

6 Some pupils are given this question to do.

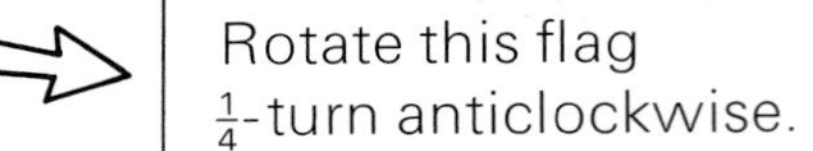

Rotate this flag $\frac{1}{4}$-turn anticlockwise.

Use C as the centre of rotation.

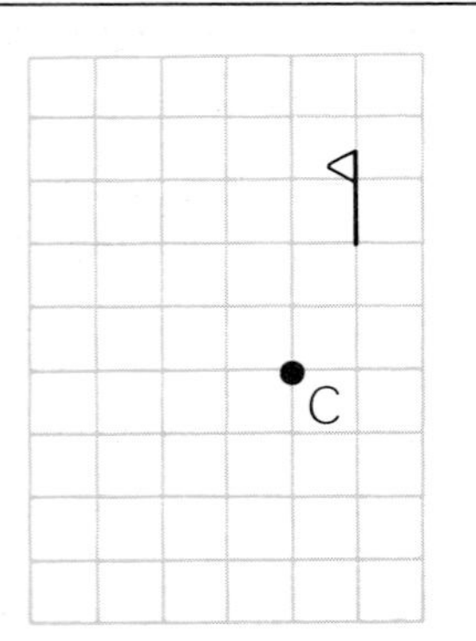

Here are their answers.

Discuss these questions with your friend. Each of you write down your answers.

a) Whose answer is correct?

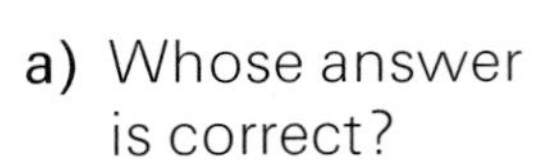

b) Describe what the others have done wrong.

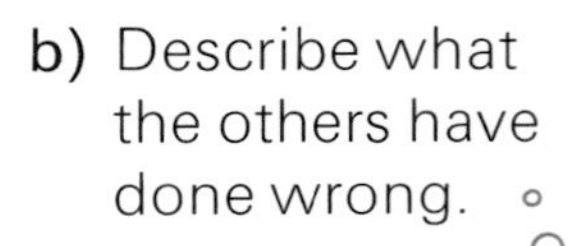

for example, 'not rotated about the centre', or 'not rotated through 90° anticlockwise', . . .

c) Mark the answers out of 10. Each of you write down what you decide.

Jamal

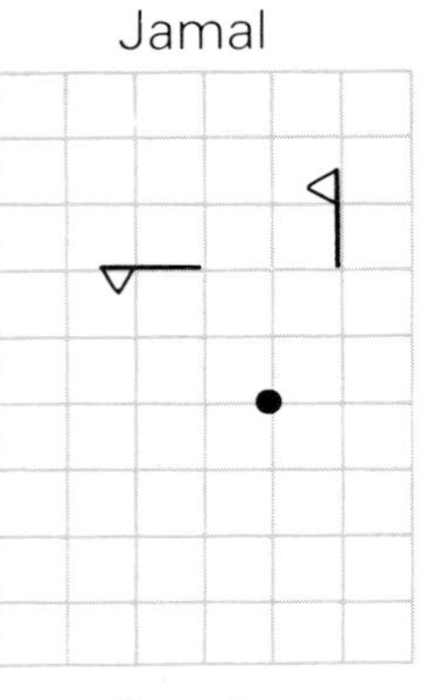

Jane

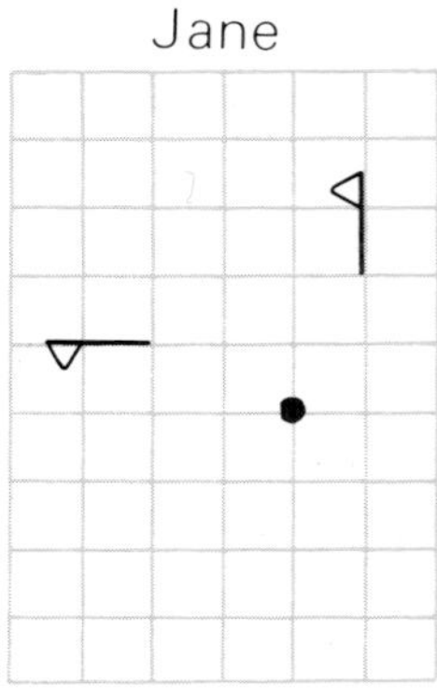

Maureen

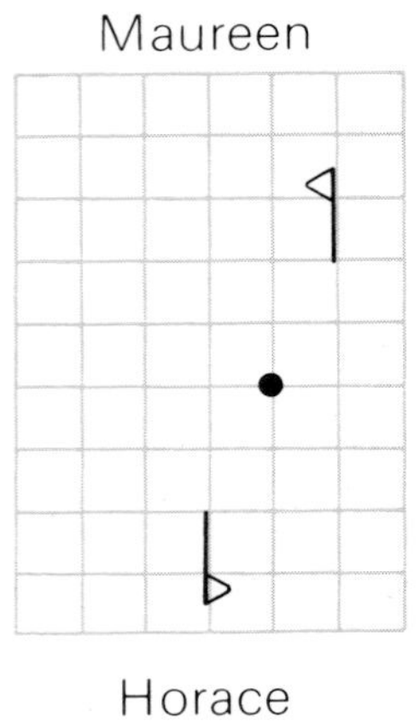

Stanley

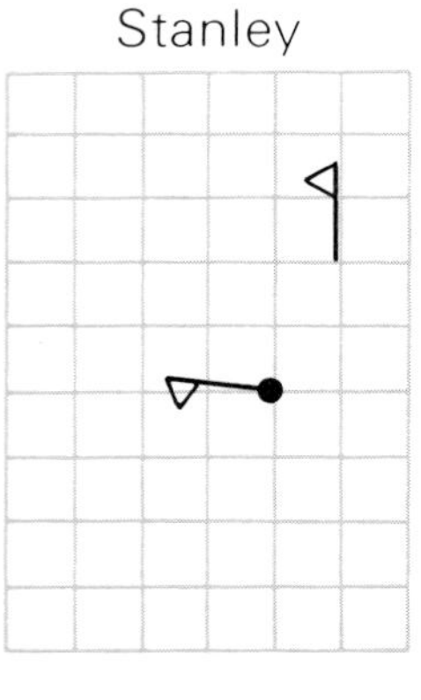

Seema

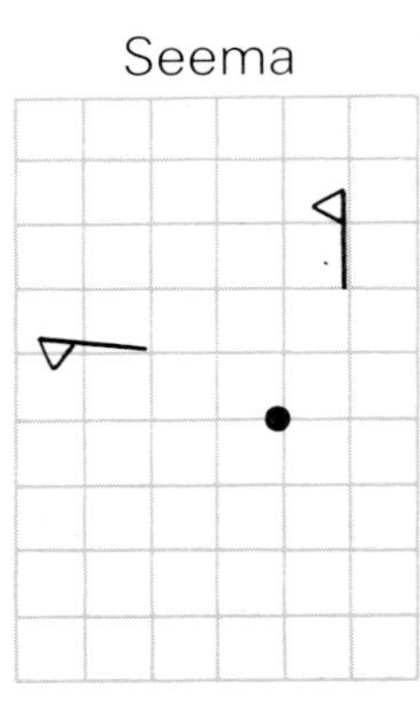

Horace

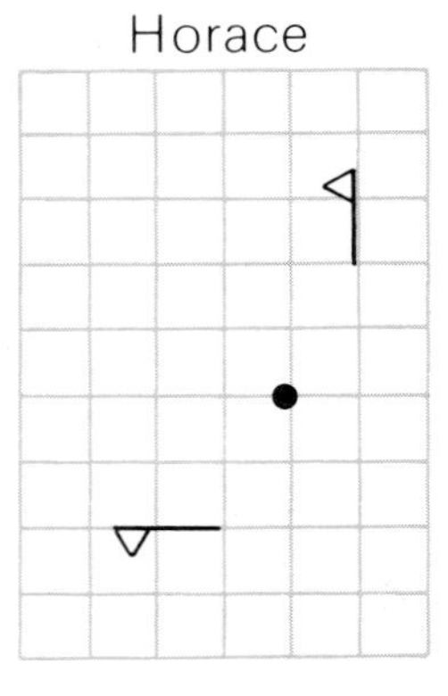

7 Copy these onto squared paper.

Rotate each flag through $\frac{1}{4}$-turn clockwise.
Use C as the centre of rotation.

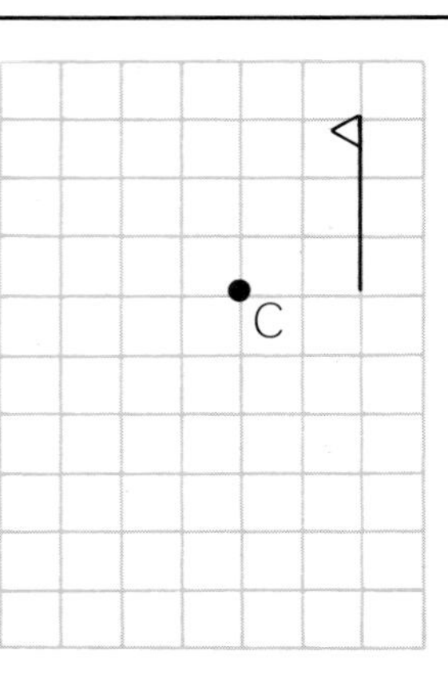

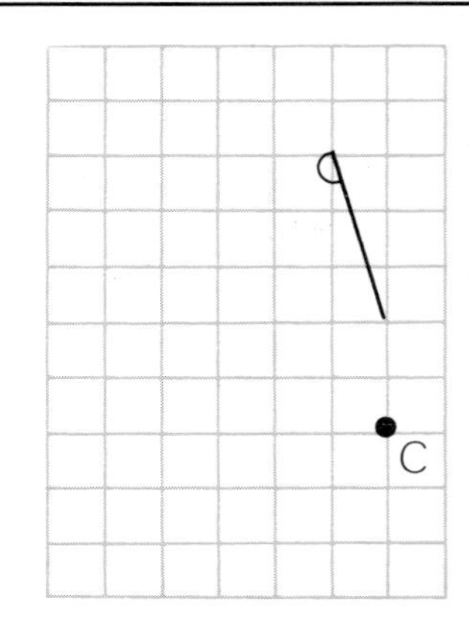

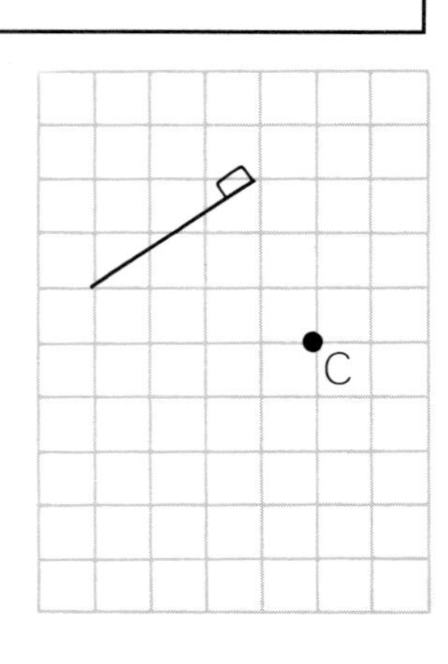

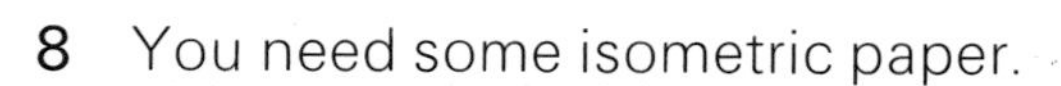

8 You need some isometric paper.

a) Make a larger version of this drawing.

b) Find the centres that take F onto the other flags. Use tracing paper to help.

9 a) Make larger versions of these drawings.

b) Find the centres of rotation. (Use tracing paper only as a check.)

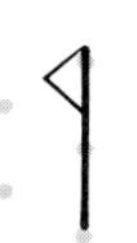

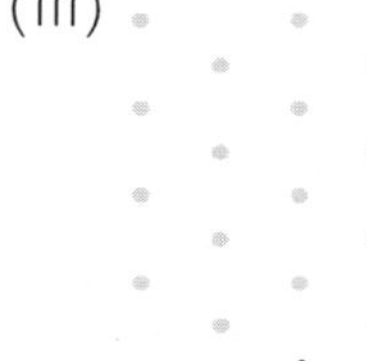

(i) (ii) (iii)

10 Here are Glenda's (G) and Horace's (H) answers to question **9**. Write down what each of them is doing wrong.

a) b) c)

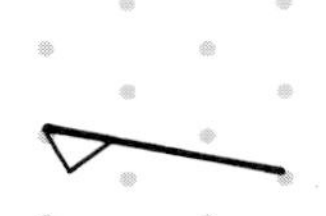

11 Which of these points are centres of rotation?

a) b) c)

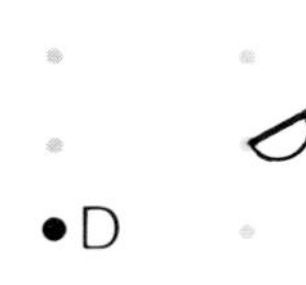

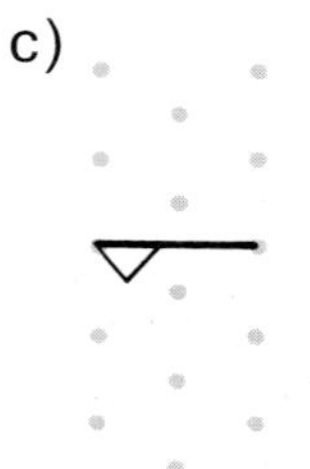
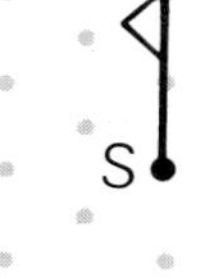

D Think it through

1 Get a plain sheet of A4 paper

Mark two points A and B.

Draw a flag on point A.

Draw four exact copies of the flag. Each one must have its stem ending on B.

Trace the flag on A.

Find a centre that takes the flag to one of the flags on B.

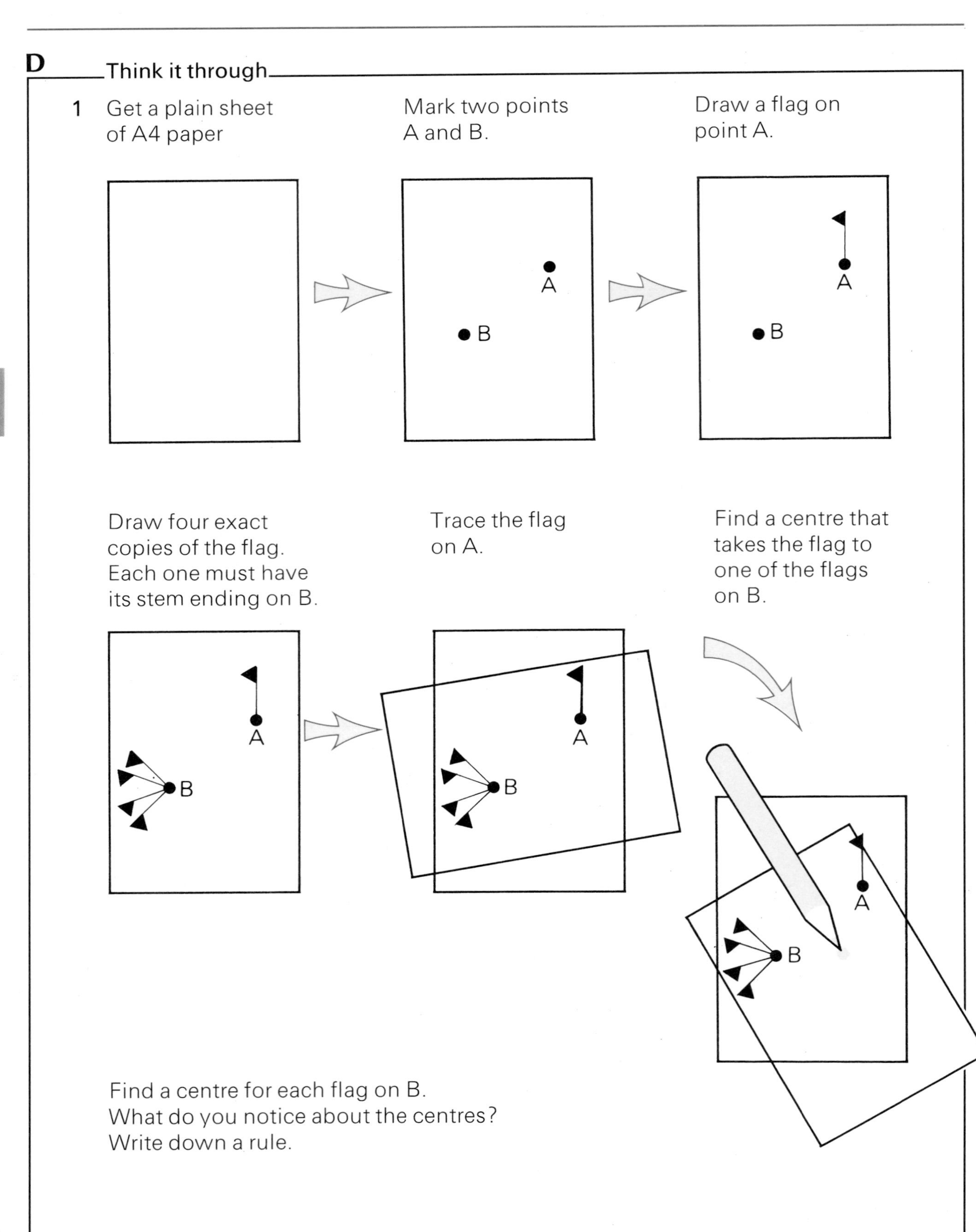

Find a centre for each flag on B.
What do you notice about the centres?
Write down a rule.

2 a) Make a very careful copy of this drawing on squared paper.
Also trace the flag.

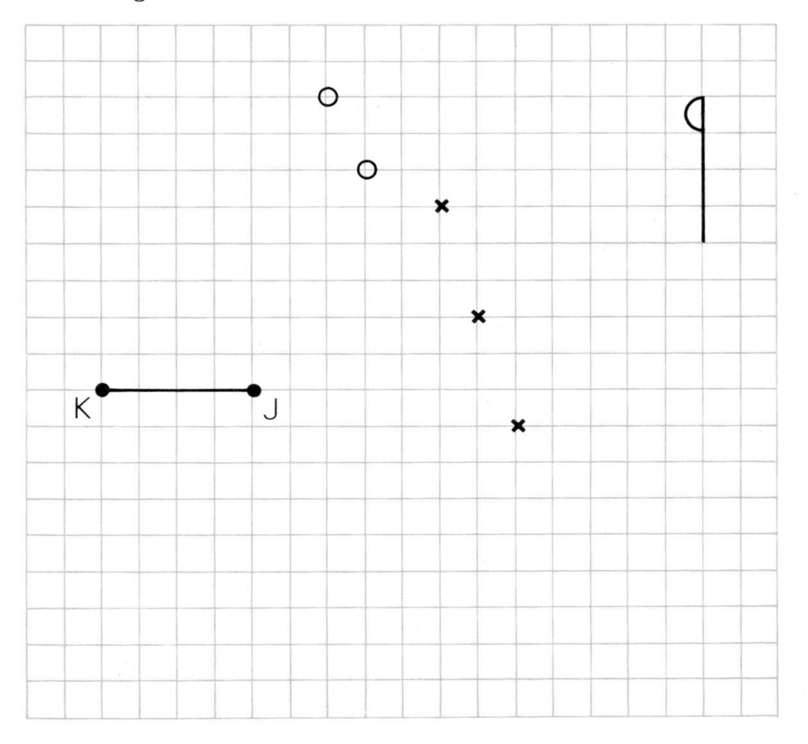

b) What can you say about the centres marked × ?
Draw three more centres of this sort.

c) What can you say about the centres marked O ?
Draw four more centres of this sort.

d) Mark the centre that takes the flag onto line JK.

3 a) Make a careful copy of this drawing on squared paper.

b) F′ is the image of F.
Use the method of question **2** to find the centre of rotation.
(But do **not** use tracing paper.)

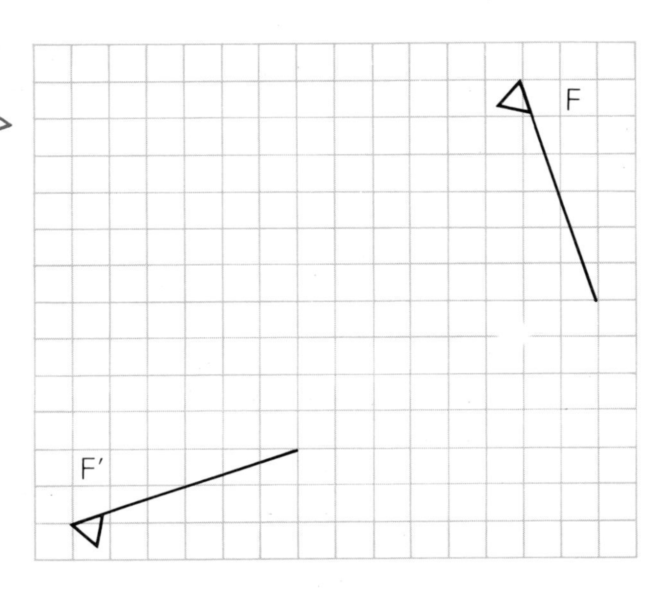

23

4 Horace wants to find a centre of rotation for this pair of flags.

He uses the method of question 2. What happens?

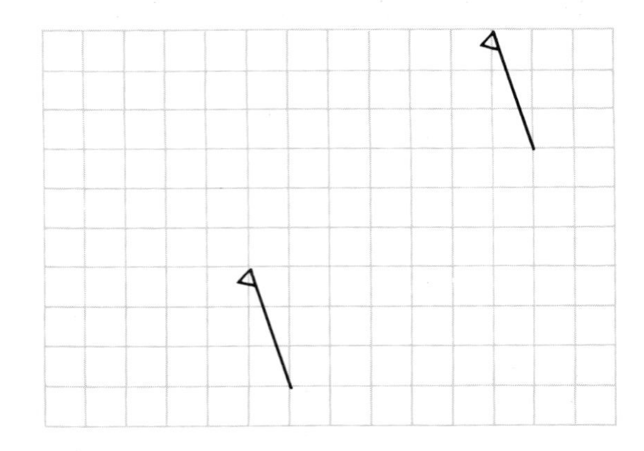

23

5 Horace wants to find a centre of rotation for this pair of triangles.

Midge says there isn't one.

Horace says there are three. Who is right?

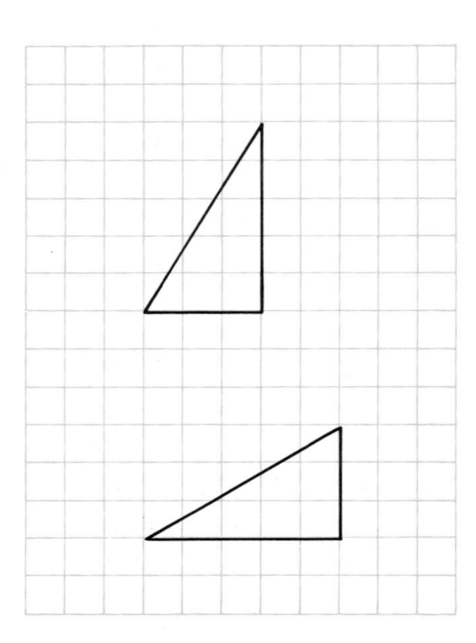

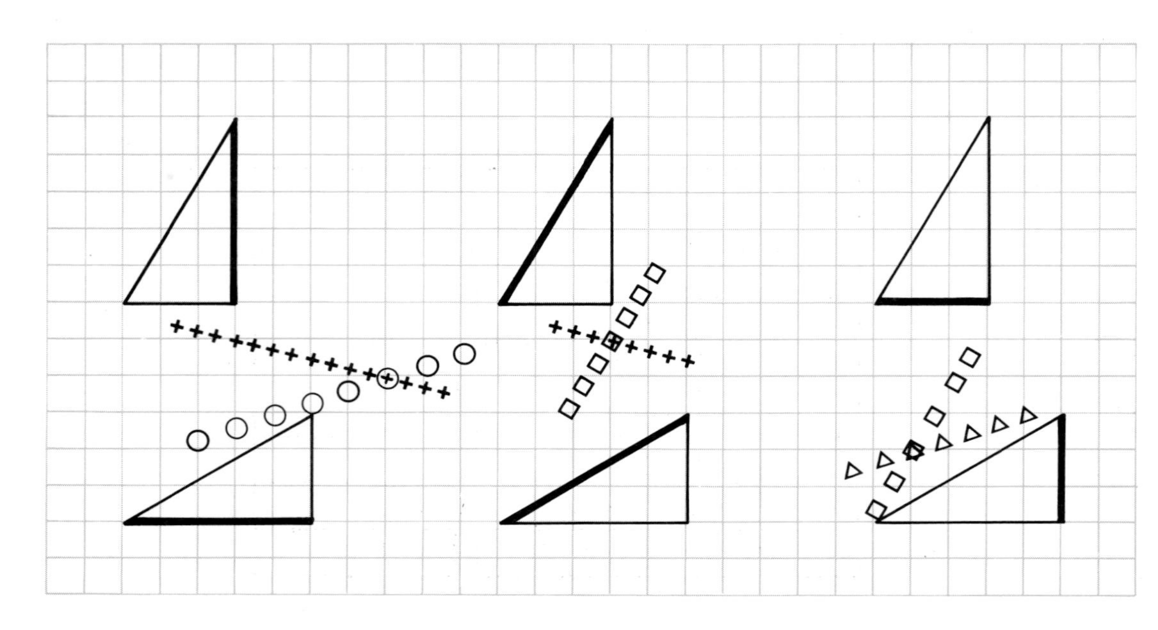

Rotations and coordinates

E 1 a) Copy this diagram onto squared paper.

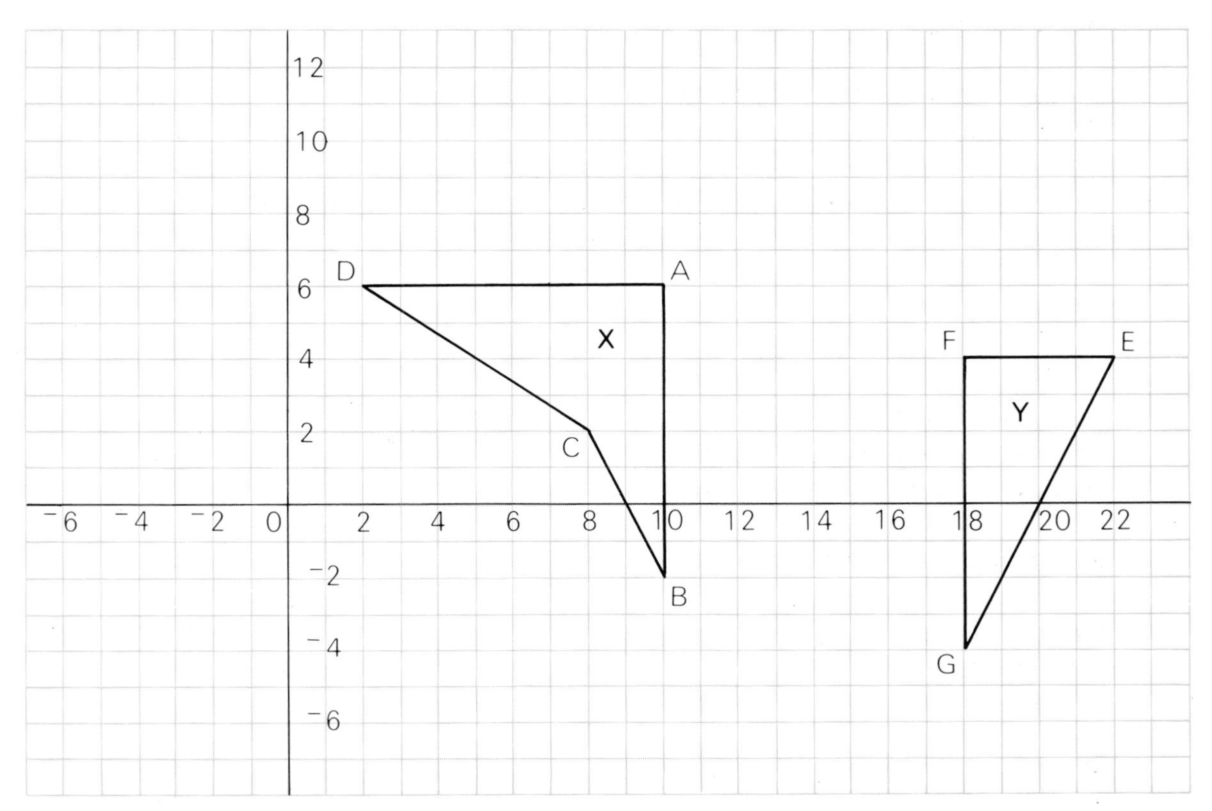

b) Figures **X** and **Y** are to be rotated through a quarter-turn anticlockwise.
The centre of rotation is at (0, 0).

(i) Draw the image of **X**.

(ii) Copy and complete this table.

Point	Original position	Position of image
A	(10, 6)	
B	(10, ⁻2)	
C	(8, 2)	
D	(2, 6)	
E	(22, 4)	
F	(18, 4)	
G	(18, ⁻4)	

Turning solids

F 1 These shapes are all made from five 1 cm cubes stuck together.

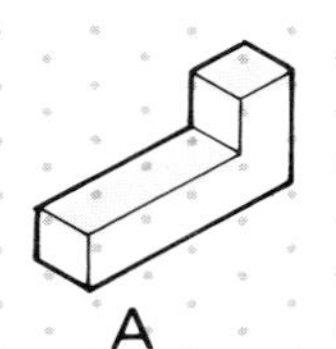
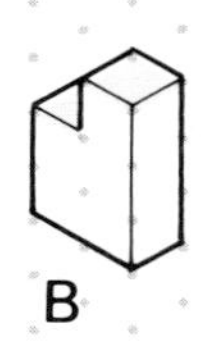
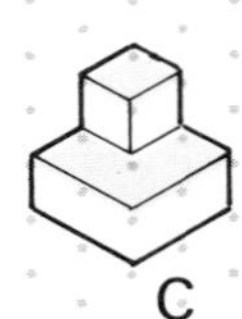
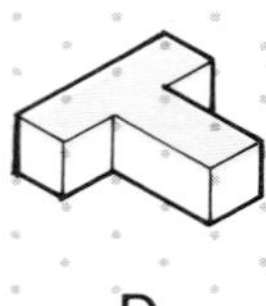
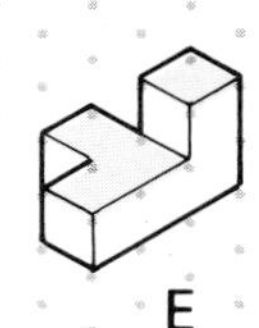

Shape **A** is moved from this position to this position.

The same move is applied to all the other shapes.

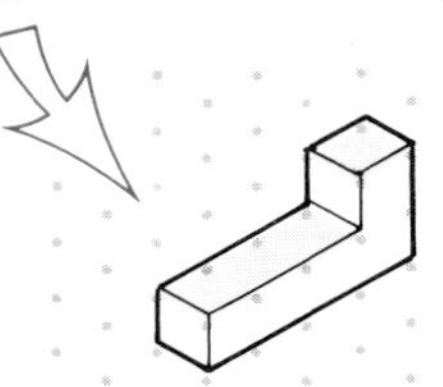

a) Which of these drawings shows the new position of shape **B**?

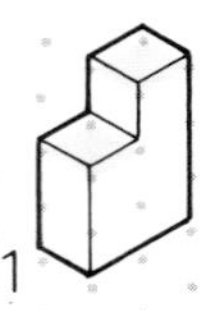
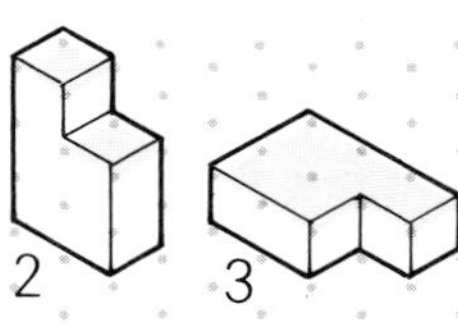
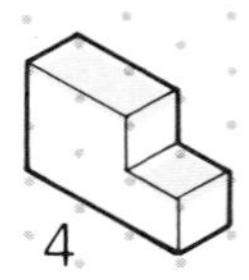
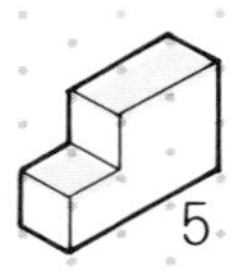
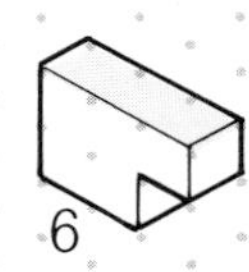

b) Which of these drawings shows the position of shape **C**?

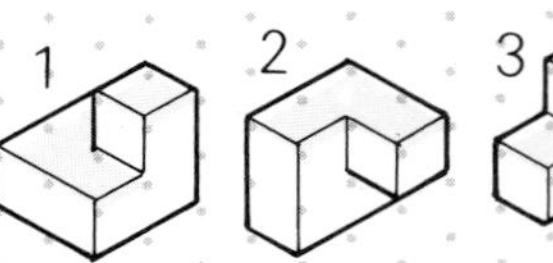
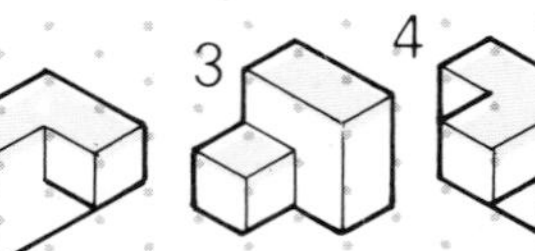

c) Draw the new position of
(i) shape **D**.
(ii) shape **E**.

d) The same move is also applied to this shape. (The shape is made of four 1 cm cubes.)
(i) Draw its new position.
(ii) What do you notice?

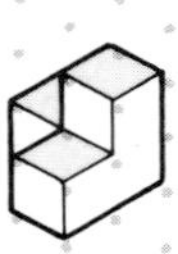

Challenge

(iii) Draw another, different shape that behaves in the same way.

Challenge

2 This shape is made from five 1 cm cubes.
A move is applied to the shape that **appears** to leave it in the same position.

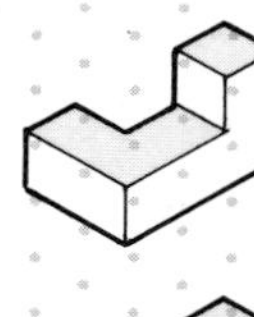

Apply the same move to this shape. Draw its new position.

24 Using percentages

A 1 Do you remember . . . ?

71% means 71 out of 100 . . . or $\frac{71}{100}$.

Read these clippings and advertisements.

Daily Blab
Tea consumption down to 71% of 1961 figures.

In 1978, 54% of all British households had central heating.

In 1979, 15% of all British households consisted of a pensioner living alone.

Gower worktops
20% off all units after the first one.

MEGASIZE TISSUES
Now 25% stronger

7UP
10% extra free
□litres for the price of 1.5 litres

a) (i) On average, how many British households in every 100 had central heating in 1978?
(ii) How many is this in every 5 million?

b) On average, how many British households in every half million consisted of a pensioner living alone in 1979?

c) The Gower Worktops advertisement means that you pay full price for the first unit and 20% less than that price for the second, third, . . . and so on.
What would you pay for three units whose normal price is £23.15 each?

d) Copy and complete the 7UP advertisement.

With a friend

e) Decide between you what the tissues advertisement means.
Write down what you decide.

f) Imagine that you have one of the old tissues and one of the new, 'stronger' tissues.
Design an experiment to test the advertiser's claim.
Write a short paragraph describing what you would do, and what '25% stronger' would mean.

more difficult to tear, . . . stands up to a stronger 'blow', . . . supports more weight, . . . ?

Think it through

2 For a promotional campaign, a hair shampoo manufacturer plans to sell 750 ml bottles of shampoo for the price of a standard 600 ml bottle.

like the one for 7UP in question 1

Design an advertising label to be stuck on the bottles.

3 There are three ways to complete this statement with one of these fractions.

$\frac{1}{2}$ $\frac{1}{4}$ $\frac{1}{5}$ $\frac{20}{100}$ $\frac{1}{20}$ $\frac{2}{10}$

20% of a bar of chocolate is $\frac{\square}{\square}$ of it.

Write down all three ways.

4 a) Some of these percentages can be paired off with some of the fractions. *(for example, 75% is $\frac{3}{4}$)*

12% 25% $\frac{1}{12}$ 50% $\frac{1}{4}$ 15% 60% $\frac{4}{5}$ 33% $\frac{3}{5}$ 30% $\frac{3}{10}$ $\frac{1}{3}$ $\frac{3}{20}$ $\frac{5}{6}$ 83% $\frac{1}{2}$ 80%

Match up as many of the percentages as you can with fractions.

b) Arrange the remaining percentages and fractions in order, starting with the smallest.

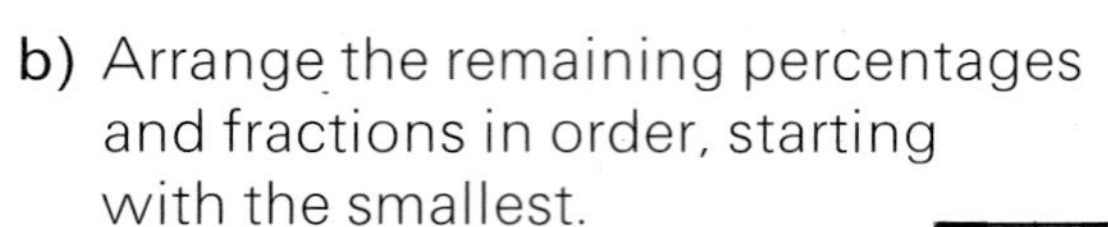

5 What percentage of the tiles in each tiled rectangle are (i) red? (ii) white?

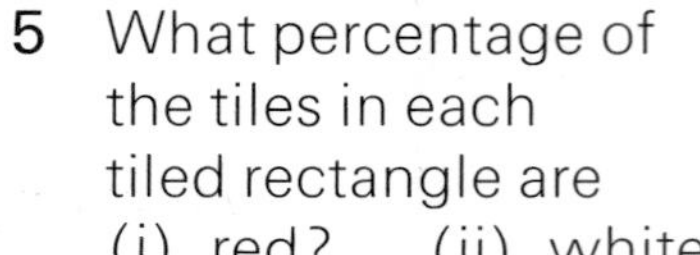

(i)

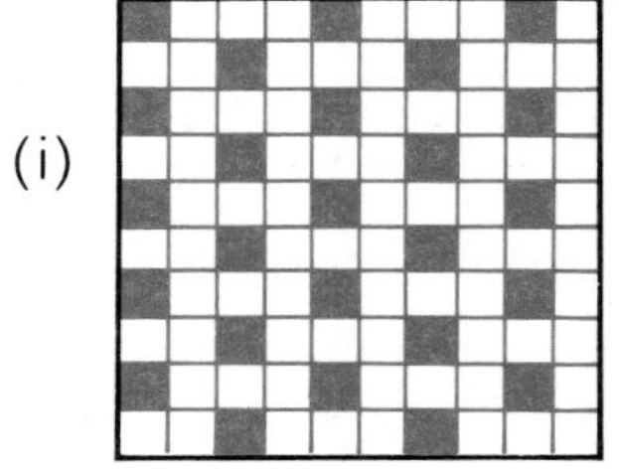

(ii)

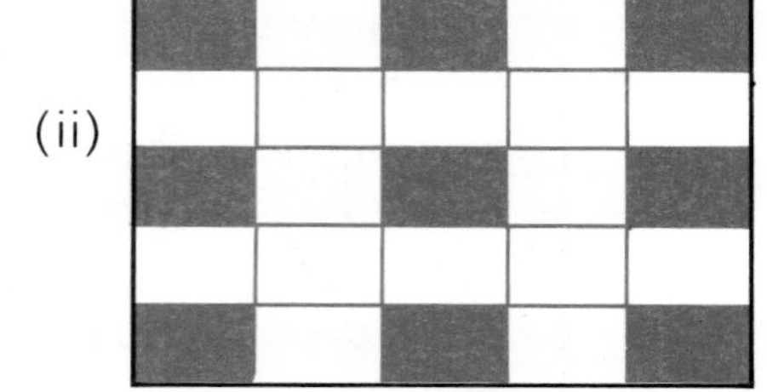

Activity *Ask a question*

6 Choose **one** of these questions. Ask 20 people in your class to answer it.

A Which person is second in line to the British throne?
B What is the capital of Australia?
C Who is the Secretary of State for Education and Science?
D Which mountain is the tallest in Great Britain?

a) Use your results to estimate what percentage of pupils in your year know the correct answer. *(Your teacher will tell you what it is.)*

b) Write one or two sentences to explain how you arrived at your answer in **(a)**.

7 Carol gets a 7% pay rise. She used to earn £650 a month. She starts to work out her new monthly pay on her calculator:

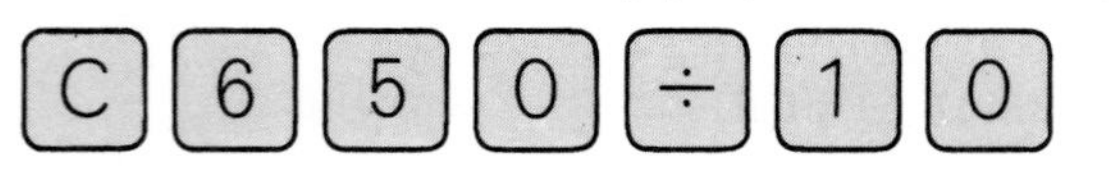

. . .

a) Copy and complete the record of the keys Carol presses.

b) How much will she earn after the pay rise?

c) Write one or two sentences to explain how Carol's method works.

Percentage changes

B 1 The Bike Shop is having a sale.
Copy and complete this table.

Usual price	Sale price
£100	
£110	
£180	
£250	
	£264

think and test

2 The DIY shop shows all its prices before VAT.
The amount of tax to be added is 15% of the price on the label.

a) How many pence must be added in every £1?

b) Copy and complete this table.

Item	Ticket price	VAT	Total price
Power drill	£50		
Electric saw	£76		
Pick-axe	£24.60		
Pincers	£1.80		

3 Work out the sale price of each of these items:

a) Washing machine £350; 20% off

b) Stepladder £8.20; 10% off

c) Microcomputer £420; 15% off

d) Set of glasses £24.50; 5% off

e) Extension cable £8 + VAT; 10% off

at 15%

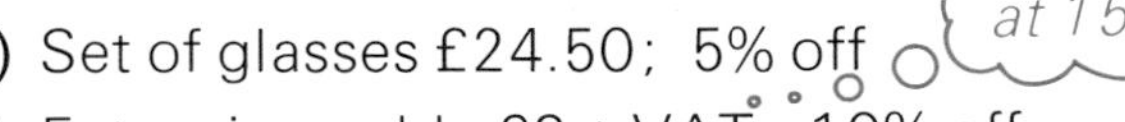

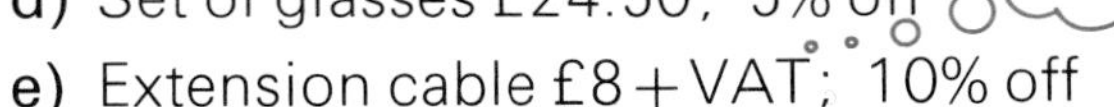

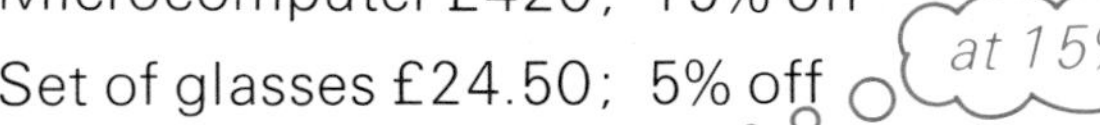

4 The Hole in the Wall Restaurant tells its customers that it automatically adds a service charge of 12% and VAT at 15%.

a) Does it matter which of the following methods of calculation is used?

A Add 12% of the bill to the bill.
Then add 15% of the result to get the total.

B Add 15% of the bill to the bill.
Then add 12% of the result to get the total.

C Add 12% of the bill to the bill.
Then add 15% of the bill to get the total.

b) Give an example to illustrate your answer.

c) Which method, **A**, **B** or **C**, do you think is normally used?

What percentage?

C 1 a) Estimate what fraction of each 24 hours you spend sleeping.

b) Is this nearest to 10%, 20%, 30% or 40%?

2 Write a sale card like this, using percentages, for each of these items. 

Children's shoes slashed!
30% off

a) Plants: save 15p in the pound.

b) 500 cc motorbike: usual price £800, save £80.

c) Paint: usual price £4, save 80p.

d) Table lamp: usual price £10, sale price £8.

e) Denims: usual price £20, save £15.

f) Refrigerator: usual price £600, sale price £576.

24

3 The sale price of the single quilt set is £12.99.

Executive Collection
—MAYFAIR DESIGN
Save at least $\frac{1}{3}$
Quilt cover and pillowcase sets in grey/black.
M.R.P. Sale
Single £19.99 £12.99
Double £29.99 £19.99

a) How much do you save if you buy it in the sale?

b) Roughly, what percentage saving is this?

c) Which is the best saving 'percentagewise', the single quilt set or the double quilt set? Explain why you made this choice.

4 Calculate the percentage increase in pay for each of these.
Give your results to the nearest 1%.

a) John Allen (teacher): before the pay rise – £12 000
after the pay rise – £12 500

b) Mary Rose (doctor): before the pay rise – £18 500
after the pay rise – £21 000

5 What percentage (by weight) of the Ski yogurt is

a) protein?

b) carbohydrate?

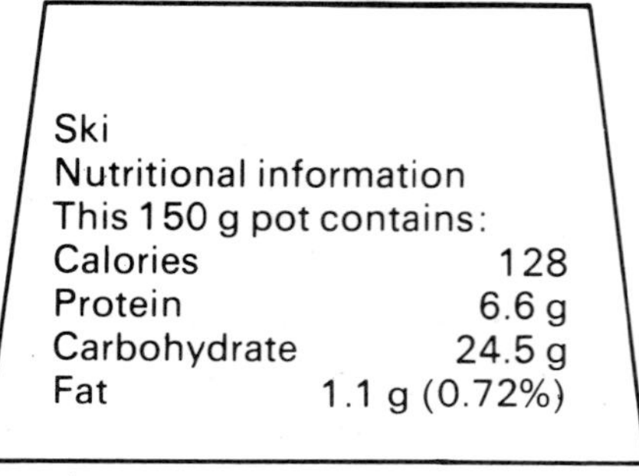

Ski
Nutritional information
This 150 g pot contains:

Calories	128
Protein	6.6 g
Carbohydrate	24.5 g
Fat	1.1 g (0.72%)

6 In 1979, the population of the United Kingdom was 55.9 million; 16.1 million men and 10.3 million women had a job, and 1.3 million workers were unemployed.

a) What percentage of the UK population were
(i) working men? (ii) working women? (iii) unemployed?

b) Why do the percentages in **(a)** not add up to 100%?

c) What percentage of UK workers were unemployed?

Metropolitan Police members 1829-1970
Figures in thousands

1829 3

1860 6

1880 11

1900 15

1940 18.5

1970 25

Left The first London police were set up under the Middlesex Justice's Act of 1792. The force was to become the Metropolitan Police after Peel's Act of 1829.

Number of amateur and professional police in London 1790-1832

Amateur: 1000 (1790), 2000 (1821), 600, 3000 (1832)
Professional: 0 (1790), 250

1790 1821 1832

Above Before Peel's Act most of the police were watchmen or parish constables. The 1829 Act effectively made most of the capital's police force professional.

7 The charts give information about the London (Metropolitan) police force between 1790 and 1970.

a) Between 1880 and 1900, the police force increased from 11 000 to 15 000.
What percentage increase is this? *(to the nearest 1%)*

b) The bar chart shows that between 1790 and 1821 the number of amateur police officers increased from 1000 to 2000.
What percentage increase is this?

c) Calculate some more percentages from the charts.
Use your results to write a newspaper article about how the police force changed between 1829 and 1970.
You might like to copy some of the drawings to illustrate your article.

Calculators and percentages

D 1 A DIY store buys electric sabre saws at £18 each.
So that the store will make a profit, the saws are put on sale at a price that is 27% higher than this.

27% is called the 'mark-up percentage.'

a) The pricing clerk works out the mark-up on her calculator like this:

[C] [0] [.] [2] [7] [×] [1] [8] [=]

(i) Why will this give the correct mark-up?
(ii) The clerk then adds the mark-up to the original cost of a sabre saw.
What is the sales price of an electric sabre saw at this store?

b) At the cash desk, a sales assistant has to add VAT at 15%.
She calculates the total cost of an electric sabre saw like this:

[C] [1] [.] [1] [5] [×] [2] [2] [.] [8] [6] [=]

(i) Why will this give the correct total cost?
(ii) What is the total cost of an electric sabre saw at this store?

24

c) The pricing clerk could have avoided having to do an addition by calculating the sales price directly.
What keys would she have had to press to achieve this?

d) What keys could you press to calculate the **total** cost directly from the original cost of £18?

e) Calculate (i) the sales price
(ii) the total price including VAT
of a set of vanadium steel socket wrenches which the store buys for £24 and marks up by 32%.

at 15%

Round results to the nearest 1p, if necessary.

2 Janet is training for the 400 m.
Yesterday she ran it in 54.6 s; today she managed it in 52.9 s.
On her calculator, she presses

[C] [5] [2] [.] [9] [÷] [5] [4] [.] [6] [=]

a) Write down what result this gives her.

b) 'That's just over a 30% improvement,' she claims.
Is she correct? Why?

3 Write descriptions of the shortest and easiest ways you know of using a calculator

a) to find a certain percentage of a certain amount.

b) to find the result when a certain amount is increased by a certain percentage.

c) to find the result when a certain amount is decreased by a certain percentage.

d) to find out what percentage one amount is of another.

Example: 12% of 12m

Example: 12m is increased by 15%.

Example: 12m is decreased by 20%.

Example: what percentage 5 s is of 14 s.

▲ 229
● Next chapter

Percentages in the news

E 1 These bar charts appeared in a newspaper to illustrate an article about when Britons do their shopping.

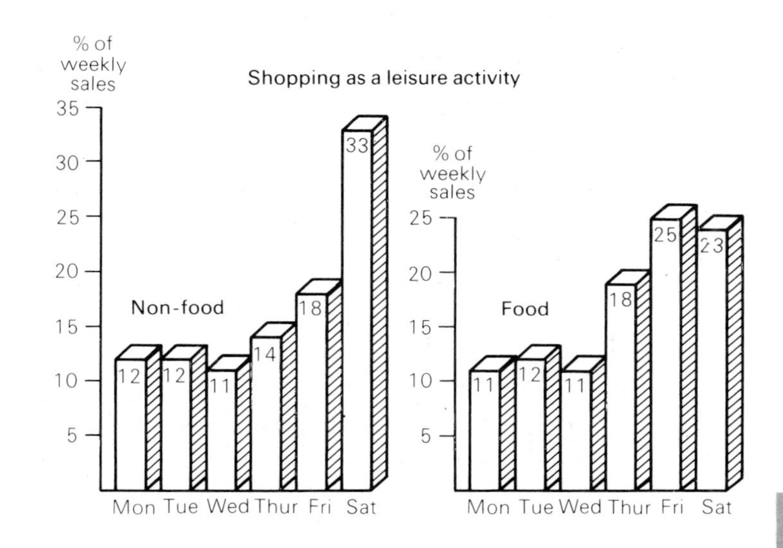

a) Check that in each chart the total of the percentages represented by the bars is 100%. Why should this be so?

b) Write a headline for the newspaper article which accompanied the bar charts. Then write an introductory paragraph for the article.

2 A survey of MOT tests revealed that 50% of the cars that failed had faulty brakes.
A company that manufactures brake linings heard about this information and, in their next advertising campaign, claimed that '50% of people drive on faulty brakes'.
Do you think this is a fair thing to say? Why?

3 You need a protractor.
This is called a pie chart.

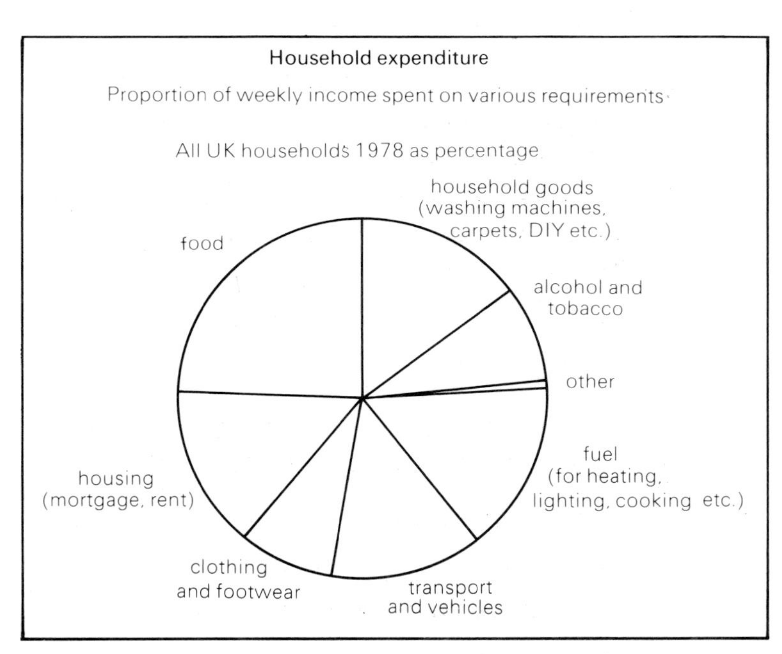

a) Roughly, what fraction of the whole circle is used to represent spending on food?

b) Write a short newspaper article to go with the chart. In your article mention these things:
food
clothing
alcohol and tobacco.
Say what percentages of weekly incomes are spent on these items.

c) Roughly, how much would a household with a weekly income of £200 spend each year on each of the items in **(b)**?

Savings and depreciation

F With a friend

10.75% interest | 6.25% interest

1 Here are two ways in which interest on savings can be calculated.

Simple interest	**Compound interest**
You receive interest only on the lump sum you invest. The interest is calculated at regular intervals (say, every year) and added to your account.	You receive interest on the total amount in your account – your original lump sum plus any interest you have earned so far. The interest is calculated at regular intervals (say, every year) and added to your account.

a) Discuss the two ways of paying interest.
Make sure you understand what they mean.
Write down which method of calculating interest you would prefer for your own savings.

b) You have £500 to invest.
Together work out how much you would have in your savings account after 2 years for each of these methods.

SAVE WITH HOMELY SAVINGS SIMPLE INTEREST OF 11% PAID EVERY YEAR

SAVE WITH SAVE-IT COMPOUND INTEREST OF 11% PAID EVERY YEAR

24

2 £2000 is invested in a savings account.
What is the total amount in the account after 4 years, if interest is calculated at

a) 10% simple interest? b) 10% compound interest?

Challenge

decreases

3 The value of a car depreciates by 15% in the first year and by $12\frac{1}{2}$% of its previous year's price, every following year.
Tim buys a new car for £8700.
After how many years will it be worth less than £4000?

4 £p is invested in a savings account which pays simple interest at r% per annum.

Write an expression involving p and r for the total amount in the account after

a) 1 year. b) 2 years. c) 3 years. d) n years.

25 Dealing with information

A 1 20 people work for Mason's Car Tyres. This is what they earn each year.

Management			
J M Mason (Director)	£30 000		
J J Sparks (Director)	£25 000		
J D Smith (Manager)	£20 000		
Shop floor			
K Dempsey	£6000	L Smith	£6000
D Stone	£6000	N Kite	£6000
N Star	£5000	D Trevors	£5000
B Samson	£5000	L Fingle	£5000
S Marriot	£5000	L Fortune	£5000
G Megs	£5000	M Malik	£5000
Secretarial			
A Mulak	£4000	T Vine	£3000
S Star	£3000	N Miles	£3000
T Train	£3000		

a) Which of these do you agree with?

(i) Management's mean earnings is £25 000.

(ii) Most shop floor workers earn £5000.

(iii) The mean wage of a shop floor worker is about £5300.

(iv) Most secretaries earn £3000.

(v) The mean earnings of the 20 staff is £7750.

b) This advertisement appeared in the local press.

Do you think it is fair? Write one or two sentences to explain why.

> Good wages at Mason's. Average earnings more than £7750 per year.

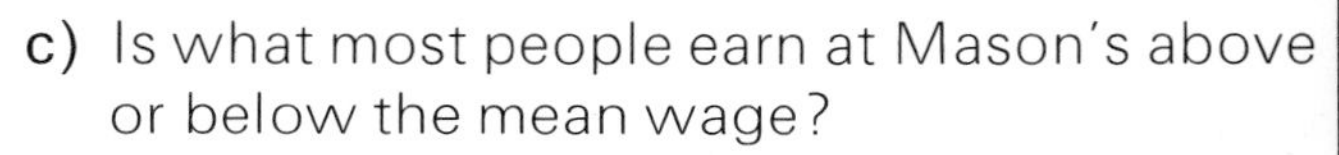

c) Is what most people earn at Mason's above or below the mean wage?

d) Sketch a bar chart to show employees' earnings at Mason's.

2 The three graphs show earnings from three different firms.

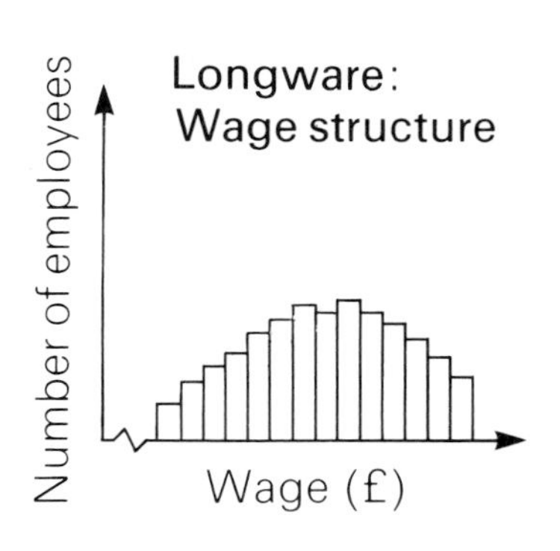

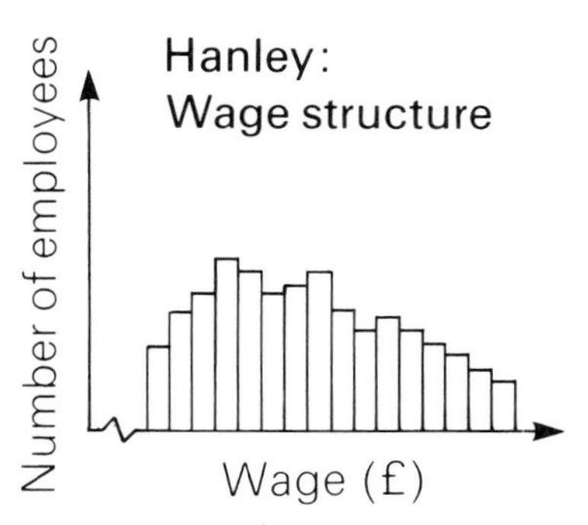

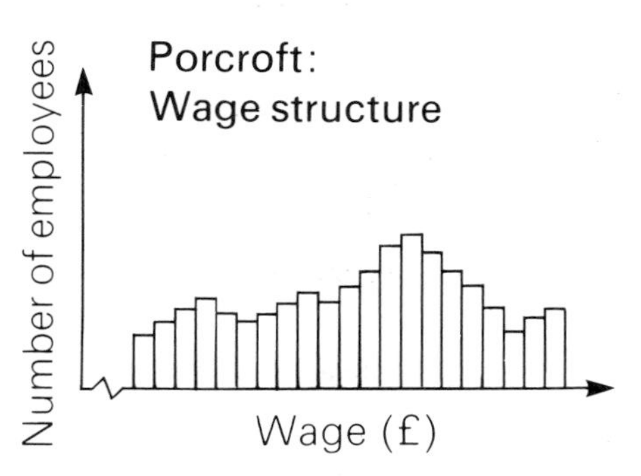

a) In which firm do most people earn more than the mean?

b) In which firm do most people earn less than the mean?

c) In which firm are there approximately the same number of people who earn more than the mean as earn less than the mean?

Take note

The mean does not tell us anything about individual measurements (*e.g. earnings*), or about how the measurements are 'distributed' (*spread out*).

Activity

3 a) Choose three people in your classroom.
Choose them so that two of them are shorter than the mean of all three.
Write down their names, their approximate heights and the approximate mean height.

ask them (or measure them)

b) Now choose three people so that two of them are taller than the mean of all three.
Write down their names, their approximate heights and the approximate mean height.

4 The dot diagram represents the ages of five children in a ballet class.
a) Guess their mean age.
b) Calculate their mean age.
How close was your guess?
Was it *good*, *OK* or *abysmal*?

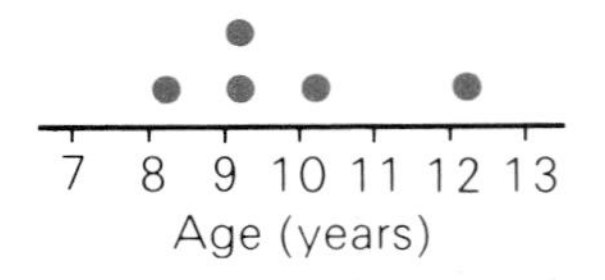

5 These charts represent the ages (in years) of six different groups of children.

A
7 8 9 10 11 12

B
7 8 9 10 11 12

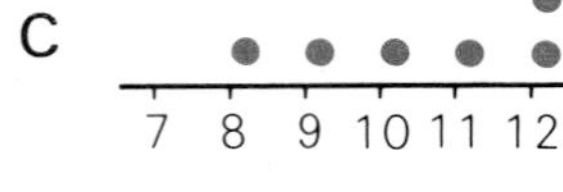

D
7 8 9 10 11 12

E
7 8 9 10 11 12

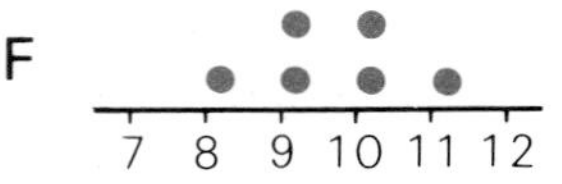

a) Without doing any calculations, decide whether the mean age for each group is greater than 9, equal to 9 or less than 9.
b) Check your results in **(a)** by calculating each mean.

Take note

The ages in chart **A** vary from 7 years to 11 years.
We say that the **range** of the distribution is 7–11 years.

c) What is the range of each of the other distributions?

Think it through

6 A survey of 1000 teenagers revealed that the mean number of hours they spend watching television each week is 15.
A report of the survey results mentions that the range is 8 hours – 20 hours.
Which of these do you think is most likely to represent all the information correctly?
Explain why.

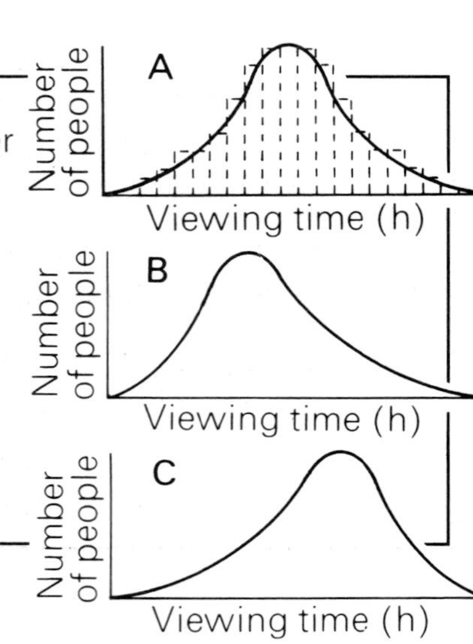

Back to nature

B

With a friend

1 These two charts record the results of two sunflower experiments. They show how tall two sets of 25 sunflowers grew using different horticultural methods.

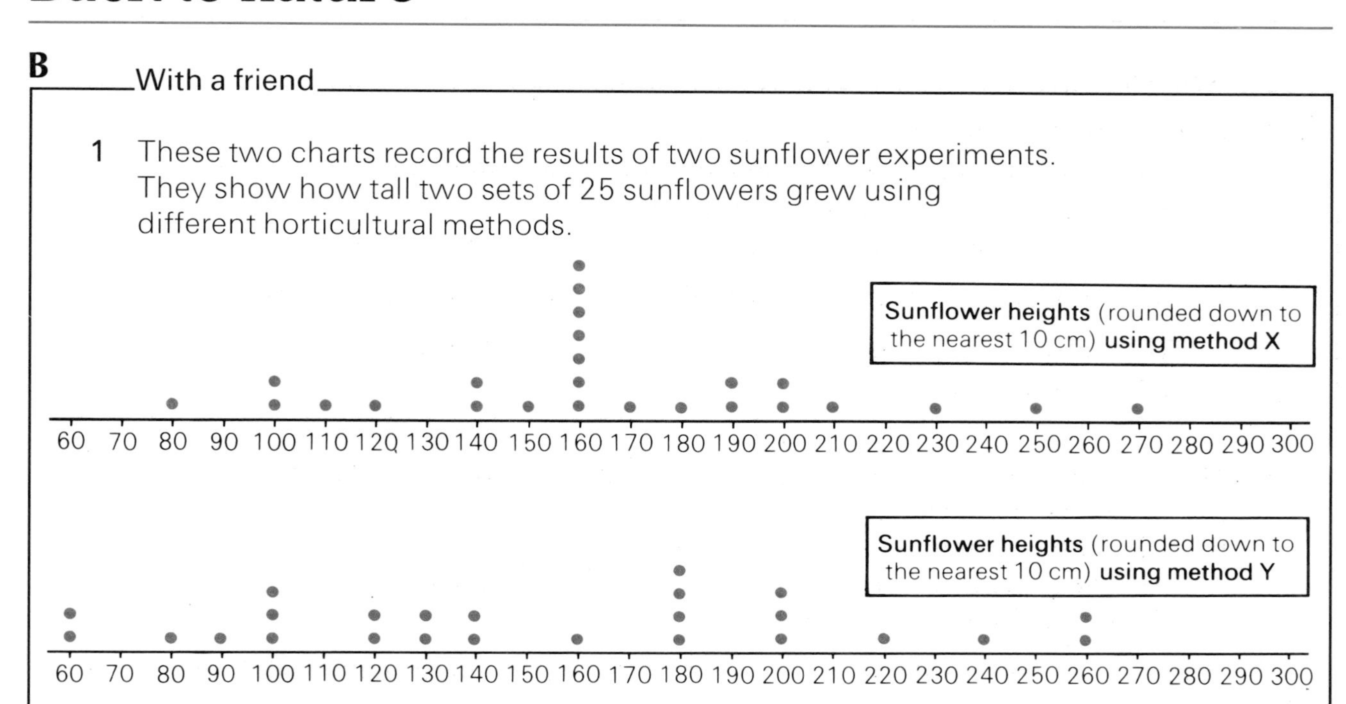

You are entering a 'Grow a sunflower' competition.
You are given one seed and must sow it in a specially marked pot.
Discuss between you which horticultural method you would choose, method X or method Y.
Each of you write down what you decide and why.

Take note

Each of the diagrams show how the heights of the sunflowers are **distributed**.
Each set of heights is called a **frequency distribution**.

80 cm, 100 cm, 100 cm, 110 cm, 120 cm, 140 cm, 140 cm, 150 cm, . . .

2 The two bar charts show the distribution of the weights of 500 11-year-olds from two different parts of the UK. Without doing any detailed calculations,

a) decide which group has the greater mean weight.

b) estimate the mean weight for each group.

Each of you write down what you decide.

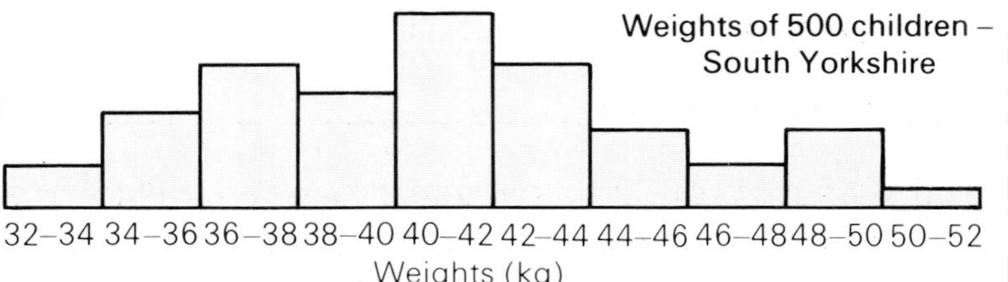

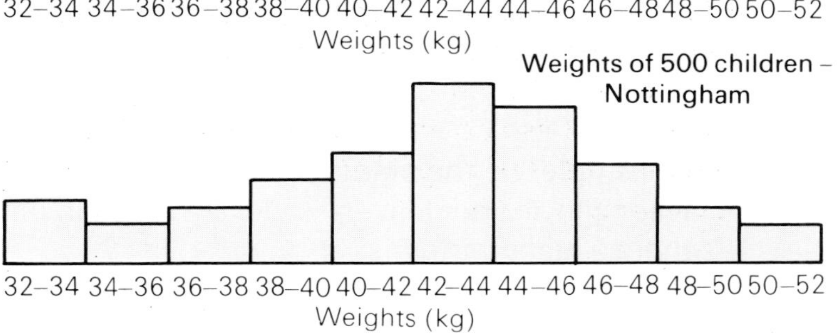

25

Assignment

3 Collect some leaves from a tree or bush, or petals from flowers.
Measure their lengths and their widths.
Draw frequency charts to show the distribution of the lengths and widths.

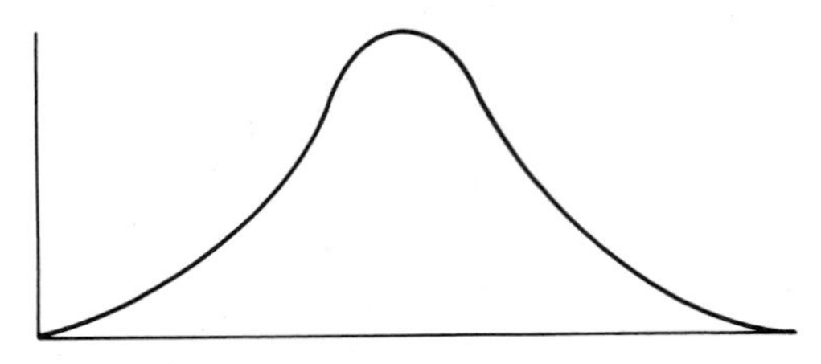

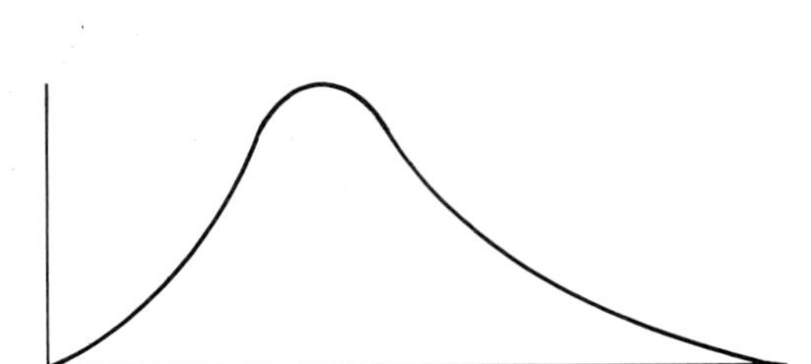

Do there tend to be
proportionally more longer leaves,
proportionally more shorter leaves,
or equal proportions of longer and shorter leaves?

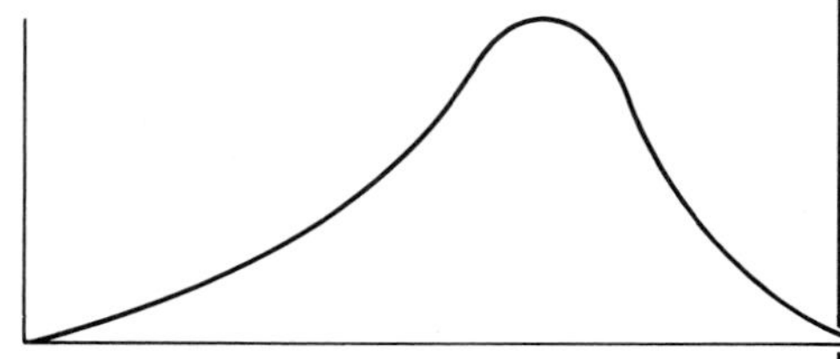

Do there tend to be
proportionally more wider leaves,
proportionally more narrower leaves,
or equal proportions of wider and narrower leaves?

Investigate leaves or petals from different trees, bushes or flowers.

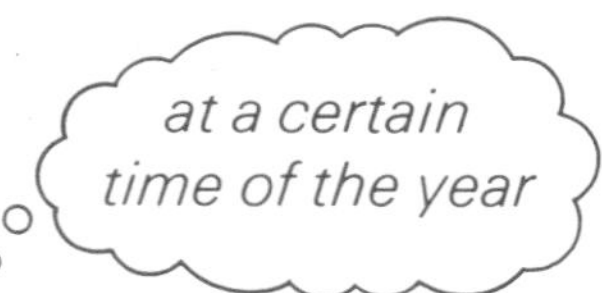

Write a report to say how the ranges of width and length vary.

For example, do the plants tend to have a greater proportion of longer leaves than shorter leaves, wider leaves than narrower leaves, . . .?
Is there a reason for this?
Do the patterns change with the seasons of the year?

Three averages

C

Challenge *Mean potatoes*

1 You want to make baked (jacket) potatoes for lunch (for three people). There are two bags of potatoes you can buy.

A good size jacket potato weighs about 340 g.

But we are not allowing you to view the potatoes before you buy!

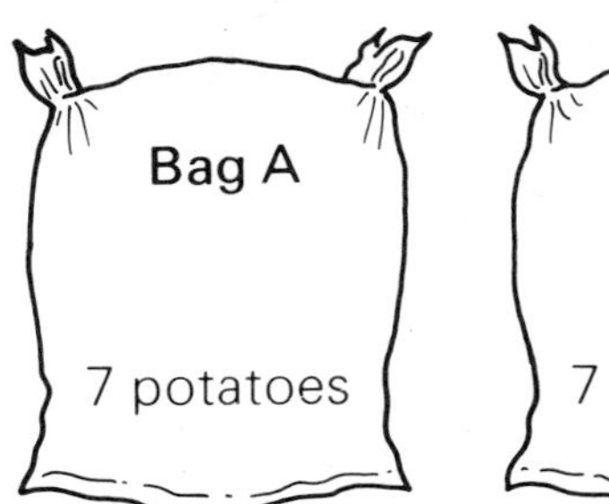

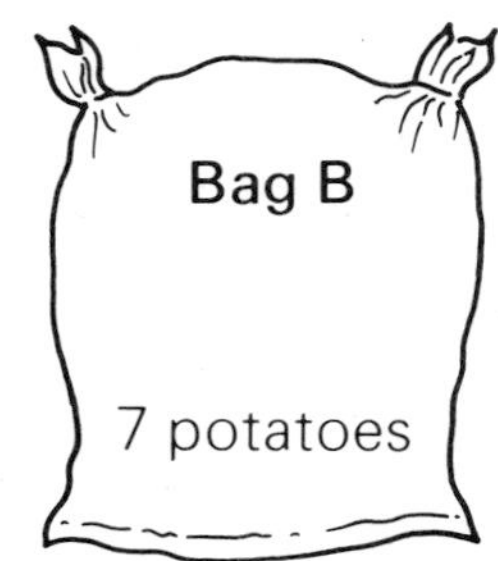

a) The mean weight of potatoes in bag **A** is **320 g**.

The mean weight of potatoes in bag **B** is **280 g**.

On this information alone, which bag would you be tempted to buy? Why?

b) Now you are told that if all the potatoes in each bag are set out in a line, end to end, starting with the smallest . . .

     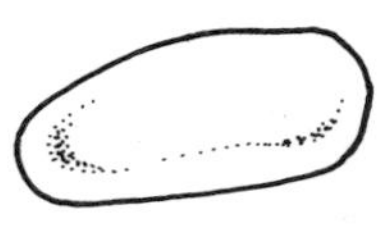 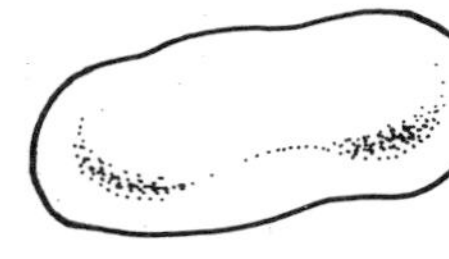

. . . the middle one in bag **A** would weigh **200 g** and the middle one in bag **B** would weigh **300 g**.
Does this make you want to change your mind about which bag to buy?
Why?

c) On the information so far, you should have some idea of the relative weights of the potatoes in the two bags.
Make a sketch of the contents of the bags, giving your 'best guess' at the weights of the potatoes.

d) You are now told that the most common weight in bag **A** is 200 g (4 potatoes) and the most common weight in bag **B** is 300 g (4 potatoes).
Which bag would you finally decide upon?
Change your 'best guess' picture (if necessary) to show what you now think the contents in each bag look like.

25

Take note

Different information about distributions might lead to different conclusions.
The 'mean' is not always the best information to have.

For example, the 'best' information we could have about the potatoes is how many weigh about 340 g.

The **middle** measurement of a distribution is called the **median**.
In this example it is 110 g.

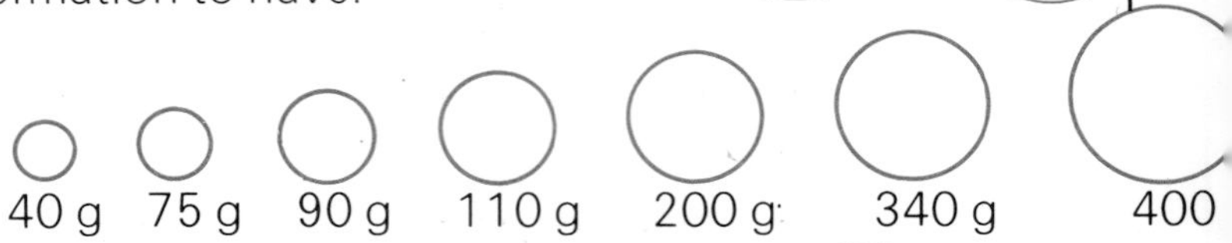

40 g 75 g 90 g 110 g 200 g 340 g 400

The **most common** measurement is called the **mode**.
In this example it is 80 g.

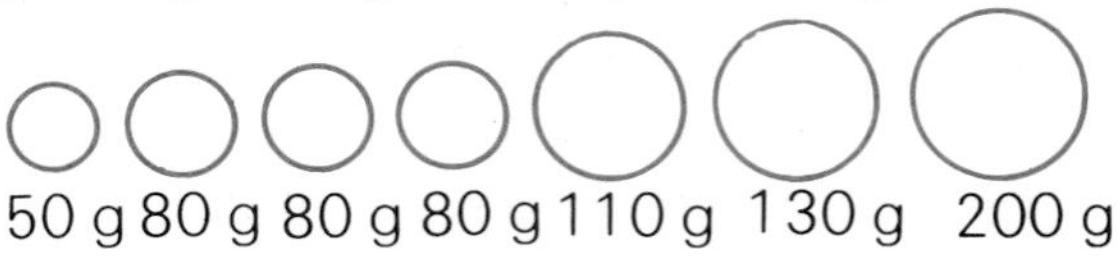

50 g 80 g 80 g 80 g 110 g 130 g 200 g

2 The three dot charts show the ages (in years) of three different groups of pupils.
What is
(i) the mean (to 1 DP)?
(ii) the median?
(iii) the mode for each distribution?

a) 7 8 9 10 11 12

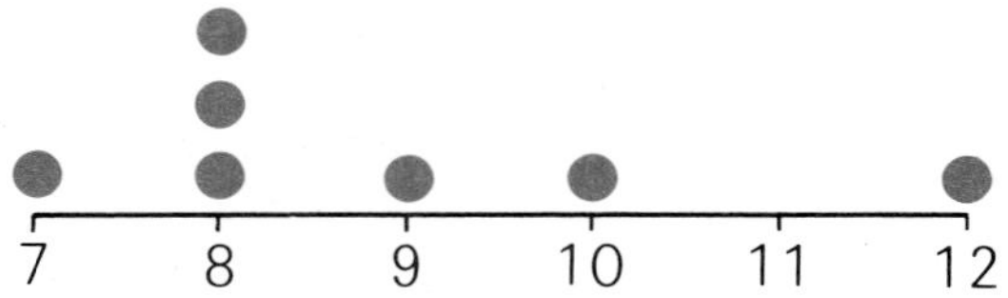

b) 7 8 9 10 11 12

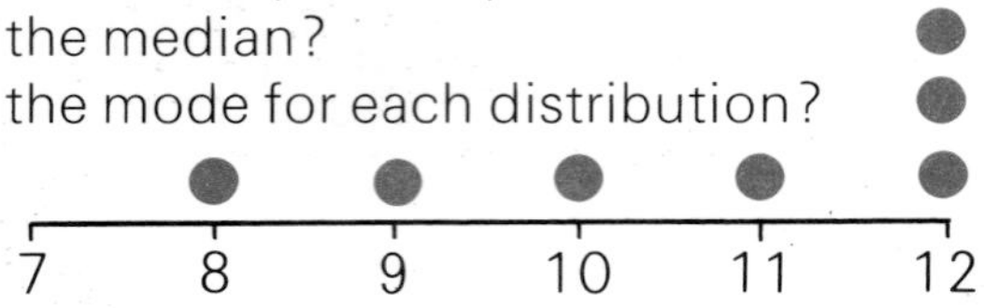

c) 7 8 9 10 11 12

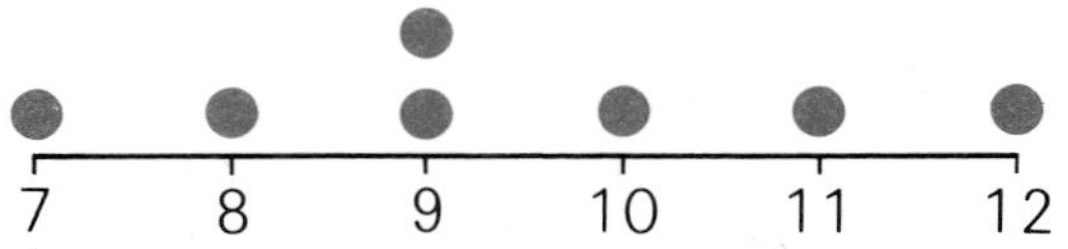

Take note

Sometimes there may be two or more modal values.

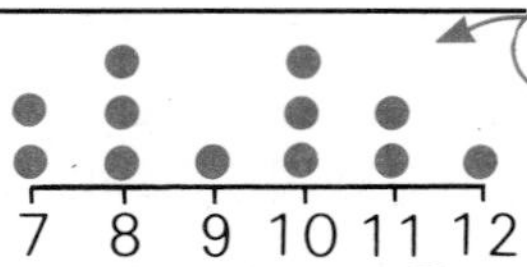

two modes: 8 and 10

7 8 9 10 11 12

When there is an even number of pieces of information, the median is the mean of the middle two measurements.

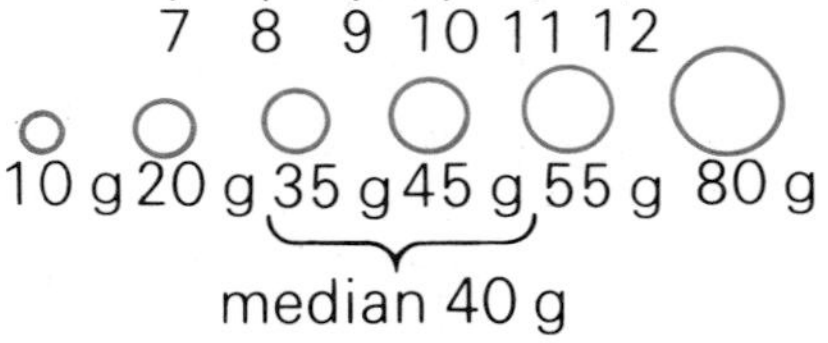

10 g 20 g 35 g 45 g 55 g 80 g

median 40 g

Activity

3 Ask five people in your class
- their age in years
- their shoe size.

Write down the mean, mode and median for each survey.

4 These are the numbers of children in 15 families:

Number of children in the family	1	2	3	4	5	6
Number of families	4	3	4	1	2	1

How many children does the 'middle family' have?

a) There are two modes. What are they?

b) What is the median family size?

5 a) In each of these, write down whether 'average' means 'median', 'mode' or 'mean'.

A *I earn an average wage. As many people earn more than me as earn less than me.*

B *The average woman's dress size must be size 12. We sell more of that size than we do of any other.*

C *Our average pocket money is £3.50 a week. If we pooled our money each week and shared it out fairly we'd have £3.50 each.*

b) Write down an example of your own where 'average' means
(i) median. (ii) mode. (iii) mean.

6 a) You own a sports shop, and you have run out of sports shirts.
Which of these pieces of information do you think would be of most value to you to decide upon your next order?
Why?

the mean size of shirt sold last month

the modal size of shirt sold last month

the median size of shirt sold last month

b) Think of an example of your own where it would probably be most useful to know
(i) the mean value rather than the modal value.
(ii) the modal value rather than the mean value.

7 In the semi-finals of the hockey competition there are four teams.
The mean number of goals scored by each team is 2.
The modal number of goals scored is 3.
How many goals did each team score?
Write down all the possibilities.

8 You are starting a new paper round.
The chart shows the daily wages of the other paper round girls and boys at the same shop.
The shopkeeper says, 'You can have a round with the modal wage, the median wage or one which pays closest to the mean wage'.

You want one which pays as much as possible.
Which round would you choose?

9 An advertising company carries out a survey on colour preferences.
1000 people are asked which colour they think is most appropriate for packaging green beans.
The choices are a range of nine colours.
From the results which of these could be arrived at, and which could not?
(i) a mean value (ii) a modal value (iii) a median value
Explain your answer.

10 In a game of chance you have to predict the total score on three spinners.
Which of these would be the most useful information to know before you begin? Why?
A the mean total score for 1000 throws **B** the median total score for 1000 throws
C the modal total score for 1000 throws

Take note

The mean, median and mode are all different types of 'average'.
The mean is a 'fair shares for all' average.
The mode is what occurs most often.
The median is the middle value of a range.

11 The members of Great Nettleford Bowls Club are very proud – they won this year's regional tournament.
The local paper is writing about their success, and the reporter wants to know the average age of the members of the club.
Imagine you are the Club Secretary.
You will have to decide whether to tell the reporter the mean, median or modal age.

The members would like to give bowls a 'youthful image' so would prefer you to use the smallest figure.

The club records show this distribution of ages:

Age	58	59	60	61	62	63	64	65	66	67	68	69	70	76	80	88
Number of members	2	1	4	3	8	1	4	1	6	1	3	1	2	1	1	1

a) How many club members are there?
b) What is the modal age?
c) Imagine all the members lined up in order of age.
What is the median age?
d) The combined total of the ages of the 58-year-olds is 116. *(2×58)*
What is the combined total of (i) the 58- and 59-year-olds?
(ii) the 58- to 60-year-olds? (iii) the entire membership?
e) What is the members' mean age?
f) What will you tell the reporter is the 'average' age of the club members?

Challenge

12 a) This is a set of information about two groups of nine 14-year-olds. If you only have this information, which group would you choose to form a 'tug of war' team? Why?

	Mean weight (kg)	Median weight (kg)	Modal weight (kg)
Group **A**	60	65	60
Group **B**	60	55	60

b) What about these groups? Why?

Group **C**	60	65	55
Group **D**	60	55	65

c) Sketch what you think the distribution of the winning team would look like.
Say which one of the four groups it is.

Finding averages for grouped information

D 1 This table shows the age distribution of male convicts in British prisons in 1981.

Age	14–18	19–23	24–28	29–33	34–38	39–43	44–48	49–53	54–58	59+
Number of men	6300	10 100	6500	4000	3500	1900	1500	750	550	375

a) We cannot calculate the exact mean age of the prisoners. Why not? Write one or two sentences to explain.

b) The Prison Service can **estimate** the mean age like this: In this estimate, everyone in the 14–18 age group is assumed to be 16 years old; everyone in the 19–23 age group is assumed to be 21 years old, and so on.

Copy and complete the calculation.

Age group	'Middle' age of the age group	Number of prisoners	'Middle' age × number of prisoners in the group
14–18	16	6300	100 800
19–23	21	10 100	212 100
24–28	26	6500	169 000
⋮	⋮	⋮	⋮
54–58	56	550	30 800
59+	say 61	375	22 875
TOTALS		□	□

call this T call this M Estimated mean $= \frac{M}{T}$

Challenge

c) We could find the **lowest** possible value that the mean can have, and the **greatest** possible value that the mean can have. How? Write one or two sentences to explain.

assuming the 'upper' age limit is, say, 70

d) Think about how we might find the median age. Why is it impossible to find the exact median age?

e) Imagine that all the prisoners are lined up according to age. In which age group is the 'middle' prisoner?

Take note

This age group is called the median age group.

f) Think about how we might find the modal age.
(i) Why is it impossible to find a particular modal age?
(ii) In which age group are there most prisoners?

Take note

This age group is called the modal age group.

g) Suppose you are the prison governor, and the Press asks you these questions. How would you reply?
(i) How old is the average male British prisoner?
(ii) How old are most male British prisoners?

2 Choose **one** of the activities **A** to **D**.

Activity A

Group the heights or weights of people in your class, like this

height . . . 150 cm – 154 cm, 155 cm – 159 cm, 160 cm – 164 cm, . . .

weight . . . 40 kg – 44 kg, 45 kg – 49 kg, 50 kg – 54 kg, . . .

Make out a frequency chart, and from your chart calculate the mean, mode and median height or weight.

Assignment

How do the mean, mode and median vary with age and/or sex in your school?

Activity B

Choose a newspaper, colour magazine or a book.
Investigate the number of words used in each sentence.
Make out a frequency chart for these groups:

Less than 5, 5–9, 10–14, 15–19, . . .

Find the mean, mode and median length of a sentence.

Assignment

Do different newspapers, magazines, authors, . . . use different length sentences?

Activity C

Make a paper aeroplane. Test its flight range.
Make out a frequency chart of its test flights for a large number of trials (say 100).

0 – 1 m, 1.1 m – 2 m, 2.1 m – 3 m, 3.1 m – 4 m, . . .

What is the mean distance, modal distance and median distance for your aircraft?

Assignment

Do different designs, sizes, . . . of paper aeroplane have different powers of flight?

Activity D

Play 'shove ha'penny' with a 2p piece.
How close can you get the 2p piece to a line or to the edge of the table?
Make out a frequency chart of the distances for a large number of trials (say 100).

0 – 4 cm, 5 – 9 cm, 10 cm – 14 cm, 15 cm – 19 cm, . . .

What is the mean, mode and median of your attempts?

Assignment

Are you better with one type of coin than another? . . . better over a short distance than over a longer distance? What about your friends?

Design a survey

3 Design your own survey where you need to collect 'grouped data' . . . and calculate the mean, mode and median.
Write a short report about your survey, and explain the different benefits of knowing the three measures of average – if there are any.

26 Squares and square roots

A 1 You need some 2 mm squared graph paper.

a) Write down the area of a square with sides of length
(i) 1 cm. (ii) 2 cm. (iii) 3 cm.
(iv) 4 cm. (v) $2\frac{1}{2}$ cm.

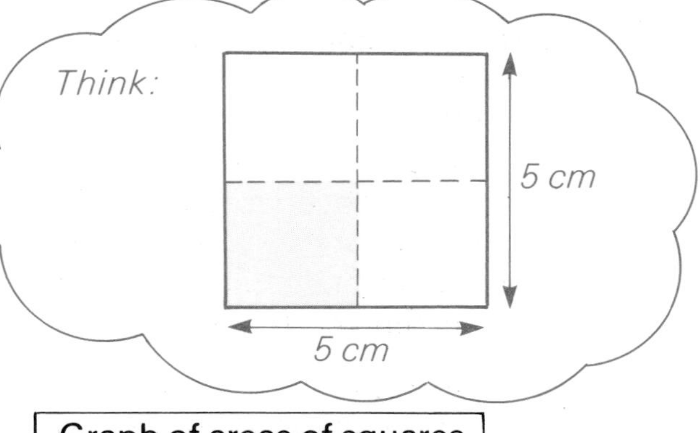

b) Copy and complete the graph, using your results from **(a)** and other calculations you need to make.

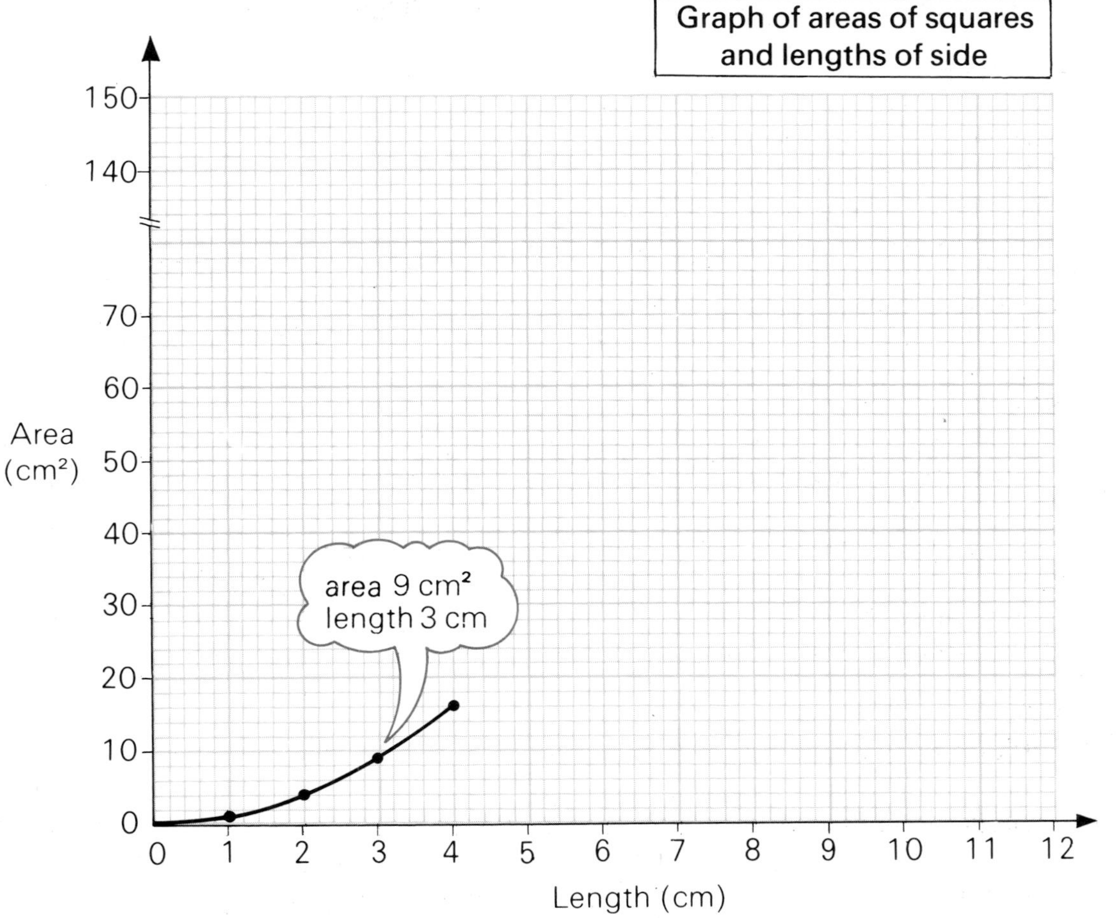

c) Use your graph to estimate
(i) the area of a square with sides 5.7 cm long.
(ii) the value of 8.3×8.3.

d) Use your calculator to check your estimates in **(c)**.

e) Use your graph to estimate
(i) the length of side of a square with area 85 cm².
(ii) the number which when multiplied by itself gives 60.

f) Use your calculator to check your estimates in **(e)**, and to find a closer estimate to each result.

Shorthand writing

B Do you remember . . . ?

7^2 is a shorthand way of writing 7×7.

We say 'seven squared' or 'seven to the power of 2'.

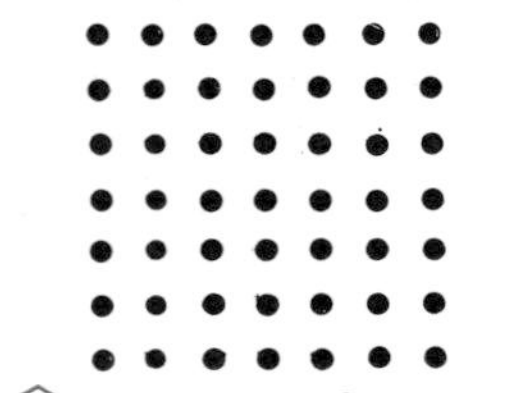

1 Copy and complete:

a) $18^2 = \square \times \square = \square$. b) $3.4^2 = \square \times \square = \square$.

c) $\square^2 = \square \times \square = 196$. d) $\square^2 = \square \times \square = 72.25$.

both $\square$s replaced by the same number

2 What can you say about $\square^2$ and $2 \times \square$; are they

A always equal, **B** sometimes equal, or **C** never equal?
Explain your answer.

3 a) Copy and complete:
(i) $2^2 \times 3^2 = \square^2$. (ii) $1.2^2 \times 3.8^2 \times 25^2 = \square^2$.

b) Write a sentence of your own like those in **(a)**, and complete it.

26

4 Think of squaring each **whole** number. *0^2, 1^2, 2^2, . . .*

a) How many of the results are

(i) less than 100. (ii) less than 1000.
(iii) between 101 and 1000. (iv) between 1001 and 1 000 000?

b) For how many whole numbers is this true: $\square^2 = \square$?

That is, the square of the number is the number itself.

Think it through

5 Study these sentences:
$\sqrt{9} = 3$ $\sqrt{1.44} = 1.2$ $\sqrt{12.25} = 3.5$
a) Write down what the symbol $\sqrt{\ }$ tells us to do.

b) Copy and complete:
(i) $\sqrt{100} = \square$. (ii) $\sqrt{6.25} = \square$. (iii) $\sqrt{\square} = 0.8$.

both $\square$s replaced by the same number

6 What can you say about $\sqrt{\square}$ and $\frac{1}{2} \times \square$; are they
A always equal, **B** sometimes equal, or **C** never equal?
Explain your answer.

7 a) Copy and complete:
(i) $\sqrt{5^2 \times 7^2} = \square$. (ii) $\sqrt{1.2^2 \times 15^2} = \square$.

b) Explain an easy way to find results like those in **(a)**.

The x^2 and √ keys

C 1 You need a calculator with an x^2 key.

a) Press C 1 2 x^2 and write down the result.

b) Try some examples of your own.
Explain what the calculator does when you press x^2

2 a) Press C 2 × 3 x^2 = and write down the result.

b) Is the result equal to $(2 \times 3)^2$ or $2 \times (3^2)$?

Take note

We say 'x squared' or 'x to the power of 2'.

x^2 is a shorthand way of writing $x \times x$.

The value of x which is multiplied by itself when you press x^2 is the number on the display just before you press x^2.

*In question **2**, the number 3 was on the display just before x^2. So only 3, and not 2 × 3, was squared.*

3 Midge enters a number, and then presses x^2
The result is 5.29.

a) **Guess** which number she entered.

b) Use your calculator to find the number she entered.

c) Here are some more examples.
For each one,
(i) guess the number Midge entered.
(ii) use your calculator to find the number.

A 9.61 B 331.24 C 2.89 D 6.76 E 0.49 F 0.01 G 0.0144

Exploration

4 Enter any number and press x^2.
The result is one one of these:

smaller than the original number	or	greater than the original number	or	equal to the original number

Investigate which numbers give which result.
Write down what you find out.

Challenge

5 a) Press C 2.
How many times do you need to press x^2 for the display to show the result of 2 × 2 × 2 × 2 × 2 × 2 × 2 × 2?

2 × 2, 2 × 2 × 2 × 2 × 2, . . .

b) Which products of 2s can be calculated by pressing

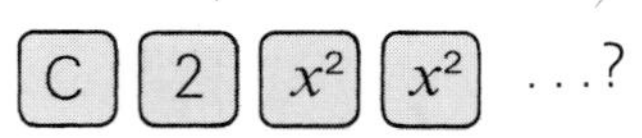

. . . ?

6 You need a calculator with a [√] key.

a) Press [C] [1] [6] [9] [√] and write down the result.

b) Try some more examples of your own.
Record each example like this: $\sqrt{169} = 13$.

c) Explain what the calculator does when you press [√].

*Notice we press [√] **after** entering the number, but we write the √ sign **before** the number.*

7 a) Press [C] [9] [+] [1] [6] [√] [=] and write down the result.

b) Is the result equal to $\sqrt{9+16}$ or $9+\sqrt{16}$?

Take note

We read $\sqrt{16}$ as 'the square root of sixteen'.

the number which when multiplied by itself gives 16

When you press [√] the calculator finds the square root of the number on the display just before you press [√]

or, more often, an approximation to the square root

. . . so [C] [9] [+] [2] [5] [√] [=] gives 14, that is, $9+\sqrt{25}$.

Exploration

8 When we press [√] the result is one of these:

greater than the number entered or equal to the number entered or smaller than the number entered

Investigate which numbers give which result. Write down what you find out.

Think it through

9 a) **Do not use a calculator.**
What do you think the result will be if you press

(i) [C] [1] [6] [√] [x^2]? (ii) [C] [1] [6] [x^2] [√]?

b) Now check your results in **(a)** with your calculator.

c) **Do not use a calculator.**
What do you think the result will be if you press

(i) [C] [1] [.] [2] [√] [x^2]? (ii) [C] [1] [.] [2] [x^2] [√]?

d) Now check your results in **(c)** with your calculator.

e) On Harold's 8-digit calculator, the display behaves like this:

Key	Display
C	0
1	1
.	1.
2	1.2
√	1.0954451
x^2	1.2

Actually,
$1.0954451^2 = 1.0954451 \times 1.0954451$
$= 1.199\,995\,967\,114\,01$, **not** 1.2.

Explain what you think the calculator is doing.

10 a) **Do not use a calculator.**
Copy and complete:

(i) $\sqrt{81} = \square$	(ii) $\sqrt{121} = \square$	(iii) $6^2 = \square$	(iv) $\square^2 = 49$
(v) $\sqrt{\square} = 8$	(vi) $\sqrt{\square} = 12$	(vii) $\square^2 = 225$	(viii) $\sqrt{\square} = 16$
(ix) $0.1^2 = \square$	(x) $\sqrt{0.01} = \square$	(xi) $0.4^2 = \square$	(xii) $\sqrt{0.16} = \square$

b) Now check your results in **(a)** with your calculator.

11 a) **Do not use a calculator**; make a sensible guess.
Is the area of this square window
- nearer to 1 m²,

or • nearer to 0.6 m²?

b) Calculate the exact area with your calculator.
Were you correct in **(a)**?

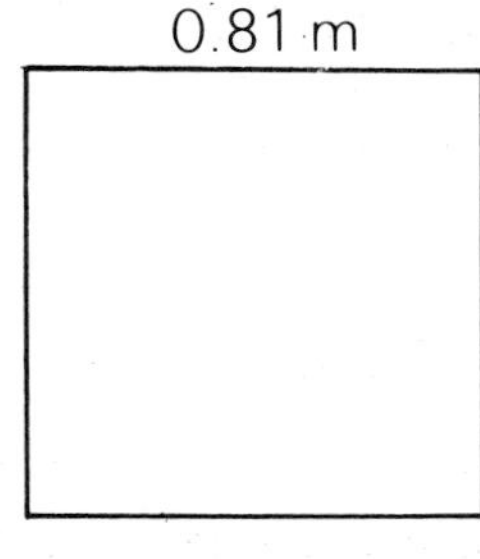

12 a) **Do not use a calculator**; make a sensible guess.
Is each side of this square window
- nearer to 0.3 m long,

or • nearer to 1 m long?

b) Calculate the exact length with your calculator.
Were you correct in **(a)**?

area = 0.9216 m²

13 Use your calculator.
Find the number which k represents in each of these equations.

a) $\sqrt{k} = 64$ b) $k^2 = 64$ c) $\sqrt{k} = 0.04$ d) $k^2 = 0.04$

14 **Do you remember . . . ?**

The area of a circle with radius r cm is $\pi \times r^2$ cm²

The circumference of a circle with radius r cm is $2 \times \pi \times r$ cm.

a) Use the x^2 key if you have one.
Approximately, what is the area of a circle with radius 8 cm?

Use $\pi \approx 3.142$

b) Use the $\sqrt{\ }$ key if you have one.
Approximately, what is the radius of a circle with area 75 cm²?

Challenge

c) This is a plan for a new 400 m running track.

Use $\pi \approx 3.14$.

Approximately,
(i) what is the length of each straight section of the track?
(ii) what area, in hectares, will the track enclose?

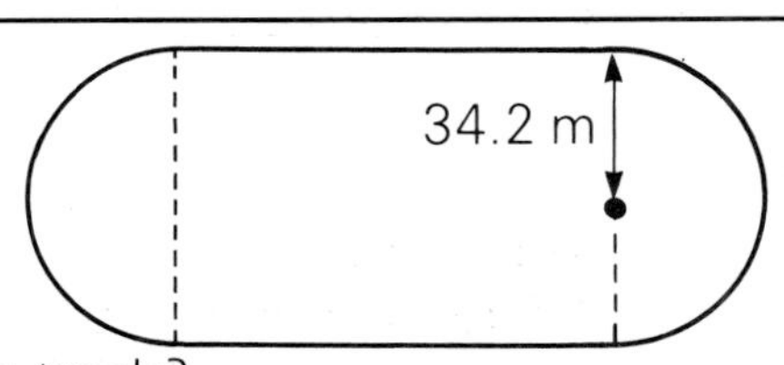

Investigating squares and square roots

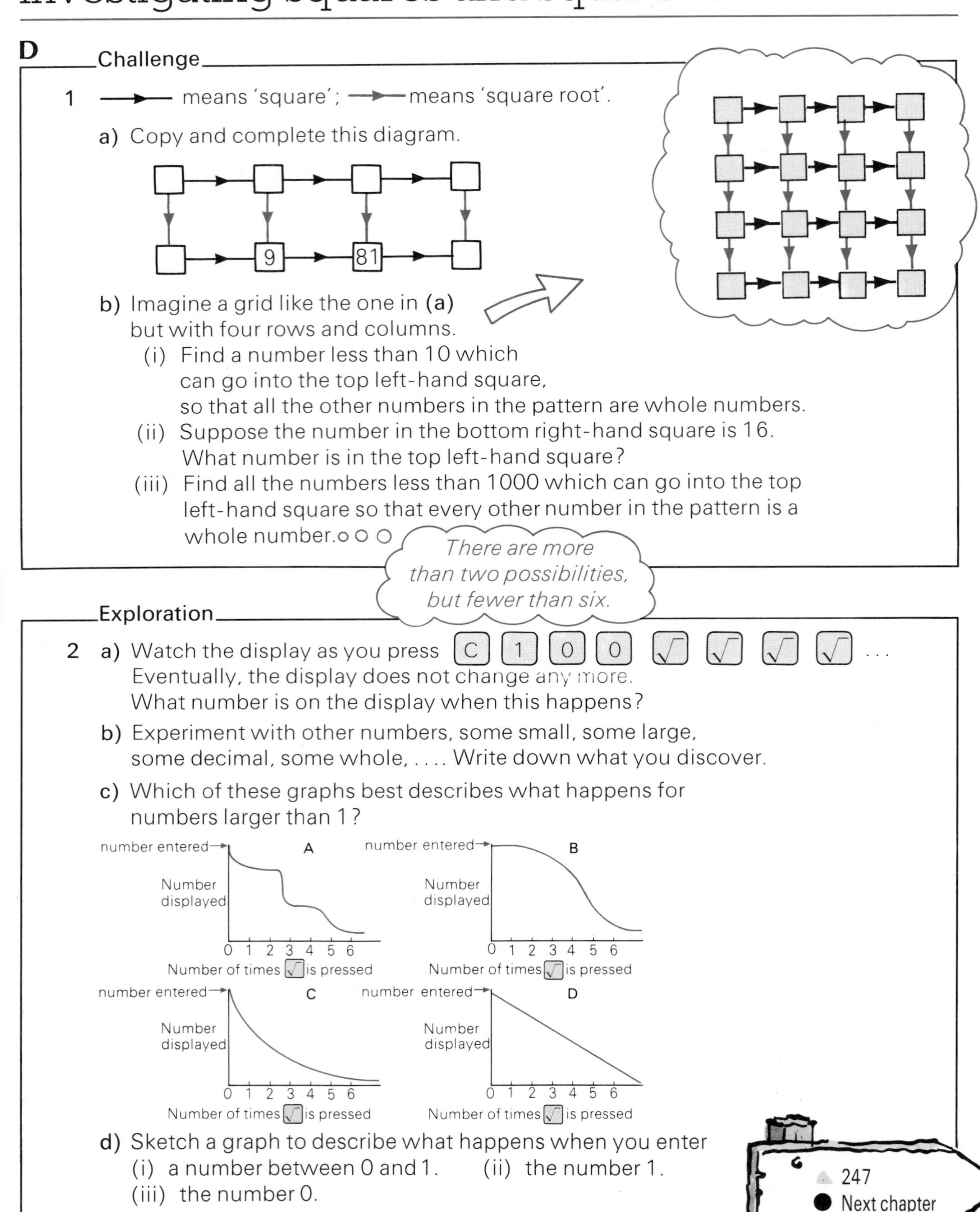

D

Challenge

1 ⟶ means 'square'; ⟶ means 'square root'.

a) Copy and complete this diagram.

b) Imagine a grid like the one in **(a)** but with four rows and columns.

(i) Find a number less than 10 which can go into the top left-hand square, so that all the other numbers in the pattern are whole numbers.

(ii) Suppose the number in the bottom right-hand square is 16. What number is in the top left-hand square?

(iii) Find all the numbers less than 1000 which can go into the top left-hand square so that every other number in the pattern is a whole number.

There are more than two possibilities, but fewer than six.

Exploration

2 a) Watch the display as you press [C] [1] [0] [0] [√] [√] [√] [√] . . .
Eventually, the display does not change any more.
What number is on the display when this happens?

b) Experiment with other numbers, some small, some large, some decimal, some whole, ... Write down what you discover.

c) Which of these graphs best describes what happens for numbers larger than 1?

d) Sketch a graph to describe what happens when you enter
(i) a number between 0 and 1. (ii) the number 1.
(iii) the number 0.

▲ 247
● Next chapter

Naturally occurring square roots

E Activity *Simple pendulum*

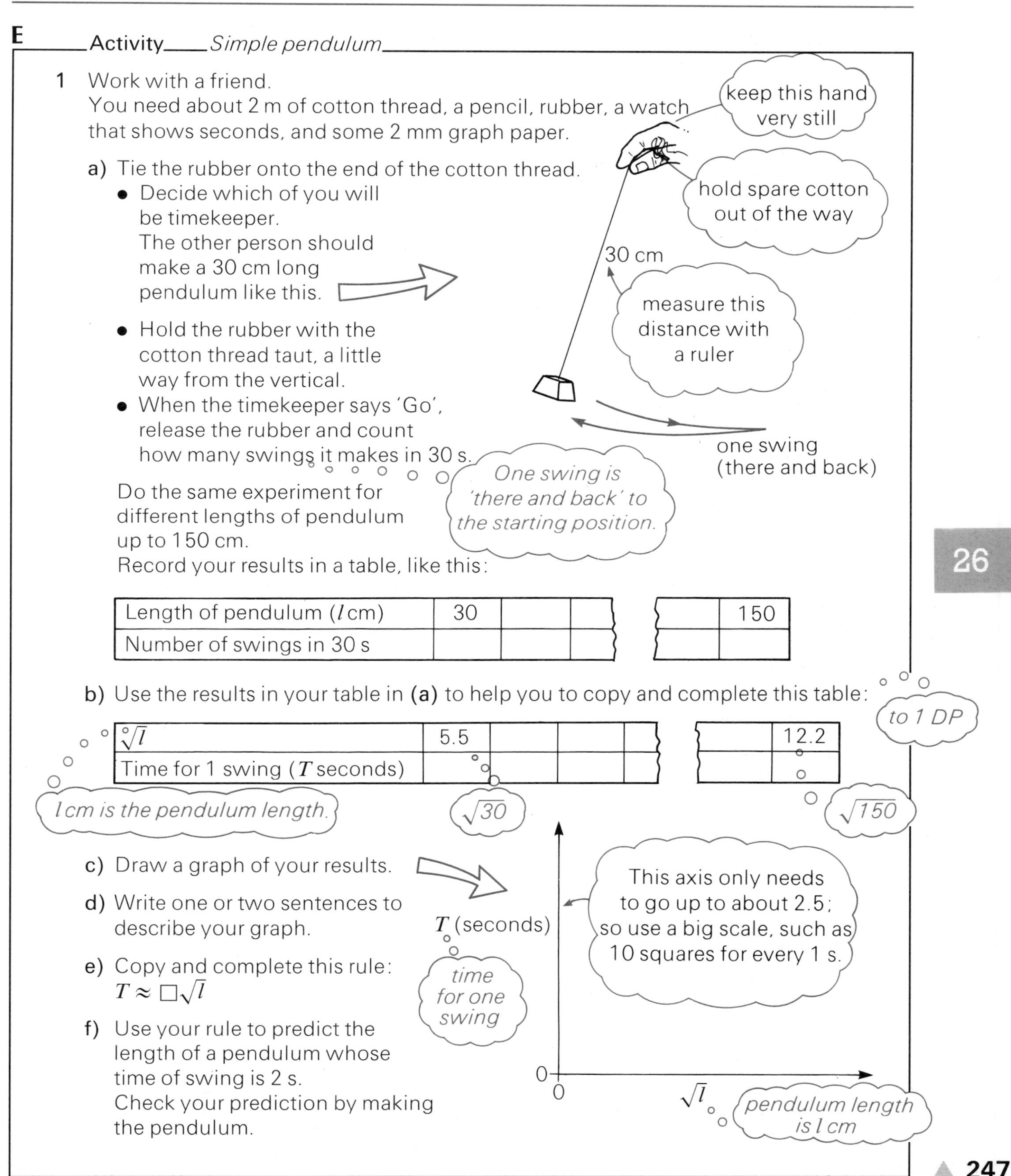
keep this hand very still
hold spare cotton out of the way
30 cm
measure this distance with a ruler
one swing (there and back)
One swing is 'there and back' to the starting position.
to 1 DP
l cm is the pendulum length.
$\sqrt{30}$
$\sqrt{150}$
This axis only needs to go up to about 2.5; so use a big scale, such as 10 squares for every 1 s.
T (seconds)
time for one swing
0
0
$\sqrt{l}$
pendulum length is l cm

1 Work with a friend.
You need about 2 m of cotton thread, a pencil, rubber, a watch that shows seconds, and some 2 mm graph paper.

a) Tie the rubber onto the end of the cotton thread.
- Decide which of you will be timekeeper.
  The other person should make a 30 cm long pendulum like this.
- Hold the rubber with the cotton thread taut, a little way from the vertical.
- When the timekeeper says 'Go', release the rubber and count how many swings it makes in 30 s.

Do the same experiment for different lengths of pendulum up to 150 cm.
Record your results in a table, like this:

Length of pendulum (l cm)	30				150
Number of swings in 30 s					

b) Use the results in your table in **(a)** to help you to copy and complete this table:

$\sqrt{l}$	5.5					12.2
Time for 1 swing (T seconds)						

c) Draw a graph of your results.

d) Write one or two sentences to describe your graph.

e) Copy and complete this rule:
$T \approx \square\sqrt{l}$

f) Use your rule to predict the length of a pendulum whose time of swing is 2 s.
Check your prediction by making the pendulum.

Greek roots

F 2500 years ago, the Ancient Greeks discovered how to draw a line exactly $\sqrt{7}$ cm long.
On this page, you will find out how they did it.

1 You need a pair of compasses.

a) Draw a semicircle with a triangle ABC in it, like this.

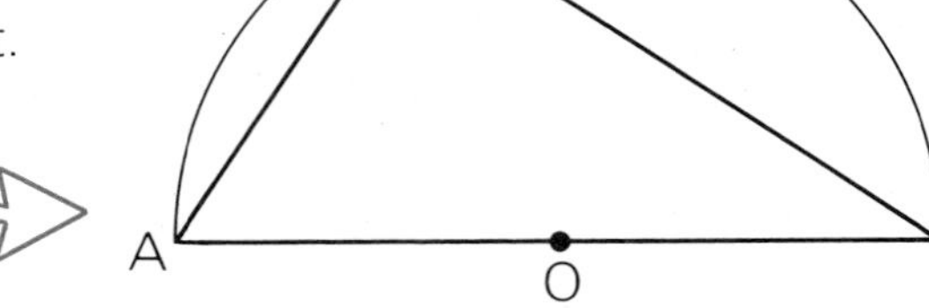

b) Join C and O.

c) Suppose that angle OBC measures $x°$ and angle OAC measures $y°$.
Explain why angles OCB and OCA have the sizes shown in the diagram.

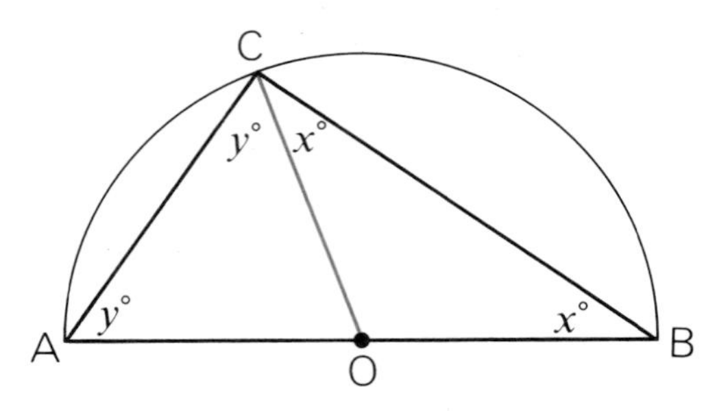

d) Look at triangle ABC.
Explain why $x + y$ must be 90.

e) You need a pair of compasses.

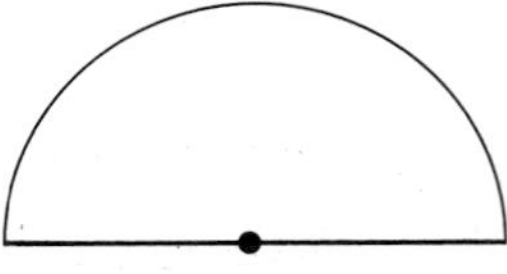

Draw a semicircle.

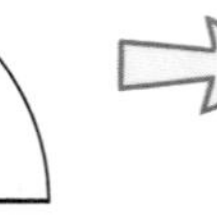

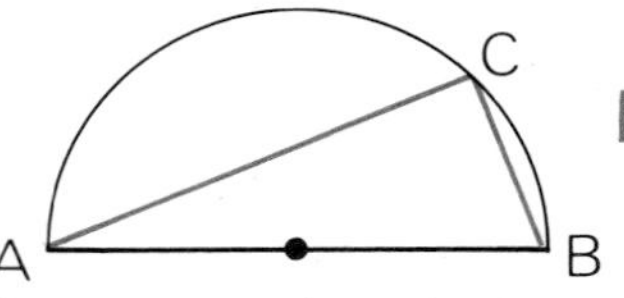

Draw any triangle ABC in it.

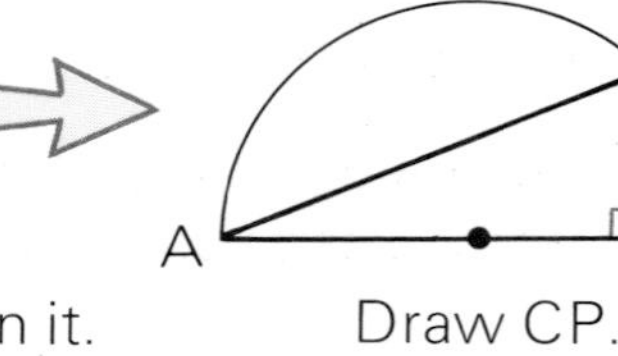

Draw CP.

f) In **(d)** you discovered that angle ACB must be 90°.
Now explain why triangles APC and CPB must be similar.

Remember . . . ?
Similar triangles have corresponding angles equal, and corresponding sides in the same ratio.

g) Copy and complete: $\frac{\text{AP}}{\text{CP}} = \frac{\text{CP}}{\square}$

So AP × □ = □²

h) Study these drawings:

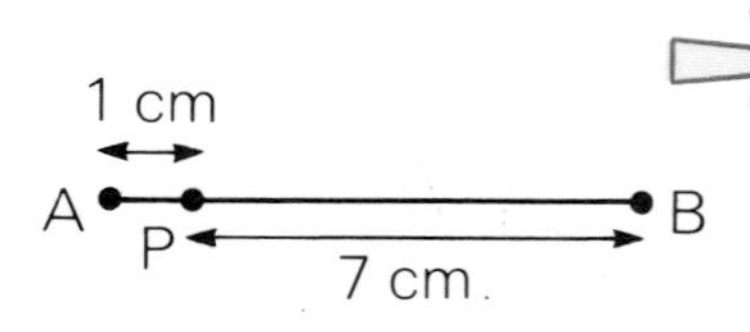

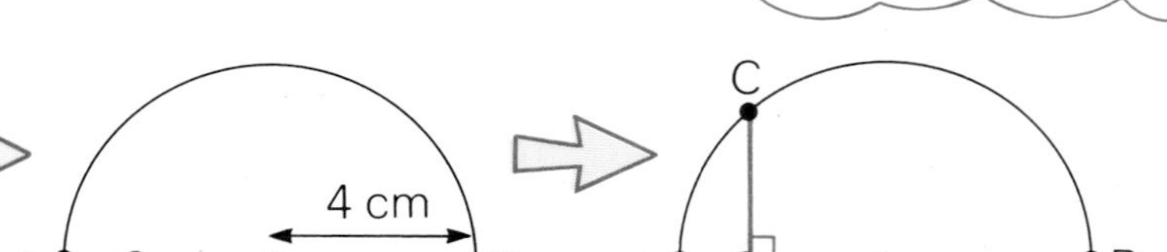

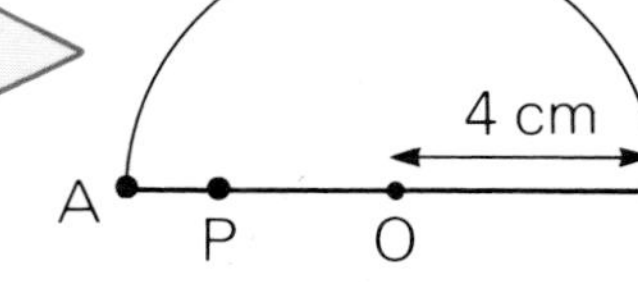

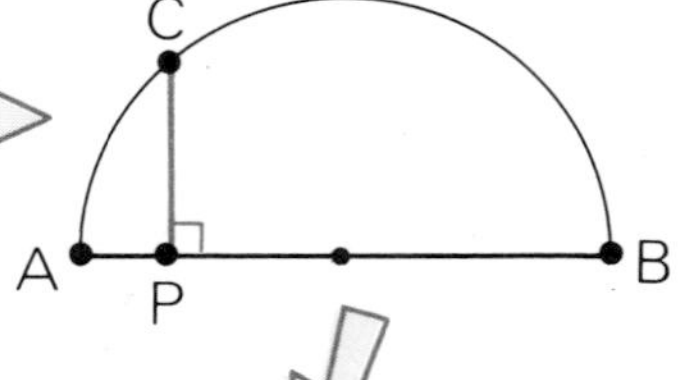

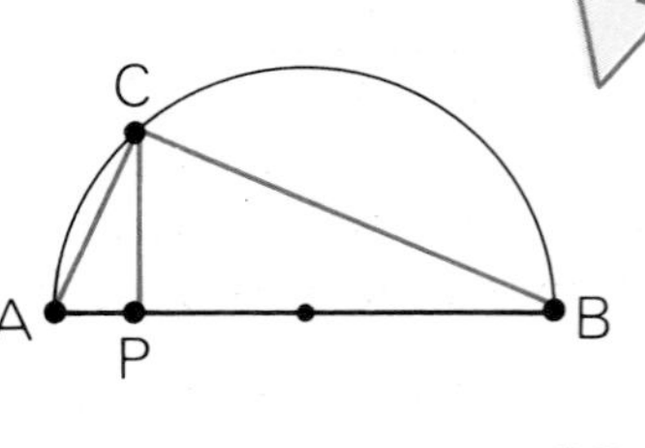

. . . then explain why CP is $\sqrt{7}$ cm long.

Now you know how the Ancient Greeks could draw a line $\sqrt{7}$ cm long.

i) Use the method to draw a line $\sqrt{11}$ cm long.

● Next chapter

27 Working with graphs

A With a friend

Discuss these situations with your friend.
Each of you write down what you decide for each situation.

1 Water is flowing from the tap at a constant rate.

a) One of the three graphs describes accurately how the water level in the container changes. Which one?

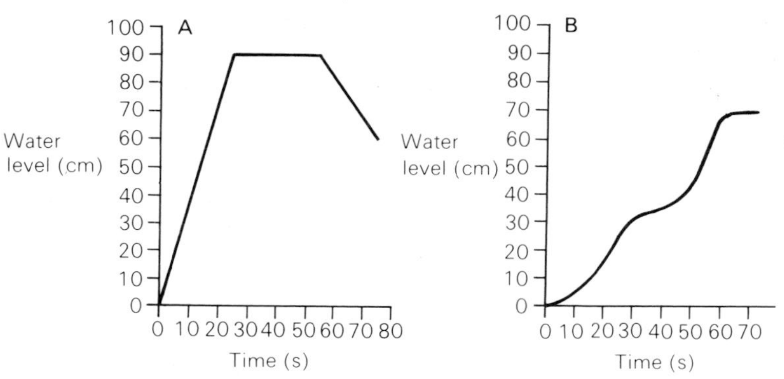

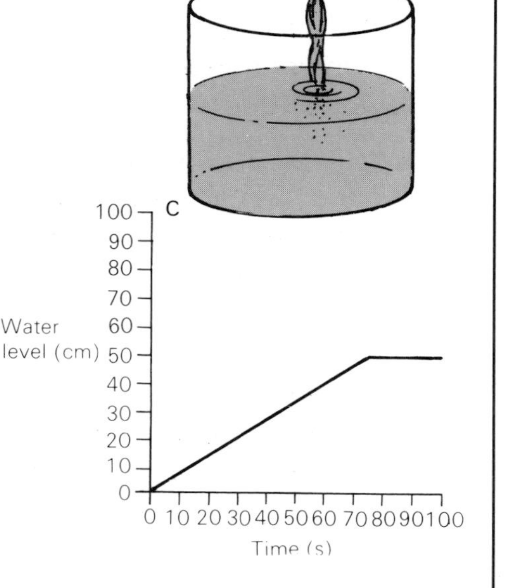

b) How many centimetres deep is the container?

c) Write one or two sentences for each of the other graphs to explain why they cannot be correct.

2 a) Water is leaking at a constant rate through a hole in the bottom of the bucket.
One of the three graphs describes accurately how the water level in the bucket changes.
Which one?

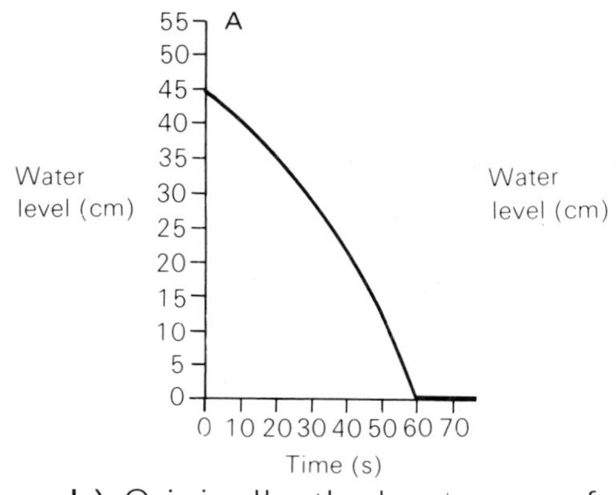

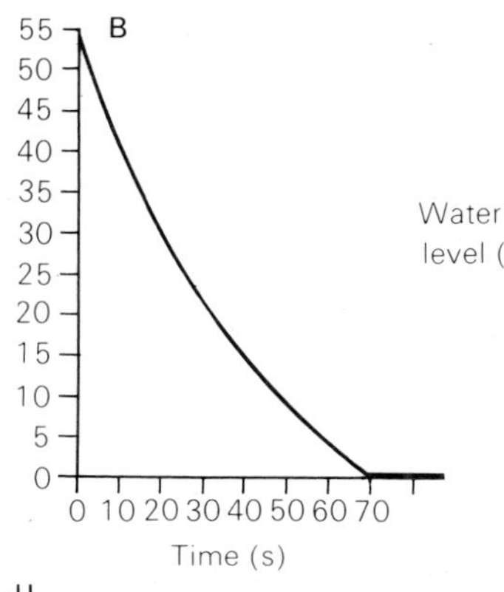

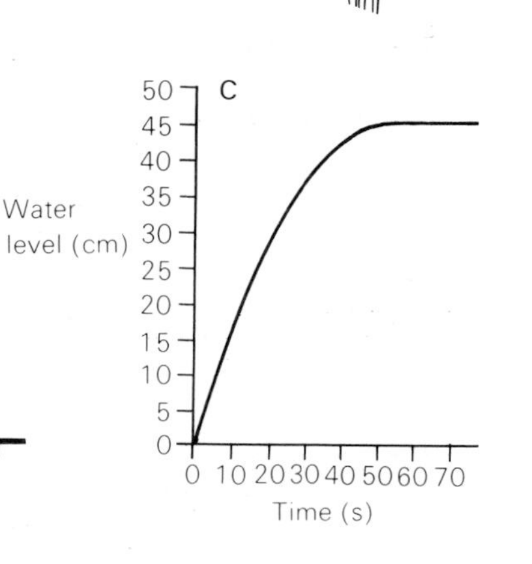

b) Originally, the bucket was full.
How many seconds did it take to empty?

c) Write one or two sentences for each of the other graphs to explain why they cannot be correct.

Challenge

3 The cannonball is rolling down the ramp.

a) One of these graphs shows its speed at various times.
Which one?

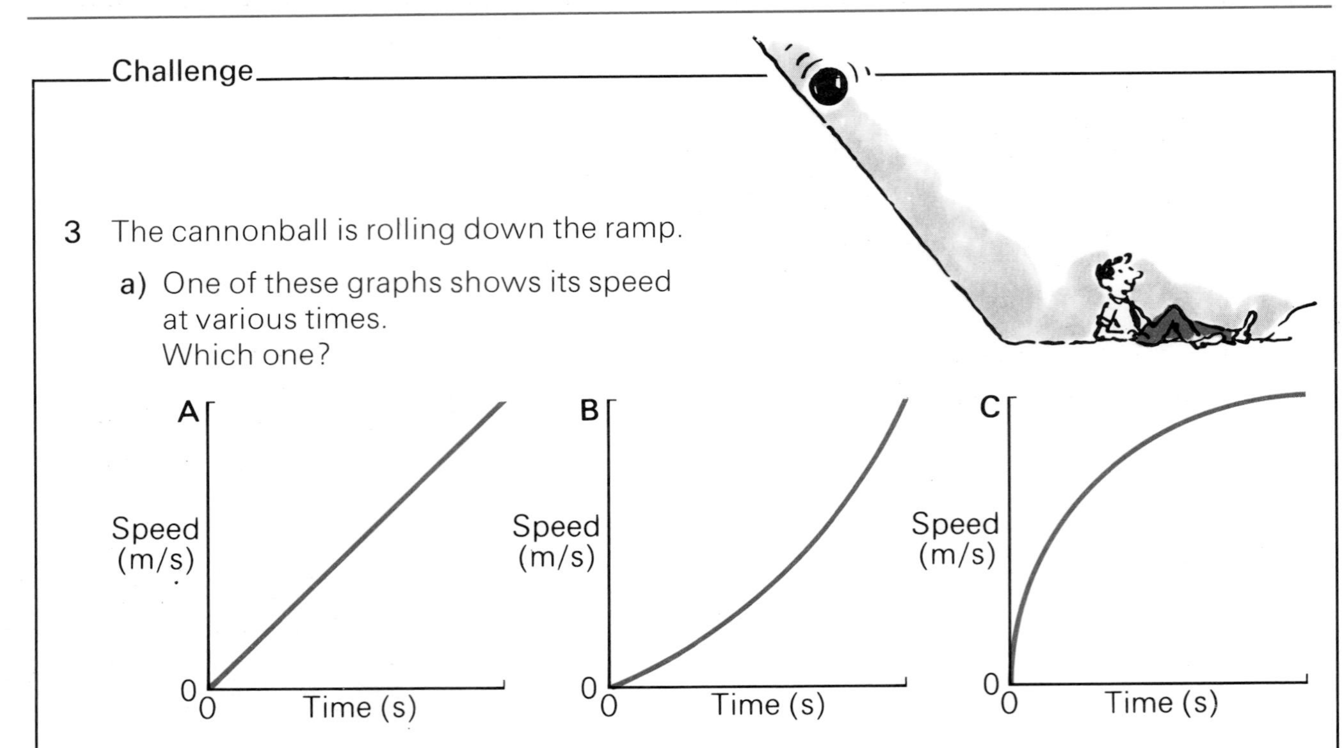

b) One of these graphs shows the height of the cannonball above ground level.
Which one?

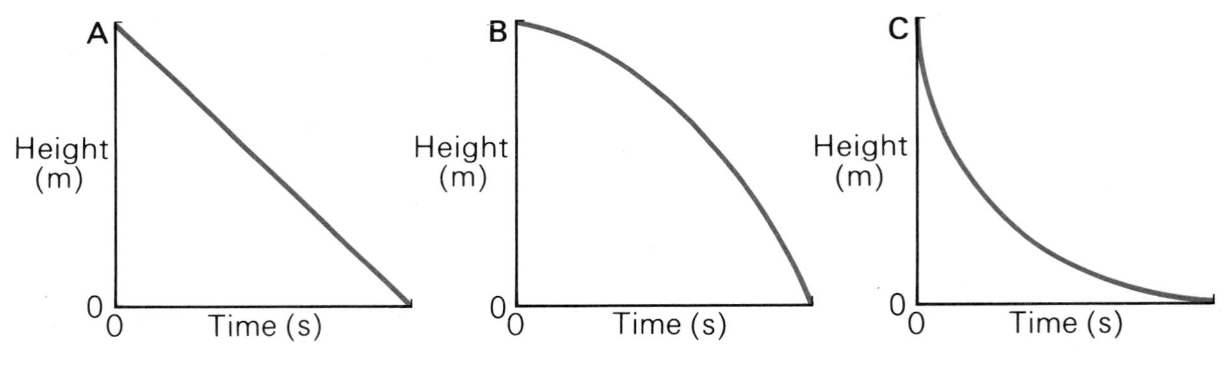

c) One of these graphs shows the distance the cannonball has rolled down the ramp.
Which one?

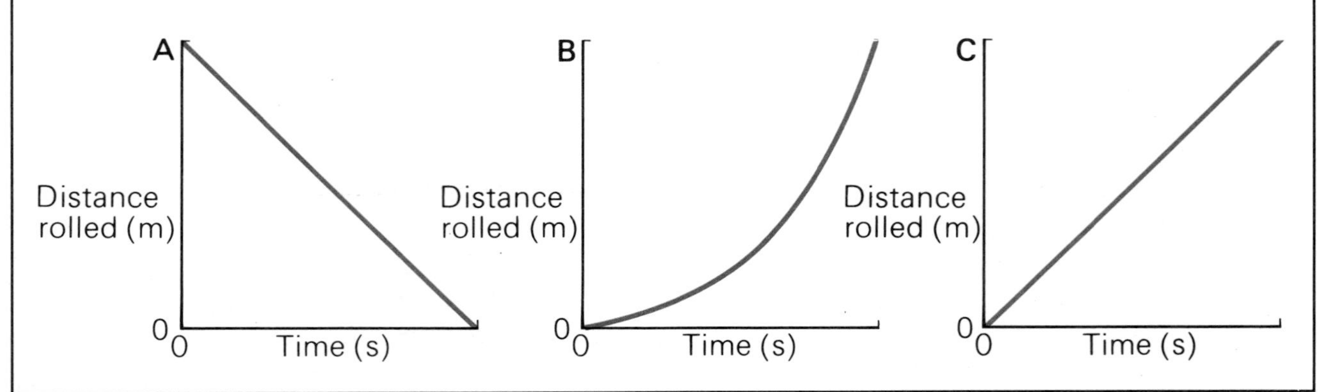

27

Drawing and interpreting graphs

B 1 You need 1 cm squared paper.

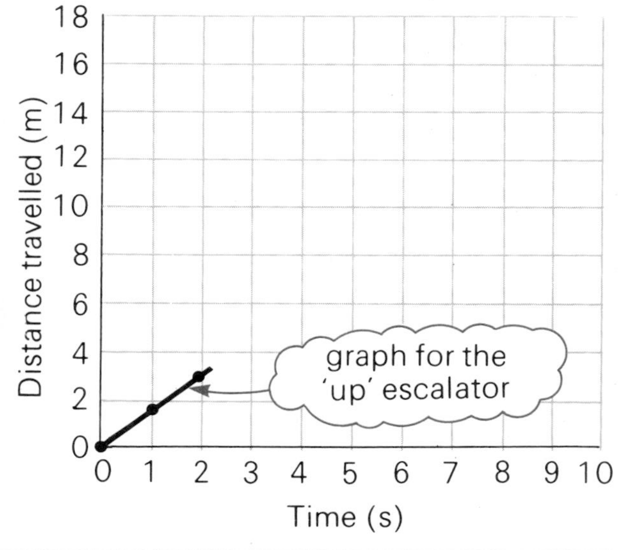

The 'up' escalators travel at different speeds to the 'down' escalators.

This table shows the distances you travel on an 'up' escalator:

Time (s)	0	1	2	3	4	5	6	7	8
Distance (m)	0	$1\frac{1}{2}$	3	$4\frac{1}{2}$	6	$7\frac{1}{2}$	9	$10\frac{1}{2}$	12

a) How many metres do you travel each second?

b) Copy and complete the graph for the 'up' escalator.

c) The 'down' escalator travels at 2 m/s.
Draw a graph for distances travelled on the 'down' escalator, on the same axes as the ones you used in **(b)**.

d) Which graph has the greater slope?

e) On the same axes draw a graph for a moving pavement which carries you 20 m in 20 s.

f) How can we tell immediately from a graph which escalator is travelling fastest?

2 The four lines in the diagram are the graphs for
- a cruise liner sailing in the Mediterranean
- a camel crossing the Sahara desert
- Concorde travelling at cruising speed across the Atlantic
- the Orient Express on its way to Vienna.

The ship, the animal, the aircraft and the train are all travelling at steady speeds.
Which graph is which?

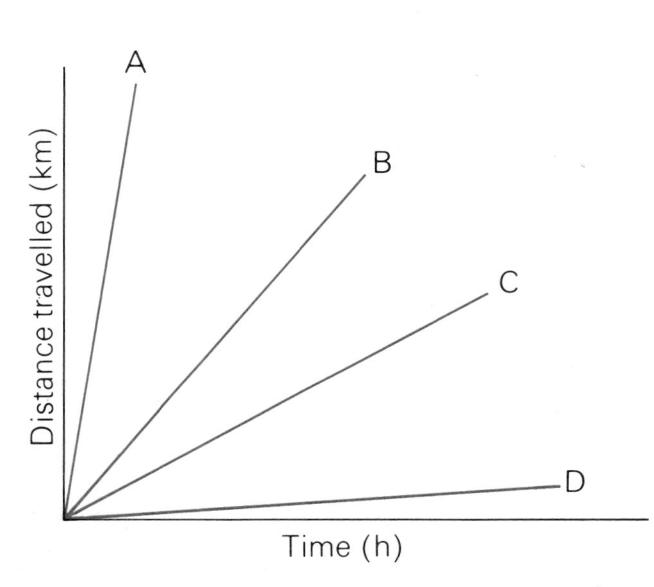

27

3 You need 2 mm squared paper.
A stone is dropped down a mine shaft.
The graph shows the distance it has fallen after 1, 2, 3 and 4 s.

a) From the graph we can see that the stone fell about 35 m during the 4th second.
Roughly, how far did it fall during
(i) the 1st second?
(ii) the 2nd second?
(iii) the 3rd second?

b) Make a rough guess of how far the stone will fall during the 5th second.

c) As the graph continues, its slope gets steeper and steeper.
What do you think this tells us about the speed of the stone?
Write one or two sentences to explain why.

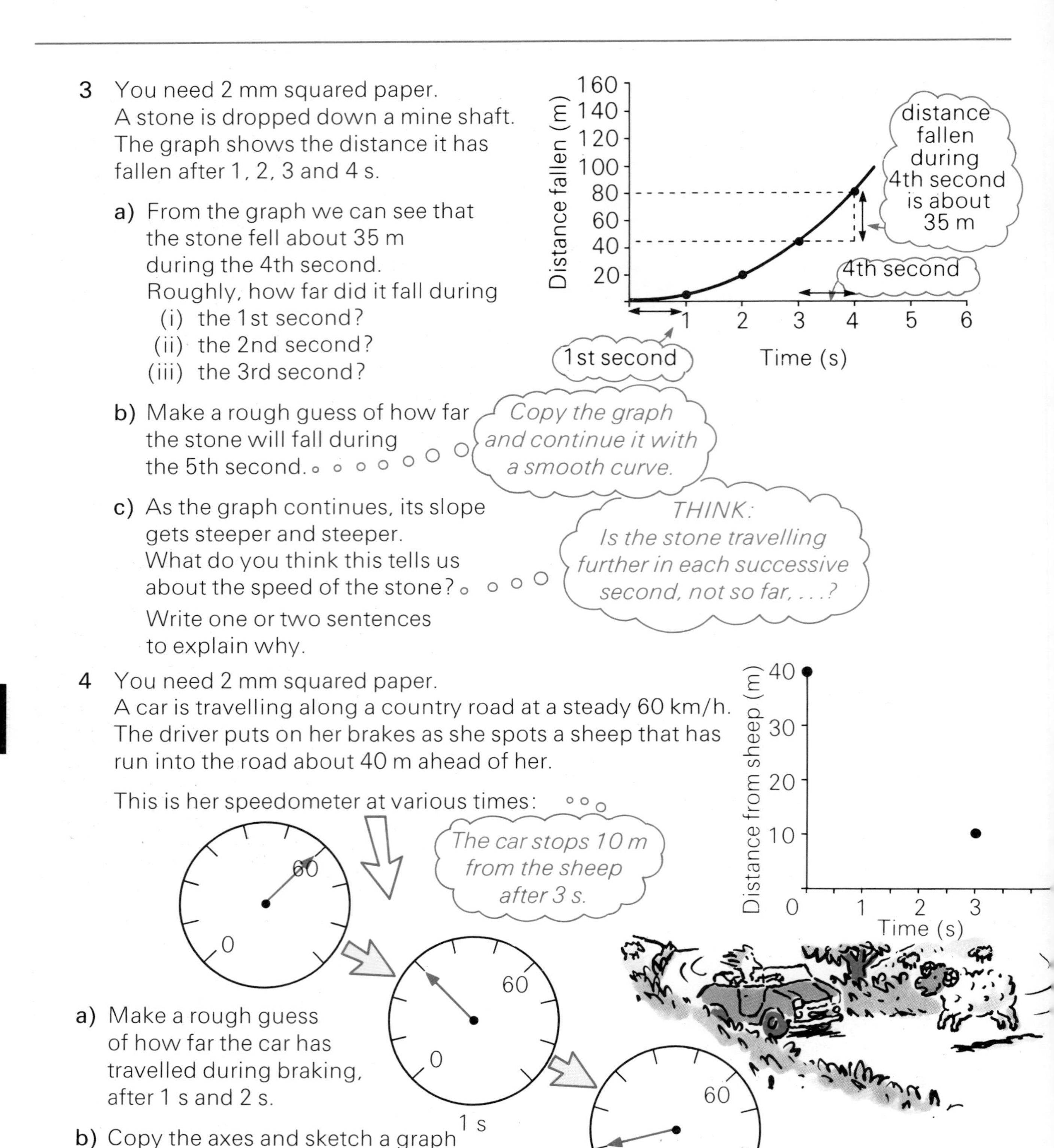

4 You need 2 mm squared paper.
A car is travelling along a country road at a steady 60 km/h.
The driver puts on her brakes as she spots a sheep that has run into the road about 40 m ahead of her.

This is her speedometer at various times:

a) Make a rough guess of how far the car has travelled during braking, after 1 s and 2 s.

b) Copy the axes and sketch a graph to show what you decided in **(a)**.

c) Does your graph get steeper or less steep as time passes?

What do you think this tells you about the speed of the car?
Write one or two sentences to explain.

Take note

On distance–time graphs, the greater is the slope then the greater is the speed.

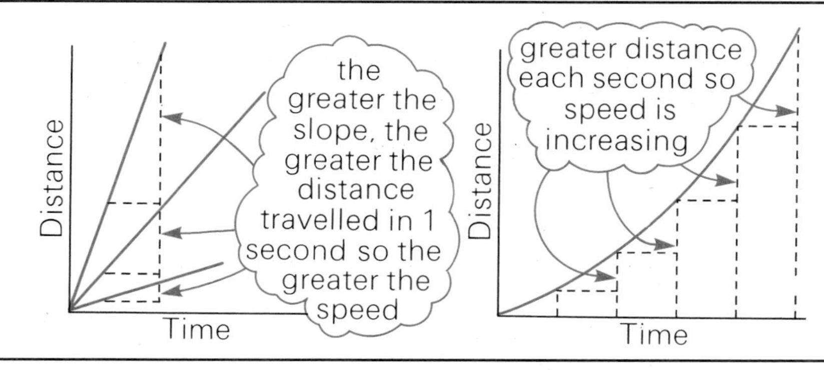

5 The graph represents the progress of two athletes in a 100 m race.

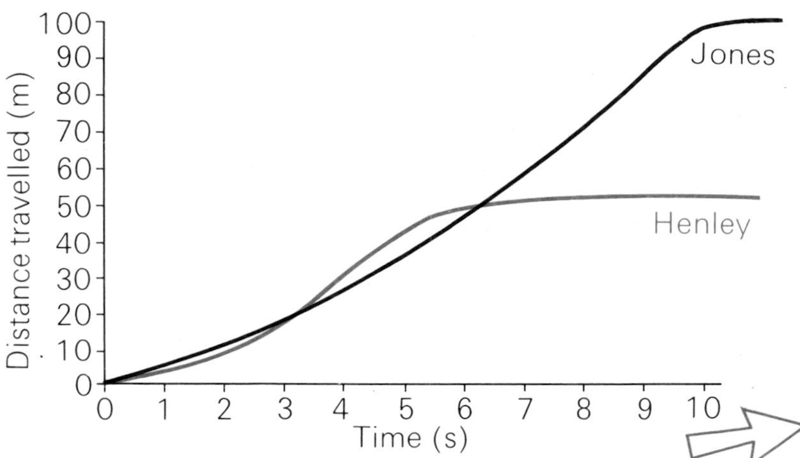

A wins 100 m Challenge

B was fast out of the starting blocks, but **C** was faster still. At the 15 m mark **D** was up by about **E** m with **F** struggling to stay with him. Then a sudden surge by **G** rapidly closed the gap until, by the time they had reached the **H** m mark, the two athletes were neck and neck. With about **I** m to go, **J** looked as though he would win easily, as he pulled away. But then disaster struck. Clutching his right calf muscle, he limped helplessly onto the central grass area, leaving **K** to win the race in **L** seconds.

a) This is a newspaper report about the race. Write down all the missing information. *A to L*

b) During the race, who reached the faster speed? Explain how you know.

Think it through

6 This graph tells the story of Moira's bath time.

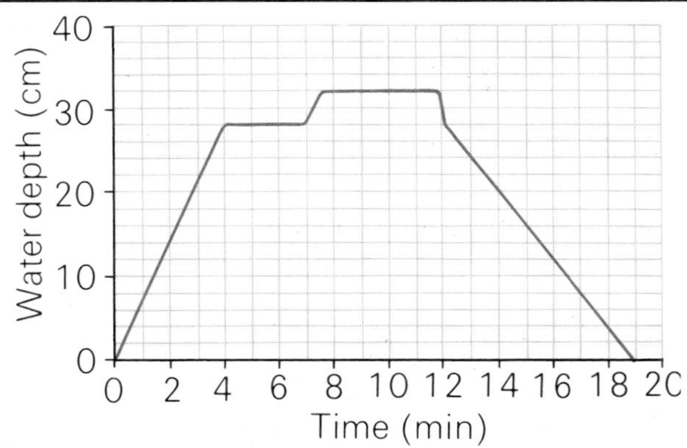

a) For how many minutes did Moira leave the taps running?

b) How many more minutes passed before Moira stepped into the bath?

c) By how much did Moira's presence in the bath increase the water level?

d) After how many more minutes did she step out of the bath?

e) Which was faster, the rate at which water flowed into the bath, or the rate at which water flowed out of it?

7 This graph shows the petrol used by a car for journeys during one week in Holland.

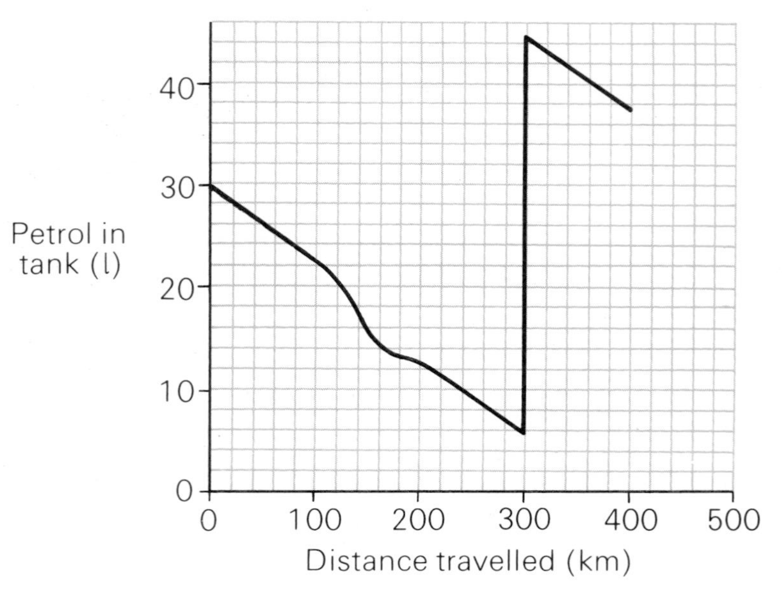

Holland is very flat!

a) At the start of the week there are 30 l of petrol in the tank.
How many litres are there in the tank after the car has travelled 100 km?

b) Do you think the car travelled at a steady speed during the first 100 km?
Explain why you think this.

c) Do you think the car travelled at a steady speed for the next 150 km?
Explain why you think this.

d) What do you think happened after the car had travelled 300 km?

e) How much petrol did the car use for the last 100 km of its travels?

f) During the last 100 km, how many kilometres did the car travel on each litre of petrol?

g) When do you think the car was travelling fastest:

A after 50 km? **B** after 100 km? **C** after 150 km? **D** after 250 km?

Explain why you chose the answer you did.

Think it through

8 Draw graphs showing the different rates at which you think a car will use petrol in these three situations.

a) during a 'racing start' from a set of traffic lights

b) cruising along a main road at 90 km/h

c) cruising along a motorway at 115 km/h

9 It is 12 noon.
Two ships, the *Tiger Bay* and the *Atlas* are sailing along the Suez Canal.
The *Tiger Bay* leaves Port Said at a steady 6 km/h.

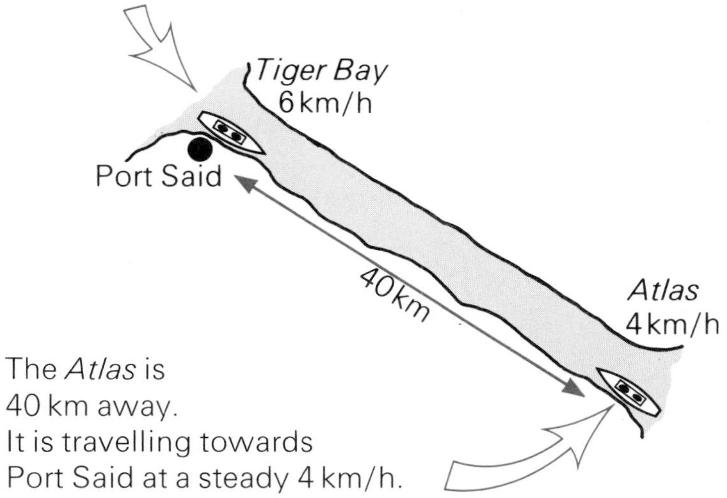

The *Atlas* is 40 km away.
It is travelling towards Port Said at a steady 4 km/h.

a) Have a guess at where they will meet and when.

? km from Port Said at ? o'clock

b) Drawing a graph can help you find out where the ships meet.
Copy the graph and complete it.

Use 1 cm squared paper.

Show the graph for each ship until 17:00.

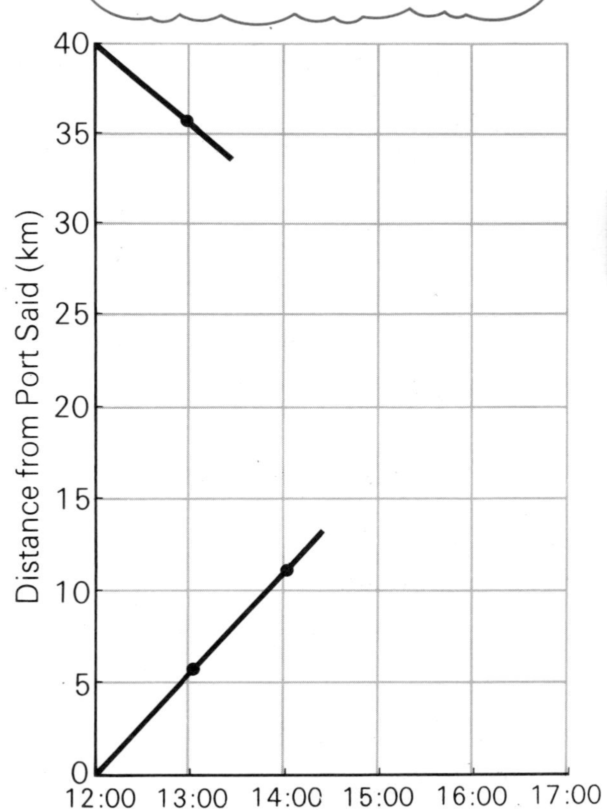

c) From your graph find out
- where the two ships meet

and
- at what time they meet.

How good were your original guesses?

d) Try to complete these rules for the two graphs:

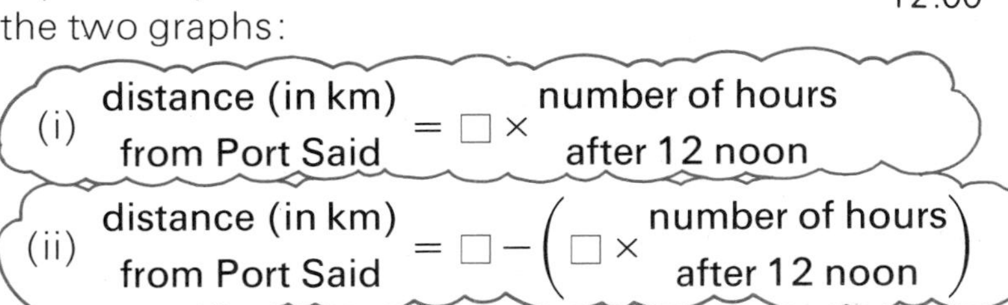

(i) distance (in km) from Port Said = □ × number of hours after 12 noon
(ii) distance (in km) from Port Said = □ − (□ × number of hours after 12 noon)

Using graphs to solve problems

C 1 a) What is the area of a square with length of side
(i) 3 cm? (ii) 3.5 cm?

b) What is the length of the side of a square with area
(i) 25 cm^2? (ii) 1.44 cm^2?

c) Choose two more squares of your own.
Write down the length of side and the area of each one.

d) One of these graphs represents the rule
area of a square = (length of side)²
Which one?

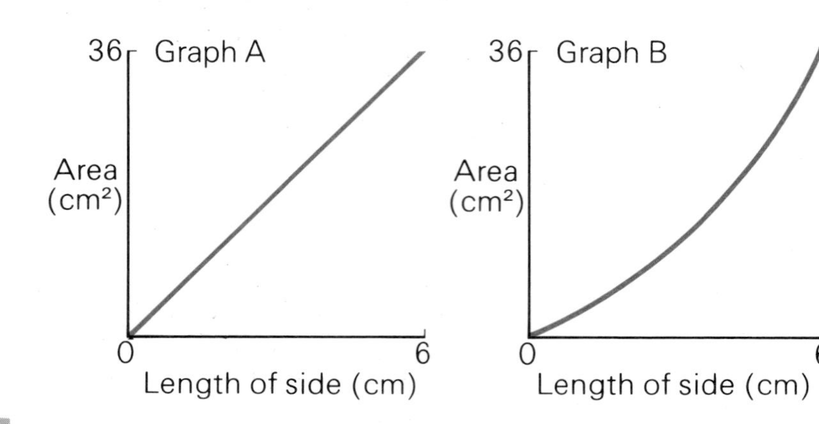

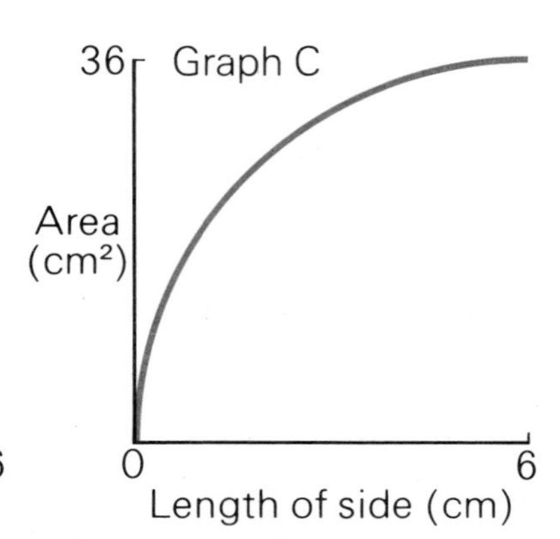

e) Sketch the graph you chose and mark on it dots for the six squares in **(a)**, **(b)** and **(c)**.

f) On the same set of axes you used for **(e)**, sketch the graph for the rule
perimeter = 4 × length of side

g) Is there a square whose perimeter and area have the same number of units?

for example, perimeter = 12 cm
area = 12 cm^2

If so, give as much information about it as you can.

Think it through

2 Is there a cube whose total edge length and volume have the same number of units?
If so, give as much information about it as you can.

You may find the √ *key of your calculator, and a graph, useful.*

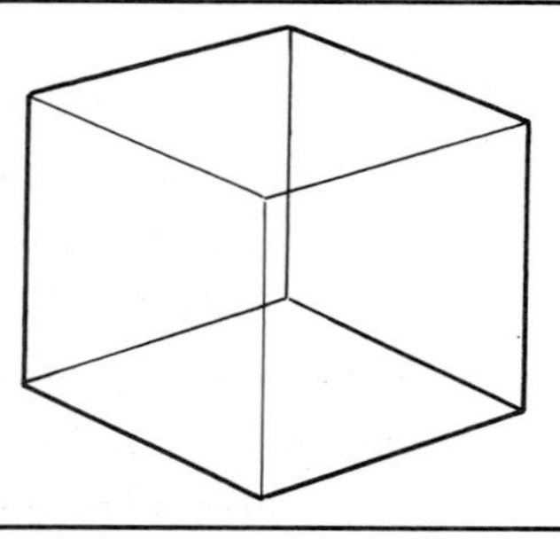

3 Most people sell their houses through an estate agent.
Estate agents charge fees that depend upon the selling price of the house.
Here are three of the methods which estate agents use to calculate how much to charge for selling a house.

Method 1	Fee = $1\frac{1}{2}$% of the selling price.
Method 2	Fee = 2% of the first £20 000 of the selling price + $1\frac{1}{4}$% of the remainder of the selling price.
Method 3	Fee = £150 + 1% of the selling price.
On top of the fee you have to pay VAT at 15%.	

a) An agent using method **1** calculates her fee for selling a £48 000 house like this:

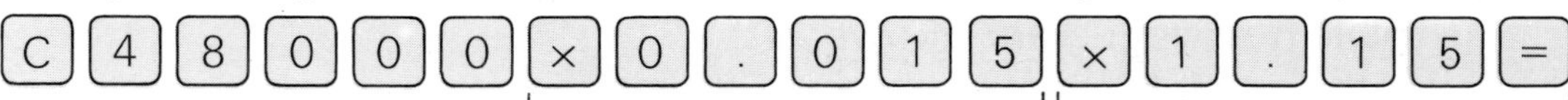

(i) What does this part of the calculation represent? . . .
(ii) . . . and this part?
(iii) What is the agent's total fee, including VAT?

b) Make a table of fees for calculation method **1** like this:

Selling price	10 000	20 000	…	80 000	90 000	100 000
Fee, including VAT						

c) An agent using method **2** calculates his fee for selling a £48 000 house like this:

C 2 0 0 0 0 × 0 . 0 2 + (1)

(i) Explain each of the four numbered parts of the calculation.
(ii) What is the agent's total fee, including VAT?

d) Make tables like the one in **(b)** for calculation methods **2** and **3**.

e) On a single pair of axes, draw graphs for the three methods of calculating fees.

f) Use your graphs to estimate, for each method, the selling price of a house for which the estate agent's fee would be £1000.

including VAT

Agent's fee (£): 0, 200, 400, 600, 800, 1000, 1200, 1400, 1600, 1800
Selling price (£000): 0, 10, 20, 30, 40, 50, 60, 70, 80, 90, 100

thousands of pounds

g) Write a short newspaper article explaining which fees are cheapest for various price ranges.

You may have to do some extra calculations to find all the information you need. A simplified version of your graphs would make a good illustration for your article.

D 1 This table gives the population of the world between 1250 and 1980.

Year	1250	1500	1650	1700	1750	1800	1850	1900	1920	1950	1970	1980
Population (millions)	375	420	570	615	720	900	1125	1625	1862	2533	3704	4410

a) Using a graph we can see more clearly how the population has changed. Draw the graph yourself.

b) The world population in 1250 was about 375 million.

(i) Roughly, how many years went by before the population doubled?

(ii) Roughly, how many more years went by before the population doubled again?

(iii) . . . and again?

c) Is the rate at which the world population is growing increasing or decreasing? Explain how your graph shows this.

d) Use the graph to estimate the year in which the population will be double what it was in 1970.

e) The official United Nations estimate of the world population for the year 2000 is 6200 million. Add this information to your graph.

*You might wish to revise your estimate in **(d)** now.*

2 Here is some information about the population of the United Kingdom.

Year	1700	1750	1800	1850	1900	1951	1971	1975	1976	1977	1978	1979
Population (millions)	6.8	8	11	22	38	50.3	55.6	55.9	55.9	55.9	55.9	55.9

a) Draw a graph to show the information more clearly.

b) Write a short paragraph describing how the population of the UK has grown. Say when it was growing slowest and when it was growing fastest.

c) Compare the way in which the world population and the population of the UK have grown since 1800.

Do this by designing a poster, or writing a report, or writing an illustrated magazine article, or . . .

d) What similarities and differences do you see between the growths of the populations of the world and of the UK? What do you think the reasons are?

Challenge

3 This table gives information about the changes in the populations of different age groups in England and Wales during the years 1950 to 1975.

Write a report, illustrated by graphs, about what seems to be happening to the relative sizes of the various age groups as the overall population gradually stops increasing.

Population in millions

Age group	Year					
	1950	1955	1960	1965	1970	1975
65 upwards	4.59	5.11	5.46	5.85	6.41	6.97
45–64	10.46	11.16	11.80	11.87	11.89	11.54
25–44	13.22	12.60	12.09	12.10	11.78	12.42
15–24	5.73	5.54	5.93	6.79	7.04	6.95
5–14	5.91	6.74	6.93	6.82	7.58	7.91
0–4	2.82	3.28	3.54	4.09	3.97	3.44

28 Working with numbers

A

Challenge *Division families*

1 These divisions have something common.

a) Use your calculator to find out what it is.

$21 \div 8$ $14.7 \div 5.6$ $1.05 \div 0.4$ $126 \div 48$

b) Find two more divisions which belong to the same 'family'.

c) Find three divisions which belong to the same family as $3 \div 8$.

d) Investigate some more divisions. Starting with any division, how can we find other divisions which belong to the same family?

Write one or two sentences to explain any methods you discover.

Do you remember . . . ?

A whole family of divisions gives the same decimal number. Examples:

0.1 family $\{1 \div 10,\ 2 \div 20,\ 0.5 \div 5,\ 1.2 \div 12,\ 11 \div 110, \ldots\}$

0.375 family $\{3 \div 8,\ 6 \div 16,\ 0.3 \div 0.8,\ 0.15 \div 0.4,\ 33 \div 88, \ldots\}$

2 Find two more divisions for each of these families.

3 Find three members of

a) the 0.36 family. b) the 1.013 family.

4 Copy and complete these members of the 0.8 family.

a) $8 \div \square$ b) $0.8 \div \square$ c) $\square \div 4$ d) $\square \div 4.4$

With a friend

5 Discuss this question with your friend.
Are there such things as
- addition families
- subtraction families
- multiplication families?

Each of you write a short report to explain what you decide.

6 a) Which of these rules give divisions which belong to the same family?

Rule A Multiply both numbers by the same number.

Example: $6 \div 18 \rightarrow 36 \div 108$ (×6, ×6)

Rule B Add the same number to both numbers.

Example: $6 \div 18 \rightarrow 12 \div 24$ (+6, +6)

Rule C Divide both numbers by the same number.

Example: $6 \div 18 \rightarrow 1 \div 3$ (÷6, ÷6)

Rule D Subtract the same number from both numbers.

Example: $6 \div 18 \rightarrow 0 \div 12$ (−6, −6)

b) Use rule **A** to find three divisions which belong to the same family as $7 \div 3$.
Use your calculator to check that the divisions really do belong to the same family.

c) Use rule **C** to find three divisions which belong to the same family as $24 \div 18$.
Use your calculator to check that the divisions really do belong to the same family.

d) Use rule **A** or **C** to help you copy and complete these:

(i) $7 \div 12 = 14 \div \square$ (ii) $8 \div 7 = 56 \div \square$ (iii) $9 \div 8 = \square \div 48$

(iv) $3 \div 13 = \square \div 52$ (v) $58 \div 26 = \square \div 13$ (vi) $100 \div 84 = 25 \div \square$

28

Think it through

7 Meg is finding divisions which belong to the same family as $24 \div 16$.
This is what she writes:

$\cancel{24}^{3} \div \cancel{16}^{2} = 3 \div 2$

$\cancel{24}^{12} \div \cancel{16}^{8} = 12 \div 8$

$\cancel{24}^{48} \div \cancel{16}^{32} = 48 \div 32$

a) To get $3 \div 2$ Meg divided 24 and 16 by 8.

*She used rule **C**.*

Which rule did she use to get
(i) $12 \div 8$? (ii) $48 \div 32$?

b) Use Meg's method to find five divisions which belong to the same family as $24 \div 18$.

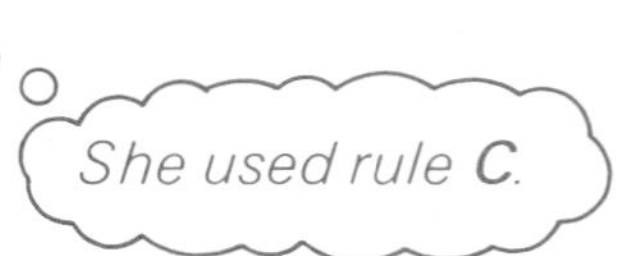

c) Here is another example of Meg's work.
Find three more divisions for the same family.

Equal ratios

B 1 These are two pieces of copper tubing.

Vince cuts the 2 m piece into five equal lengths.
He wants to cut the 8 m tube into pieces of the same length as the five he has already cut.

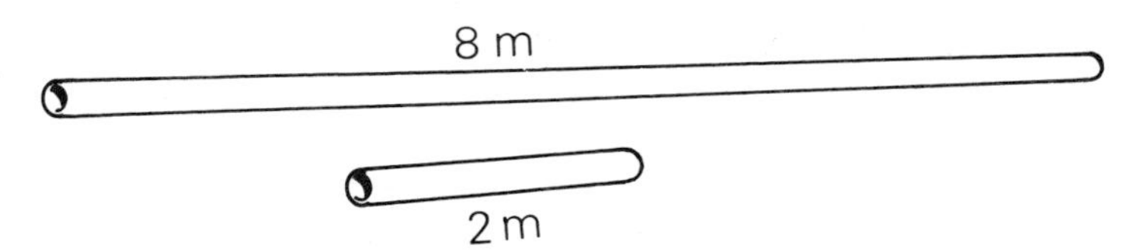

a) Into how many pieces should he cut it?

b) This number sentence represents Vince's problem.
Copy and complete it.

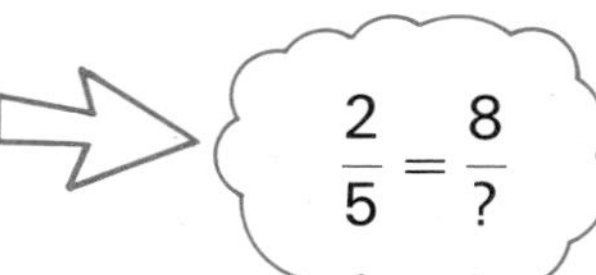

2 At a wedding reception the hostess shares a 1 l bottle of wine equally into ten wine glasses.
Now she wants to share out a $2\frac{1}{2}$ l bottle.
She wants the portions to be the same size as those from the first bottle.

a) How many glasses should she use?

b) This number sentence represents the wine problem.
Copy and complete it.

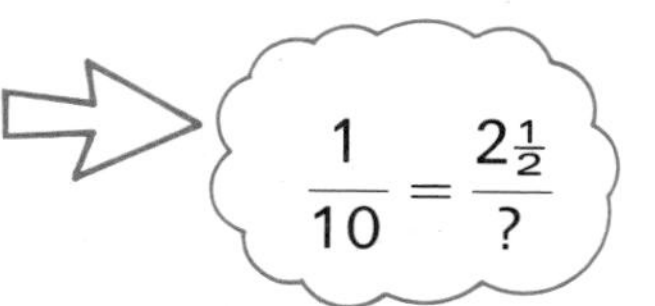

3 a) Write your own number sentence for this situation.
Use **?** for the number of piles made from a 4 kg bag.

b) Use your number sentence to help you to find the number of piles into which the 4 kg bag is divided.

A 4 kg bag of rice is divided into a number of equal piles.
A 6 kg bag is divided into nine equal piles.
All the piles are the same size.

4 a) Write your own number sentence for this situation.

b) What is the circumference of the first circle?

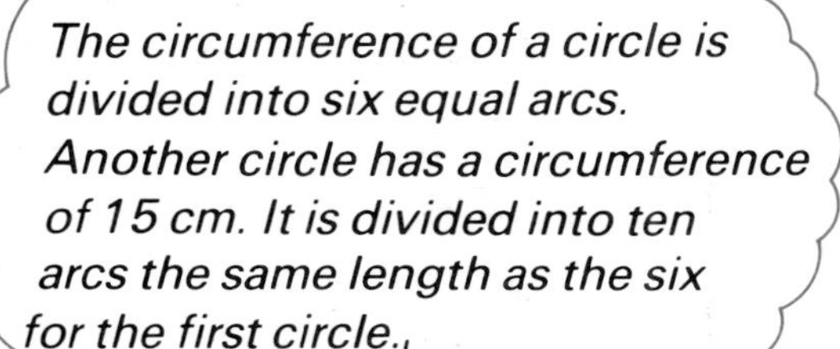

The circumference of a circle is divided into six equal arcs. Another circle has a circumference of 15 cm. It is divided into ten arcs the same length as the six for the first circle.

5 The ratio of red beads to black beads in a 'Tudor' necklace is always the same, no matter how long the necklace. In one necklace there are 8 red beads and 12 black beads. In another there are 20 red beads.

a) Write down a number sentence for this situation.
Use **?** for the number of black beads in the second necklace.

b) How many black beads are there in the second necklace?

Think it through

c) Why can't a 'Tudor' necklace have 15 red beads?

6 a) Write your own story for this number sentence:

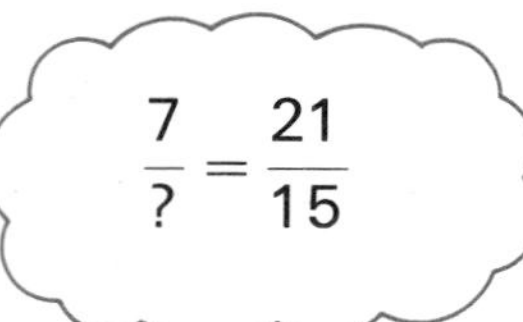

b) Write down the number which **?** represents in your story.

c) Repeat **(a)** and **(b)** for this number sentence.
Make sure that the replacement for **?** makes sense in your story:

$$\frac{?}{6} = \frac{5}{12}$$

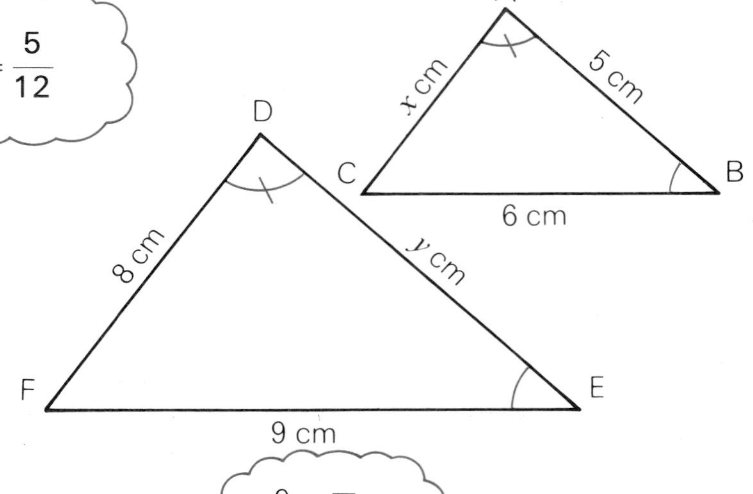

7 The two triangles are similar.

a) These number sentences tell us about ratios of corresponding sides. Copy each sentence and replace □ by the correct number from the diagrams.

A 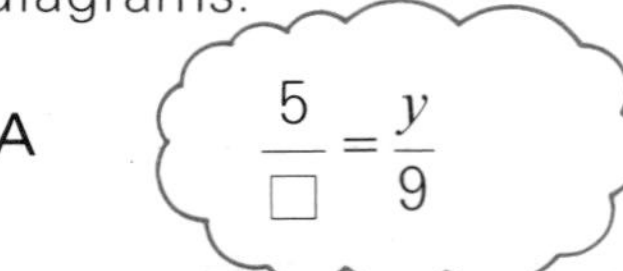

B $\frac{9}{6} = \frac{\square}{x}$

b) Use your sentences to help you to find the distance
(i) DE. (ii) AC.

8 ABCDE and A′B′C′D′E′ are similar.

a) Find
(i) B′C′. (ii) CD.
(iii) E′D′. (iv) AE.

b) What is the scale factor for the enlargement?

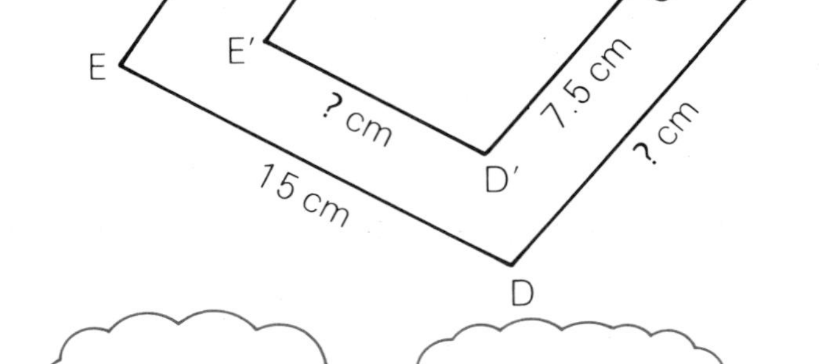

9 These two number sentences are for two similar triangles.

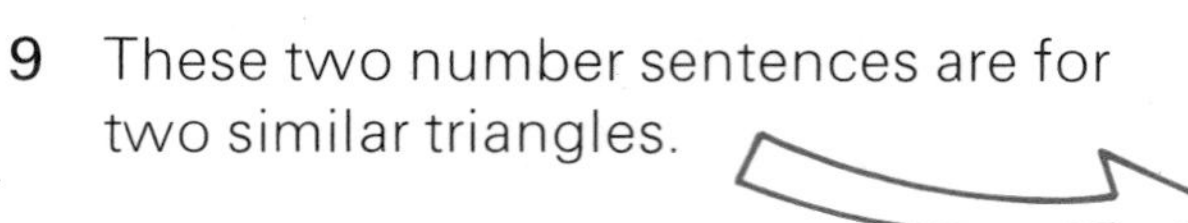

$$\frac{2}{9} = \frac{6}{27} \qquad \frac{6}{27} = \frac{5}{22.5}$$

a) **Sketch** two possible triangles.
Mark in their corresponding sides and angles.

b) Is there another possible pair of triangles?
If so, sketch them.

28

Divisions, ratios and fractions

C **Take note**

$2 \div 5$, $4 \div 10$, $6 \div 15$, $8 \div 20$, ... all belong to the same division family.

$2 \div 5 = 0.4$ $4 \div 10 = 0.4$ $6 \div 15 = 0.4$ $8 \div 20 = 0.4$

$\frac{2}{5}$, $\frac{4}{10}$, $\frac{6}{15}$, $\frac{8}{20}$, ... all belong to the same fraction family.

$$\frac{2}{5} = \frac{4}{10} = \frac{6}{15} = \frac{8}{20} = \ldots = 0.4$$

1 a) Use your calculator to check that these fractions belong to the same fraction family: $\frac{2}{9}$, $\frac{14}{63}$, $\frac{22}{99}$

b) Find another fraction which belongs to the same family.

2 Which of these rules give fractions which belong to the same family?

Rule A
Multiply the numerator and the denominator by the same number.
Example: $\frac{2}{9} \xrightarrow{\times 7} \frac{14}{63}$

Rule B
Add the same number to the numerator and the denominator.
Example: $\frac{2}{9} \xrightarrow{+4} \frac{6}{13}$

Rule C
Divide the numerator and the denominator by the same number.
Example: $\frac{12}{27} \xrightarrow{\div 3} \frac{4}{9}$

Rule D
Subtract the same number from the numerator and the denominator.
Example: $\frac{12}{27} \xrightarrow{-6} \frac{6}{21}$

3 Use rule **A** to find three fractions which belong to the same family as $\frac{3}{7}$.
Use your calculator to check that your fractions really do belong to the same family.

4 Use rule **C** to find three fractions which belong to the same family as $\frac{18}{24}$.
Use your calculator to check that your fractions really do belong to the same family.

5 a) Use rule **A** or **C** to help you to complete these:

(i) $\frac{4}{9} = \frac{\square}{27}$ (ii) $\frac{5}{16} = \frac{15}{\square}$ (iii) $\frac{4}{\square} = \frac{12}{15}$ (iv) $\frac{9}{36} = \frac{3}{\square}$ (v) $\frac{15}{45} = \frac{1}{\square}$

(vi) $\frac{\square}{7} = \frac{21}{49}$ (vii) $\frac{120}{1240} = \frac{12}{\square} = \frac{60}{\square} = \frac{\square}{310} = \frac{\square}{62} = \frac{15}{\square} = \frac{\square}{248}$

b) All the fractions in **(a)** (vii) give the same decimal.
What is it?

28

Think it through

6 Meg is finding fractions which belong to the same family as $\frac{60}{80}$.
This is what she writes:

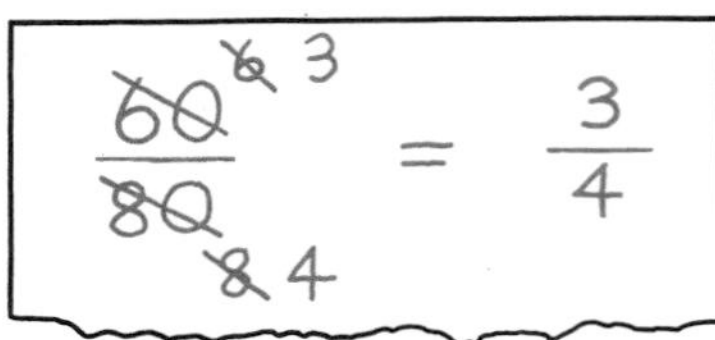

a) Can you explain how she arrived at $\frac{3}{4}$?

b) Find the fraction from the same family as $\frac{72}{180}$ which has the smallest possible whole-number numerator.

c) Find the fractions belonging to the same families as the following, which have the smallest possible whole-number numerator.
(i) $\frac{16}{24}$ (ii) $\frac{100}{125}$ (iii) $\frac{64}{88}$ (iv) $\frac{84}{120}$

7 Find three fractions which belong to

a) the 0.8 family
b) the 0.72 family
c) the 0.1111 ... family
d) the 1.25 family.

8 Copy and complete the diagram by writing **six** fractions for **each** decimal shown on the line.

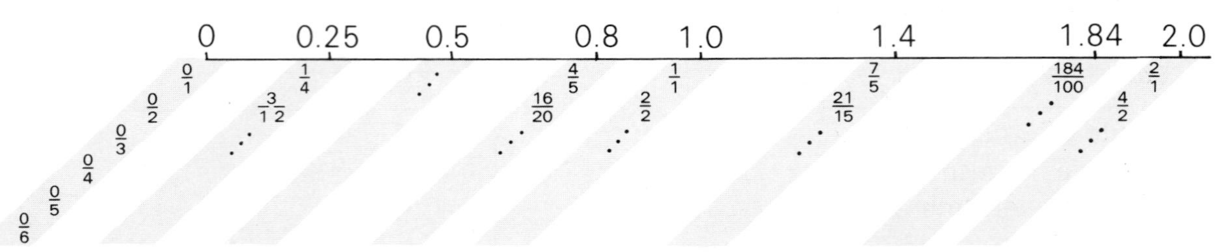

9 The coloured parts of the rectangles represent two fractions from the same family: $\frac{12}{20}$ and $\frac{6}{10}$.

Draw your own diagrams to represent $\frac{7}{8}$ and $\frac{21}{24}$.

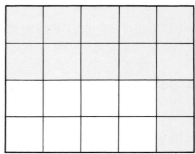

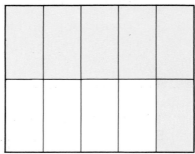

10 Two of these shapes have the same fraction of their area coloured. Which two are they?

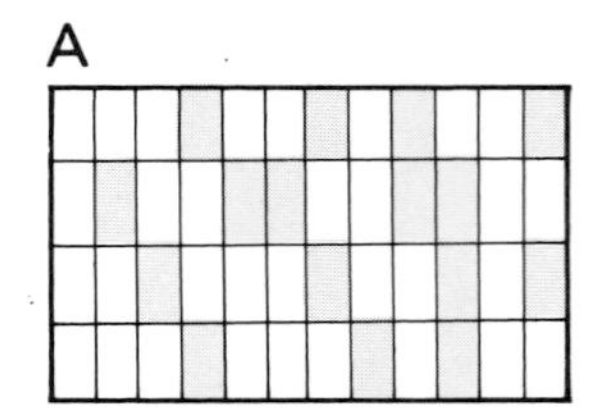

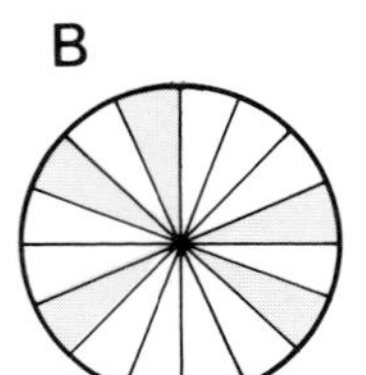

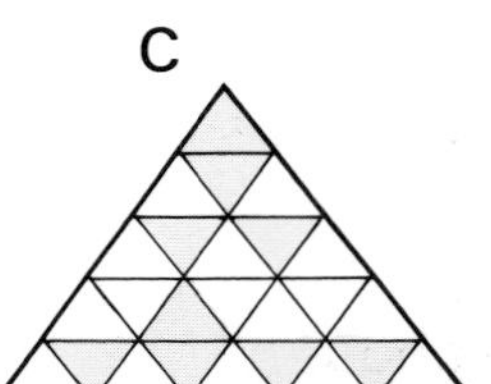

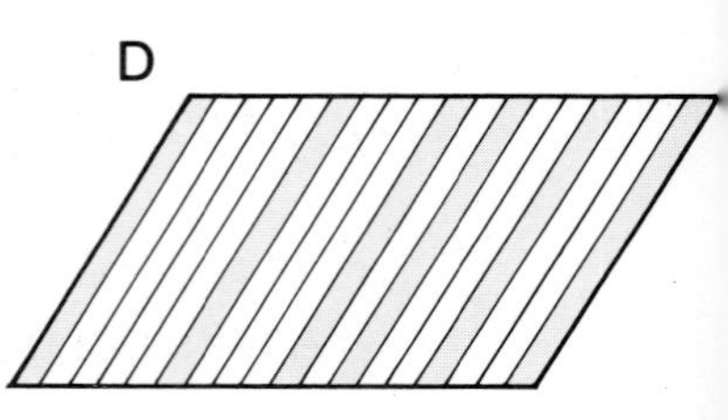

28

11 a) Use your calculator to arrange these fractions in order, smallest first.

$\frac{1}{2}$ $\frac{2}{3}$ $\frac{3}{7}$ $\frac{5}{8}$ $\frac{4}{9}$ $\frac{7}{10}$ $\frac{5}{7}$ $\frac{8}{9}$

b) 0 ——— $\frac{1}{2}$ ——— 1

Sketch the number line.
Mark each fraction on it, roughly in its correct position.

Do not use a calculator on the rest of this page.

Challenge

12 Arrange these four fractions in order, smallest first.

$\frac{3}{8}$ $\frac{1}{3}$ $\frac{5}{12}$ $\frac{7}{24}$

13 a) Write $\frac{5}{8}$ and $\frac{3}{5}$ as fractions with denominator 40.

b) Which is the larger fraction?

14 a) Write $\frac{9}{14}$ and $\frac{5}{8}$ as fractions with denominator 56.

b) Which is the larger fraction?

15 Which is the larger fraction, $\frac{7}{12}$ or $\frac{11}{18}$?

Think it through

16 Find five fractions which lie between $\frac{3}{13}$ and $\frac{4}{13}$.
Write your fractions in order, smallest first.

17 Find t in each sentence.

Think and test until you find t.

a) $\frac{t}{3} = \frac{4}{12}$ b) $\frac{5}{t} = \frac{15}{18}$ c) $\frac{1}{2} = \frac{t}{8}$

d) $\frac{2}{3} = \frac{10}{t}$ e) $\frac{4}{5} = \frac{28}{t}$ f) $\frac{t}{4} = \frac{18}{8 \times t}$

18 $\frac{t}{5}$ is larger than $\frac{9}{25}$.
What can you say about t?

19 $\frac{t}{8}$ is larger than $\frac{5}{11}$.
What can you say about t?

Ratio squares

D

With a friend

1 These are special arrangements of numbers.

a) Discuss between you what is special about them.

Each of you write down what you decide.

b) Find two more examples of arrangements like them.

2	12
3	18

0.4	1.6
0.8	2.4

7	49
8	56

$1\frac{1}{2}$	$7\frac{1}{2}$
$\frac{1}{2}$	$2\frac{1}{2}$

2

3	18
5	30

In this arrangement 3, 5, and 18, 30 are in the same ratio: $\frac{3}{5} = \frac{18}{30}$

Look for other equal ratios among the four numbers.
Write down each pair you find.

Take note

We will call this kind of number arrangement a **ratio square**.

5	20
3	12

Notice:

$\frac{5}{3} = \frac{20}{12}$ $\frac{3}{5} = \frac{12}{20}$

$\frac{5}{20} = \frac{3}{12}$ $\frac{20}{5} = \frac{12}{3}$

$5 \times 12 = 3 \times 20$

3 Write down four pairs of ratios and a multiplication sentence for these ratio squares.

a)

7	28
21	84

b)

0.2	1.0
1.6	8.0

4 Which of these are ratio squares and which are not?

a)

1	3
2	4

b)

1	4
$\frac{1}{2}$	2

c)

3	12
1	10

d)

0.1	10
1.1	11

e)

99	100
9	10

28

Challenges

5 a) Check that this ratio sentence is correct: $\frac{2}{3} = \frac{10}{15}$.
Can the numbers 2, 3, 10, 15 be arranged to make a ratio square?
If so, show how.

b) If you start with any correct ratio sentence, can the numerators and denominators be arranged to make a ratio square?
If you say **no**, give an example to explain.

6 Starting with any two numbers, how can you form a ratio square?
Give examples to explain for different starting positions.

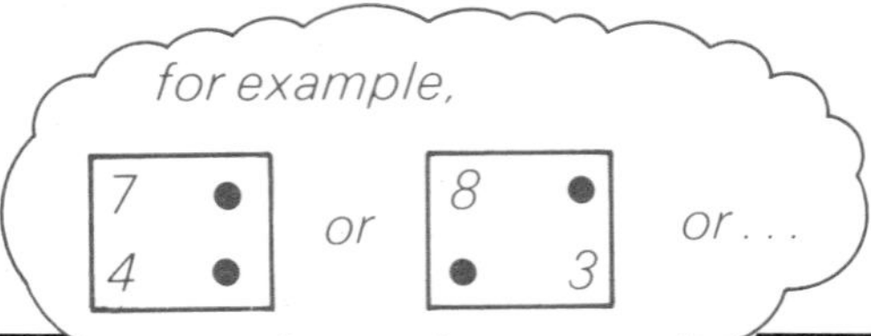

7 This is a multiplication sentence from a ratio square: $7 \times 8 = 4 \times 14$.

a) Complete the ratio square.

b) Write down four equal ratios for the square.

c) Repeat **(a)** and **(b)** for this sentence: $0.4 \times 1.8 = 1.2 \times 0.6$.

7	?
?	?

8 This is a ratio square in which all four numbers are whole numbers.

7	?
?	8

a) Write down a possible pair of numbers for the missing values.

b) How many different pairs are possible?

c) What happens if numbers other than whole numbers are allowed in **(b)**?

9 These are ratio squares. How can we find the missing numbers? Write an explanation for each square.

a)

?	15
4	20

b)

3	12
?	60

10 a) These are ratio squares. For each one, find the pair of smallest possible whole numbers which fits the square.

28	84
?	?

?	100
?	36

b) Find the two numbers with the smallest **sum** which fit this square.

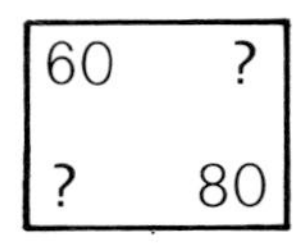

60	?
?	80

11 a) Check that

5	15
12	36

is a ratio square.

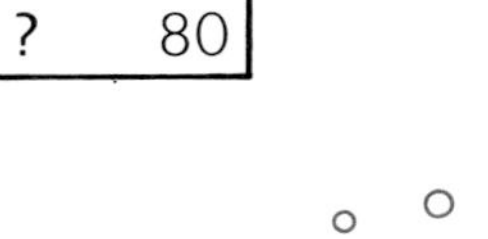

b) Think about rearranging the numbers in different ways.

Do we always get a ratio square?
If you say **no**, list all the arrangements which are **not** ratio squares.

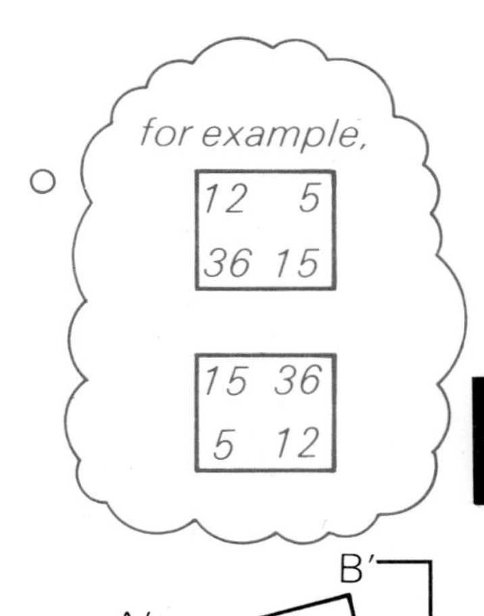

for example,
12 5
36 15
15 36
5 12

Think it through

12 ABCD and A′B′C′D′ are similar.

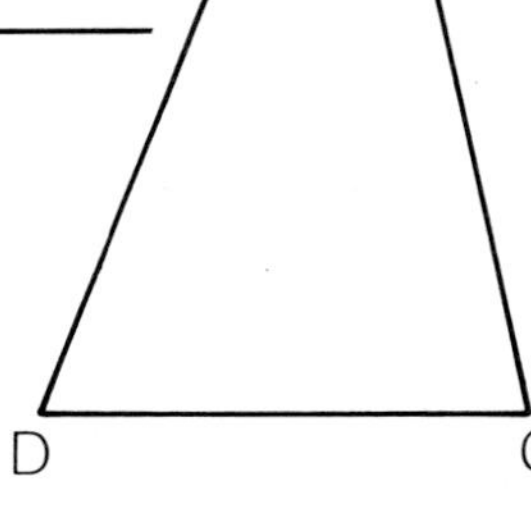

a) So

AB	A′B′
AD	A′D′

must be a ratio square.

Write down two more ratio squares.

b) AB = 2 cm, BC = 5 cm,
A′D′ = 7 cm, C′D′ = 7 cm,
A′B′ = 3 cm.

(i) What is the scale factor for enlargement?

(ii) How long are these: CD, AD, B′C′?

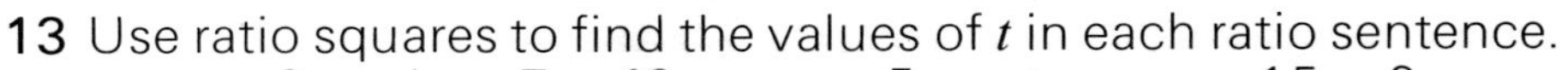

13 Use ratio squares to find the values of t in each ratio sentence.

a) $\frac{t}{4} = \frac{9}{12}$ b) $\frac{7}{t} = \frac{49}{63}$ c) $\frac{5}{11} = \frac{t}{66}$ d) $\frac{15}{t} = \frac{3}{8}$

▲ 268
● Next chapter

Letters in ratio sentences

E 1 a) Start with $\frac{a}{b} = \frac{c}{d}$.

Explain why $a \times d$ **must** equal $b \times c$.

b) Start with $p \times q = r \times s$.

Is this sentence always true, sometimes true or never true? ⇨ $\frac{p}{r} = \frac{q}{s}$

Explain why.

Think about this: $\frac{5}{8} = \frac{20}{32}$.

So $\frac{5}{8} \times \frac{8}{5} = \frac{20}{32} \times \frac{8}{5}$.

But $\frac{5}{8} \times \frac{8}{5}$ is $\frac{5 \times 8}{8 \times 5}$, which is 1.

So what does this tell us about 20×8 and 32×5?

2 $\frac{a}{b} = \frac{c}{d}$. Which of these must also be true?

a) $a \times d = b \times c$ b) $a \div c = b \div d$ c) $b \div a = d \div c$

d) $a + c = b + d$ e) $c \div b = d \div a$ f) $a - c = b - d$

For each one which is not necessarily true, give an example to explain.

3 Find the number which k represents in each sentence.

a) $\frac{k+1}{4} = 2$ b) $\frac{8}{k+1} = 4$ c) $\frac{4}{5} = \frac{12}{k+1}$ d) $\frac{15}{18} = \frac{k}{k+1}$

4 BE is parallel to CD.

a) Explain why triangles ABE and ACD are similar.

b) AB = 5 cm, BC = 3 cm, AE = 7 cm, ED = k cm. Find k.

Hint: Write a ratio sentence involving k.

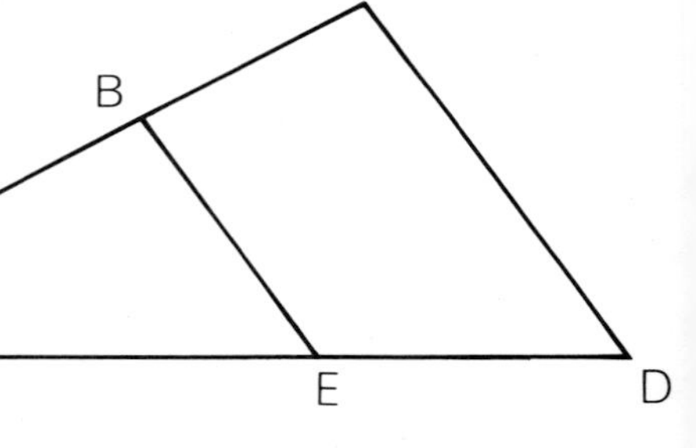

5 a) In these diagrams, do you think the two points are lined up with the origin?

b) Write a short section for a textbook of your own to explain to your readers how to use ratio to decide if two points are lined up with the origin. Include the two pairs here as worked-out examples. Also include this 'aircraft problem' as a worked-out example.

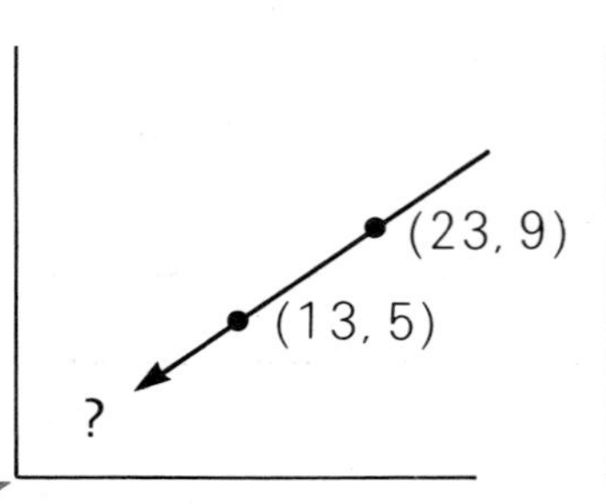

An aircraft is on its final approach to an airport. From here on, it will fly in a straight line. The aircraft is 348 km west and 44 km north of the airport. Will it fly directly over a hospital that is 56 km west and 5 km north of the airport?

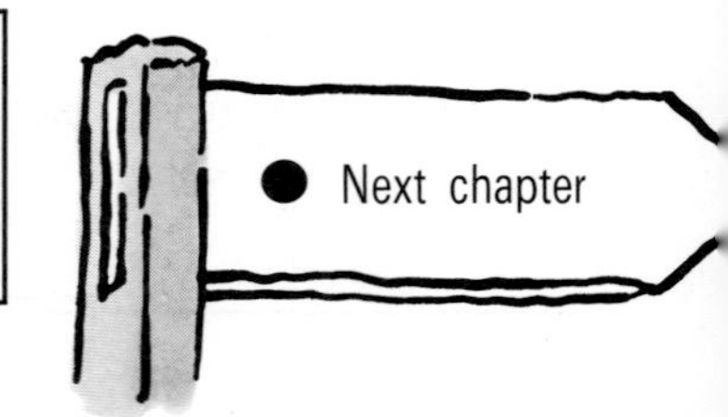

28

29 Angles and distances

A

Do you remember . . . ?

We say that a shape and its enlargement are similar. Corresponding angles are equal.
The lengths of pairs of corresponding sides form a ratio square:

AB	A′B′
BC	B′C′

1 Which of these are pairs of similar shapes?
If the shapes are not similar, say why.

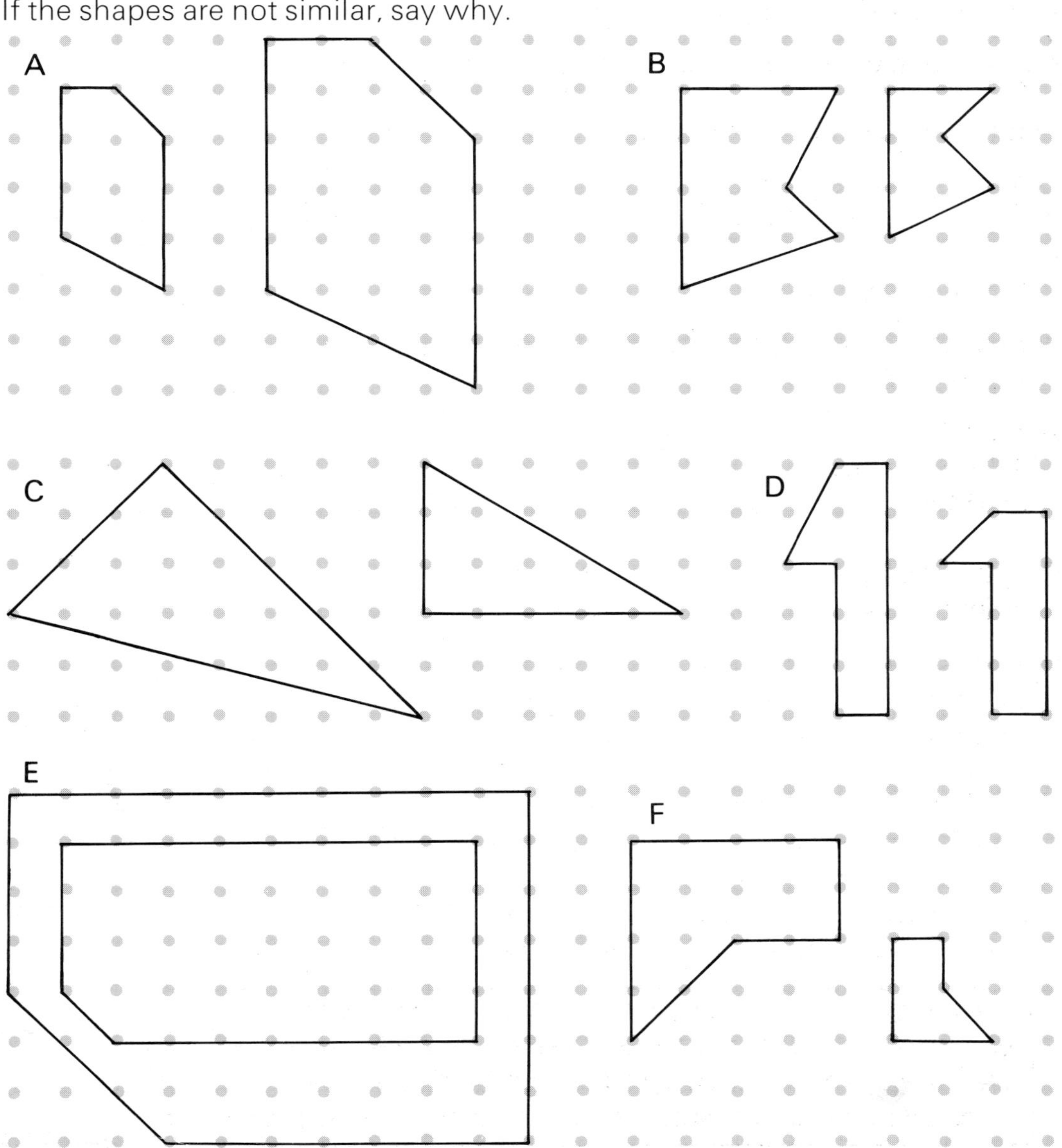

Think it through

2 Use 1 cm dotted or squared paper.

a) Draw two quadrilaterals which have the same sized angles but which are not similar.
Write one or two sentences to explain why they are not similar.

b) Draw two quadrilaterals whose corresponding sides are in the same ratio, but which are not similar.
Write one or two sentences to explain why your shapes are not similar.

3 **Do you remember . . . ?**

Triangles with the same sized angles are always similar.
Pairs of corresponding sides form ratio squares.
Also, **triangles** with corresponding sides in the same ratio are **always** similar.
Their corresponding angles are always equal.
This is true **only** for triangles.

Equal angles . . . so sides form ratio squares, for example,

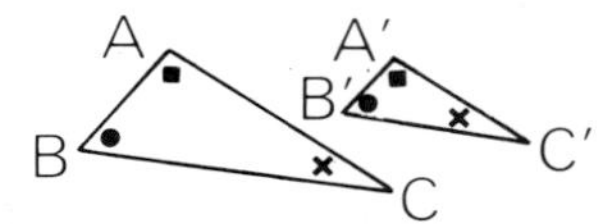

AC	A′C′
AB	A′B′

Sides form ratio squares, for example,

XY	X′Y′
XZ	X′Z′

X Y Z X′ Y′ Z′

. . so the corresponding angles are equal.

Do not use a protractor.
The drawings are **not** to scale.
Which of these pairs of triangles are similar?
Write **similar** or **not similar** for each pair.

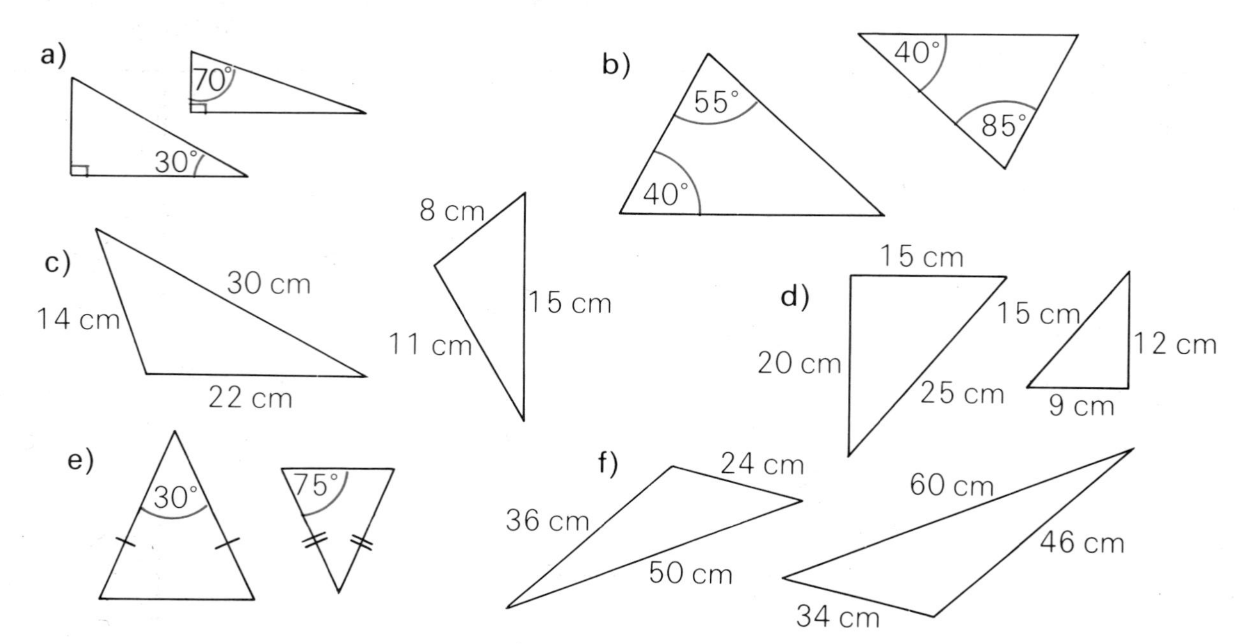

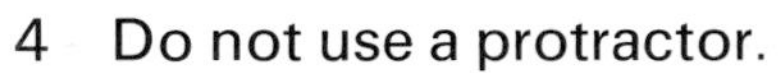

4 **Do not use a protractor.**
The two hexagons are similar.

a) What is the scale factor for enlargement from ABCDEF to A′B′C′D′E′F′?

b) What is the scale factor from A′B′C′D′E′F′ to ABCDEF?

c) What is the size of angle AFE?

d) What is the size of angle A′B′C′?

e) How long is AF?

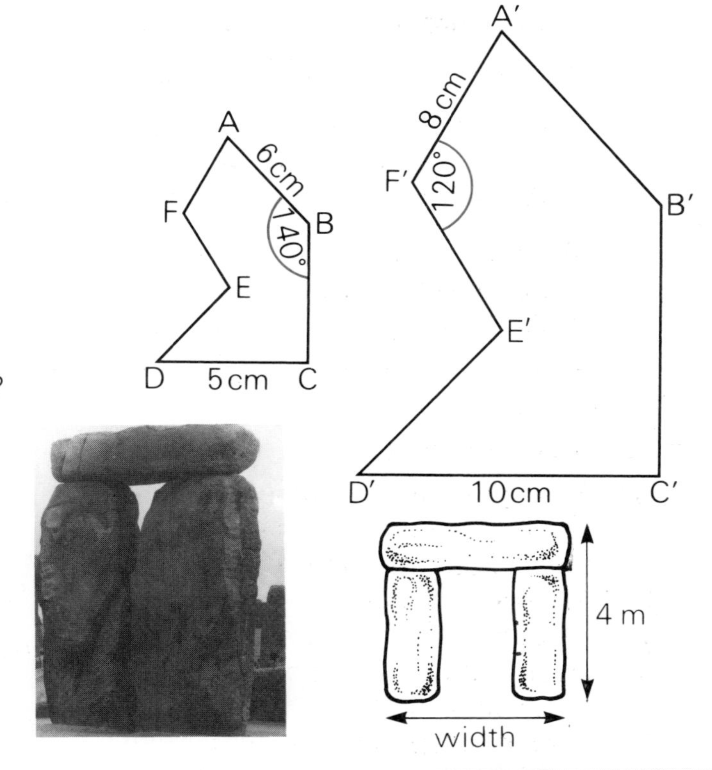

5 These are photographs of Stonehenge.
The height of the arch in real life is 4 m.
Estimate, as accurately as you can,

a) the width of the arch.

b) the height of each of the stones that form the uprights of the arch.

c) the total distance across the Stonehenge circle.

distance across Stonehenge

Think it through

6 These three triangles are similar.

What is the value of

a) x?

b) p?

c) t?

d) q?

e) y?

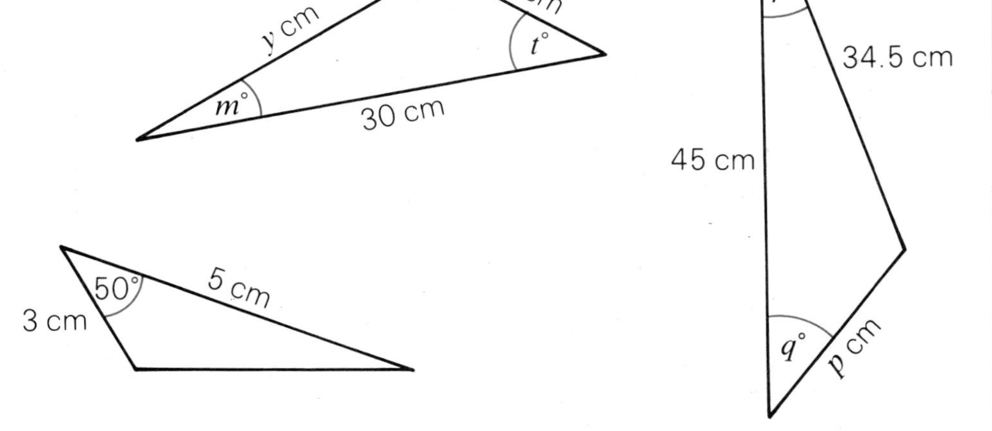

Working with ratios of lengths

B With a friend

1 a) Discuss Bert's problem with your friend.
Both of you write down what you decide.

Bert needs to know the height of the fir tree.
Fortunately it is a sunny day, and he has
a tape measure and a friend.
Bert is almost 2 m tall.
How can Bert and his friend use the
tape measure to get a good estimate
of the height of the tree?

He is making telegraph poles.

without chopping it down or climbing it!

b) Now discuss this ladder problem.
Both of you write down what you decide.
Use squared paper if you think it will help.

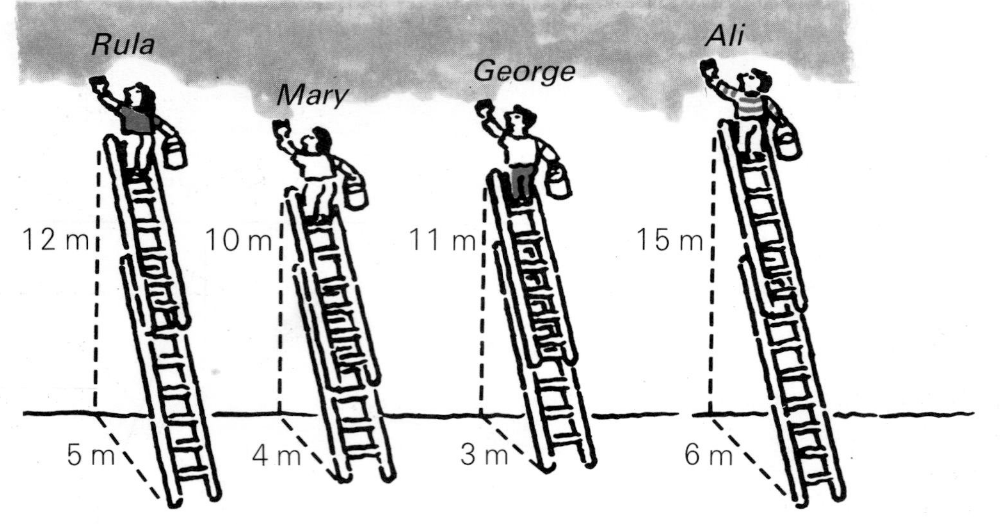

Rula, Mary, George and Ali are painting the wall of the gymnasium.

(i) Which two ladders are sloping at the same angle?

(ii) Which of the four ladders is the steepest?

29

With a friend

2 You need 1 cm squared paper.
Both of you draw two **different** right-angled triangles with one angle measuring 40°.

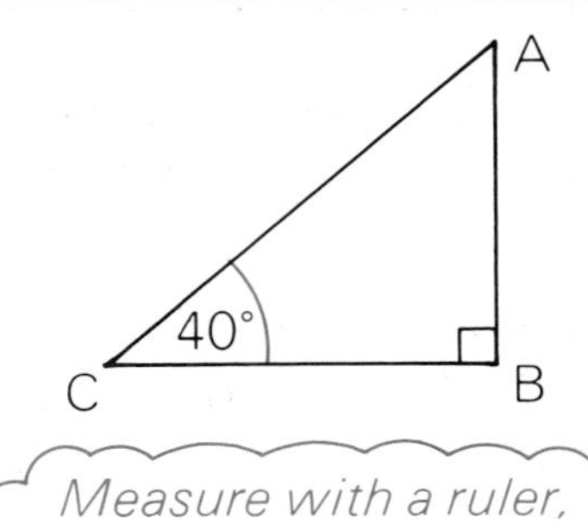

a) For all four triangles, calculate

the ratios $\frac{AB}{BC}$, $\frac{AB}{AC}$ and $\frac{BC}{AC}$.

Measure with a ruler, then use your calculator.

Each of you write down the three sets of four ratios.

four for $\frac{AB}{BC}$, four for $\frac{AB}{AC}$ and four for $\frac{BC}{AC}$

b) Why should the ratios in each set be equal?
Decide between you, then write one or two sentences to explain.

3 a) Draw a right-angled triangle with an angle of 35°.

b) Use your triangle to work out the height of this pole.

c) Write one or two sentences to explain how your method works.

35°
15 m

4 In the diagram of the triangles, x is 29.
Use a ruler (but **not** a protractor) to find out whether a, b, c, d and e are
smaller than 29,
equal to 29
or larger than 29.

5 Triangles LMN and PQR are similar, and LM : MN = RQ : PQ.
Angle MNL = 42°.
Which angle in triangle PQR is 42°?

6 Triangles ABC, DEF, GHI and JKL are similar and length BC ÷ length AB = 0.75.

a) What is length DE ÷ length EF?

b) Write down two more divisions which give the value 0.75.

c) Name three angles equal to angle BAC.

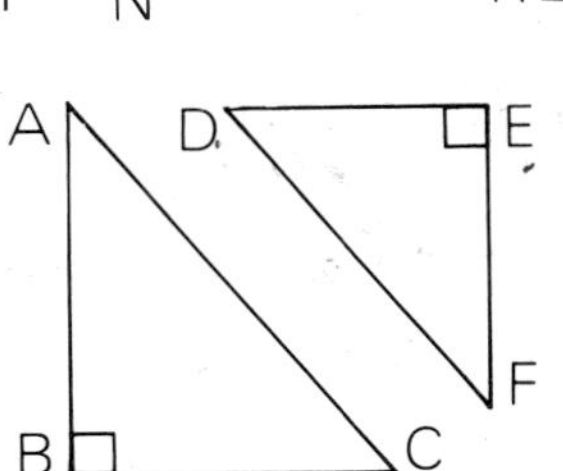

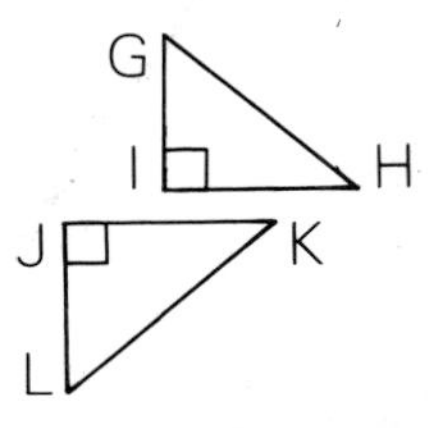

7 Meg drew these triangles and measured AB and BC in each one.

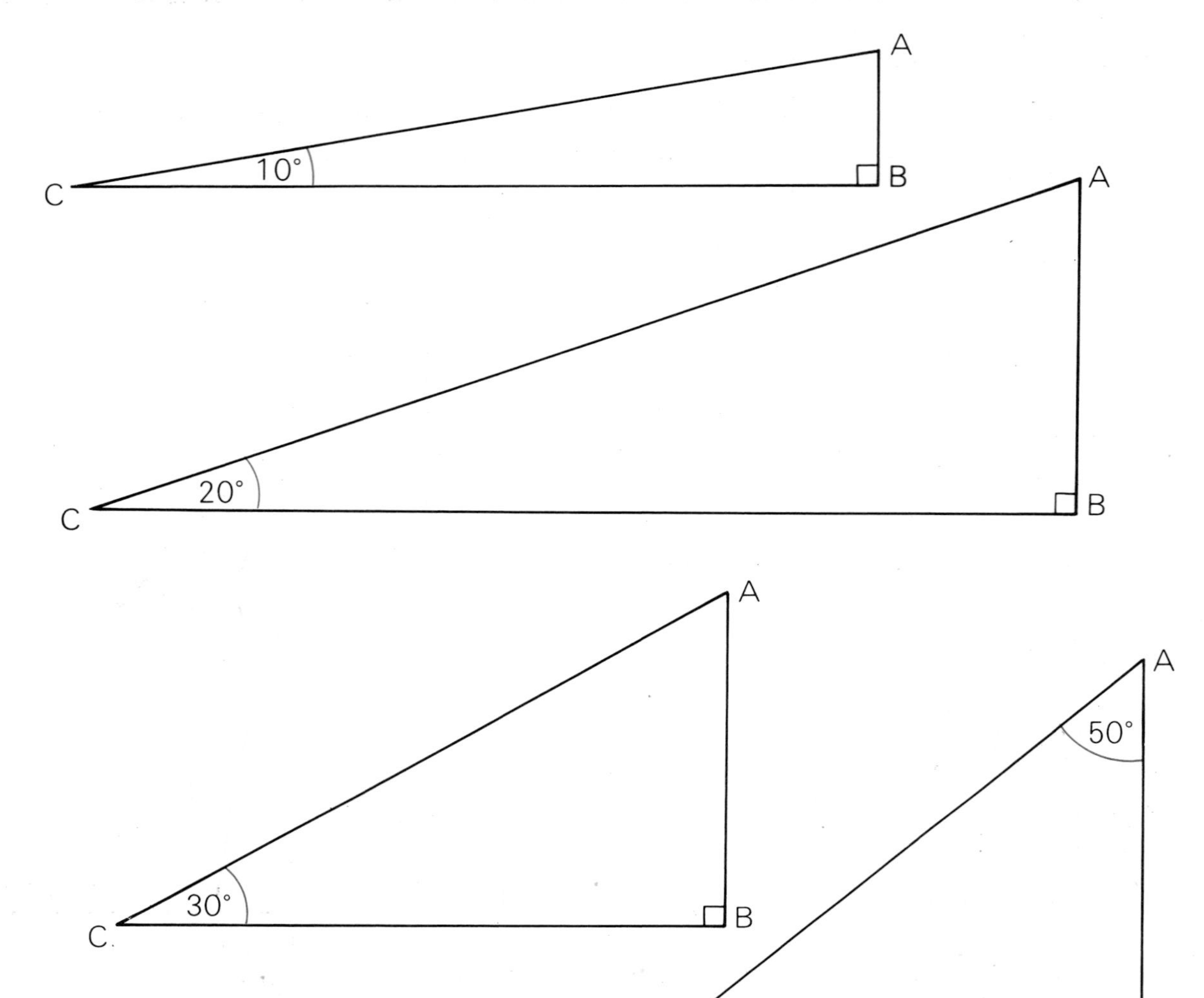

She produced this table.

Angle ACB	10°	20°	30°	40°	50°	60°	70°	80°
$\frac{AB}{BC}$ (to 2 DP)	0.18	0.36	0.58	0.84				

Copy Meg's table.
Complete it by making your own measurements.
Keep your table somewhere safe.
You need it for question **8**.

Think it through

8 You need your table from question **7**.

a) In each of these situations, the angle marked * measures a multiple of 10°. Use only your calculator and your table from question **7** to help you find the size of each one.

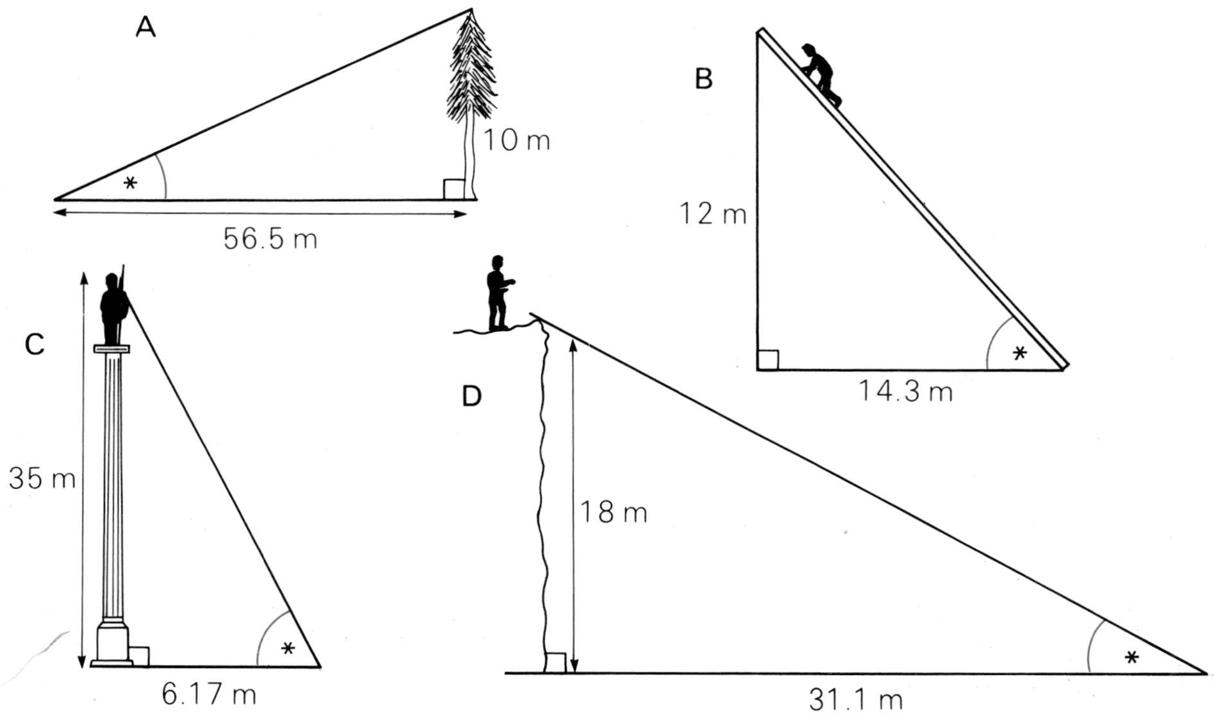

b) Use only your calculator and your table from question **7** to help you find the distances marked ? m in each of these situations.

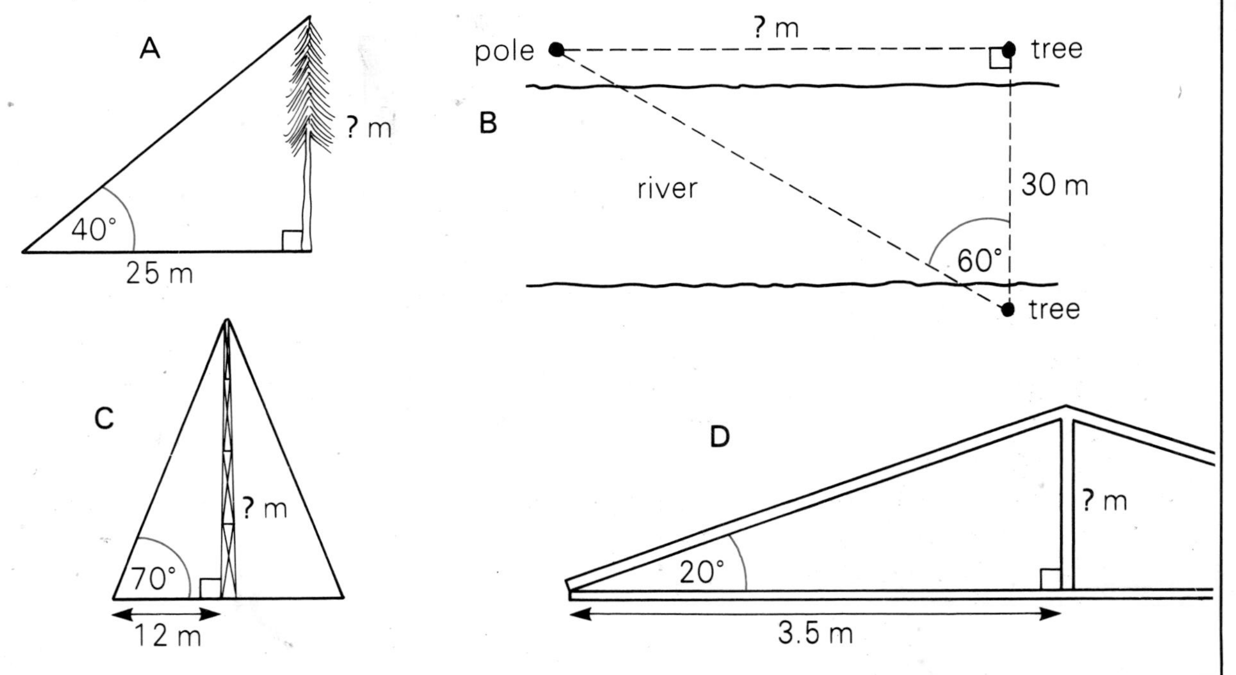

Find out for yourself

9 You need a **scientific calculator** which has these keys: sin cos tan.

Make sure it is in **degree mode**.

Press C 6 0 tan.

Write down the result.
Try with the numbers 10, 20, 30, ..., 80.
Compare your results with your table from question **7**.

Write one or two sentences to explain what the tan key tells us about right-angled triangles.

Think it through

10 You need a scientific calculator in degree mode.

a) Measure AB and AC, and find the ratio $\frac{AB}{AC}$.

Press C 6 0 sin

What do you think the sin key tells us about right-angled triangles?
Write one or two sentences to explain.

b) Draw some right-angled triangles of your own, with different sized angles at C.
Measure the sides, then use the sin key to find the angles at C.

c) Use your triangles in (a) and (b).
Investigate the behaviour of the cos key.
Explain what you discover.

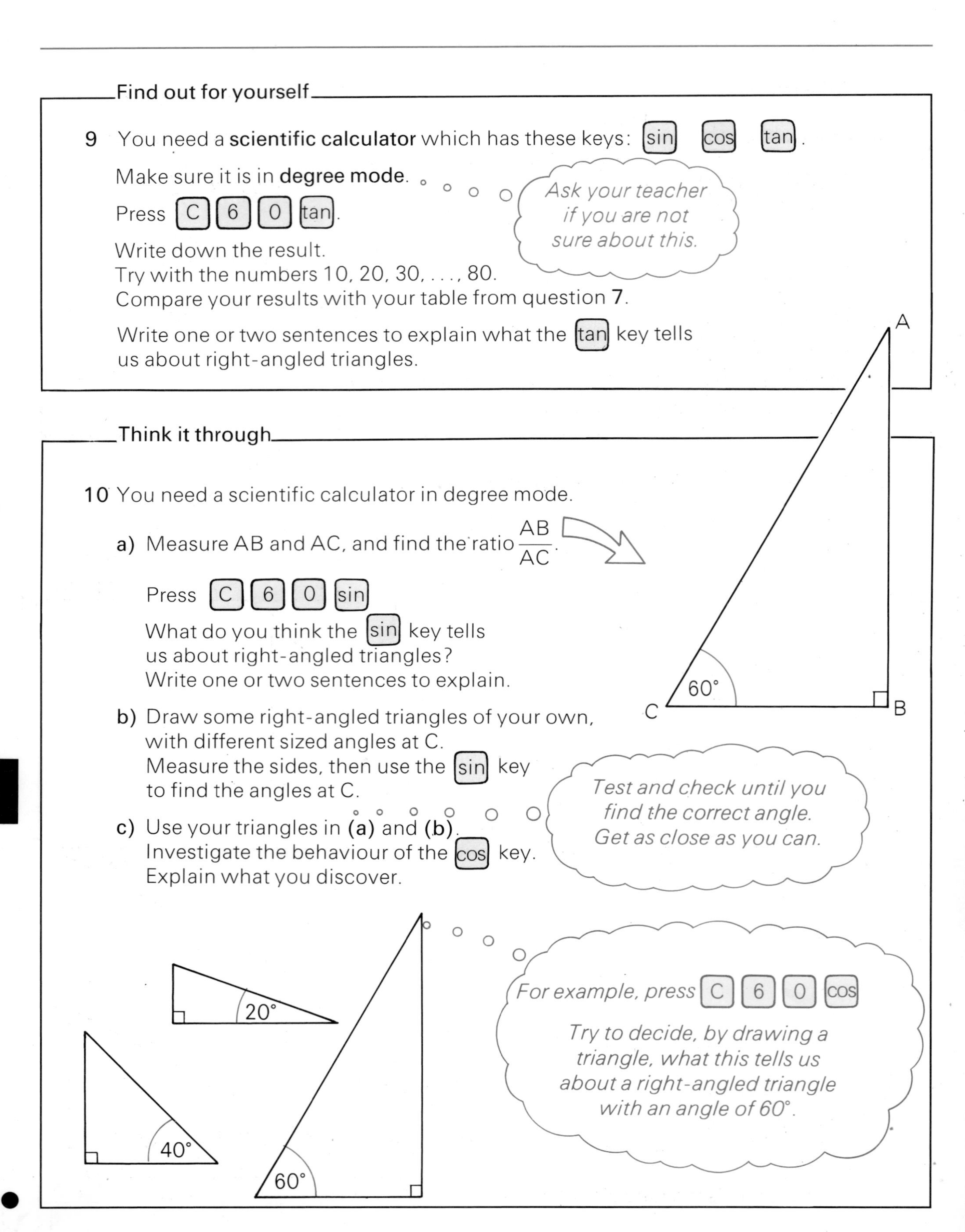

Ask your teacher if you are not sure about this.
A
B
C
60°
Test and check until you find the correct angle. Get as close as you can.
20°
40°
60°
For example, press C 6 0 cos
Try to decide, by drawing a triangle, what this tells us about a right-angled triangle with an angle of 60°.

Take note

When we press . . .   . . . the calculator displays

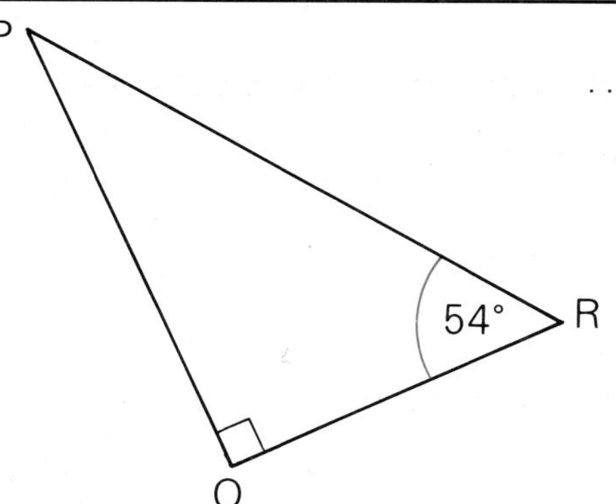

the ratio $\frac{PQ}{QR}$

the ratio $\frac{PQ}{PR}$

the ratio $\frac{QR}{PR}$

These ratios are the same for all 54°, right-angled triangles, no matter how big or how small they are.

11 You need a scientific calculator in degree mode.

a) Use your calculator to find these ratios:

(i) $\frac{LN}{NM}$ (ii) $\frac{LN}{LM}$ (iii) $\frac{NM}{LM}$

L, N, M, 38°

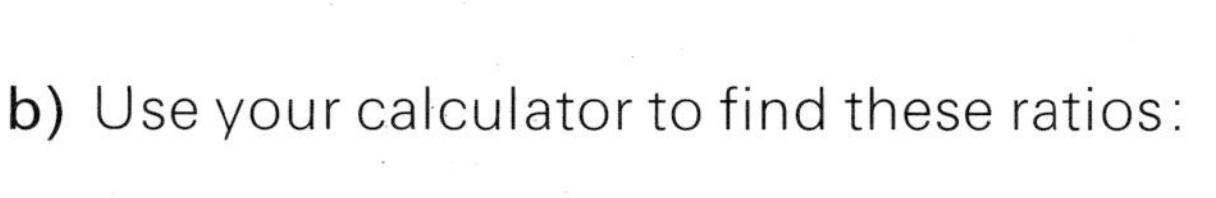

b) Use your calculator to find these ratios:

(i) $\frac{PQ}{PR}$ (ii) $\frac{PQ}{RQ}$ (iii) $\frac{RQ}{PR}$

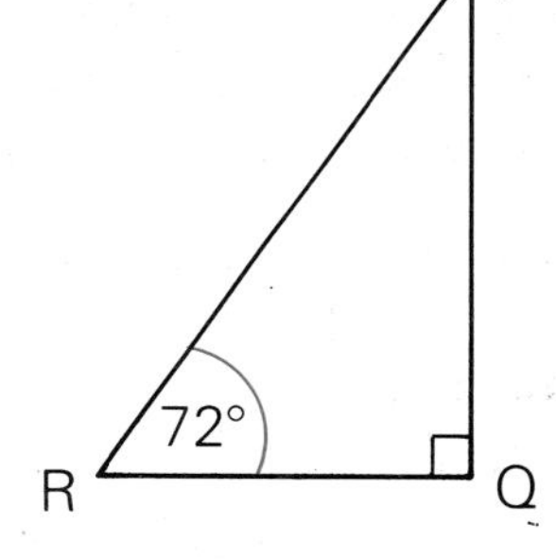

c) Use your calculator to find these ratios:

(i) $\frac{AB}{BC}$ (ii) $\frac{AC}{BC}$ (iii) $\frac{AC}{AB}$

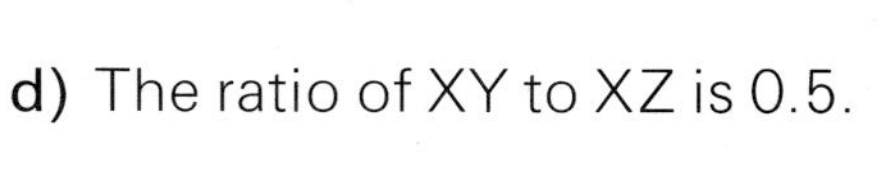

d) The ratio of XY to XZ is 0.5.

(i) What is the angle XZY?

(ii) What is the angle ZXY?

(iii) What is the ratio $\frac{ZY}{XY}$?

(iv) What is the ratio ZY : ZX?

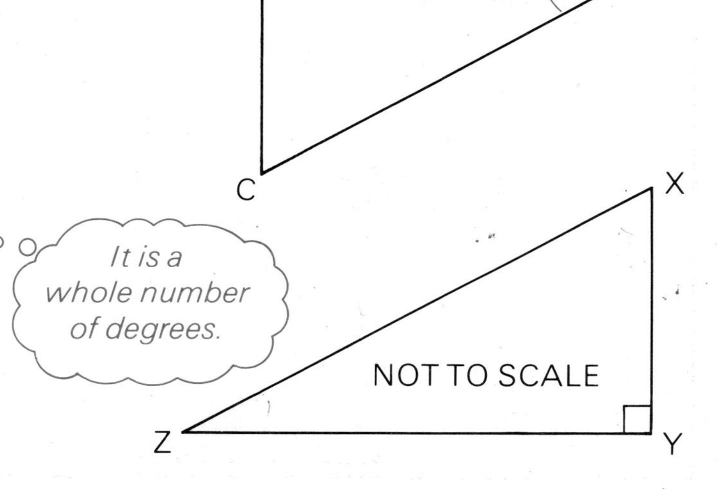

29

Challenge

12 This table gives the ratios *to 2DP*

tan $\frac{PQ}{RQ}$ sin $\frac{PQ}{PR}$ cos $\frac{RQ}{PR}$

for different angles *.

Angle *	10°	20°	30°	40°	50°	60°	70°	80°
tan ratio: $\frac{PQ}{RQ}$	0.18	0.36	0.58	0.84	1.19	1.73	2.75	5.67
sin ratio: $\frac{PQ}{PR}$	0.17	0.34	0.50	0.64	0.77	0.87	0.94	0.98
cos ratio: $\frac{RQ}{PR}$	0.98	0.94	0.87	0.77	0.64	0.50	0.34	0.17

a) Use the table and your calculator to help you to find the angles marked *. *Each one is a multiple of 10°.*

b) Use the table and your calculator to help you to find the lengths marked ? m.

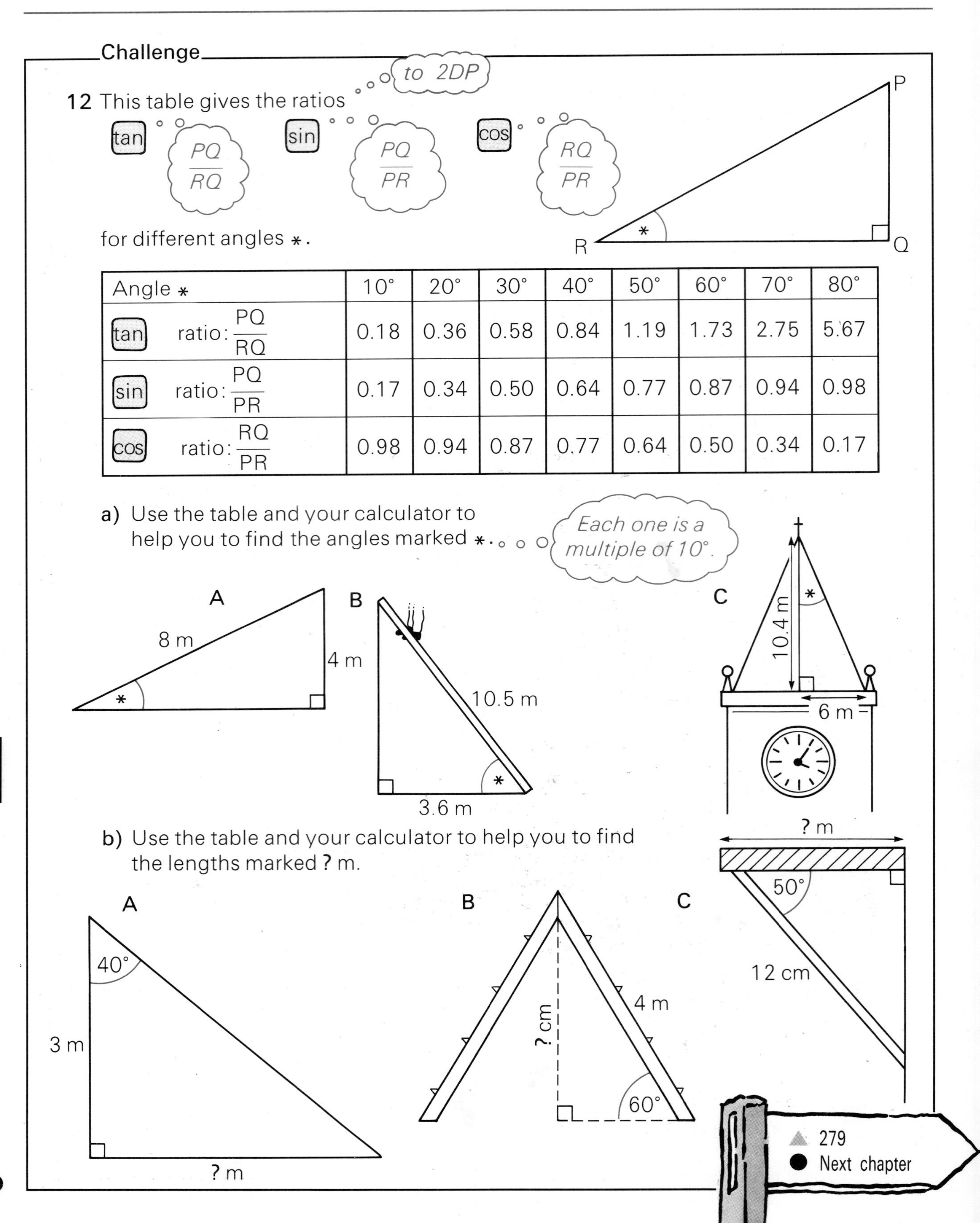

279
Next chapter

C 1 You need a scientific calculator in degree mode.

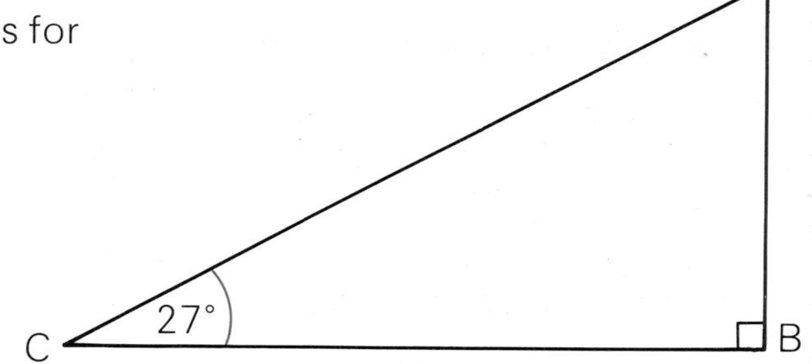

a) Use your calculator to find these ratios for the triangle ABC.

(i) $\frac{AB}{BC}$ (ii) $\frac{AB}{AC}$

(iii) $\frac{CB}{AC}$ (iv) $\frac{CB}{AB}$

b) **Imagine** you press [C] [6] [3] [tan]

The result will be one of the results in **(a)**.
Which one?
Why?

c) Press [C] [6] [3] [tan] and check your answer in **(b)**.

2 The diagram shows the path of the ship SS *Coalton*.

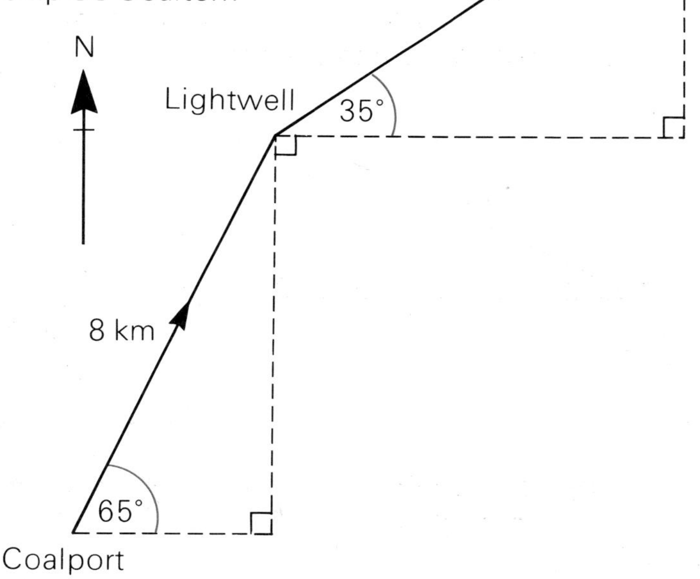

Use your calculator to find

a) how many kilometres north of Coalport the ship is when it arrives at Lightwell.

b) how many kilometres east of Coalport the ship is when it arrives at Lightwell.

c) how many kilometres north of Lightwell the ship is when it arrives at Mycen.

d) how many kilometres east of Lightwell the ship is when it arrives at Mycen.

e) how many kilometres north of Coalport the ship is when it arrives at Mycen.

f) how many kilometres east of Coalport the ship is when it arrives at Mycen.

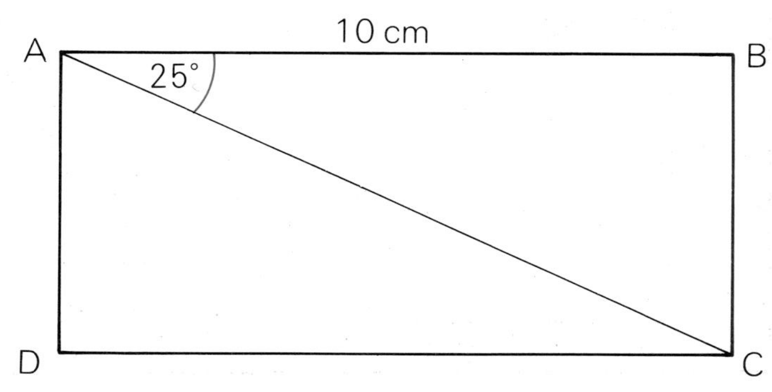

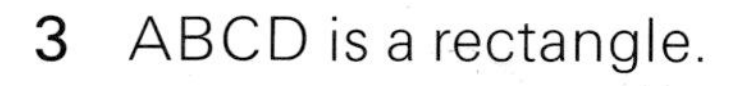

3 ABCD is a rectangle.

a) How long is BC?

b) How long is AC?

29

Challenges

4 This is a 30° spiral.
How long is KO?

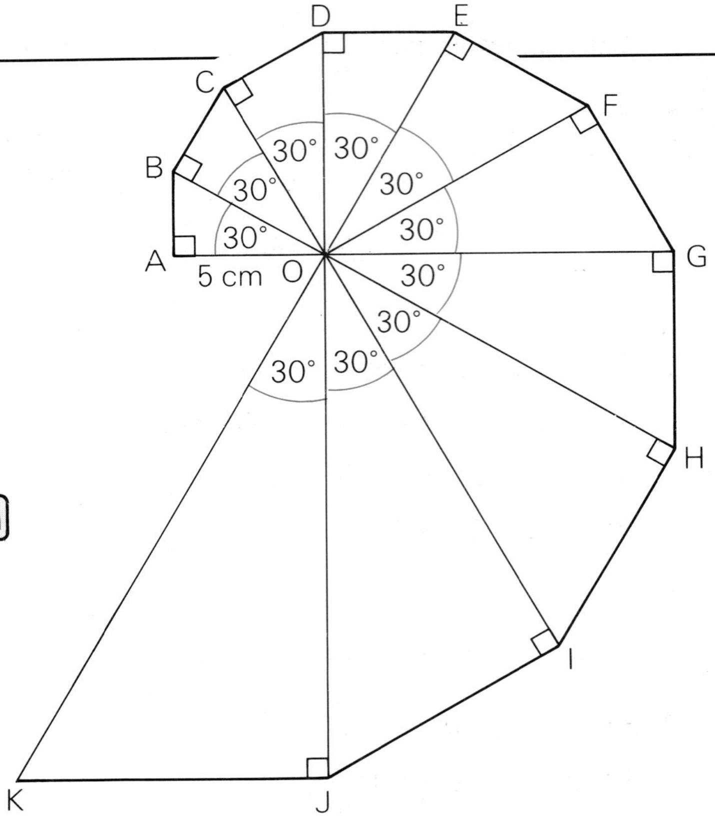

5 a) Press [C] [3] [0] [sin]

Write down the result.

Press [C] [.] [5] [INV] [sin]

Write down the result.

b) Press [C] [4] [5] [tan]

Write down the result.

Press [C] [1] [INV] [tan]

Write down the result.

c) Think carefully about the answers in **(a)** and **(b)**, then use your calculator to find the angle which has a sin ratio of 0.56.

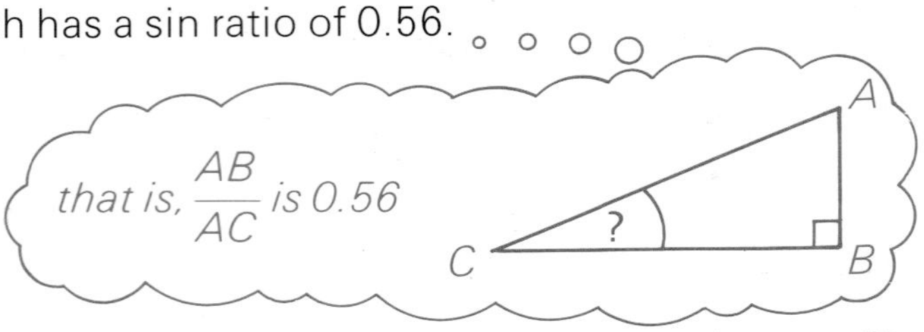

6 What is the longest wooden pole that could fit inside this chest?

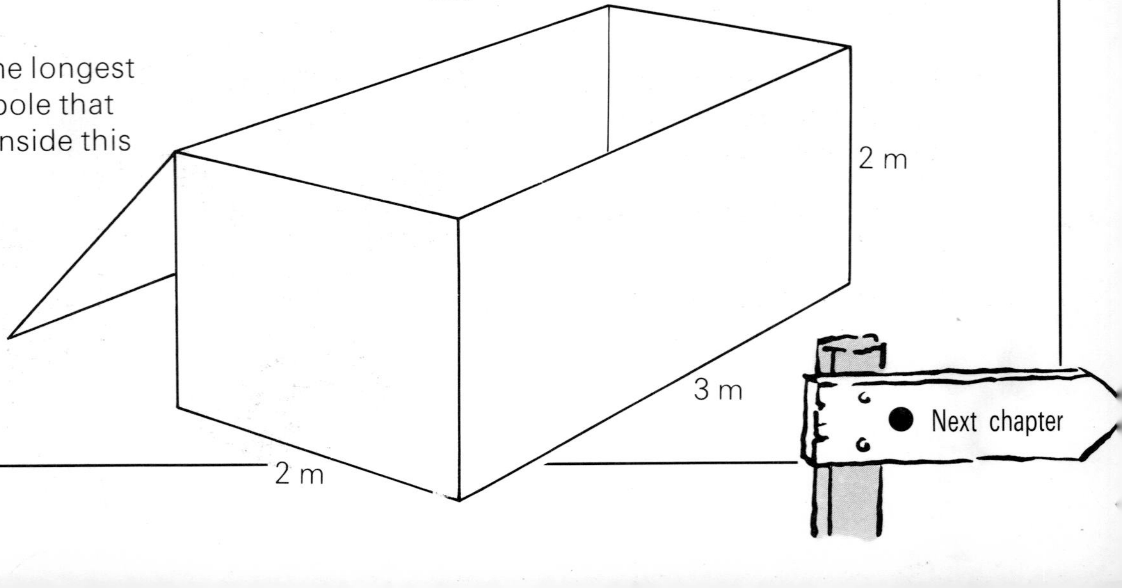

29

30 Using indices

A Do you remember . . . ?

We can write 5×5 more simply as 5^2.

We say, '5 to the power 2' or '5 squared'.

and $7 \times 7 \times 7$ more simply as 7^3.

We say '7 cubed' or '7 to the power 3'.

1 Here is a shorthand way of writing $9 \times 9 \times 9 \times 9 \times 9 \times 9 \times 9 \times 9$:

9^8

We say, 'nine to the power 8'.

Write each of these in a similar way.

a) $3 \times 3 \times 3 \times 3$ b) $10 \times 10 \times 10 \times 10 \times 10 \times 10$

c) $1.2 \times 1.2 \times 1.2$

2 Write these as single numbers.

a) 4^2 b) 7^2 c) 2^3 d) 3^4

3 a) Is 3^2 equal to 2^3?

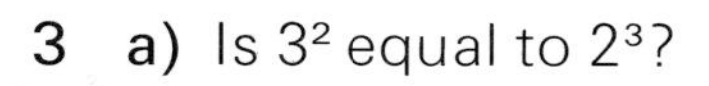

b) Is 3×4 equal to 3^4?

c) Which is larger, 2×100 or 2^{100}?

4 a) Guess the size of this number:

About 20?
. . . about 200?
. . . about . . . ?

b) Now use your calculator to find the number.
Was your guess an overestimate or an underestimate?

c) Do the same for each of these.
(i) 3^{10} (ii) 9^5 (iii) 18^5

5 **$2^{\square} \approx 1$ million.** How close can you get to $\square$?

Write down what you discover.

6 There are six eggs in a tray, six trays in a carton and six cartons in a box.

How many eggs are there in a box?

Write your result like this: $\square^{\square}$.

30

Take note

Numbers like 5^3 and 7^4 are called **powers**.

We say the power 5^3 has **base** 5 and **index** 3. *plural: indices*

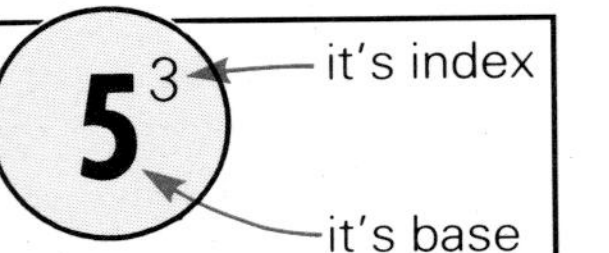

7 We can write 32 as a power like this: 2^5.

a) Write each of these as a power. (i) 125 (ii) 1000 (iii) 81

b) Is there more than one way to answer any of the parts in **(a)**? If so write down the alternatives.

c) Write 729 as a power in three different ways.

8 The Greek astronomer Ptolemy (about 90–168 AD) made the first recorded classification of stars. He classified them into various **magnitudes**.

He thought the brighter a star is, the bigger it must be.

A star of the first magnitude sheds about 2.5 times as much light as a star of the second magnitude.

A star of the second magnitude sheds about 2.5 times as much light as a star of the third magnitude . . . and so on.

a) How much more light is shed by a star of the first magnitude than a star of (i) the third magnitude? (ii) the fifth magnitude? (iii) the tenth magnitude?

Write your results as powers.

Challenge

b) A star of the first magnitude sheds as much light as 100 stars of a certain magnitude. Which magnitude?

c) To shed about as much light as **one** star of the first magnitude how many stars do you need of
(i) the fourth magnitude? (ii) the seventh magnitude? (iii) the ninth magnitude?

Exploration

9 We can write 81 as a power in these three ways:

81 → 81^1, 9^2, 3^4

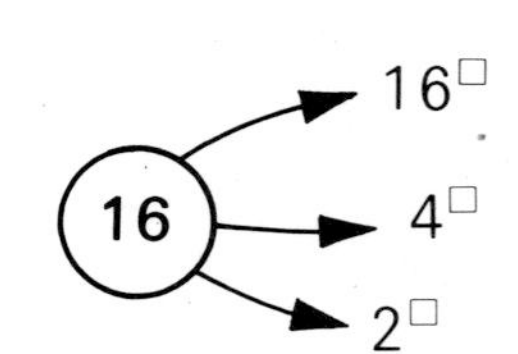

a) Copy and complete these three ways of writing 16 as a power.

b) Find four different ways of writing 64 as a power.

c) Which numbers less than 100 can be written as a power in more than one way?

d) Which numbers less than 1000 can be written as a power in the greatest number of different ways?

Powers of 10

B 1 About 56 million people live in the UK.
56 million is 56 000 000.

> One million is 1 000 000 which is 10^6.
> One billion is one thousand million.
> One trillion is one thousand billion
> or one million million.

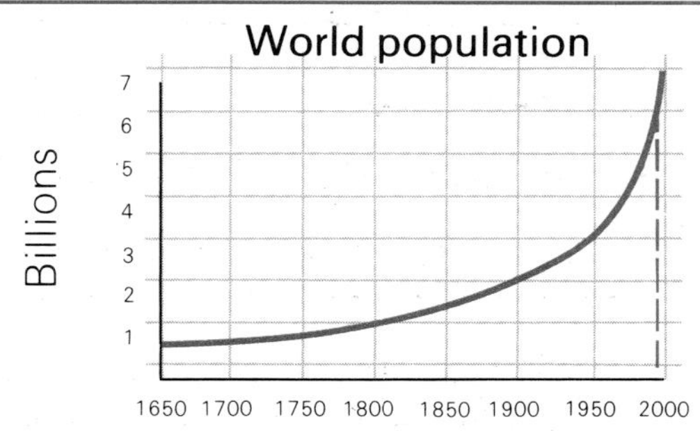

a) The population of the world passed the 5 billion mark in 1987.
Write 5 billion as a single number.

b) Write these large numbers as powers of 10.
(i) 10 million (ii) 1 billion (iii) 100 billion (iv) 1 trillion

c) Write these numbers in words.
(i) 10^5 (ii) 10^8 (iii) 10^{10} (iv) 10^{13}

2 To go from one line to the next in this table, we divide by 10.

except for the last line

a) Copy and complete the table.

b) What power of 10 would continue the pattern on line 7?

Number	Power of 10	In words
1 000 000	10^6	one million
100 000		
10 000		
	10^3	
		one hundred
10		
		one

STOP

Take note

> We agree to let 10^0 mean 1.

c) Think about adding another line to the table.
What should line 8 of the table be?
What power of 10 continues the pattern?

Take note

> We agree to let 10^{-1} mean $\frac{1}{10}$.

d) Write out lines 7 and 8 of the table, and then add four more lines.
Continue the pattern for each line.
Finish with the number 'one hundred-thousandth'.

e) Write one billionth as a power of 10.

f) We can write 10^{-1} in decimal form, like this: 0.1.
Write these small numbers in decimal form.
(i) 10^{-2} (ii) 10^{-3} (iii) 10^{-4} (iv) 10^{-5} (v) 10^{-6}

3 We can write 10^{-3} as $\frac{1}{10^3}$ or $\frac{1}{1000}$; we can write 3^{-2} as $\frac{1}{3^2}$ or $\frac{1}{9}$.

Write these numbers as fractions.

a) $5^{-3} = \frac{1}{5^3} = \square$ b) $20^{-1} = \square = \square$ c) $4^{-2} = \square = \square$ d) $3^{-4} = \square = \square$

Think it through

> 4 Write each of these as a power.
>
> a) $\frac{1}{10\,000}$ b) 0.000 001 c) $\frac{1}{343}$ d) 0.25

30

Numbers large and small

C 1 a) The mean distance of the planet Venus from the sun is 108 000 000 km.
We can write this distance using indices in many ways.
For example,
108 000 000 km = 108 × 1 000 000 km = 108 × 10^6 km

Copy and complete: **108 000 000 km = 108 × 10^6 km**
= 10.8 × $10^{\square}$ km
= $\square$ × 10^8 km

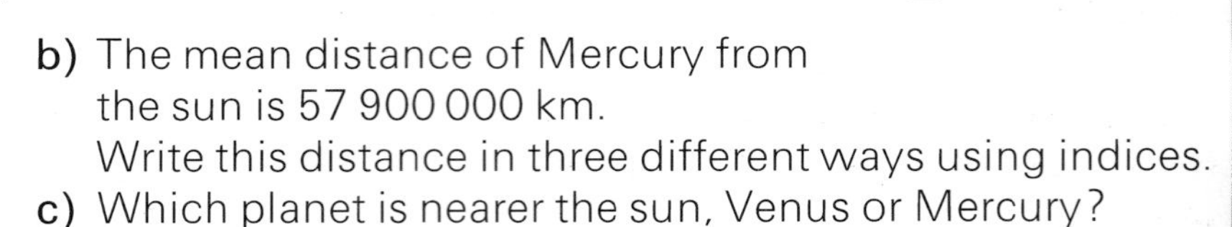

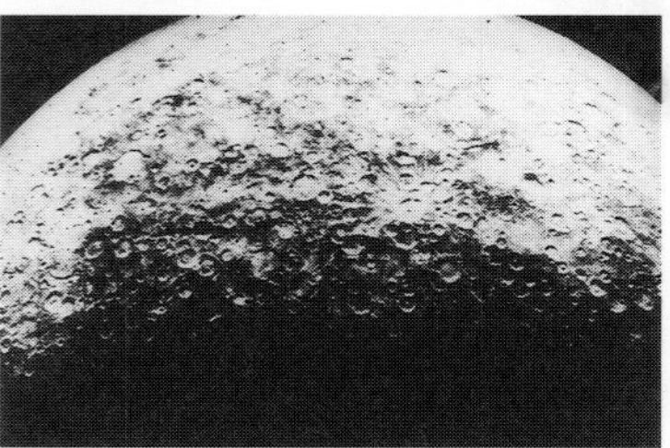

b) The mean distance of Mercury from the sun is 57 900 000 km.
Write this distance in three different ways using indices.

c) Which planet is nearer the sun, Venus or Mercury?

Take note

Indices help us to write large numbers in a simpler way:
2 590 000 000 000 = 259 × 10^{10} = 25.9 × 10^{11} = 2.59 × 10^{12}, . . . and so on.

2 Write down which number in each pair is the larger.

a) 2.64×10^8; 2.6×10^8 b) 1.74×10^6; 1.74×10^5

c) 3.76×10^6; 37.6×10^6 d) 0.415×10^8; 41.5×10^5

3 a) Here are the mean distances of the planets from the sun.

Planet	Distance
Venus	10.8×10^7 km
Mars	228×10^6 km
Saturn	14.3×10^8 km
Earth	150×10^6 km
Jupiter	7.78×10^8 km
Mercury	579×10^5 km
Uranus	2870×10^6 km
Neptune	450×10^7 km
Pluto	5.9×10^9 km

Write each distance like this: $\square.\square \times 10^{\square}$ km.

A number between 1 and 10, . . . including 1, but not 10.

Examples: 6.7×10^4 km, 1.25×10^7 km

b) Use your results in **(a)** to write the planets in order, starting with the one whose mean distance from the sun is greatest.

4 'Light' is the name we give to the electromagnetic waves that can be detected by the human eye.
The wavelengths of 'light' lie between 400×10^{-9} m and 760×10^{-9} m.
Write these two limits using decimals.

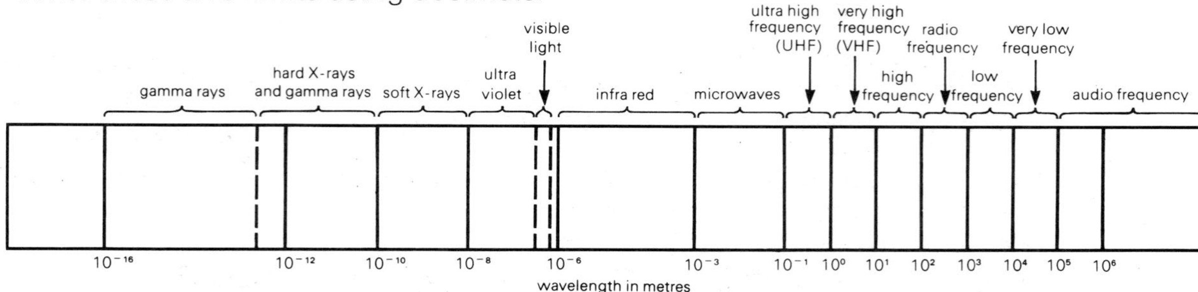

The electromagnetic spectrum

5 There is an enormous variety of living organisms, ranging from a virus (with a mass of only about 10^{-21} kg) to a blue whale (with a mass of about 1.4×10^5 kg).

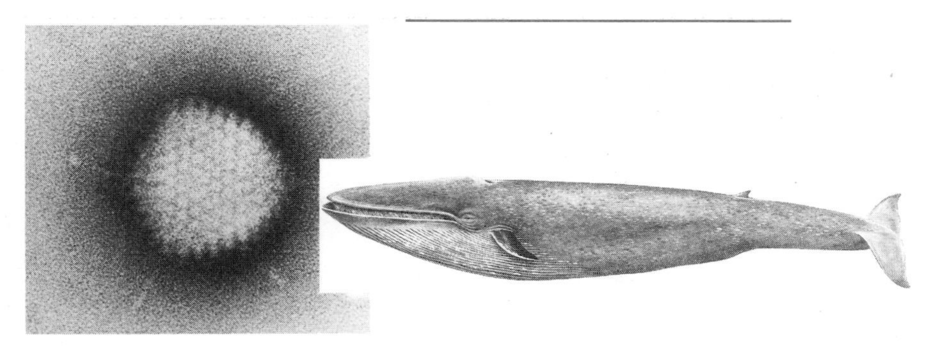

a) Which of these best describes the mass of a blue whale relative to that of a virus?

A about the same B quite a bit bigger

C much bigger D millions of times bigger

Explain your choice.

b) A house spider weighs about 10^{-4} kg;
5 spiders weigh about 5×10^{-4} kg; 10 weigh about 10^{-3} kg.
Roughly, what is the weight of (i) 8 spiders? (ii) 100 spiders?

c) The total mass of two viruses is about 2×10^{-21} kg.
How many viruses are needed to make a total mass of
(i) 10^{-20} kg? (ii) 10^{-19} kg? (iii) 10^{-14} kg?

d) Here are the masses of some more things.

Humming bird: 2×10^{-3} kg

Average man: 7.29×10 kg

Molecule of water: 2.99×10^{-26} kg

African elephant: 6.3×10^3 kg

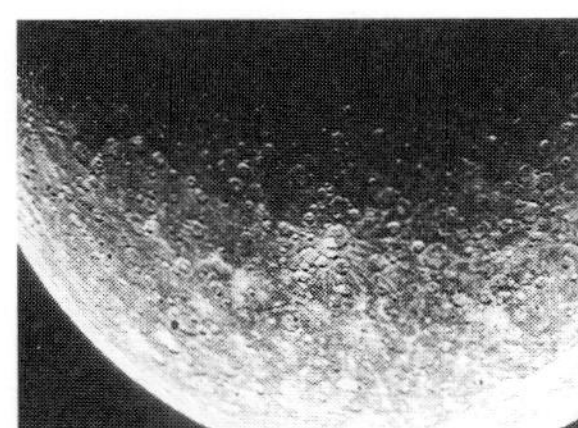

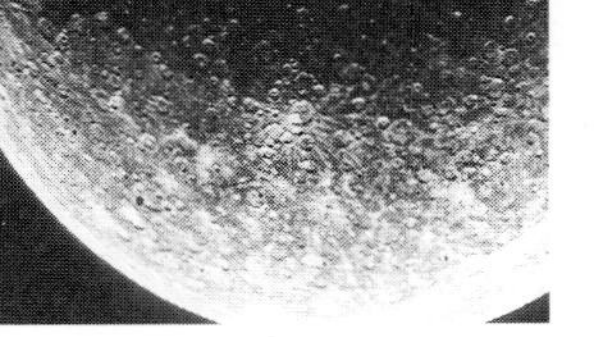

Planet mercury: 3.31×10^{23} kg

Parasitic wasp (the smallest insect): 5×10^{-9} kg

Approximately, how many times heavier is

(i) an average man than a parasitic wasp?
(ii) an African elephant than the average man?
(iii) the planet Mercury than a molecule of water?

Write your results using indices: $\square \times 10^{\square}$.

e) If a humming bird ate its own weight in parasitic wasps how many wasps would it consume?
Is it about
A 4000? B 40 000? C 400 000? or D 4 000 000?

Challenge

6 Study this photograph of a stylus travelling through the grooves of a record. The magnification is × 200. As a record turns, the stylus vibrates and a signal is transmitted to the speakers of the hi-fi system.
Use a ruler to estimate the mean width of the grooves of an LP. Write your estimate, using indices, in millimetres.

7 a) Use a ruler to estimate the measurements indicated for each of the three bacteria.
Write your estimates, using indices, in millimetres.

(i) *Legionella* bacillus

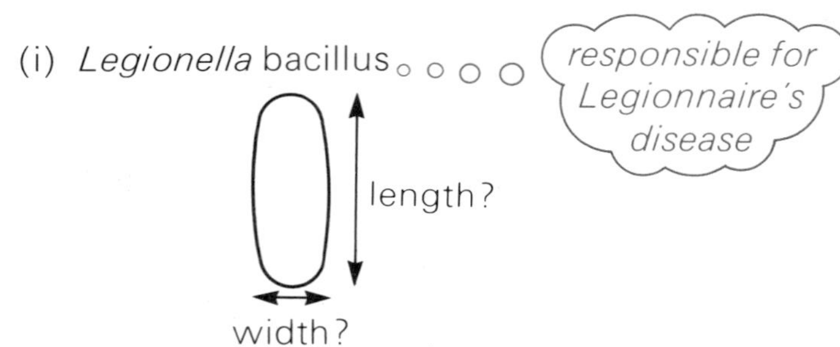

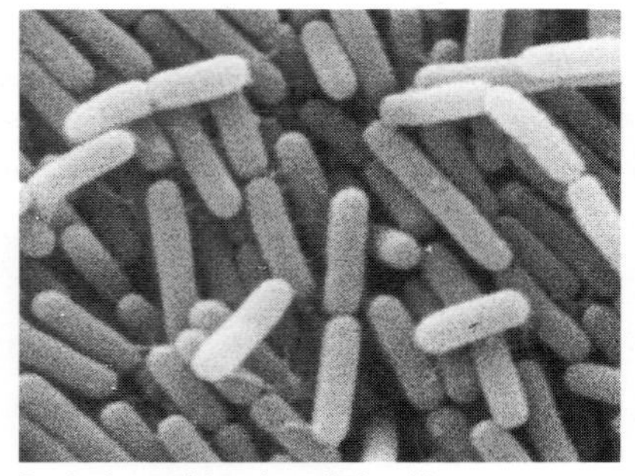
Rod-shaped *Legionella* bacillus magnified 3000 times

(ii) *Pseudomonas fluorescens* bacterium

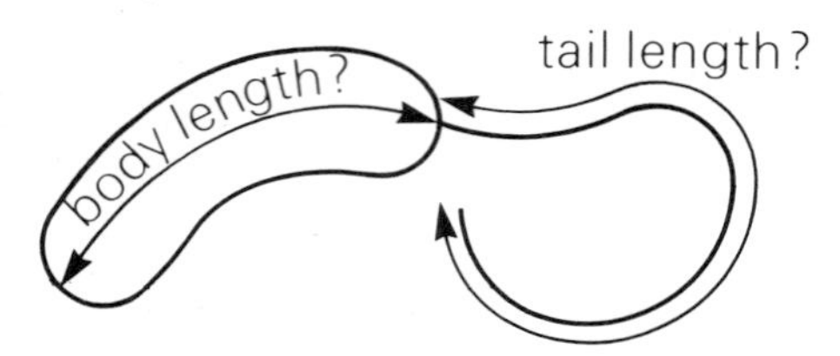

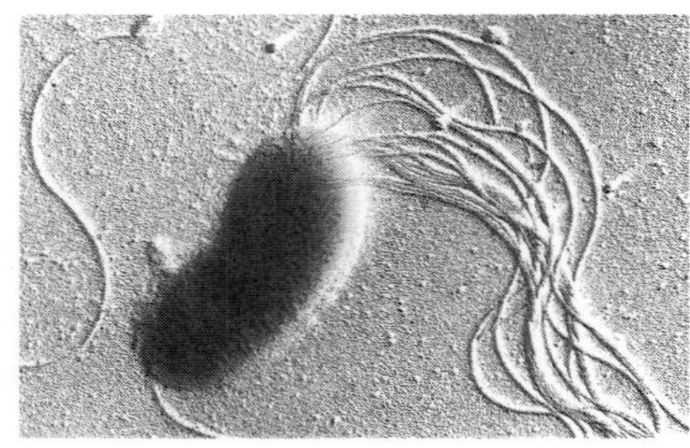
Pseudomonas fluorescens bacterium (× 10 000)

(iii) *Staphylococcus aureus* bacterium

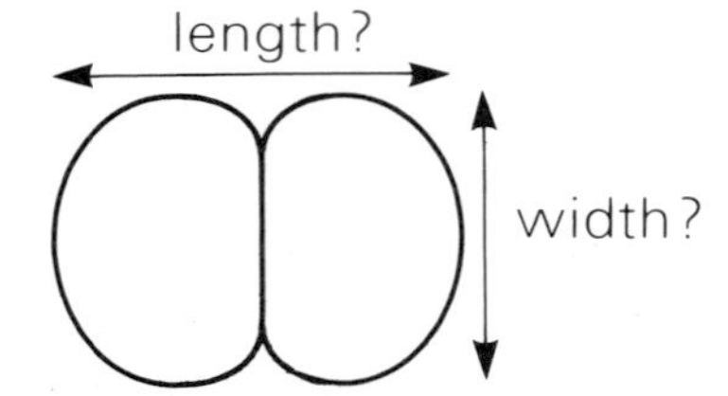

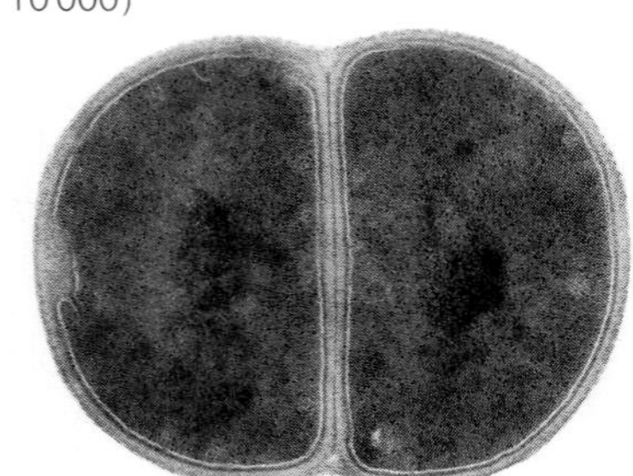
Staphylococcus aureus bacterium (× 50 000)

b) Which of the three bacteria is
(i) the longest? (ii) the shortest?

30

Standard form

D 1 Scientists, engineers and astronomers often have to use very large or very small numbers.

By using powers of 10, they can make these numbers much easier to read and to compare.

A woman running for an hour uses about 1.5×10^6 joules of energy.

The wingbeat of a honeybee needs about 8.0×10^{-4} joules of energy.

For example, 700000000000 is 7×10^{11} 7000000000000 is 7×10^{12}

Here are several ways of writing the amount of energy that the woman athlete uses: 1500×10^3 joules 150×10^4 joules 15×10^5 joules 1.5×10^6 joules 0.15×10^7 joules

a) Write down in four different ways the amount of energy a honeybee uses to beat its wings. Start with 800×10^{-6} joules and end with 0.8×10^{-3} joules.

Take note

Numbers written like this are said to be **in standard form** or **in scientific notation.**

1.5×10^6
8.0×10^{-4}

a number between 1 and 10, including 1 but not 10

a power of 10

Standard form helps us to compare large numbers and small numbers more easily.

b) Which is (i) the largest (ii) the smallest of these numbers?
A 2.7×10^8 B 3.1×10^7 C 2.7×10^9 D 3.1×10^8 E 2.1×10^7

2 Write the number in each of these situations in standard form.

a) In a day's heavy manual labour you would use 17×10^6 joules of energy.

b) A burning match uses 4000 joules of energy.

c) The earth receives 560×10^{22} joules of energy each year from solar radiation.

d) The mass of a carbon atom is about 0.2×10^{-22} g.

e) The sun-grazing comet Ikeya-Seki passed within half a million miles of the sun's surface in October 1965. *not this number!*

f) A cubic millimetre is 0.000 000 001 m^3.

3 a) Write these distances in order, shortest first, in standard form.
13 000 000 m 0.000 001 m 12 000 000 000 000 m
0.0001 m 760 000 000 000 000 000 000 m 0.005 m

b) Match the distances in (a) with these: **the length of an insect**
the diameter of our galaxy the diameter of the earth the diameter of an atom
the diameter of a grain of sand the diameter of the solar system.

Calculators and powers

E For this page you need a calculator with a 'power' key (or function key).

It may be labelled [x^y] or [y^x]

On this page we use [x^y]

1 One of these sequences displays 5^3 and the other displays 3^5.

A [C] [3] [x^y] [5] [=]

B [C] [5] [x^y] [3] [=]

Which is which?

$1^3, 2^3, 3^3, \ldots 30^3$

2 a) Use your calculator to make a list of the first 30 cube numbers.
b) Underline the last digit each time.
What do you notice about the last digits?
c) What is the chance that a cube number chosen at random ends in
(i) 6? (ii) 7? (iii) 5?

Challenges

3 What is the chance that the fourth power of a whole number chosen at random ends in
a) 6? b) 7? c) 5?

4 a) Which of these numbers are the squares of whole numbers, and which are cubes?
3249 729 262144 884736 5206 5929 6085

b) Are any of them both cubes and squares?

5 a) Work out 40 000 × 30 000 in your head and write down the result.
b) Now press [C] [4] [0] [0] [0] [0] [×] [3] [0] [0] [0] [0] [=].
Copy the display.
c) Write one or two sentences to explain how your calculator display and your result in **(a)** are related.

30

Think it through

6 Astronomers deal with extremely large distances, so they use an extremely long unit of length – a light year.
A light year is the distance that light travels in a year, travelling at 300 000 km/s.
Use your calculator to find, as accurately as you can, how long a light year is in kilometres.
Write your result in standard form.

apart from our own sun, that is

7 The earth's speed as it orbits the sun is about 107 220 km/h.
If the earth could be used as a spaceship to travel from star to star, how many years would it take to reach the closest star, Proxima Centauri, 4.22 light years away?

▲ 289
● Next chapter

Constant change

F 1 Pepper's uncle gave her £100.
She invested it in a Building Society account which paid 8% interest each year.

Interest 8% per annum

a) How much was the interest after 1 year?

b) Pepper left the £100 and the interest in the account.
What was the interest on this new amount at the end of the second year?

c) If Pepper continues to take no money out of the account, n years after she was given the £100 she will have a total of £A, where $A = 100 \times (1.08)^n$.
How much will she have after 5 years?

Think it through

d) Will Pepper ever have more than £300 in her account?
If so, after how many years?

e) Is it possible for the account ever to pass the £1000 mark?
Explain why.

2 A roller is used to make a very thin strip of gold foil.
The roller passes many times over the gold foil.
A small piece of gold 10 cm long is placed on the slab.
Each pass of the roller lengthens the piece of gold by 20%.

a) Each pass of the roller multiplies the length of the piece of gold by how much?

b) Copy and complete this rule giving the length, l cm, of the piece of gold after n passes of the roller:

$$l = 10 \times \square^n$$

c) The final length of foil must be greater than 2 metres.
How many passes of the roller are needed in order to produce a piece that long?

3 The school office gets a new carpet, and immediately it begins to fade in the sunlight.
When it had just been laid it had a measured degree of 'brightness' of 16 units.
Each year the brightness fades by 4%.

a) Each year the brightness is multiplied by how much?

b) Copy and complete this rule giving the brightness, b units, of the carpet after n years:

$$b = 16 \times \square^n$$

c) The Local Authority will allow the carpet to be replaced if its brightness falls below 8 units.
After how many years can the carpet be replaced?

Growth

G 1 Imagine that you have a new paper round.
Your boss must be losing his mind because he has offered you

Plan A £10 000 for each day's delivery in July (31 days).
or **Plan B** 1p for 1 July's delivery,
2p for 2 July's delivery,
4p for 3 July's delivery,
8p for 4 July's delivery, and so on, up to 31 July.

a) You have to make a quick decision – in 5 seconds; which plan would you choose?

b) Now work out how much you would earn in July
(i) under plan **A**. (ii) under plan **B**.

c) Would you have made a good choice in **(a)**?

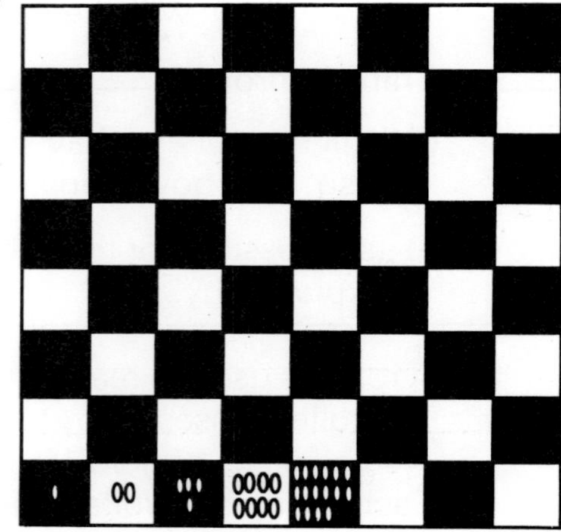

2 Legend tells us that the inventor of the game of chess asked to be paid in rice.
He wanted
1 grain of rice on the first square,
2 grains on the second square,
4 grains on the third square, and so on.

a) How many squares are there on a chess board?

b) How many grains of rice would there be on the last square of the board?

Write your answer as a power.

c) Explain why the total number of grains of rice would be
$1+2+2^2+\ldots+2^{62}+2^{63}$

d) Study this argument:

$$\begin{aligned}
\text{Total} &= 1 + 2 + 2^2 + \ldots + 2^{62} + 2^{63} \\
\text{Total} &= 1 + 2 + 2^2 + \ldots + 2^{62} + 2^{63} \\
\hline
\text{Add:}\quad 2\times\text{Total} &= (2\times 1) + (2\times 2) + (2\times 2^2) + \ldots + (2\times 2^{62}) + (2\times 2^{63}) \\
&= 2 + 2^2 + 2^3 + \ldots + 2^{63} + 2^{64} \\
&= (1+2+2^2+\ldots+2^{62}+2^{63}) + 2^{64} - 1 \\
&= \text{Total} + 2^{64} - 1
\end{aligned}$$

So $\boxed{\text{Total} = 2^{64}-1}$

Use your calculator to find an approximation to this number in standard form.

e) 50 grains of rice weigh about 1 g.
Find the approximate weight of the inventor's desired payment.

in tonnes

Is this more or less than the total world production of rice in 1985, 465 970 000 tonnes?

Radioactive decay

H Assignment

1 Some substances such as uranium are 'unstable' and are said to be 'radioactive'.

a) Find out what you can about radioactive substances.
- Who first discovered them?
- What are alpha, beta and gamma rays?
- Who first suggested alpha, beta and gamma rays existed?
- What are the most well-known radioactive substances?
- How are they detected?
- What are they used for?
- What is 'half-life'?

b) Sketch a graph for radioactive decay against time.

${}^{238}_{92}U \rightarrow {}^{4}_{2}He + {}^{234}_{90}Th$	α-decay
${}^{234}_{90}Th \rightarrow {}^{0}_{-1}e + {}^{234}_{91}Pa$	β-decay
${}^{234}_{91}Pa \rightarrow {}^{0}_{-1}e + {}^{234}_{92}U$	β-decay

2 The half-life of a substance is 2 seconds.
How long will it be before all the radioactivity is gone?

3 The half-life of radium is 1620 years.

a) You start with 1 g of radium. How long will it be before only 0.125 g remains?

b) In the process of radioactive decay, substances pass through various stages.
For example, uranium, as it decays, eventually turns into radium, and finishes as lead, which is a stable element.
There are other stages along the way.
The half-life of uranium is 4.5×10^9 years.
Uranium exists naturally on earth.
Why do you think this is?

4 Some archaeologists use carbon-dating to help them determine the age of the objects they discover.
Living organic matter, such as plants, absorbs radioactive carbon from the atmosphere.
When a plant dies it takes in no more radioactive carbon.
The half-life of radioactive carbon is about 5730 years.
The roof beam of a house is found with an activity of 7.5 counts per second.
A similar amount of living wood from the same kind of tree has an activity of 15 counts per second.
About how old is the roof beam?

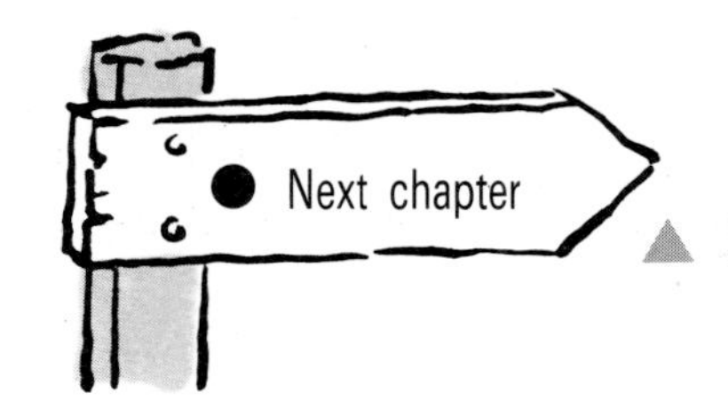

31 Thinking about letters

A 1 This is a side view of a hinge.

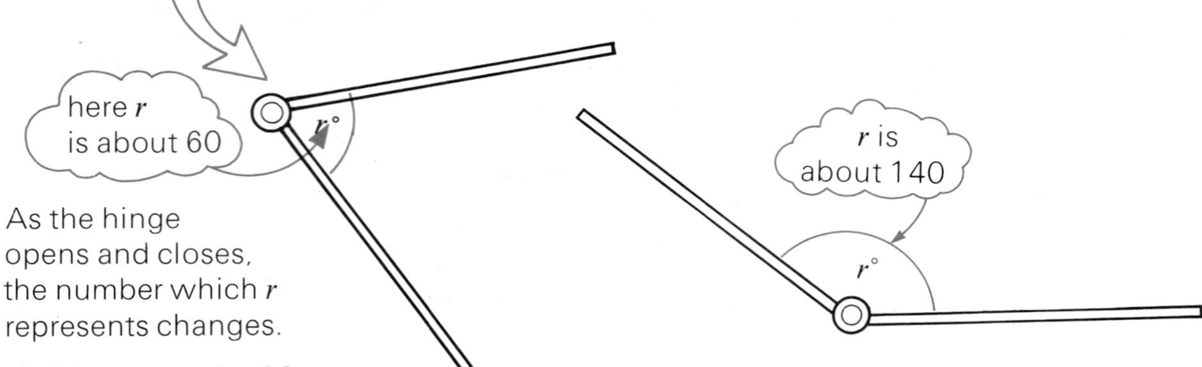

As the hinge opens and closes, the number which r represents changes.

a) Choose r to be 90.
Make a rough sketch of the hinge.

b) Make a rough sketch of the hinge where r represents a number between 180 and 270.
Write down the approximate value of r.

c) This is the hinge opened out to its full extent.
Roughly, what number does r represent here?

Take note

In the hinge the value of r changes.

r represents all the numbers from 0 to about 360.

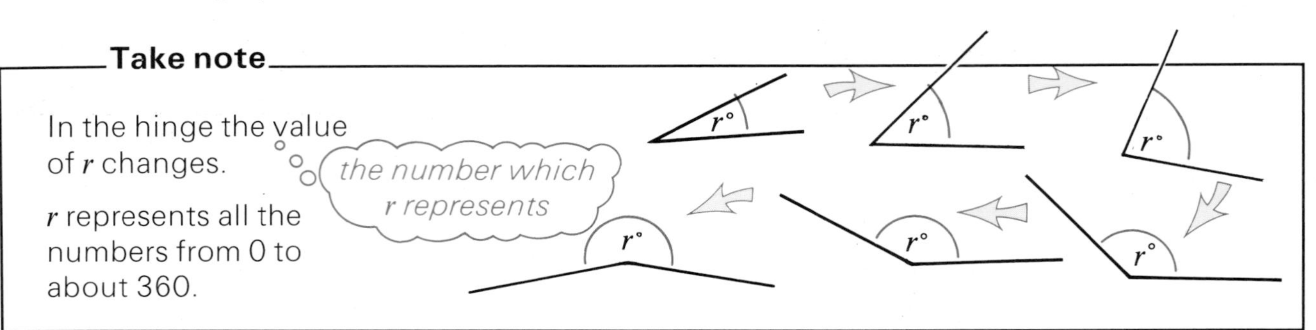

2 These are two pieces of Meccano joined by a piece of elastic.

Think about the Meccano strips opening and closing.

a) Is angle ABC a **constant** or a **variable** angle?

b) Is AB a **constant** or a **variable** length?

c) Is AC a **constant** or a **variable** length?

d) Does r have a **constant** (single) or a **variable** value?

e) Does k have a **constant** (single) or a **variable** value?

f) Does b have a **constant** (single) or a **variable** value?

nut and bolt (not tightened)
B
t cm
Meccano strips
$r°$
b cm
C
k cm
elastic
A

3 a) Winston used these instructions to draw this set of quadrilaterals.

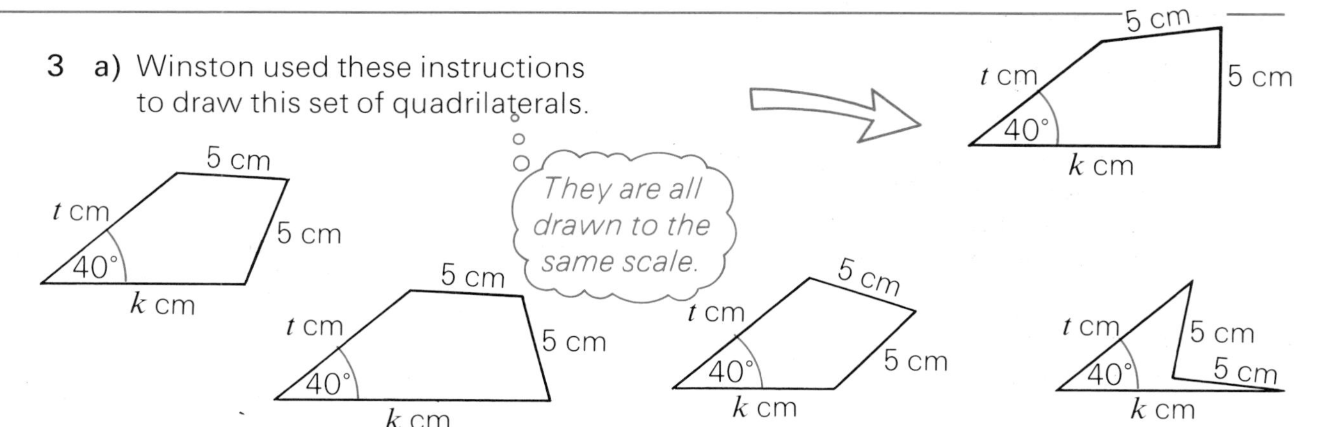

He chose a different value for k in each quadrilateral. Does the value of t in the set of quadrilaterals increase, decrease, or increase and decrease?

b) Make your own freehand sketch of four quadrilaterals for which t remains constant while k increases.

4 This is a set of instructions for drawing triangles.

4 cm, 6 cm, $r°$, k cm

They are all drawn to the same scale.

These are sketches of some possible triangles.

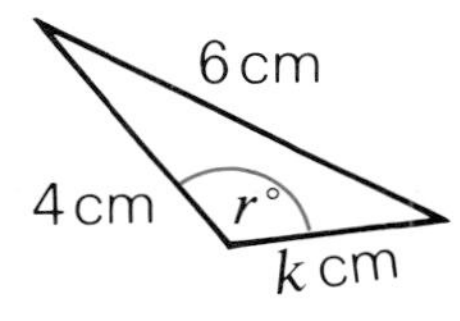

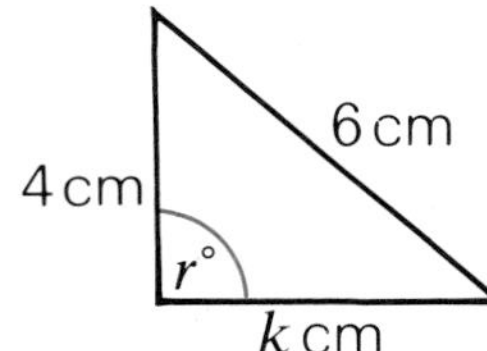

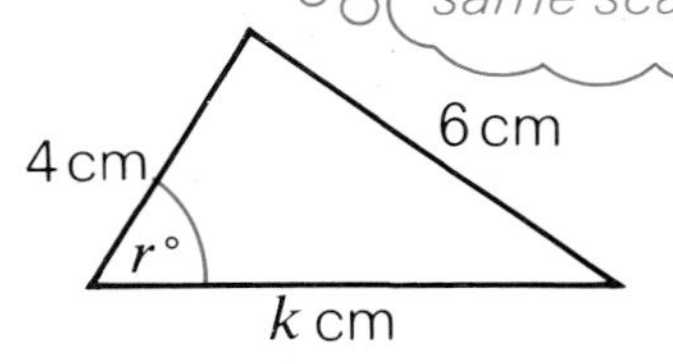

4cm, 6cm, $r°$, k cm

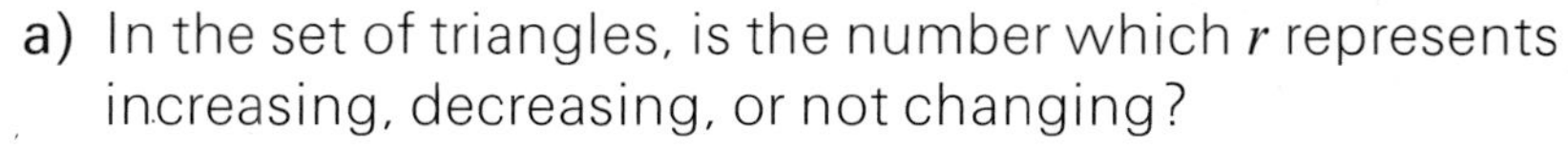

a) In the set of triangles, is the number which r represents increasing, decreasing, or not changing?

b) Is the value of k increasing, decreasing, or not changing?

c) What is the largest possible number which k can represent? Make a sketch of the triangle when this happens.

d) What is the smallest possible number which k can represent? Make a sketch of the triangle when this happens.

Challenge

5 You have four Meccano strips, two long and two short, and four nuts and bolts.
Describe how to make a framework including two angles measuring $r°$ and $s°$ such that as the value of r increases
the value of s decreases,
and as the value of r decreases
the value of s increases.
Make sketches to illustrate what you have in mind.

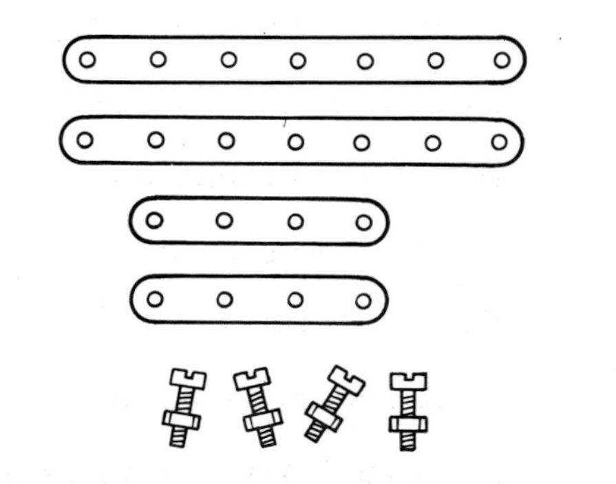

B

Take note

Letters can be used to represent **constant** values for angles, distances, . . . or **variable** values for angles, distances, . . .

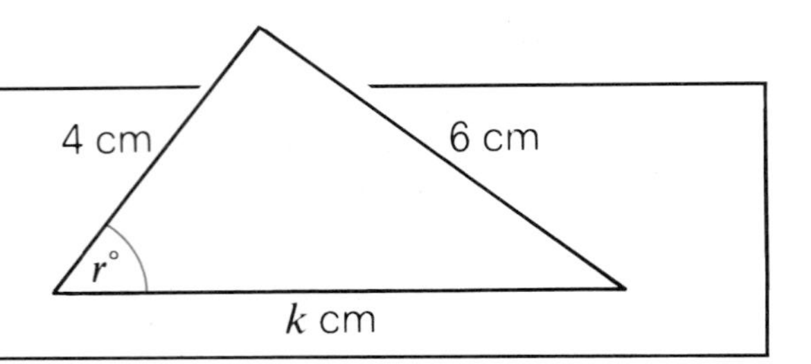

1 In the triangle ABC, angle BAC is a variable angle and length BC is a variable length.

a) What is the largest possible value that b can take?

b) What is the smallest possible value that b can take?

c) What is the value of r when b has its largest value?

d) What is the value of r when b has its smallest value?

e) What range of values can b take?

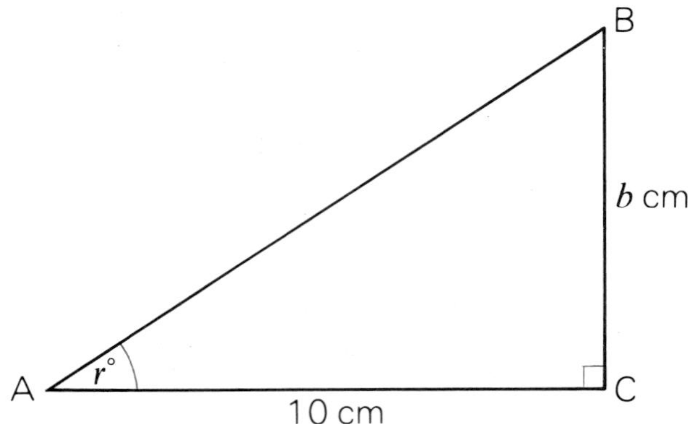

With a friend

2 Study these figures carefully.
Some lengths and angles are represented by letters.
You are not told if the letters represent variable values or constant values . . . but you should be able to decide if a variable value or only a single value is possible.
Discuss each figure.
Each of you write down what you decide for each letter.

For example, m represents a constant value . . r represents . . .

a)

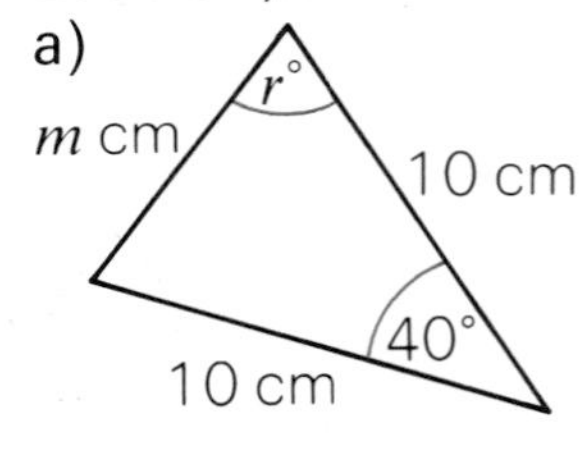

b)

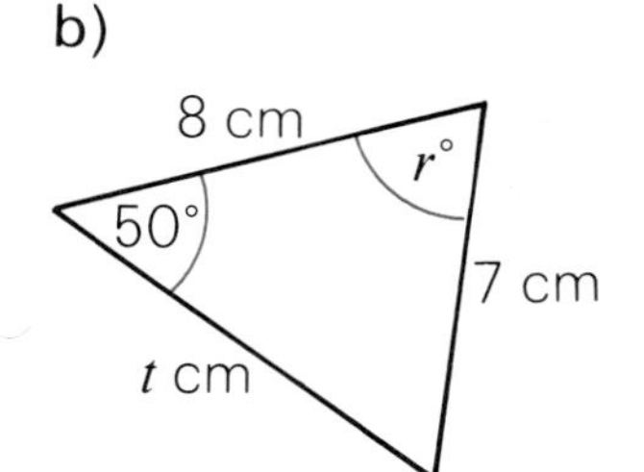

c)

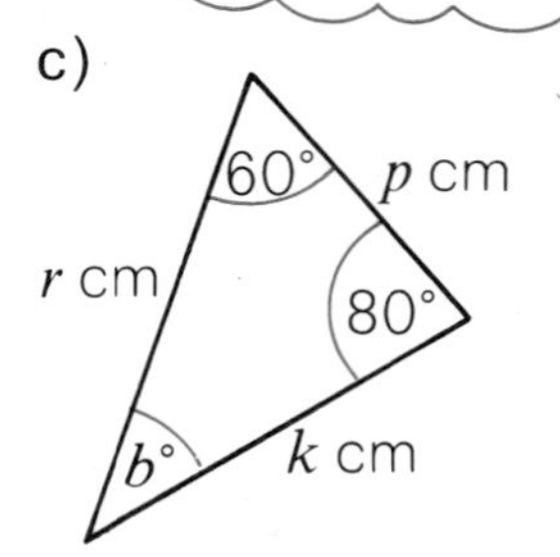

d)

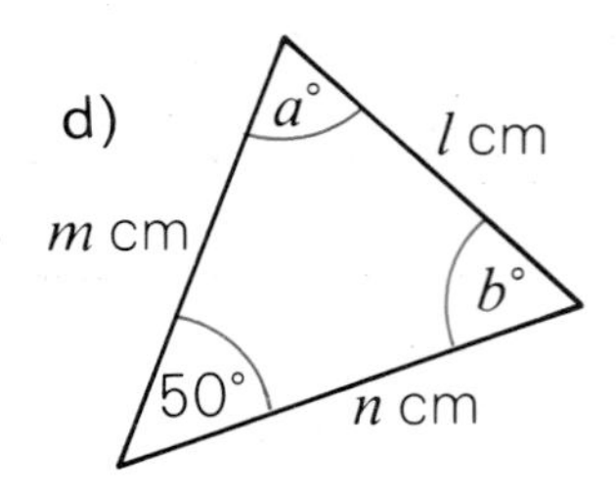

e)

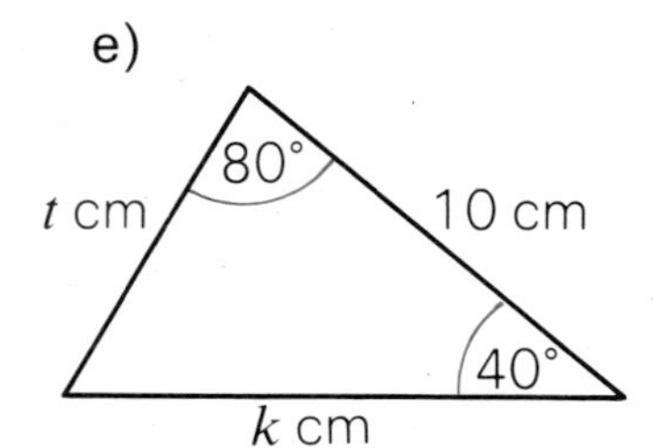

f)

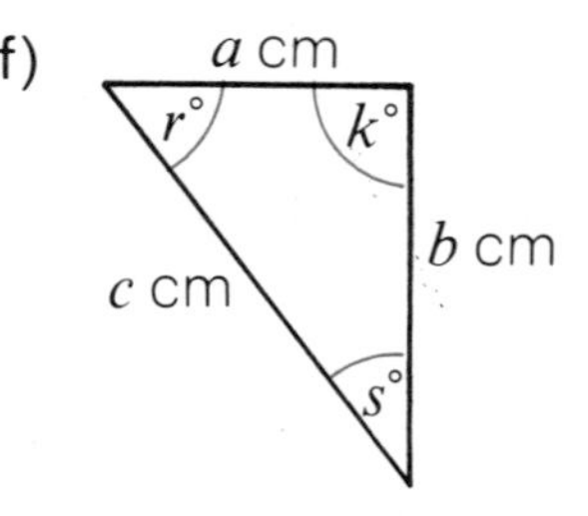

Stay with your friend for the next question.

With a friend

3 Describe an experiment which you could carry out for each of these. Explain exactly what you would do, and write down what result you think you would get.

A T seconds represents the time for one full swing (there and back) of this pendulum. l metres represents the length of the string. As the length of the string varies, is T a constant or a variable?

B W kg represents the mass of the bob on the end of the string. Keep the length of the string constant, but vary the mass of the bob. Is T a constant or a variable?

C $S°$ represents the angle of swing. Keep the length of string and the mass of the bob constant. Vary the angle of swing. Is T a constant or a variable?

D D seconds represents the time a sugar lump takes to dissolve in water. H°C represents the temperature of the water. As the water temperature varies, is D a constant or a variable?

E F seconds represents the time it takes for ice to form on the surface of a pond. T°C represents the air temperature; d metres represents the depth of water in the pond. If the air temperature remains the same, but the depth of water varies, is F a constant or a variable?

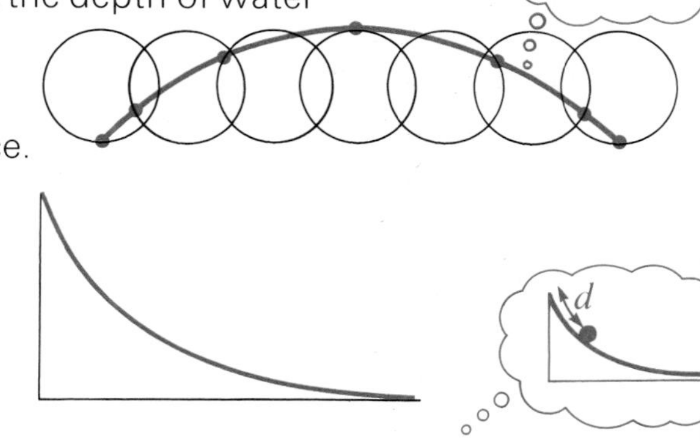

F This is the path that a point on a wheel follows as the wheel rolls along a flat surface. The curve can be used as a children's slide. T seconds represents the time taken for a child to complete the slide. W kg represents the mass of the child. For children of different weights, is T a constant or a variable?

G d metres represents the distance down the slide from which the child starts. As the same child starts from different parts of the slide, is T a constant or a variable?

Assignment

4 Choose one or more of the situations in question **3**. Collect or make the apparatus you need, and carry out the experiment(s) you suggested. Draw a graph to represent your results. Write a report about what you find out.

C 1 The stepladder is gradually slipping!

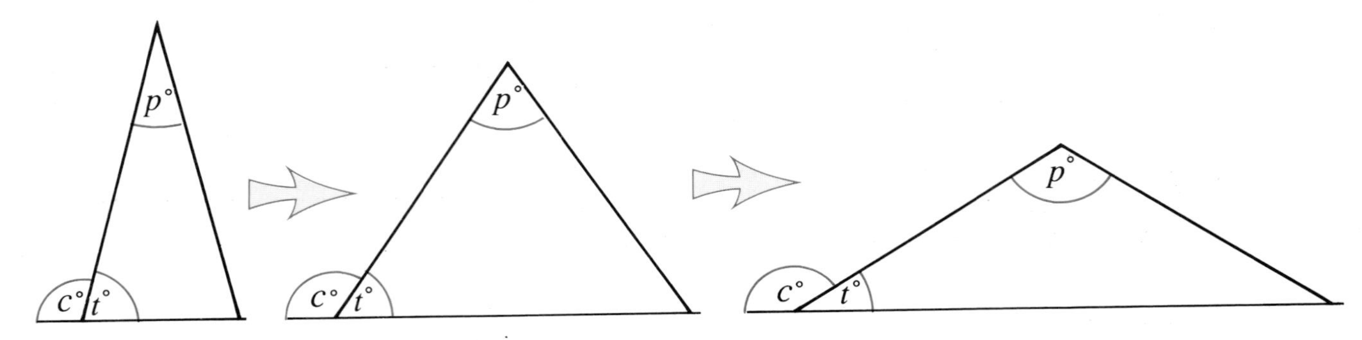

a) Write down whether the numbers which each letter represents are increasing, decreasing or remaining constant.

b) This is a rule connecting c and t for all possible positions of the ladder. Copy and complete it.

$c + t = \square$

c) Copy and complete this rule for all possible positions of the ladder:

$p + (2 \times \square) = 180$

d) . . . and this rule:

$c = \square + t$

2 The ladder is sliding down the wall!

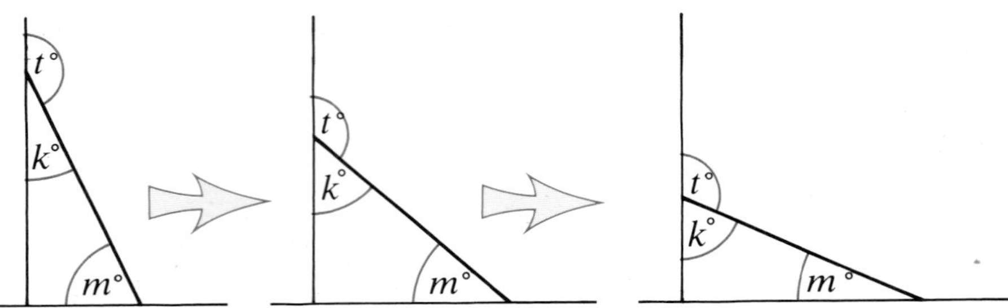

These rules are correct for each diagram.
Copy and complete them.

a) $k + m = \square$ b) $t = m + \square$

31

With a friend

3 Think about these two rules.

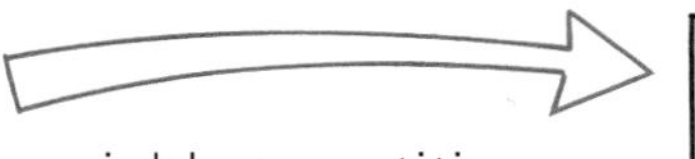

A	$100 \times k = p$
B	$(2 \times k) - 1 = p$

a) Assume p and k both represent variable quantities. Think of a situation which each rule might represent. Each of you write down what you decide.

b) Try to think of a situation in which p and k represent constant quantities. Each of you write down what you decide.

4 This is the plan of a lawn.

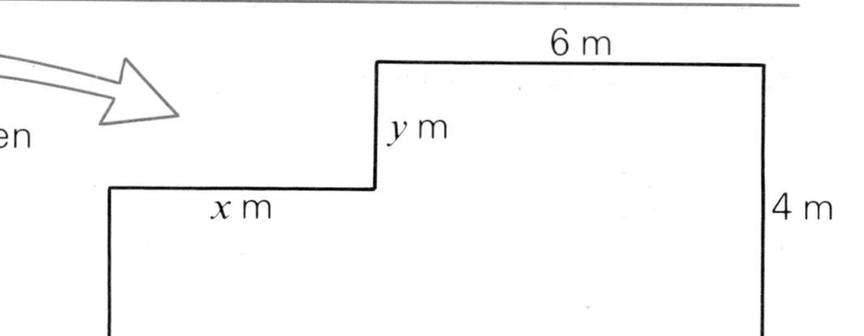

a) Make a rough sketch of the lawn when
 (i) x is 2 and y is 2.
 (ii) x is 4 and y is 0.
 (iii) x is 0.

b) What is (i) the largest
 (ii) the smallest
value which x and y can take in this situation?

5 This is a set of instructions for drawing triangles.

p°
5 cm
a cm
q°
k cm

a) Sketch, roughly to scale, **two** possible triangles for which a is 5.
What special kinds of triangles are they?

b) Sketch, roughly to scale, all the possible triangles for which a and k are both 5.
Say what special kind of triangle each possibility is.

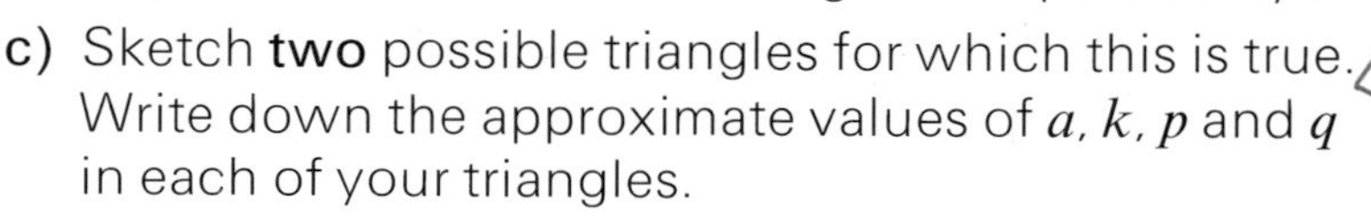

c) Sketch **two** possible triangles for which this is true.
Write down the approximate values of a, k, p and q in each of your triangles.

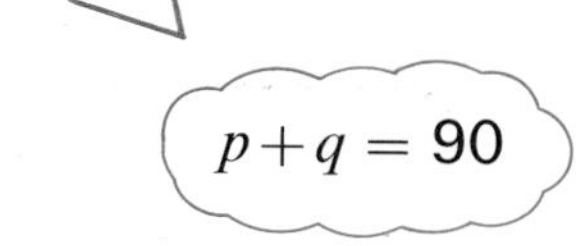

d) Choose a to be 8 and k to be 15.
What happens?
Write one or two sentences to explain.

6 This is a set of instructions for drawing kites.

a) Sketch, roughly to scale, **two** possible kites for which k is 7.

b) Sketch, roughly to scale, all the possible kites for which a is 130.
Say what special kind of kite each possibility is, and write down the value of k.

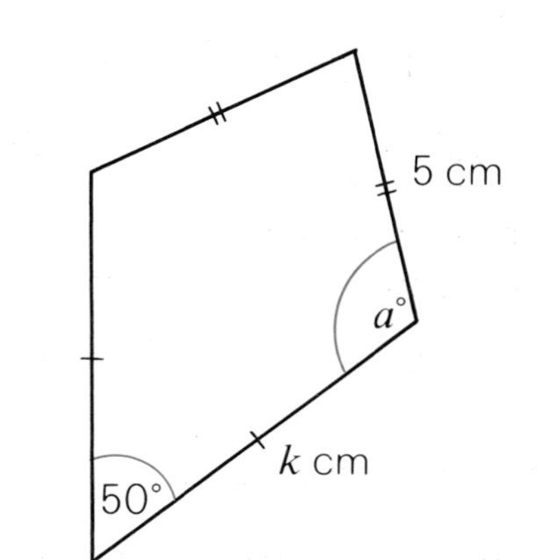

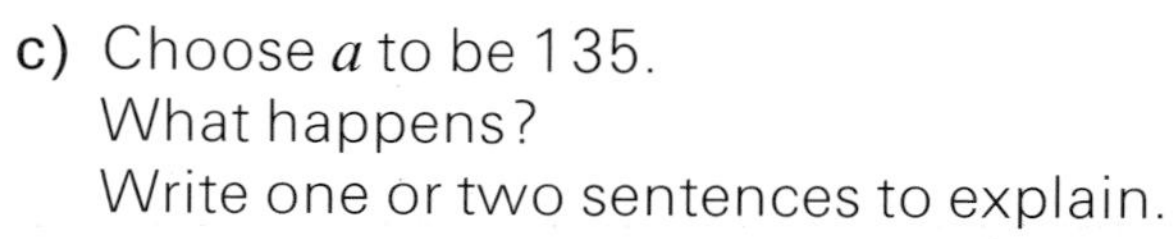

c) Choose a to be 135.
What happens?
Write one or two sentences to explain.

d) Choose k to be 15.
What happens?
Write one or two sentences to explain.

Challenge

e) By making some accurate drawings, find the greatest possible value which k can take for the kite.

31

7 These are instructions for drawing quadrilaterals.

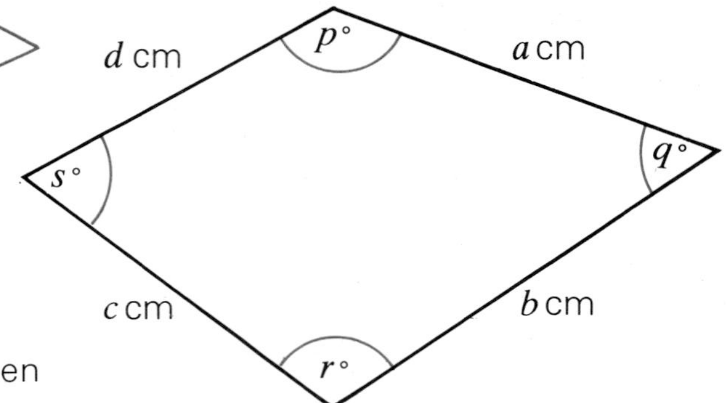

a) Sketch a possible quadrilateral when the values of a, b, c and d are all equal, and the values of $p, q, r,$ and s are all equal.
What special kind of quadrilateral is this?

b) Sketch **two** possible quadrilaterals when a, b, c and d all have the same value.
In one of your quadrilaterals choose r to be 90.
In the other, choose r to be less than 90.
Name the two types of special quadrilaterals that you have drawn.

c) Sketch **two** possible quadrilaterals for which

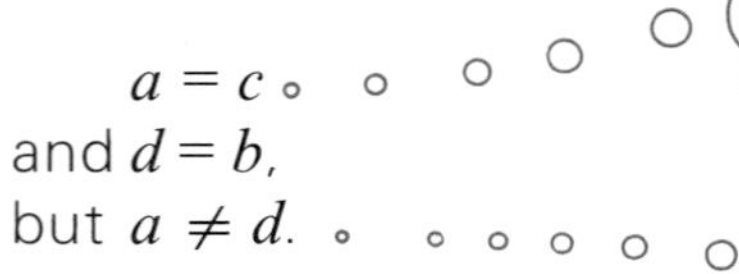

$a = c$

and $d = b$,

but $a \neq d$.

the value of a is equal to the value of c

the value of a is ***not equal*** *to the value of d*

In one of your quadrilaterals, choose p to be 90.
In the other, choose p to be greater than 90.
Name the two types of quadrilaterals that you have drawn.

d) Sketch **two** possible quadrilaterals for which this is true.

$p + q = 180$

In one of your quadrilaterals, choose r to be 90, and b to be equal to a.
In the other, choose r to be larger than 90 and b to be equal to a.
Name the two types of quadrilaterals that you have drawn.

31

Think it through

e) Draw a quadrilateral in which the letters obey all of these rules.

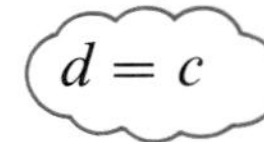

$a = b$

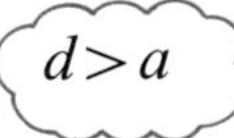

What special kind of quadrilateral have you drawn?

f) What will be true about the values of a, b, c, d, p, q, r and s if the quadrilateral is
(i) a kite? (ii) a trapezium? (iii) an isosceles trapezium?

Exploration — *Quadrilaterals*

8 a) Draw two different quadrilaterals for which both of these are true.

$$p + r = 180$$

$$s + q = 180$$

Is it possible to draw a circle which encloses each of your quadrilaterals like this?

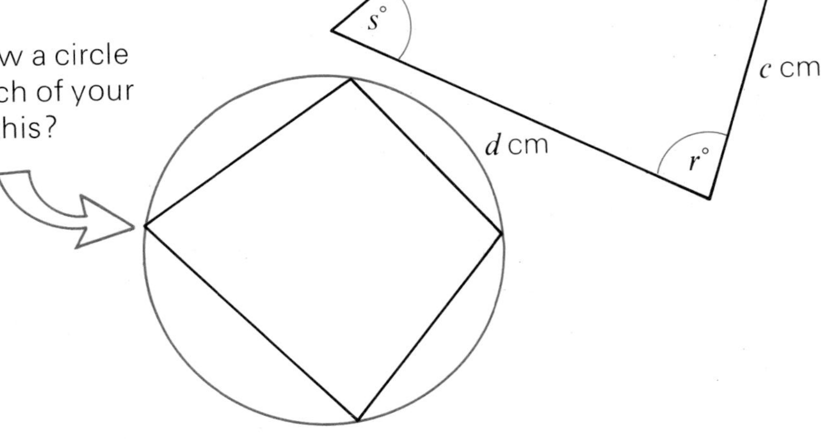

Investigate some more quadrilaterals for which the same two rules are true.
Write down what you discover.

b) Investigate quadrilaterals for which this rule is true.

$$p + q = 180$$

What rule connects s and r in each of your quadrilaterals?
What is special about all the quadrilaterals which obey the rule?

c) Investigate concave quadrilaterals for which

$$p = 360 - (2 \times r)$$

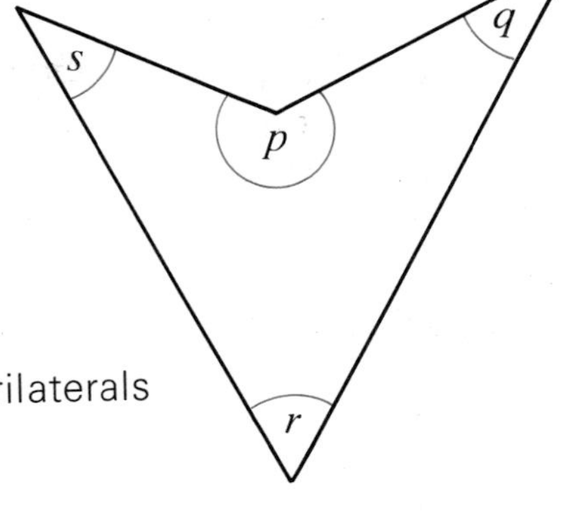

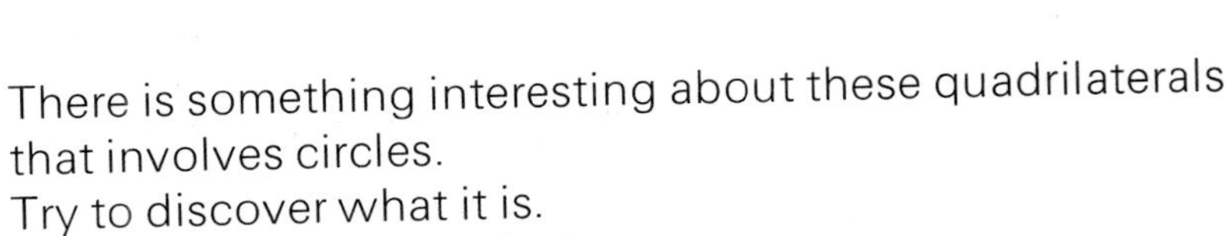

There is something interesting about these quadrilaterals that involves circles.
Try to discover what it is.
Write about what you find out.

d) Choose some rules of your own which connect the lengths of sides or sizes of angles of quadrilaterals.
Write about the special properties of the sets of quadrilaterals which these produce.

Diagrams and graphs

D 1 a) p and q are variables.
The diagram shows how the values of p and q vary together.

As the values of p increase, how do the values of q change?
Do they decrease steadily,
decrease and then increase,
or increase steadily?

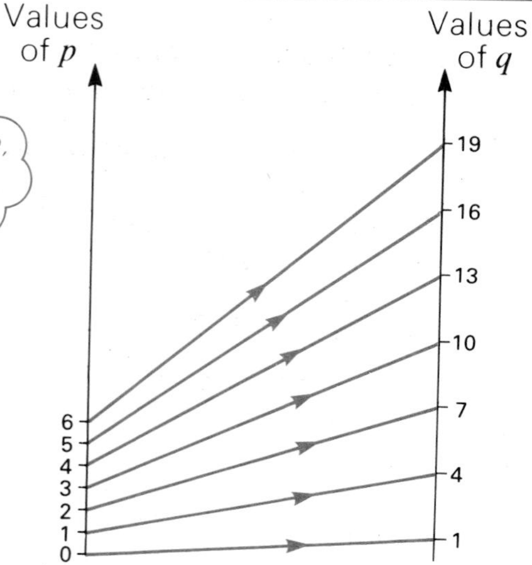

b) (i) This is the rule which connects the values of p and the values of q.
$q = (\square \times p) + \square$
Copy and complete it.
(ii) Write the rule like this: $p = \ldots$
(iii) Write the rule in a third way, different from those in (i) and (ii).
(iv) What is the value of q when p is $\frac{1}{2}$?
(v) What is the value of p when q is $\frac{1}{2}$?

c) Here is another way of representing the relationship between p and q.
Copy and complete it.

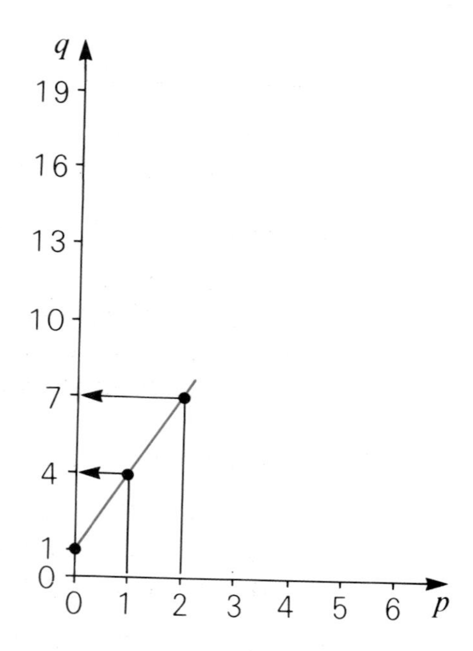

d) r and s are variables.
Here are some pairs of values of r and s.
(0, 0) (1, 1) (2, 4) (3, 9) (4, 16) (5, 25)
(i) The values of r and s are connected by a simple rule.
Sketch diagrams like those in **(a)** and **(c)** to show the relationship between r and s.
(ii) Are the values of r and s variable or constant?
(iii) What is the value of r when s is 100?
(iv) What is the value of s when r is $\frac{1}{2}$?

e) Repeat the first two parts of **(d)** for these pairs of values of r and s.
$(1, \frac{1}{2})$ $(2, \frac{1}{2})$ $(3, \frac{1}{2})$ $(4, \frac{1}{2})$ $(5, \frac{1}{2})$ $(6, \frac{1}{2})$ ...

Challenge

2 p and q are variables.
Here are two pairs of values of p and q: (1, 2) (2, 4).
The values of p and q are connected by a simple rule.
Write down **two** different possibilities for the rule.
Show that your rules really are different by drawing a graph like the one in question **1 (c)** for each one.

3 Each diagram represents the relationship between two sets of numbers, one set represented by r and the other by s.

a) For each diagram decide whether r and s are variables or constants.

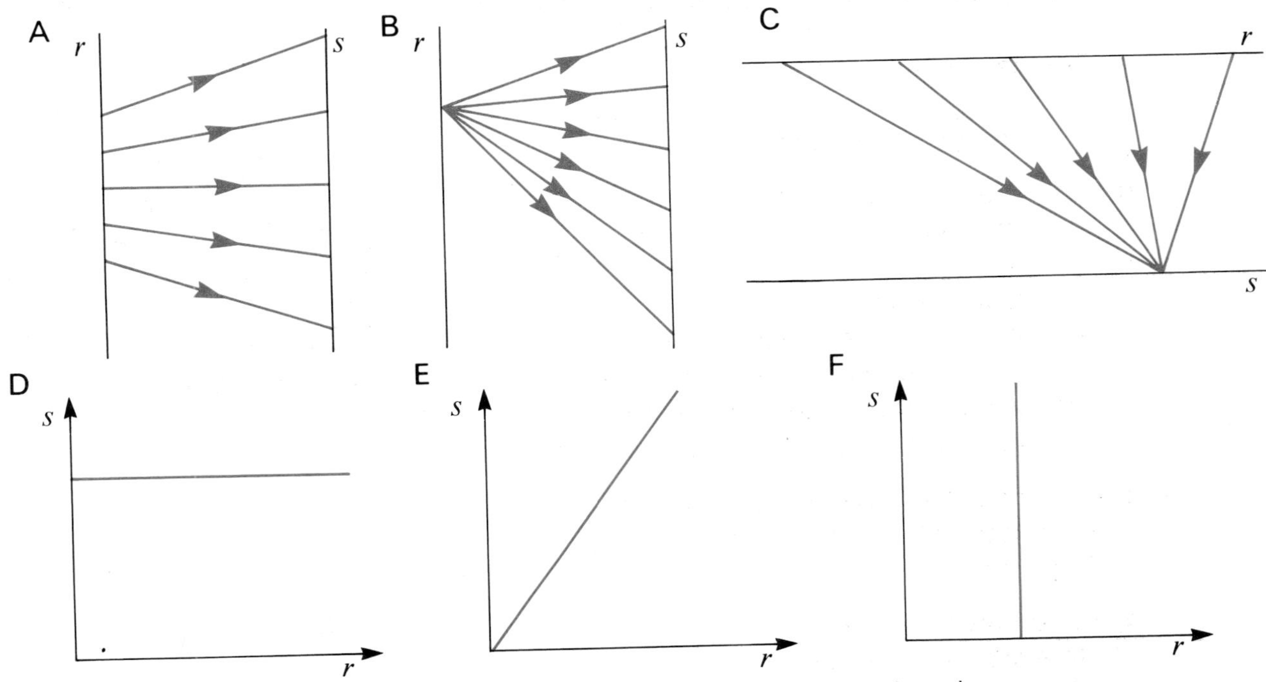

b) Sketch your own 'parallel-line' diagram to show these situations:
(i) r increases while s decreases. (ii) r and s are constant.

c) Sketch a graph for each situation in **(b)**.

d) Diagram **B** might represent the steady speed (r km/h) of a car at various times (s hours) after the start of a journey.
Write down a possible situation for each of the other diagrams.
In each situation explain what each of the letters represent.

4 A car is travelling around a circular racetrack.

a) Represent the relationship between the distance of the car from the centre of the track and the time it has been travelling on
(i) a parallel-line diagram. (ii) a graph.

b) The speed of the car is gradually increasing.
Represent these relationships on parallel-line diagrams and graphs:
(i) its distance from the centre of the circle and its speed.
(ii) the time it has been travelling and its speed.
(iii) its distance from the centre of the circle and the distance it has travelled along the track.
(iv) the time it has been travelling and the distance it has travelled along the track.

True, false or neither?

E 1 a is a variable, which can take any value.

a) One of these statements is true for all possible values of a. Which one?

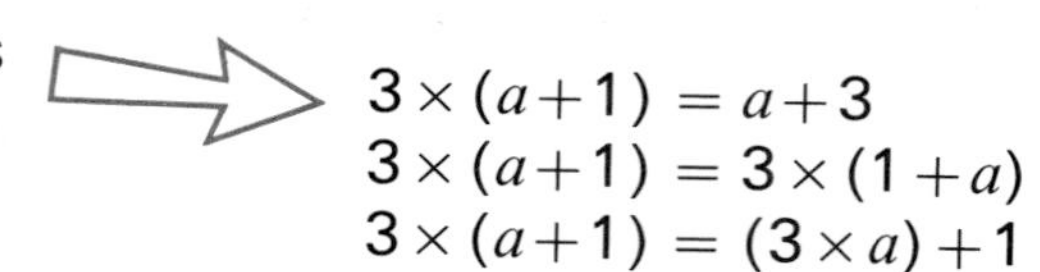

$3 \times (a+1) = a+3$
$3 \times (a+1) = 3 \times (1+a)$
$3 \times (a+1) = (3 \times a)+1$

b) Which one is false for all possible values of a?
c) Which one is true for some and false for other values of a?

2 Write down two statements of your own, involving a variable (say k), which are
a) true for all possible values of k.
b) false for all possible values of k.
c) true for some values and false for some values of k.

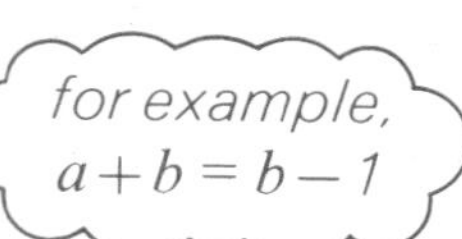

3 Use a and b as variables.
Write down a statement in which a and b both appear which is
a) true for all possible values of a and b.
b) true for some pairs of values of a and b.
c) true for no pairs of values of a and b.

Take note

Statements belong to three types:

always true	sometimes true	never true
Examples: $b+1=1+b$	Examples: $b+2=3$	Examples: $b=b+1$
$a+b+3=3+b+a$	$a+b=5$	$a+b-2=a+b$

Exploration

4 a, b and c are variables.

a) Decide which of these statements are always true.

(i) $a+b=b+a$
(ii) $a-b=b-a$
(iii) $a \times b=b \times a$
(iv) $a \div b=b \div a$
(v) $(a+b)+c=a+(b+c)$
(vi) $a-(b+c)=a-b-c$
(vii) $a \times (b+c)=(a \times b)+(a \times c)$
(viii) $a \times (b-c)=(a \times b)-(a \times c)$
(ix) $\frac{a}{b}+\frac{c}{d}=\frac{a+c}{b+d}$
(x) $\frac{a}{b} \times \frac{c}{d}=\frac{a \times c}{b \times d}$
(xi) $\frac{a}{b} \div \frac{c}{d}=\frac{a \div c}{b \div d}$
(xii) $\frac{a}{b} \div \frac{c}{d}=\frac{a}{b} \times \frac{d}{c}$

b) For each statement you reject, give an example to show why you rejected it.

31

32 Stretching and shearing

A 1 This is a drawing of Glenda's cousin Nancy. It is on a 1 cm squared grid.

Glenda decides to change the grid.

a) Copy the new grid. (Use 1 cm squared paper.)

b) Copy and complete the new drawing of Nancy.

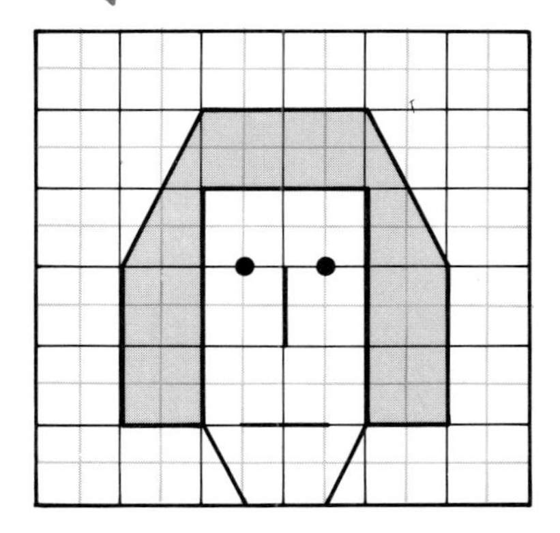

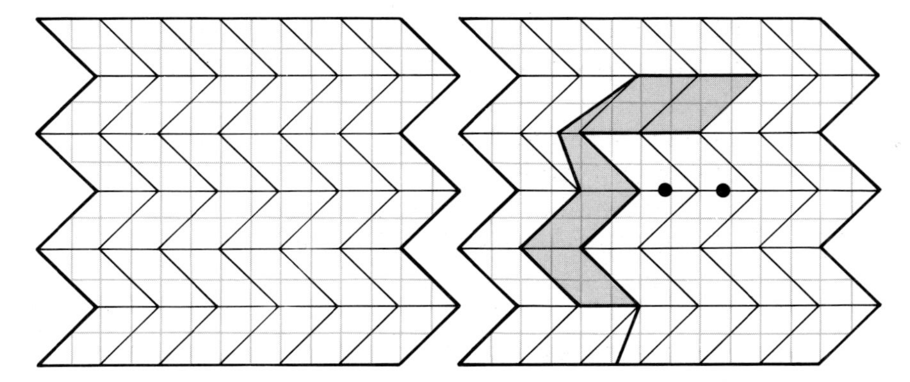

2 Glenda changes the grid again.

a) Copy this new grid.

b) Draw Nancy.

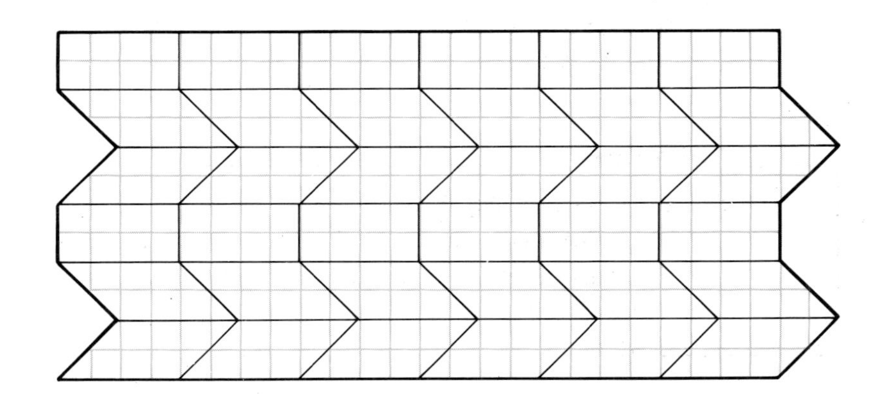

Challenge

3 Draw the grid Glenda used to get this drawing of Nancy.

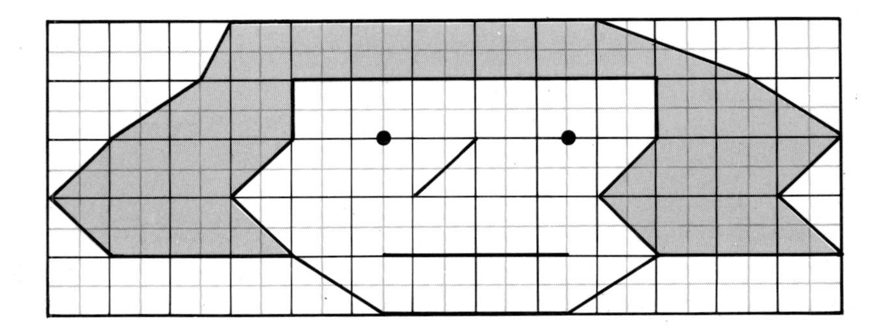

32

4 Here is a grid with a drawing of Sid (a friend of Nancy).
Here are some new versions of the grid and Sid.
Write down which goes with which.

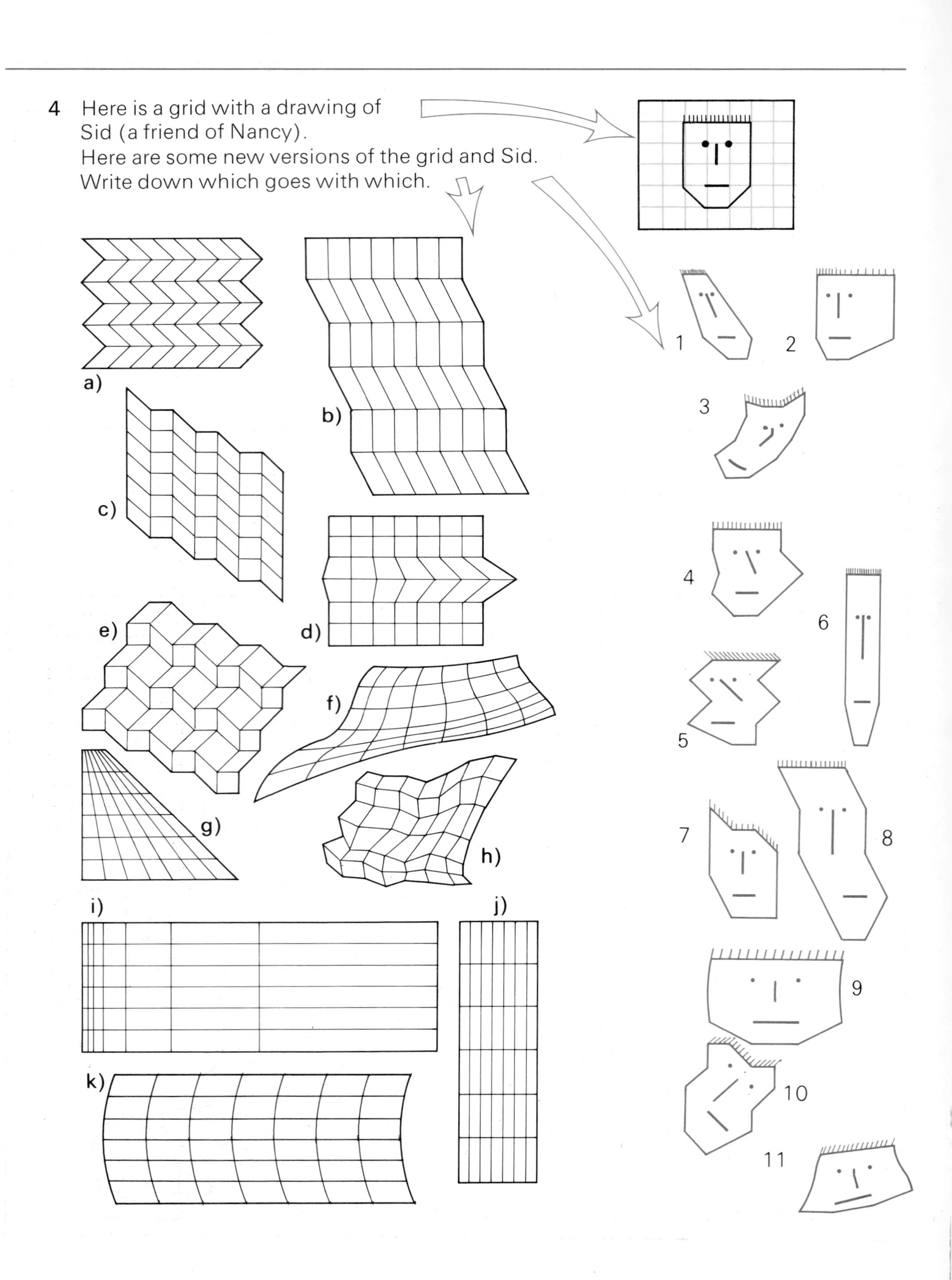

Challenge

5

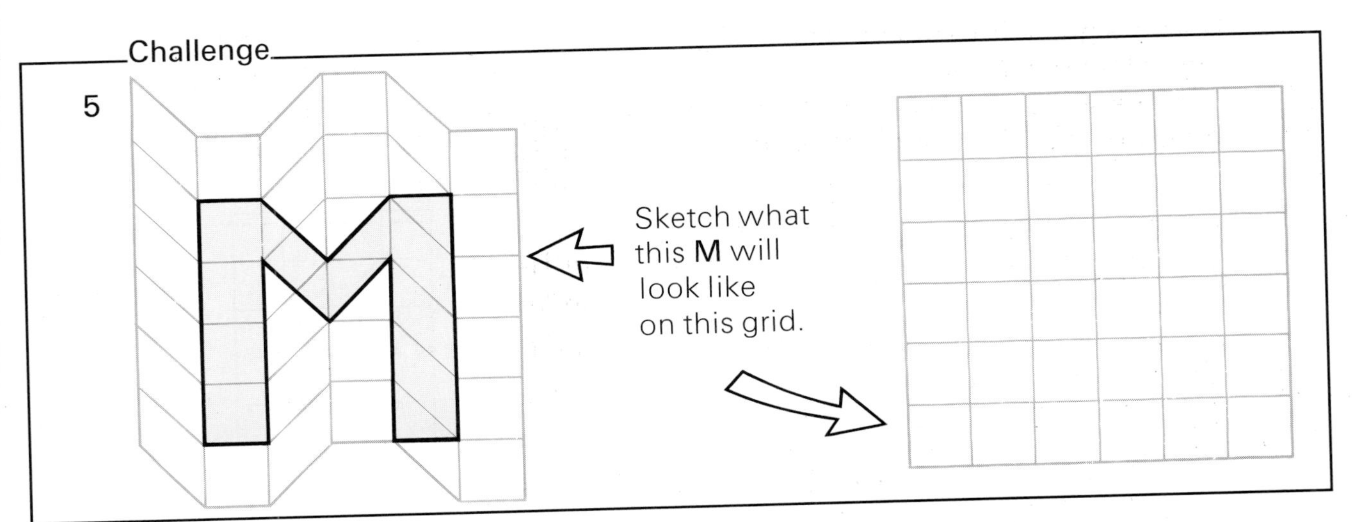

Sketch what this **M** will look like on this grid.

With a friend

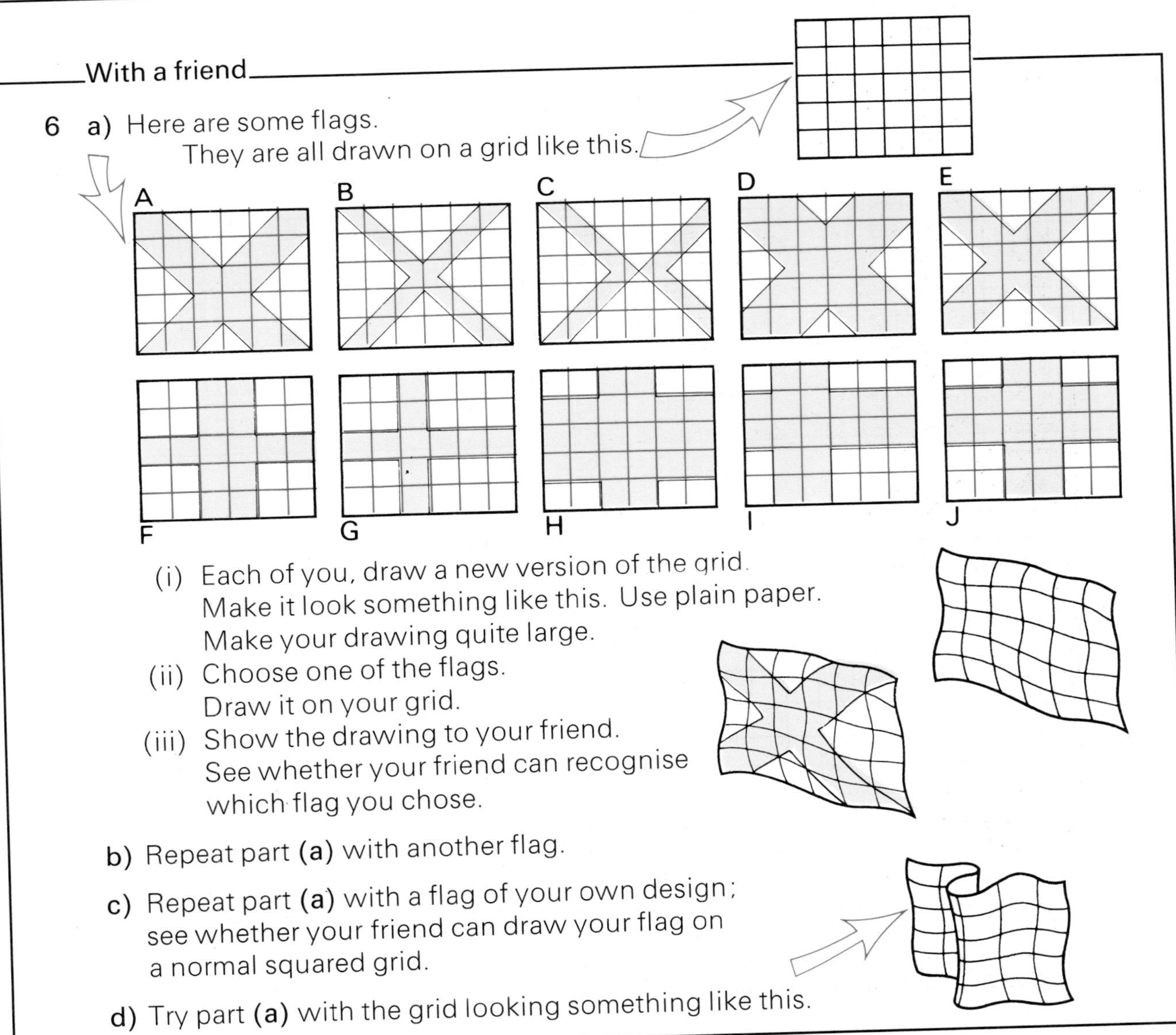

6 **a)** Here are some flags.
They are all drawn on a grid like this.

(i) Each of you, draw a new version of the grid.
Make it look something like this. Use plain paper.
Make your drawing quite large.

(ii) Choose one of the flags.
Draw it on your grid.

(iii) Show the drawing to your friend.
See whether your friend can recognise which flag you chose.

b) Repeat part **(a)** with another flag.

c) Repeat part **(a)** with a flag of your own design; see whether your friend can draw your flag on a normal squared grid.

d) Try part **(a)** with the grid looking something like this.

32

One-way stretch

B 1 Here is an **L** on a squared grid.
Here are some new versions of the grid and the **L**.
In this drawing the grid has been stretched **evenly** and in just **one** direction.
List the other **three** drawings where this has happened.

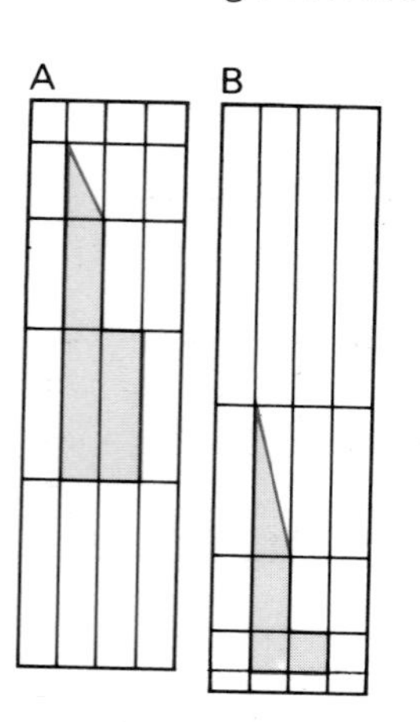

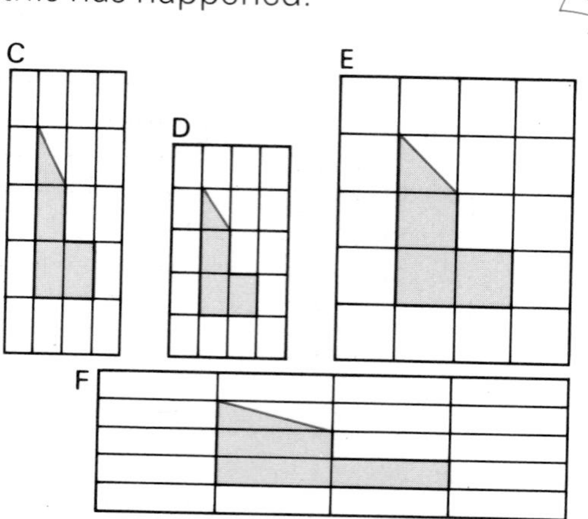

Take note

An even stretch in one direction is called a **one-way stretch**.
This is a one-way stretch. This is not: . . . neither is this.

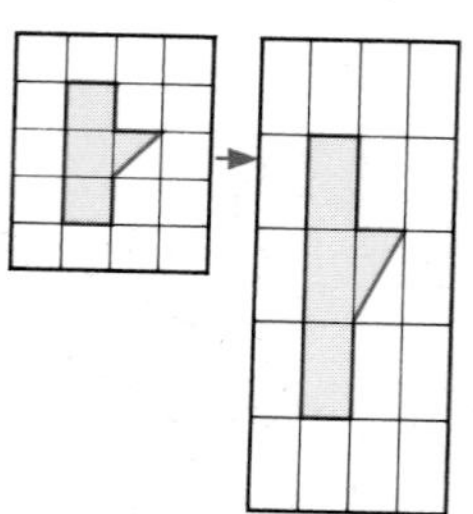

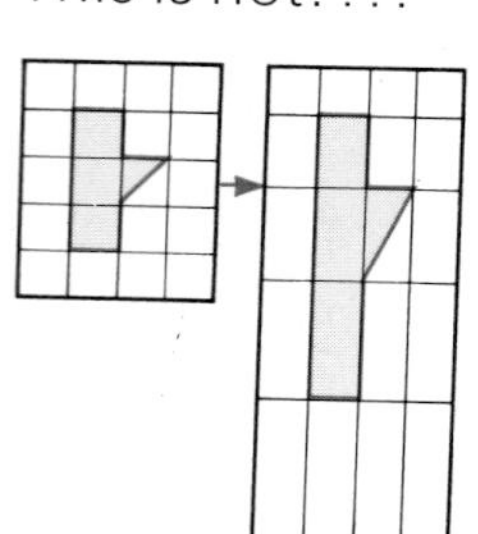

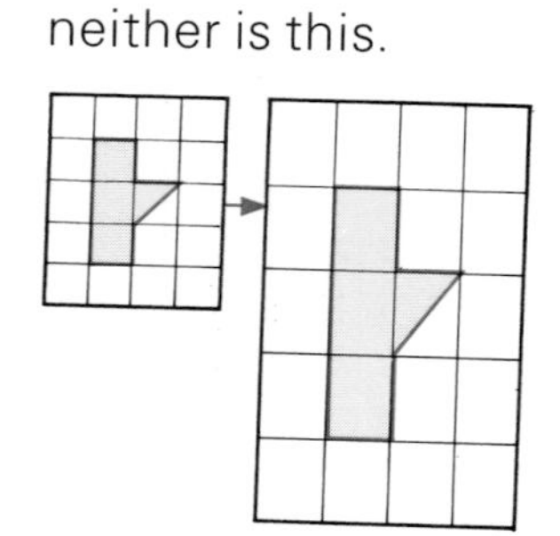

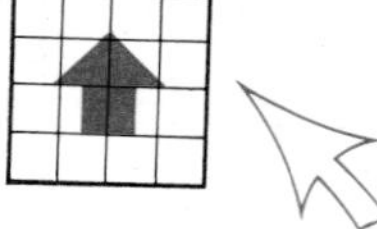

2 Here is an arrow on a grid. Here are some new versions of the grid.

A B C D

a) Which **two** grids show a one-way stretch?
b) Draw the new arrow for these two grids. Use 5 mm squared paper.

3 Here is a letter **F**.
Four of these drawings show a one-way stretch of the **F**.
List them.

4 Here are six one-way stretches of this letter **T**.
Copy and complete the drawings.
Use 5 mm squared paper.

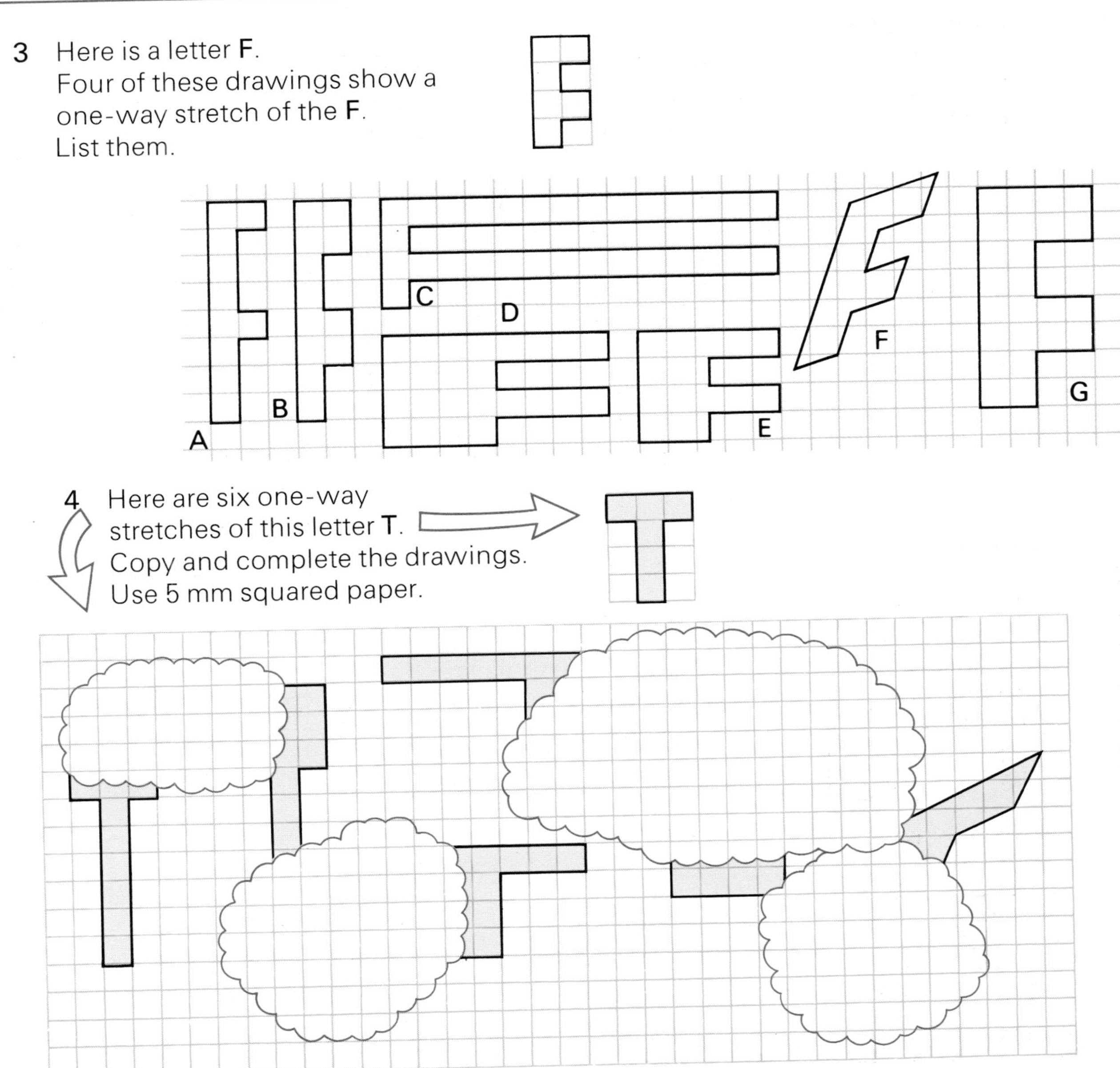

Challenge

5 Here is an arrow on a grid.
The grid is given a one-way stretch.
Draw the new shape of the arrow.

6 Here is another arrow.
Copy and complete this one-way stretch.

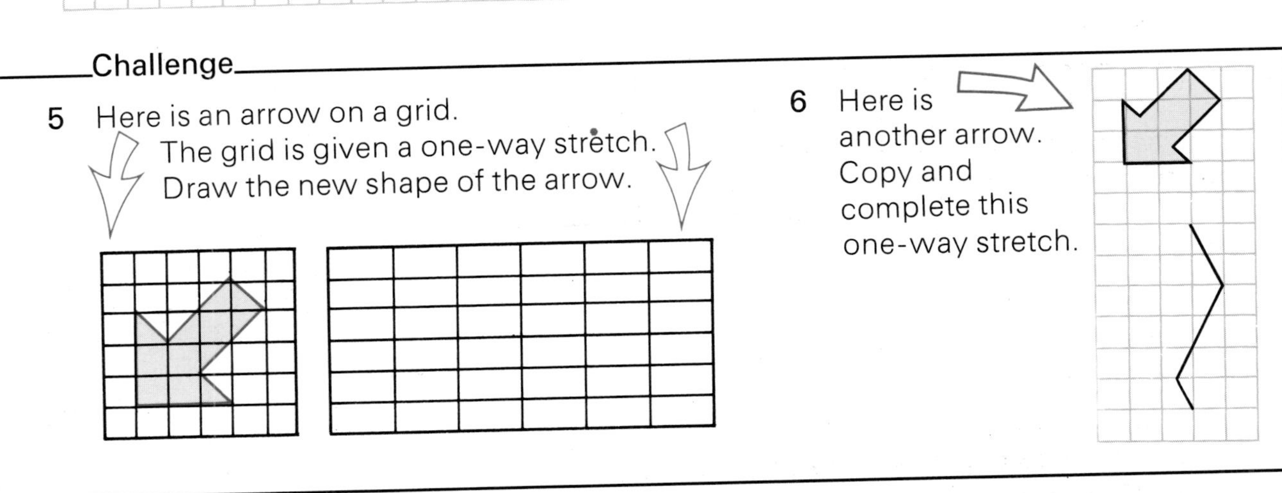

Exploration

7 You need some 5 mm squared paper.

a) A one-way stretch is applied to a 5 mm squared grid so it looks like this.

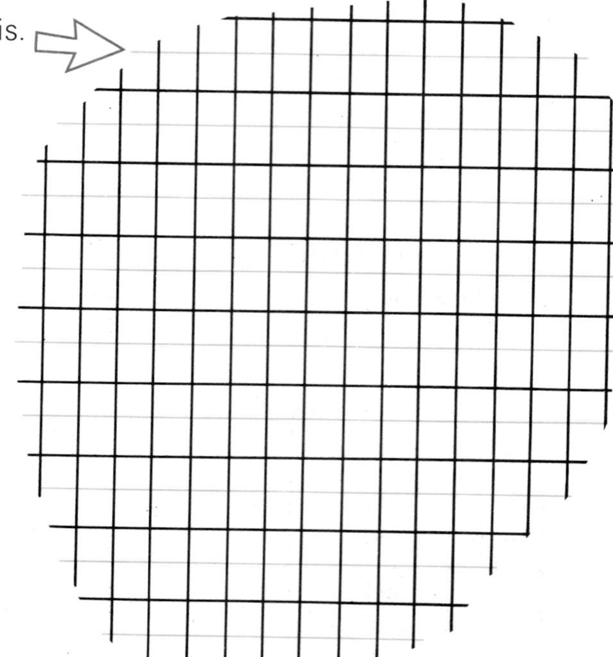

Draw the stretched grid. Make it cover about half a sheet of A4.

b) (i) Draw some squares on the original 5 mm grid.

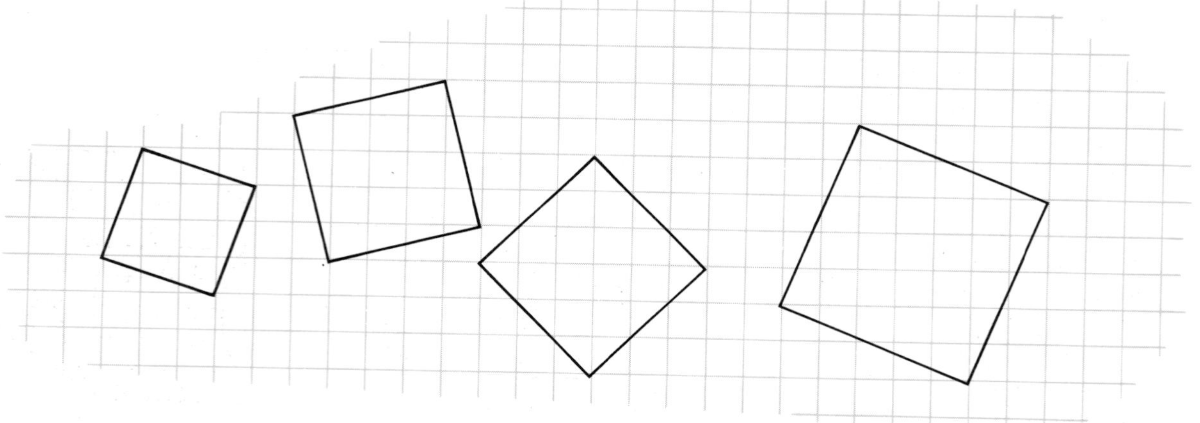

(ii) Now draw stretched versions of the squares on the stretched grid.

c) What kind of shape are your stretched squares?

d) Apply some different one-way stretches to your original squares. Do you still get the same kind of shape?

e) Instead of squares, use rectangles, parallelograms, or trapeziums as your starting shape. Name the kinds of shapes that you get in each case.

Piles of paper

C 1 Horace has a huge pile of sheets of paper on his desk.

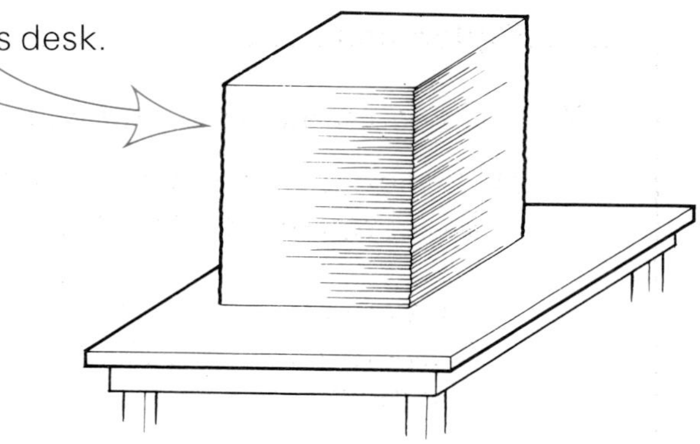

He draws a grid on a vertical face.

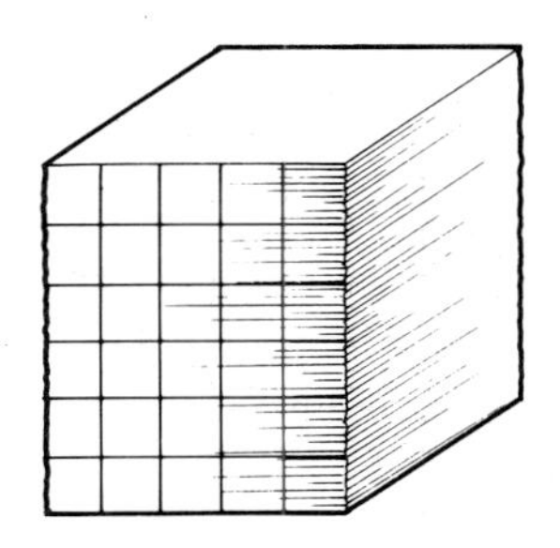

He now draws a picture of Sid on the grid.

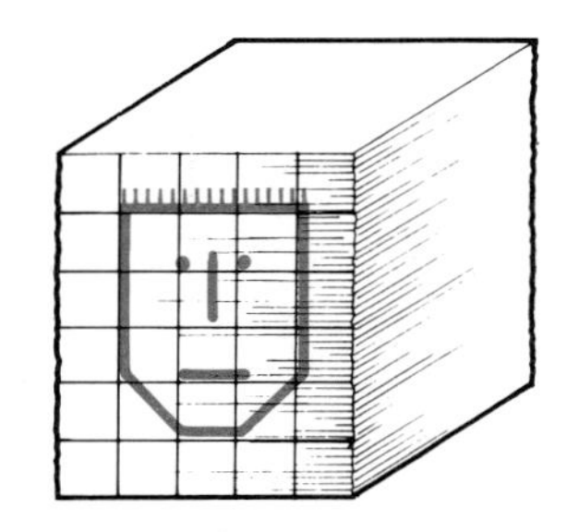

Now he slides the sheets sideways (to the left or the right, but not forwards or backwards).

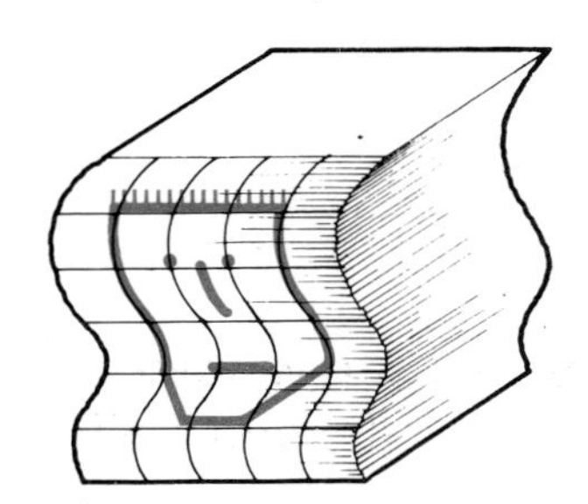

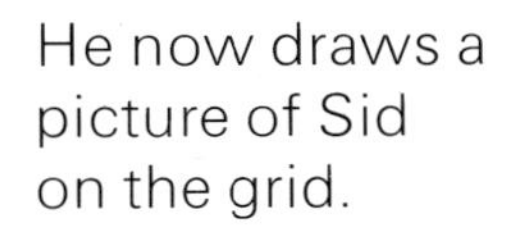

Sketch the picture of Sid when the sheets are in these positions.
Use 5 mm squared paper.

a)

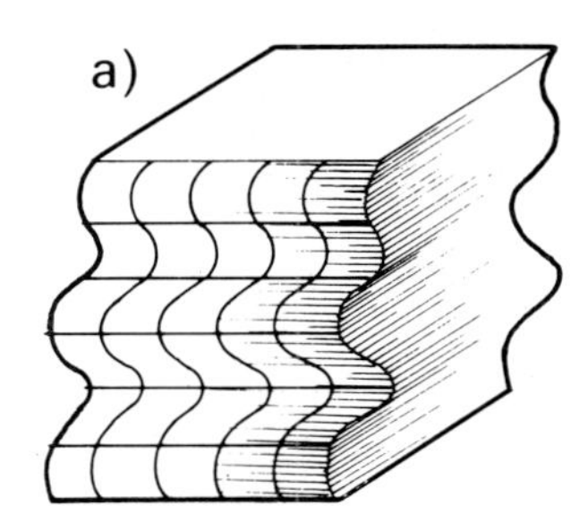

b)

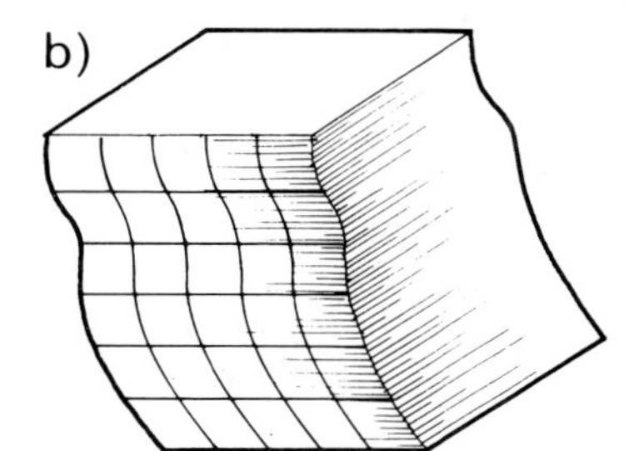

c)

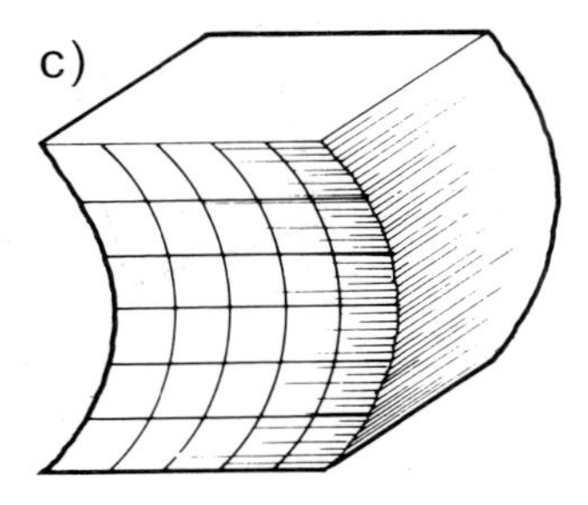

Challenge

2 Horace draws an **h** on this pile of sheets of paper.
Sketch the **h** when the pile is straightened out.

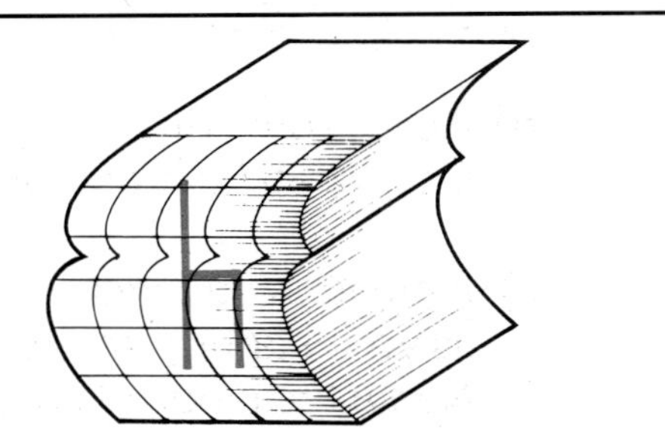

3 Here is another of Horace's piles of paper.
Again he has drawn a grid on a vertical face.

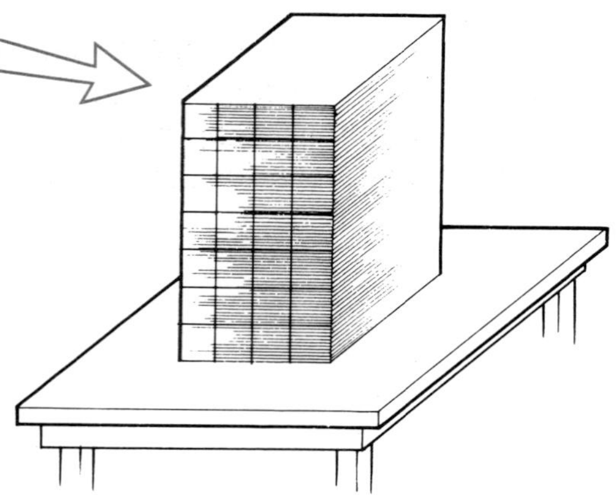

Horace slides the sheets sideways again
(left or right; not forwards or backwards).
Three of these drawings show possible positions of the grid.
List them.

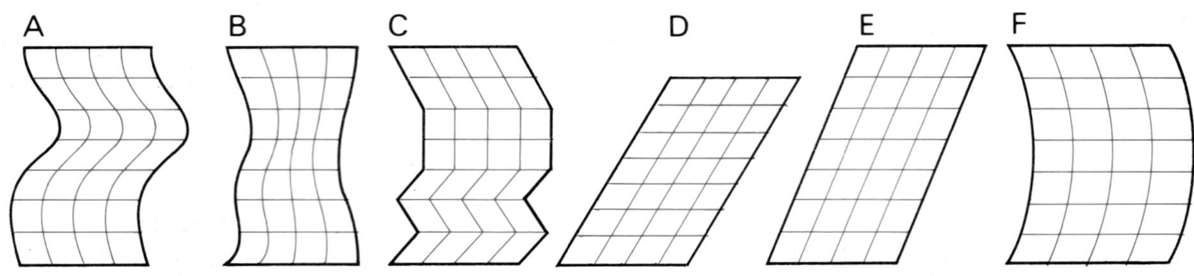

4 On this pile Horace has drawn a grid and an **H**.
Again Horace slides the sheets to the left or right.

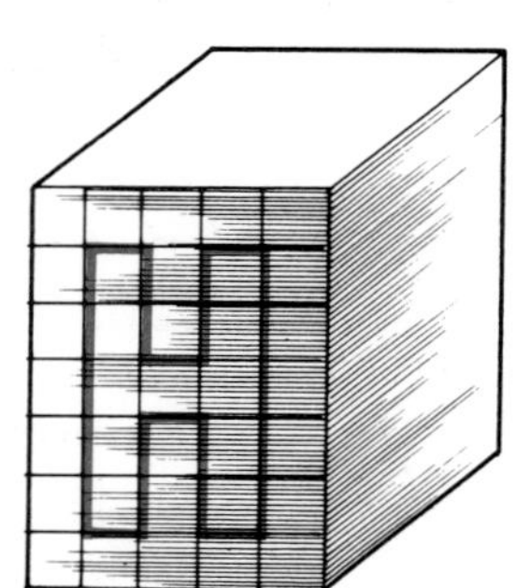

a) Three of these drawings show possible versions of the **H**.
List them.

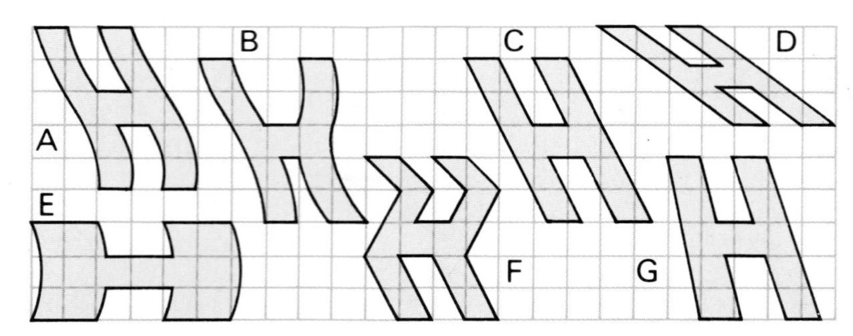

b) Horace produces this version of the **H**.
Copy and complete the drawing.
Use 5 mm squared paper.

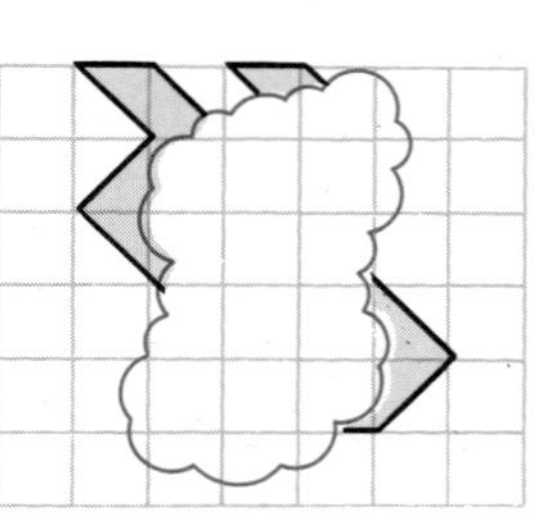

32

Shearing

D 1 Horace has another pile of paper.
Again he slides the sheets sideways.
He takes care to keep this edge a straight line.
This kind of move is called a **shear**.

One of these moves is a shear.
Which one?

A B C D

2 These moves are all made by sliding sheets of paper sideways.
Which moves are shears?

A B C D

3 Here is a letter **L**.
Copy and complete these five shears of the **L**.
Use 1 cm squared paper.

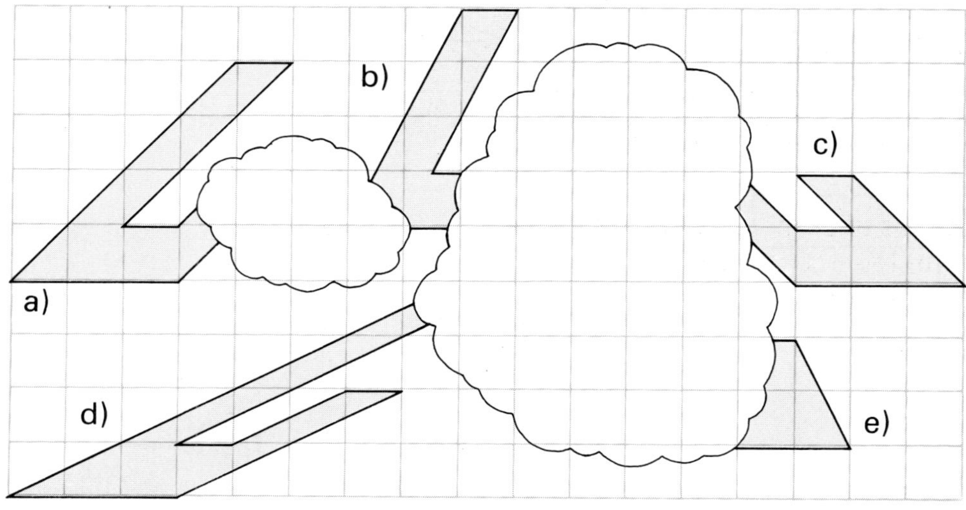

32

4 This is an example of a vertical shear.

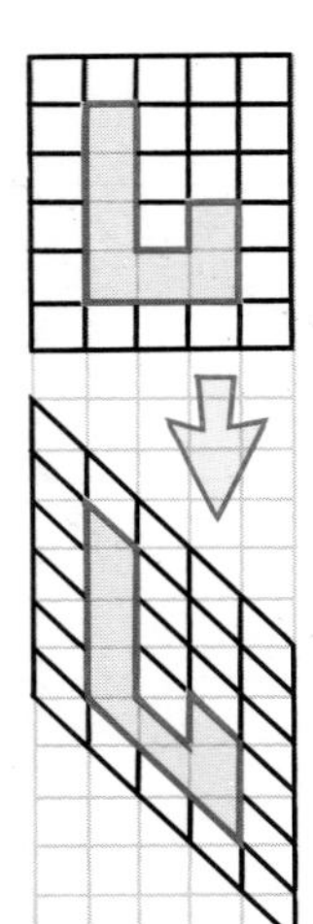

Here are three more vertical shears of the same letter **L**.
Copy and complete them.
Use 5 mm squared paper.

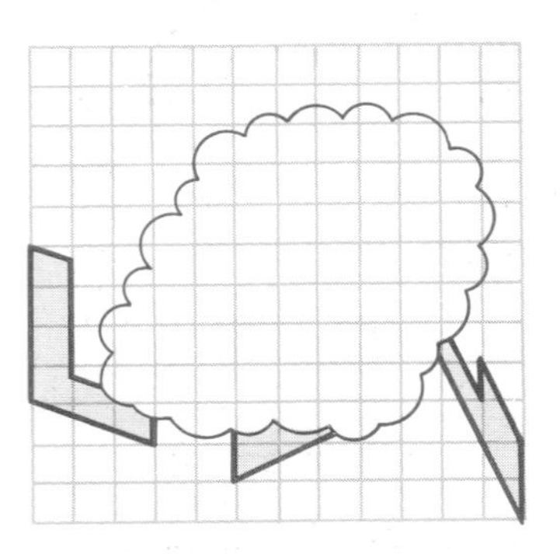

5 Copy and complete these shears.
Use 5 mm squared paper.

32

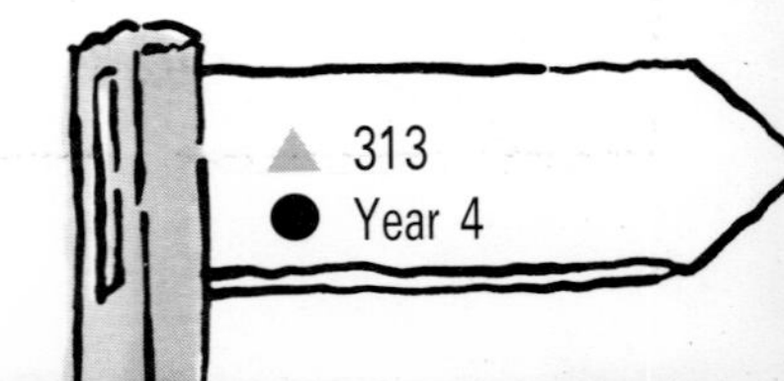

Stretch and shear

E

Think it through

1 a) A one-way stretch changes this square into this rectangle.
The same stretch is applied to this square.
Make a rough sketch of its new shape.

b) Check your answer to **(a)** by applying the same stretch to this.

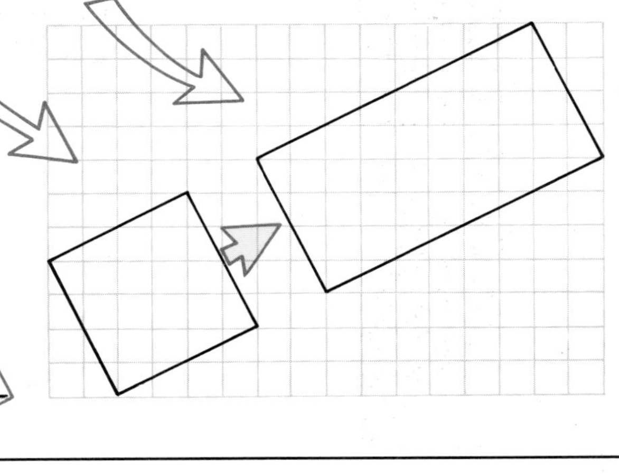

Exploration

2 a) What fraction of the red square is shaded?
Explain your answer.

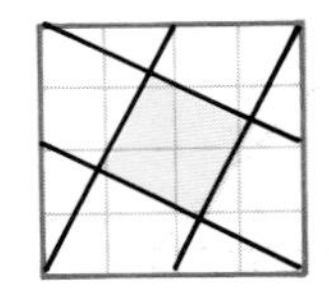

b) (i) Change the red square into a rectangle by applying a one-way stretch.
(ii) What fraction of the new rectangle is shaded?
Explain your answer.

c) Repeat **(b)** with a different one-way stretch.
Do you get the same result?
Will you get the same result for any one-way stretch?

Challenge

3 This parallelogram was produced by applying a horizontal one-way stretch to a square.
Find the original square.

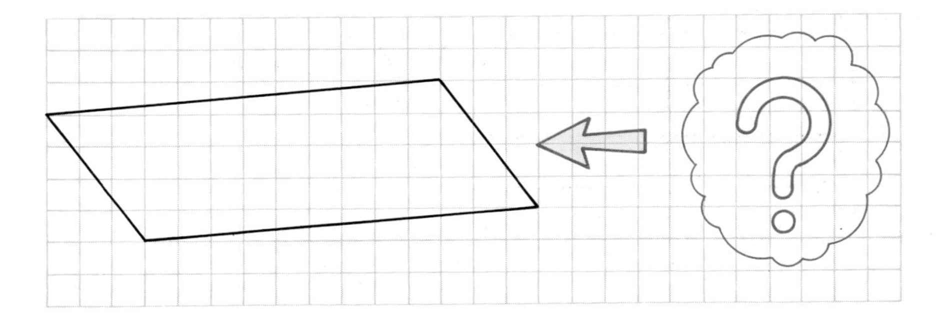

Challenges

4 Copy and complete this shear.

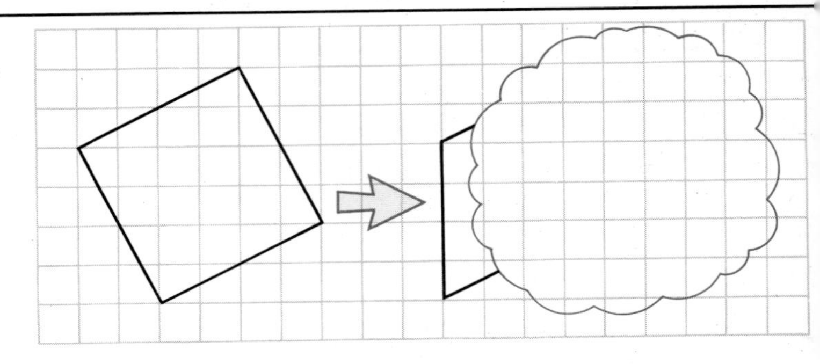

5 Copy and complete this shear.

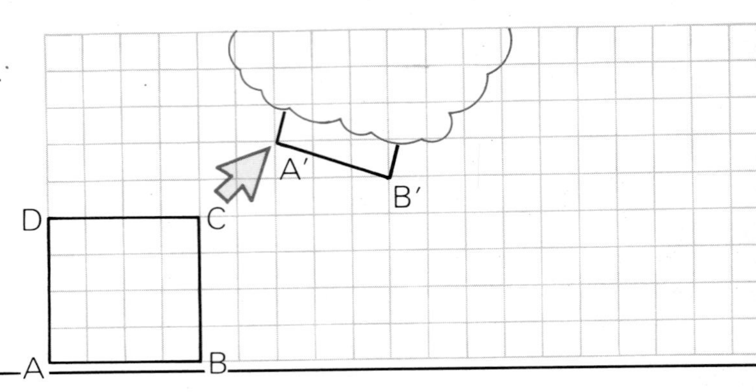

With a friend

6 Each of you

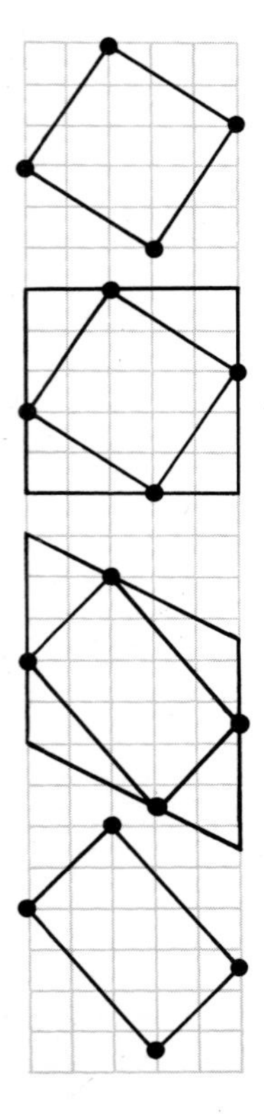

a) draw a tilted square.

b) frame it.

c) apply a vertical shear to the frame and square.

d) draw the sheared square without the frame; pass the drawing to your friend; ask your friend to draw your original square.